T0132574

History Within

History Within

The Science, Culture, and Politics of Bones, Organisms, and Molecules

MARIANNE SOMMER

THE UNIVERSITY OF CHICAGO PRESS   CHICAGO AND LONDON

MARIANNE SOMMER is professor in the Department of Cultural and Science Studies at the University of Lucerne. She is the author of *Bones and Ochre: The Curious Afterlife of the Red Lady of Paviland*.

The University of Chicago Press, Chicago 60637
The University of Chicago Press, Ltd., London

Printed in the United States of America
25 24 23 22 21 20 19 18 17 16   1 2 3 4 5

ISBN-13: 978-0-226-34732-5 (cloth)
ISBN-13: 978-0-226-34987-9 (e-book)
DOI: 10.7208/chicago/9780226349879.001.0001

Library of Congress Cataloging-in-Publication Data

Names: Sommer, Marianne, 1971– author.
Title: History within : the science, culture, and politics of bones, organisms, and molecules / Marianne Sommer.
Description: Chicago : The University of Chicago Press, 2016. | Includes bibliographical references and index.
Identifiers: LCCN 2015041267I ISBN 9780226347325 (cloth : alk. paper) | ISBN 9780226349879 (e-book)
Subjects: LCSH: Natural history—History. | Evolution (Biology)—History. | Evolutionary genetics—History.
Classification: LCC QH 15.S66 2016 | DDC 508—dc23 LC record available at http://lccn.loc.gov/2015041267

∞ This paper meets the requirements of ANSI/NISO Z39.48–1992 (Permanence of Paper).

Contents

Acknowledgments

The research that informs this book began more than ten years ago and was carried out at the ETH Zurich, the University of Zurich, and the University of Lucerne—institutions that have provided wonderful intellectual and research environments. I am particularly grateful to the Swiss National Science Foundation that financed the project for four years in the context of an SNSF Professorship. I also had the opportunity to focus on and receive responses to particular aspects of the project as senior fellow at the International Research Center for Cultural Studies (IFK) in Vienna, and as a visiting scholar at the Institute for Society and Genetics at UCLA, the Max Planck Institute for the History of Science, Stanford University, the Centre d'Estudis d'Història de la Ciència (Universitat Autònoma de Barcelona), and the Centre for the Study of Life Sciences (Egenis) at the University of Exeter, among others. I could not have written this book without the institutions that care for the published and archival materials I draw on; among the latter are the American Museum of Natural History Library in New York, the Woodson Research Center at the Fondren Library of Rice University in Houston, the New-York Historical Society, the Zoological Society of London, the Museum Archives of the Natural History Museum of Los Angeles County, the Page Museum Archives, the New York Public Library, the Wellcome Library, and the Stanford Public Libraries. I would like to express my gratitude to the members of the staff who have assisted me, sometimes during monthlong stays. The Woodson Research Center and the American Museum of Natural History Library in particular granted me the right to reproduce many of the wonderful images that characterize this book. For a project of this duration it is impossible to name individually every scholar and scientist who contributed to my knowledge and perspective, or who invited me to share my insights at conferences, and the reviewers of the manuscript for the University of Chicago Press did their great service anonymously. I would like to thank all of them, as well as the University of Chicago Press and especially Karen Merikangas Darling.

Introduction

The expression *history within* in my book title is borrowed from the global population genetic endeavor called the Genographic Project. On the basis of the analysis of the genetic variation among human populations worldwide, it reconstructs "our" modern evolutionary history of migration and diversification. This history is advertised on the project websites and told by the project director in popular books and films. It is promoted as located within our bodies, as living in a quite literal sense. As such it is sold to individual customers when they have their DNA analyzed for their personal genetic history. This book engages with that kind of history within, its science, culture, and also politics. I am interested in the role (representations of) bones, organisms, and molecules have played in the process of scientifically reconstructing deeper human pasts and in academic and nonacademic perceptions thereof. I am interested in how the evolutionary perspective has informed and informs worldviews and understandings of self and other.

My focus is on the twentieth and twenty-first century, which allows me to trace developments from "the coming of age of paleoanthropology" to the molecular approaches as implemented in the Genographic Project. And I focus on the lifeworks of three scientists: Henry Fairfield Osborn (1857–1935), Julian Sorell Huxley (1887–1975), and Luigi Luca Cavalli-Sforza (1922). I have chosen these scientists because they took part in major shifts in the history of the historical life sciences; because they undertook outstanding efforts to bring about an evolutionary perspective in the ways in which academic disciplines and people in their everyday lives understood the human past and the light it throws onto possible futures; and because they believed that their deeper histories held clues for the organization of contemporary societies, if not the world. I situate their careers in the powerful institutions and scientific networks that not only facilitated but also shaped their endeavors.

People, Objects, and Institutions

I develop my history along three parts that focus on the interpenetrating scientific, public, and popular work of Osborn (part 1), Huxley (part 2), and Cavalli-Sforza (part 3). In doing so, I follow a trend in the history of science to rekindle and reconceive the biographical genre. As Mary Jo Nye has appraised in her introduction to a focus on the topic in the journal *Isis*: "While historians of science often use biography as a vehicle to analyze scientific processes and scientific culture, the most compelling scientific biographies are ones that portray the ambitions, passions, disappointments, and moral choices that characterize a scientist's life" (2006, 322). Biographical approaches emphasize the historical agent—in my case the scientists with their practices, knowledges, emotions, desires, aims, limitations, and frustrations. While such an approach must not lose sight of the conditions of possibilities for action, belief, thought, and so forth, it is to a certain degree a perspective that renders history the product of human endeavor, success, failure, and inability even to act, rather than of systemic (r)evolution. This does not, therefore, constitute a return to the histories of "great men and great ideas." Rather, among other things, through new biographies, the careers and ideas of powerful and influential men and women have come to be seen as conditioned by material, institutional, social, economic, and cultural resources.

Osborn's family belonged to New York's dominant class and his social network opened many doors for him. He advanced to the station of a paragon of vertebrate paleontology and a powerful representative of this science's nineteenth- and early twentieth-century tradition. As curator and president (1891–1933), he was centrally involved in turning the American Museum of Natural History in New York into a hub for international science and into an attraction to large publics. In the early twentieth century, Osborn was attracted to paleoanthropology because of the discovery of spectacular remains and cave art. By bringing to bear the widely shared notions of parallel evolution and orthogenesis on the fossils of animals and humans, he revealed their phylogenies. Osborn's friend Huxley, on the other hand, who was his junior by thirty years and himself an offspring of a British family of intellectual and scientific prominence, was among the movers of the evolutionary synthesis of the 1930s and 1940s. The term *evolutionary synthesis* refers to the process by which the Darwinian variation-selection theory (modified through new insights into the processes of heredity) was (re)integrated into many biological fields. Huxley reinterpreted Osborn's hereditarily determined evolutionary trends as the result of natural selection. In Huxley's understanding, organisms in all their diversity were the basis of evolution. I will follow Huxley in his attempts at implementing and communicating his evolutionary perspective through institutions and organizations such as the London Zoo (director, 1935–1942), UNESCO (first director general, 1946–1948), and the World Wildlife Fund (WWF) (founded 1961).

If Huxley recognized great potential in human population genetics, he insisted on the holistic view from the entire organism. For those like Cavalli-Sforza, who could reap the fruits of the molecular revolution, the perspective changed. When Cavalli-Sforza was at the University of Pavia in the 1960s, his understanding of modern human genetic evolution as a Brownian motion process enabled him to capture this evolution statistically and with the help of computer programs to produce phylogenetic trees and migration maps. His success brought him a professorship at Stanford University (1970–1992), where his lab contributed significantly to the development of molecular or genetic anthropology. At the beginning of the 1990s, Cavalli-Sforza was among the leaders of the Human Genome Diversity Project, which is in many ways a precursor to the Genographic Project. It was conceived as a global initiative to secure blood samples from those indigenous peoples who were understood to carry the historically most informative genetic markers. This would enable scientists to "read the histories in the genes" for a long time, applying increasingly sophisticated methods.

Thus, Osborn, Huxley, and Cavalli-Sforza contributed to the knowledge of "our" evolutionary past via different kinds of research and with an emphasis on different objects of analysis. These organic "traces" of evolutionary history are another structuring device of this book. The objects of science, which the historian of science Hans-Jörg Rheinberger (1992, 1997) termed *epistemic things*, are not simply "pieces of nature." They are brought into being by, and influence the development of, the instrumental and theoretical inventory of a time. As objects of science, they always already embody concepts. In part 3, for example, I examine how the system of Y-chromosome DNA markers emerged. I show how in the early 1990s, the right instincts (not least those of Cavalli-Sforza), the skill different people brought to the Stanford University team, and certainly the possibilities of the young technologies such as the Polymerase Chain Reaction (PCR) played a role in constituting these DNA sequences as markers of the male phylogenetic lines. But the systematizing sciences, to which human population genetics also belongs, first of all collect things. They dislocate them from their original contexts and reappropriate them within the scientific theoretical and epistemic-practical contexts, for example, in the process of establishing cell-line and DNA repositories (Rheinberger 2006, 336).

In fact, the sociologist (of science) Bruno Latour (2008) has mused about these processes with regard to the exhibitions of fossils at the American Museum of Natural History in New York that were constructed under Osborn: how is it possible to see long extinct animals as if they were still alive? It seemed to Latour as if there were magic at work. However, he explains that rather than succumb to that magic charm, the historian of science has to bring to light the hard work of transforming fragmentary bones hidden in sediments into lifelike fossil mounts and murals of lost animals and humans. Instead of magic, there were long, expensive, competitive as well as collaborative, and dangerous processes from organizing expeditions to finding, preparing, and transporting bones. To arrive at a mount, the scientists and technicians experimented with the animals' anatomy and posture by drawing on existing knowledge. In doing so, they might have reinterpreted the animals or humans and thereby also transformed the knowledge system. Latour emphasizes that once paleontologists happened on the fossil remains, after entering Osborn's wonderfully successful network of institutions, experts, objects, and knowledge, the bones existed in a new mode: one of reference. Instead of a genealogical chain through reproduction, the animals and humans from times long gone now existed through chains of inscriptions.

Latour and his colleague Steve Woolgar have analyzed how laboratory instruments translate material substances into inscriptions such as a figure or diagram—inscriptions that can then be set in motion and combined to larger entities (Latour and Woolgar 1979). Latour's musings on the American Museum of Natural History highlight the importance of the exchange between experts as well as of the institutional infrastructures for these processes. We will see how it took such networks to negotiate and (temporarily) stabilize the knowledge gained from fossils, organisms, and molecules. At institutions like the American Museum of Natural History, "traces" of "our" histories and phylogenies could be collected, analyzed, and translated into diagrams, pictures, and texts to be published and shared with other experts. Thus, in Latour's parlance, the photographic, graphic, filmic, and textual inscriptions that resulted from the studies of fossils, organisms, and molecules can be seen as immutable and combinable mobiles, while the institutions functioned as networked centers of calculation (Latour 1987, 227–228, 232–247).

However, places like the American Museum of Natural History or the London Zoo do not serve experts only. They also address particular publics. I am especially interested in how "our" deeper histories and kinships were presented to larger audiences. I inquire about not only the scientific, but also the artistic, literary, material, and spatial technologies of historical reconstruction; I analyze the cultural topoi and schemata that helped in the translation of abstract knowledge into identity-informing exhibits, narratives, and images. And I pay attention to the aura that the bones, organisms, and molecules could gain, despite the mundane processes of inscription, as "authentic traces from our past" in the perceptions and feelings of scientists as well as different publics. We will thus see that these things are of more than epistemic function. They acquire political value in negotiations over prerogatives of interpretation and in the exposition of scientific results to specific publics. As objects that can be made to transport meaning about people's histories and identities, these bones, organisms, and molecules are also cultural and moral things.

The Circulation of Knowledge

Even if not necessarily appreciated by their expert communities, Osborn, Huxley, and Cavalli-Sforza invested great efforts and hopes in the communication of their sciences to diverse publics. Beyond the museum exhibition halls, Osborn spread his textual and visual reconstructions of human evolutionary history through popular books, magazines, newspapers, and public lectures. Huxley, too, was a public figure. He lectured and published prodigiously, partook in radio shows, produced film documentaries, and even experimented with science fiction. Finally, Cavalli-Sforza's career was marked by the endeavor to bring human population genetics to a more general readership. Besides his engagement in the Human Genome Diversity Project that became publicly visible indeed, he published educational and popular books that appeared in several languages. His career was crowned with an international exhibition. These efforts of my protagonists were informed by a classical understanding of "the popularization of science," an understanding of the communication of scientific knowledge and scientific methods as a unidirectional, enlightening, and modernizing flow of information. In its pure form, in this view the generation of scientific knowledge appears as untouched by everyday concerns. Science works in isolation from society. When no longer under expert control, knowledge may suffer not only vulgarization but even mythical distortion. The view is a legacy of the nineteenth century, when the increasing professionalization of science seemed to rip apart the worlds of experts and laypersons, while science and technology were understood to be of paramount importance for the advance of societies.

From Osborn's time, the communication of scientific knowledge involved mass publics that were often perceived as diffuse, evasive, and even alienated. Concomitantly, science communication was institutionalized, such as in the American Science Service, and professionalized, as in the case of science journalists and writers. We will see that Osborn, Huxley, and Cavalli-Sforza thought about the mass societies, as well as the quality and effect of mass communication, of their times. Vulgarization and the nonsense kind of popularization had to be substituted with their informed kind. Osborn wanted the public to get to know the true forms and ways of life of extinct hominids and animals. He set these truths in circulation against the proliferation of monstrous ape-men and against humans that were made to live with dinosaurs in popular culture. Huxley was up against lingering notions of "Lamarckian" evolution. And although both Osborn and Huxley maintained a role for spiritual feelings and religiously inspired ethics and rituals, they positioned their own phylogenetic explanations against creationist understandings. Finally, a considerable portion of Huxley's and Cavalli-Sforza's popularizations were written against "outdated conceptions of race" such as those held by Osborn.[1]

Thus, most notably in the case of Osborn, the popularization of evolutionary history could serve conservative ethics and politics. Like figures in a game of chess, Osborn positioned his "ancestors" against the new Negro, the new woman, and the degenerate white man. Furthermore, though intended to enlighten and modernize, Huxley's and Cavalli-Sforza's narratives of evolution were also meant to have the power of myths that would provide the final word on who we are and where we come from. This indicates that Osborn, Huxley, and Cavalli-Sforza did not completely subscribe to, and that their popularizations did not entirely match, the classic understanding of the popularization of science. Indeed, scholars like Andreas Daum (1998, 2002) have shown that the classic understanding does not give an accurate picture even of the spectrum of motives for the production and communication of scientific knowledge in the second half of the nineteenth century, when the noun *popularization* was brought in connection with the treatment of scientific knowledge for wider audiences. The same holds true with regard to the actors and genres involved. Osborn, Huxley, and Cavalli-Sforza were heirs to the great scientist-popularizers of the nineteenth century like Julian Huxley's grandfather Thomas Henry Huxley. But the landscape of people and places of science communication was much more diverse. Indeed, Osborn felt threatened by the proliferation of actors and genres, and he tried to control the books on evolutionary history that others wrote for children and adults, including works of fiction.

We will also encounter alternative conceptions of the relation between science and society than the simple opposition associated with the classic understanding of popularization. Huxley, for example, was among those of his generation who arrived at a notion of (the history of) science as deeply embedded in its social context and who wanted to lure citizens into becoming everyday scientists. At this time, there were already fledgling reconceptualizations of "the popularization of science." Unknown to Huxley, the Polish immunologist Ludwik Fleck ([1935] 1980) developed ideas about the history of science that would prove very influential decades later. Fleck's thought was much more radical than that of Huxley, who believed that in a favorable environment, rational and socially beneficial knowledge would flourish. Fleck not only regarded scientific thought and work as in exchange with everyday ideas, beliefs, and practices; he also saw no science-intrinsic, logical, and cumulative process at work. He described the communication of scientific knowledge from esoteric to exoteric circles as an integral part of knowledge production. He understood the process of translating knowledge for nonexpert audiences within and outside science as one of increasing generalization, hardening, and objectification. Hypotheses become facts when a language of uncertainty gradually gives way to established knowledge (Fleck 1983, 84–127 ["Das Problem einer Theorie des Erkennens," first published 1936], 92–96, 112–113).

Osborn, Huxley, and Cavalli-Sforza addressed their more accessible publications to experts in other disciplines as well as to a general readership. And as Fleck discussed, each instance of communication carried the signature not only of the conditions of production but also of the intended audience. In fact, the processes of production, communication, and even reception cannot clearly be separated, because the potential recipients already shaped the generation of knowledge, while it was only in the act of reading narratives or contemplating images and exhibits of the evolutionary past that knowledge was activated. We will see that reception is a most creative process indeed. However, knowledge was not only communicated as an end product; sometimes the ways it was achieved were also detailed. Osborn liked his staff to explain to the general reading public the intricate work necessary to arrive at a reconstruction, for example, of a prehistoric scene in a mural or of a fossil human type as bust. To a certain extent, this made him vulnerable, exactly because he did not present closed scientific facts.

Fleck's writings were at the time not widely received, and Osborn's and Huxley's more traditional understandings of science communication did not clash with the perspectives taken by historians of science. On the contrary, the classic view of popularization strongly informed the outlooks of the early institutionalizers of the history of science in Britain and the United States, with whom Huxley interacted. The notion of science as untouched by its cultural environment only began to be more systematically questioned in science studies in the 1970s, when historians of science also increasingly became interested in the publics of the sciences and in the sciences of laypersons.[2] Science communication came to be understood as a locally, temporally, and medially specific multidimensional process, which despite unequal power relations incorporates different motivations and traditions. New concepts were introduced to complement or replace "the popularization of science." Reminiscent of Fleck's ideas, Stephen Hilgartner (1990) has substituted the binary opposition between "science" and "popularization" with a continuum of milieus of communication from lab shop talk to mass media. He has shown how "the dominant view of popularization" can be employed by scientists to gain authority, but also to distance themselves from the ways in which science is covered in the media by associating popularization with distortion.[3] These observations are relevant to my analysis because I engage with a broad spectrum of media and genres, from scientific journal to institutional pamphlet and science fiction, and because accusations of distorting knowledge were certainly issued, particularly by Osborn.

Also resonating with Fleck's ideas, Terry Shinn and Richard Whitley have proposed *scientific exposition*, or *expository science*, to refer to all intra- and interdisciplinary as well as public-oriented communications of science that involve a transformation in content.[4] What I appreciate especially about these terms is that they capture the processes of performance, visualization, and narration, as well as the accompanying translations and negotiations that, in my case, were involved in rendering data gained from organic objects meaningful. To approach science as communication has also been Jim Secord's suggestion with the concept of *knowledge in transit* (2004). He demanded that every text, image, action, and object be understood as the trace of an act of communication, with producers, receivers, and modes and conventions of transmission. At the same time, *knowledge in transit* seems to suggest that knowledge is always in flux, that its communication has no clear point of origin or goal. It is in permanent transformation in negotiations between diverse producers and between senders and recipients who themselves are transformed in the process. *Transit* might thus also allude to the precariousness of knowledge and to the possibility of its being lost.

Historical analyses and qualitative field research indicate that people encounter scientific knowledge as imbued with interests that have implications for existing social relations, values, and identities. Public readiness to engage with science is fundamentally affected by the willingness to accept the knowledge's (unstated) ideological content. The reception also depends on the trust in the institutions offering the knowledge, and on whether people believe they can act on it. We know that one strategy in science communication, but also a necessary condition of it, is to combine the new with the traditional, by conveying knowledge in well-known styles and in association with widely shared ideas and arguments.[5] The scientific and public standing of the researchers, objects, and institutions involved in Osborn's, Huxley's, and Cavalli-Sforza's accounts of (human) evolution certainly facilitated their great successes. Drawing on well-established literary genres, all three were versed in the art of narration. Their offers of evolutionary meanings to various audiences were very successful, not least because these histories arose out of the historical context into which they were released. Nonetheless, in the course of reception, adaptation to prior knowledge and beliefs, as well as to personal needs and ends, always took place. And there was also resistance, rejection, and radically alternative historicizing.

Finally, it is significant that the event at which Secord gave his programmatic talk on *knowledge in transit* was the 2004 history of science conference on circulating knowledge (the British-North American joint meeting of history of science societies). In fact, under the label *Wissensgeschichte* ("history of knowledge"), the circulation of knowledge has also been revived as a unifying concern of research in German-speaking communities. *Wissensgeschichte* is interested in life-worldly contexts, in which knowledge is interactively generated, transformed, archived, and distributed. It emphasizes diversity and exchange (Speich Chassé and Gugerli 2012). Knowledge is thereby understood as decidedly material: what circulates are objects, animals, and humans, allowing for the generation of meaning and sociality. Historians of knowledge are interested in processes of appropriation and rejection by those who engage with the knowledge thus presented (Sarasin and Kilcher 2011). Therefore, although there is no clear origin from which knowledge circulates, because it is always already intersocial, intertextual, and intermedial, particular ways of passage and itineraries of its objects, the transformations in meaning they undergo, and the obstacles they meet might be reconstructed or observed.[6]

We will see how diverse actors were engaged, for example, in the production of an exhibition. We will also see that several people influenced final book products and that the authors might think of their publications as collaborative works. Furthermore, Osborn's, Huxley's, and Cavalli-Sforza's ideas were steeped in scientific and cultural traditions, as well as historical formations. While I am thus attentive to the particular power of certain institutions and scientists, I also try to do justice to the ways in which they were part of social and discursive landscapes. And throughout this book, I follow the journeys of images and narratives published under the names of Osborn, Huxley, and Cavalli-Sforza to individual readers, into different disciplines, and into newspapers and magazines. In these travels, ancestors, phylogenies, and evolutionary histories came to life, but they also came to carry diverse meanings. As Fleck has already demonstrated for some transformations of knowledge in the process of circulation, they were at times disfigured beyond recognition (Fleck [1936] 1983, 95). In engaging with the goals the researchers pursued with their evolutionary perspectives, as well as with the ways in which these were rendered meaningful by audiences, I also address notions and processes of embodiment, in how a deeper history made living was understood and experienced as a history within.

History Within and History Without

Osborn, Huxley, and Cavalli-Sforza engaged in the circulation of a particular kind of knowledge—a knowledge that was both biological and historical. All three relied also on nonorganic material and nonbiological methods. Even if they considered the biological knowledge foundational, they variously drew on archeology, ethnology, history, and linguistics. Furthermore, their syntheses were among those histories that are intended to do something in the world and that may become histories in use (*Gebrauchsgeschichte* after Marchal 2006). The scientific agendas of Osborn, Huxley, and Cavalli-Sforza and their views of "our" deeper past were informed by and drove cultural, social, and political goals. The insights into prehistory should open up future prospects. They should shape the development of their societies, and they should become, and indeed became, part of particular kinds of identity formation. As such, they partook in the landscapes of historical cultures. Throughout the time I am concerned with here, there were contesting reconstructions of pasts, and the demand for sense-imbuing narratives seemed strongest as a result of perceived crises.

Osborn answered to a sense of loss of orientation due not least to such fields as astronomy, biology, and psychology that increasingly painted an indifferent cosmos and a hapless humankind reduced to heredity and brain processes. There seemed no place left for the creator and human destiny. Osborn was among those like H. G. Wells who stepped into "this void" and provided people with a living history. Like Osborn's popularizations, Wells's stunningly successful *Outline of History* (1920) was part of a more general effort in disseminating grand sweeps of human progress for personal internalization and orientation. Wells drew heavily on Osborn's books to render humankind's deeper past, because he still perceived the beginnings of humankind in modern human life, its political, religious, and social aspects. He subtitled the work *Being a Plain History of Life and Mankind* and even ventured as far as to attribute second rank to the reconstruction of history on the basis of written sources. Those "documents" that the geologist, paleontologist, embryologist, and natural historian contributed to the project of a world history were ontologically superior. However, while Wells made the promise of a future federal world-state, Osborn rather hoped his narratives would stabilize what seemed to be eroding social structures and norms. His images, histories, and exhibits were intended to stimulate experiences of human beings long bygone within contemporary bodies. His reconstructions from stones and bones appealed to the heritage of a past when humans lived under the stern laws of nature and in awe of her beauty.

Wells's *Outline* was well received by historians who have come to be regarded as founders of public history. As early as the 1910s, James Harvey Robinson (1912) located the task of history in helping people to understand the problems and prospects of humankind. Carl Becker, Robinson's doctoral student, wrote in his foundational address "Everyman His Own Historian" (1932) that history was essential to the performance of the simplest acts of daily life. Rather than truth, Becker therefore regarded usefulness in the present as the greatest virtue in historiography. The call for public history was thus also a reaction to the disconnectedness of an increasingly professionalized and specialized academic history from the general readership, even while this potential audience had grown through the expansion of education. Furthermore, the extra-academic representation of history gained in market share, because it profited from the differentiation in popular genres and media. And interestingly enough, Robinson also considered "the new sources" from paleontology and other historical sciences to be particularly relevant in the task of creating living histories (Sommer 2012a, 227–229).

The preoccupation with and the change in perceptions of time around the turn of the twentieth century have been related to, among other things, the transformation of everyday life through the innovations in production, as well as in transportation and communication technologies, and to the development of mass and consumer cultures (Kern [1983] 2003). Certainly, the experiences of World War I changed historical consciousness. The sense of loss increased tremendously and would only be surpassed by the experience of World War II. For the historian Walter Benjamin, for example, World War I set an abrupt end to the very possibility of storytelling (Benjamin 1974, 385–410 ["Der Erzähler," first published 1936], see especially 386). On the contrary, Huxley even gave up a professorship in zoology in the 1920s to cooperate with the Wells brothers on the so-called first textbook of modern biology, *The Science of Life* (Wells, Huxley, and Wells 1929/1930/1931/1934). It was far more than a dry rendering of the latest knowledge; it suggested an evolutionary worldview. When looking back in time with a certain degree of desperation, Huxley reread the human organism as the evolutionary process become conscious of itself; the human body (including the mind) incorporated life on earth in its historical becoming. It was a living museum of evolution. Humans had to act on the insight that they were life's spirit and potential. Julian Huxley thus imbued the meaningless evolutionary process of his grandfather Thomas Henry Huxley with new moral authority.

During Cavalli-Sforza's career, some of these trends and conflicts within historical cultures had intensified. In the 1970s and 1980s, public history was institutionalized in the United States. Its concerns and practices spread throughout western Europe and in the process were adapted to local and national traditions, such as in the German *Geschichte in der Öffentlichkeit* ("history in the public"). *Public history* referred to the employment of historians and the historical method outside academia. Historians should be trained to work in government, business, research organizations, the media, historical preservation, historical interpretation (museums and societies), archives and records management, as well as in teaching. Public history was meant to meet the practical and intellectual needs of society. Its proponents ascertained a history boom that they related to processes of modernization. Public historians should analyze this public engagement with history but also encourage it for educational purposes. Popular historical practice (referred to as *popular history* or *popular historymaking*, that is productions and appropriations by nonhistorians) was perceived as facilitating an understanding of the present and of possible futures; providing legitimation for existing social structures or the ability to criticize them; and allowing for identity formation, meaning production, and orientation in rapidly changing worlds. Public history therefore also catered and still caters to the commercially and medially reinforced needs of the more consumption- and fun-oriented histotainment cultures (Kelley 1978; Howe and Kemp 1986; Cole 1994; Rauhe 2001). When referring to this particularly market-led branch, such as historical festivals, parks, PC games, reality TV shows, fiction films, and history marketing, scholars also use the term *applied history* (Hardtwig 2005, 12–13; Kühberger, Lübke, and Terberger 2007; Hardtwig and Schug 2009).

This examination of the characteristics and aims of public history may serve to situate Osborn's, Huxley's, and Cavalli-Sforza's roles and goals. They variously addressed concerns of research, government, economy, education, the media, and historical preservation and archiving. However, their sense-imbuing histories, like those of public historians, were at odds with certain trends in the historical discipline. In Osborn's time, some scholars questioned the tradition of historicism and a progressive view of history.[7] During the 1970s and 1980s, the apprehensions that were felt by some intellectuals in the early twentieth century seized the humanities more generally. In synergies with postcolonialism, feminism, the civil rights and ecological movements, there emerged a new cultural history worldwide. The past of minority and underprivileged groups, the experiences of ordinary people, and aspects of private life, as well as popular, mass, consumer, and material cultures became central foci. A new sensibility for the literality of all texts had arisen.[8] However, the distance between academic and public history had increased rather than diminished. While scholarly standards continued to demand the meticulous reconstruction of historical context, and encouraged the demonstration of interrelated processes and structures in as much complexity as possible, public histories were characterized as a simplification and shrinking of historical distance (Hardtwig 2005, 31–32).

As Huxley would experience toward the end of his career, resistance to a universal human history with the promise of salvation had increased, in particular to a biologically based one. The Enlightenment notions of an autonomous subject, of one coherent history, and of individual perfectibility and common progress were in conflict with humanities and social sciences transformed by poststructuralist thought and postcolonial consciousness. Cavalli-Sforza's syntheses of knowledge from the diverse anthropological fields on the basis of the genetic-evolutionary approach provoked controversies with regard to the very meaning of *history* and *anthropology*. By the time the Human Genome Diversity Project was officially launched in 1991, in view of the experiences of racism, sexism, colonialism, and genocide, the idea of one grand living history that may be shared by and unite all was perceived as an instrument of power. In a world changed by indigenous rights movements, the long ongoing sampling effort attracted particular criticism.

Osborn's, Huxley's, and Cavall-Sforza's evolutionary histories seem to pertain to a humanist tradition. Osborn imbued the study and knowledge of human evolution with epic grandeur. Huxley, too, abhorred the reductionist as well as vulgarizing look at the biological and historical beings that humans constituted. He confronted them with his philosophy of scientific and evolutionary humanism. Finally, exactly as a reaction to poststructuralist science and global fragmentation of worldview, Cavalli-Sforza maintained the humanist belief in a progressive history and a possibly progressive future. All three did so among other things in view of problems such as overpopulation, war, or loss of nature and natural resources. They perceived an urgent need for public education, and science free from the weight of academic standards profited from inventions in communication technologies throughout the twentieth century. The advent of audio-visual media and eventually digitization and "globalization" have vastly expanded the circle and circles of people who may participate in the production, communication, and application of histories—a process we will see at work in the age of history in the gene.

While Osborn, Huxley, and Cavalli-Sforza were contributors to particular "history booms," the social lives of histories have increasingly become a subject of study by historians. They analyze what they call *historical cultures*, with their locally and temporally specific cognitive, political, and aesthetic dimensions (after Rüsen 1994a; also Schörken 1981, 1995; Lowenthal 1985; Füßmann, Heinrich, and Rüsen 1994). At the same time, cultural studies have partly reinvented themselves as memory studies, recognizing in the more recent turn toward the past a *memory boom* (Huyssen 1995, 5). As an interdisciplinary field, memory studies span approaches from the humanities as well as social and natural sciences. The biology of memory has become a central concern in animal learning and behavior, neurobiology, cellular biology, genetics, and the neurosciences, as well as in artificial intelligence. "Memory" as a cultural phenomenon that is performed in popular culture, literature, architecture, and the arts, and that manifests itself in the growing heritage industry, is investigated from the perspectives of literary, media, and religious studies, history, sociology, educational science, and psychology, among other fields. In short, by the 1980s, "memory" had advanced to an integrative core concept (A. Assmann 2002).[9]

However, just like "the history boom," "the memory boom" is not a new phenomenon. Furthermore, *memory*, like *history*, has a contested twentieth-century development. In fact, Osborn, Huxley, as well as Cavalli-Sforza worked with a certain concept of memory. There were widespread notions in the early decades of the twentieth century that individual behavior and by inference cultural practices could be inscribed into organic matter and passed on to following generations (Sommer 2005b). Osborn had been educated in the American "neo-Lamarckian" school, but he had to grapple with the critique from new branches such as genetics. Nonetheless, he believed in "racial souls" that were the result of the environments and ways of life of the races in the evolutionary past. This notion came close to a race-specific memory that Osborn thought could be activated most effectively by visual stimuli. His offers of evolutionary exhibits, images, and narratives should trigger transformative experiences; they were tools for "racial regeneration."

At the same time, there were also nongenetic and nonracialized ways of thinking of a living history as memory. For Robinson, for example, the personal memory incorporated aspects of the history that a person was told and which he or she read. Both Robinson and Becker conceived of a living, embodied kind of history as an artificial extension of the socially mediated personal memory. In their thinking, as in the much more famous conception of the personal and collective memory of Maurice Halbwachs (1925, 1950), the term *memory* signaled a critique of historicism; it was introduced to capture the positively valued phenomena of a living history vis-à-vis what was perceived as dead academic writing. Halbwachs was interested in the role of social frameworks in personal remembering. Each act of seemingly individual recollection was in truth a collaborative process, since it drew on a collective memory that was fed by cultural transmission. It is especially Halbwachs's notion of memories as individual appropriations of collectively constructed knowledge of a distant, inexperienced past that has remained relevant for current theories of cultural memory (J. Assmann 1992). For Aleida Assmann, the cultural memory consists of archives and the living histories constituted on their basis. The first contain objects and narratives that may be obsolete or restricted to specialist research and discourse, while the second refer to those that have been imbued with meanings and values in order to add to the social life of particular groups. The first are reservoirs for future living histories, while the second are constitutive in the process of identity formation and the legitimation of societal organization (A. Assmann 1999, 130–145; 2006, 54–58).

Huxley developed a comparable notion of the cultural memory, but he wanted to cultivate it. He dreamt of what he called *a superhuman memory* that would be developed through the progress in and sharing of culture. The idea of the superhuman memory consisted in nothing less than a worldwide archive of knowledge about every sphere of life. The collection, management, and continuation of this archive would allow belief systems to evolve. In the process, humankind, through the individual appropriation of the superhuman memory, would become a superhuman organism. This utopia informed Huxley's many plans for centers of calculation at the zoo, UNESCO, and conservation organizations, for example, in the form of national parks in Africa and Britain. In these endeavors, Huxley conceptualized the organismic variety, and in particular the human biological and cultural diversity, as a panhuman heritage. Central to this notion was the concept of trusteeship. Because humans were evolution become conscious of itself, they had the responsibility for its conscious progressive steering. This progressive steering depended on the preservation and cultivation of the organismic as well as cultural diversity.

Like Osborn and Huxley, Cavalli-Sforza was driven by greater aims. He thought about ways to mathematically capture the mechanisms of cultural transmission so that intervention in cultural evolution in progressive ways may become possible. Regardless of scale, the collection and preservation of the (cultural) memory had a reason beyond a nostalgia for the past: "And if it is important to preserve the memory and to fix it so that it does not get lost, it is not only for sentimental reasons; it is also because there is a lot to learn from history. . . . It is not impossible that some might profit from it to find new ideas that allow the modification of our social behavior in positive ways" (Cavalli-Sforza 2005b, 244, my trans.).[10] The history derived from human population genetics would have a vital part to play by revealing the deep bodily connections throughout all of humanity. The evolutionary histories of humanity and human groups had to be reconstructed on the basis of large-scale comparative analyses of the genetic variation between living populations, which required a concerted effort of collecting and storing. This human genetic variability was conceived as our heritage—a gateway into a past beyond the confusing effects of modernity that was about to close. For Cavalli-Sforza, both our cultural and natural—and especially our genetic—heritage would have to be preserved.

Thus, while Osborn thought racial temperament was genetically transmitted and memories made by the race in prehistory could be rekindled in the present, Huxley thought that the sharing of (evolutionary) knowledge would produce something like a superhuman memory in which each individual brain participated. At the same time, both Huxley and Cavalli-Sforza transferred the term *memory* to objectified and externalized (historical) knowledge. And they used the term *heritage* to refer to biological as well as cultural entities of supposedly panhuman interest and meaning. In recent years, concepts such as cultural memory and heritage have been criticized, exactly because they are associated with past notions of group consciousness and infuse objects with divine presence (Klein 2000, 129–138). Exponents of today's memory studies can answer this critique with the explanation that, on the contrary, it is the processes by which group identities are constructed and objects come into being and might indeed be sacralized in certain contexts that are of interest (Sturken 2008, 74). Along these lines, I will approach Huxley's and Cavalli-Sforza's fascination with the magic of "memory" and "heritage," and look at how Osborn fetishized the remains of cherished fossil men.

As Geoffrey C. Bowker has put it in his *Memory Practices in the Sciences* (2005), "the background (our canvas) should stay stable while the foreground (human attainment of perfection) should be changing rapidly" (209). Similarly, for my protagonists, (global) archives of history within and history without should ensure the survival of the past diversity and allow future generations to continue to study and enjoy it, so that knowledge and society may progress. Memory practices "are what makes our current reality true and our future—in will if not in deed—controllable" (ibid., 229–230). For my protagonists, the deeper past in particular could rationalize or point to the weaknesses and mistakes of the present in order to think about progressive futures. And the messages they read from this past had to reach the masses to be implemented on a wide scale. Thus, even if Osborn, Huxley, and Cavalli-Sforza worked out syntheses from different kinds of knowledge, their histories were biologically founded. What might be particular about a living history that is also a history *within?*

The Phylogenetic Diagram

History within, like any other history, may become living in the sense of being appropriated to form historical narratives of "us-groups." Biologically founded pasts therefore also involve customized histories with a pronounced topicality. They are likely to have a normative aspect in that the past is seen to explain the present and is endowed with moral lessons for the future; the past might be used to enforce or undermine current privileges. This among other things raises the question of whether *biological* history, because it is associated with the authority of science and the foundational power of biology, allows less flexibility in its appropriation. In view of the present "genetic history boom," this further begs the question of whether we are facing a re-biologization of identities, or whether biological determinism has in fact never disappeared, or indeed whether the new interconnections between science and technology, publics, and markets in genetic history go along with novel understandings of kinship and of notions such as tribe, clan, nation, ethnicity, and race (Sommer and Krüger 2011).[11]

This book will reveal significant transformations in the understanding of inner-human variation in science and its communication to diverse publics. Osborn, Huxley, and Cavalli-Sforza imagined what I call *the phylogenetic diagram* of humankind in different ways. We can see a general trend in the history of evolutionary anthropology toward a perception of increased difference among what were called the human races in the early decades of the twentieth century—a trend that in the work of some anthropologists culminated in the belief that these races in fact constituted different species or even genera (Sommer 2007a, part 2; 2010b).[12] Osborn represented that trend. He believed in clearly demarcated racial types that owing to their long independent evolutionary histories differed considerably. His hominid phylogenetic diagrams had increasingly long racial branches. At the same time, such trees established a hierarchy among the living races, with the European or the Caucasian at the top. The trees thus also created or re-created entities such as "Chinese," "Hottentot," "Mongolian," or "Australian." They reified categorizations of human groups that referred to (sometimes a combination of) geography, nation, "race," ethnicity, "indigeneity," "tribe," language, or religion.

Huxley's phylogenetic diagrams stood in stark contrast to Osborn's. Huxley played a key role in the process of redefining inner-human variation in science and society in the interwar years. For Huxley, species were natural entities that were reproductively isolated from other such entities (Huxley 1938b). The concept of species was thus objectified. Simultaneously, a notion of species was defined as the basis of the modern synthetic evolutionary theory that worked best for the animal kingdom but broke down with regard to asexual reproduction and many plants. In his introduction to the influential volume *The New Systematics* (1940b), Huxley again emphasized that also with regard to humankind, the new species concept failed: "So it does in man, who exhibits a peculiar form of reticulate descent consequent upon extreme migration" (21). In a book that did not deal with humans, Huxley considered it necessary to insist that human evolution had not been mainly a process of differentiation but of convergence. The human phylogenetic diagram was a net. Not only could there be no speciation event; there were no subspecies comparable to those in the animal kingdom. Huxley thus on the one hand cautioned about biologically reifying what were in fact historically grown cultural population labels. On the other hand, he criticized the iconography of the tree and refrained from visualizing human kinship.

Again in stark contrast, the development and visualization of phylogenies on the basis of human genetic variation was a central concern of Cavalli-Sforza's lifework. It is therefore in part 3 that I engage most closely with the power of the phylogenetic diagram. While in genetics it retained the problem of portraying human diversification without intermixture and perpetuated some of the old population labels (Takezawa et al. 2014), Cavalli-Sforza, like Huxley, believed that knowledge from human population genetics could help replace xenophobia and racism with altruism and panhumanism. This discourse of human population genetics as demonstrating the scientific indefensibleness of race also informed the Human Genome Diversity Project and is continued in the Genographic Project. Despite minor setbacks, the latter project has enjoyed enormous success with a huge public. The notion of history and ancestry in the DNA has become a decisive part of some historical cultures. While scholars from the humanities and social sciences have generated a critical discourse, warning about the potentially disruptive and essentializing effects of genetic identification, increasing numbers of the public partake in the markets as well as virtual worlds of genetic ancestry and build real-worldly connections on "the history in their genes." In what I conceptualize as the genographic network, global projects, local labs, companies, and active consumers have long since connected in novel formations focused on genetic phylogeny. And we will observe instances in which identities become hardened through commercialized genetic history, as well as cases in which people playfully patch up personal remembrances, cultural myth, and genetic information to build their own stories "from genes."

PART I

History in Bones

Henry Fairfield Osborn (1857–1935) at the American Museum of Natural History

The American Museum of Natural History (AMNH) presents to the visitor the different branches of knowledge that throw light on the history of the universe, the earth, its life, and of humans and their cultures. Besides paleontological and geological exhibitions, there are astronomic, biological, and anthropological halls. The museum celebrates evolution. Its message is that "our" history is an evolutionary history and spans an immense amount of time, during a comparatively tiny part of which the earth had been populated by animals and humans that are now mostly lost. An excerpt of that evolutionary history—the history of the vertebrates—was already conveyed to a large New York, American, and international public in the decades around 1900 through the terrific Hall of Fossil Reptiles, Hall of the Age of Mammals, and Hall of the Age of Man that had been created under Henry Fairfield Osborn's aegis and "finished" in 1905, 1895, and 1924 respectively. And already Osborn emphasized that comparative anatomy and paleontology were sciences concerned with "our history" (Osborn 1927a, 146). In what follows, I focus on how the AMNH under Osborn functioned as a meaning-making machine regarding "our" evolutionary past.

As Ronald Rainger (1991) has shown in his scientific biography, Osborn belonged to the New York City elite, a network that played its part in securing Osborn the double mission of establishing a department of vertebrate paleontology at the AMNH and building up a biology divi-

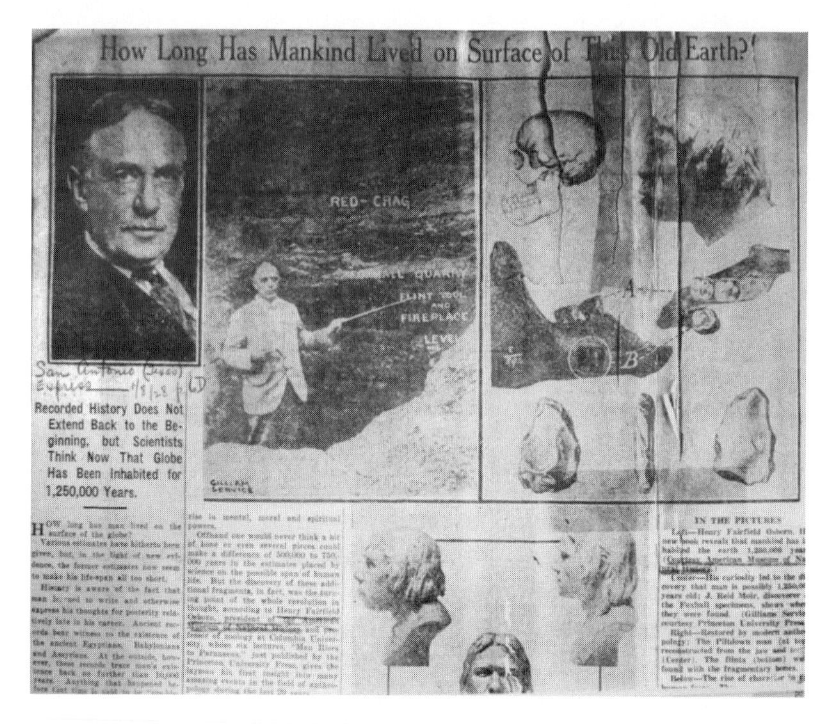

How Long Has Mankind Lived on Surface of This Old Earth?

Recorded History Does Not Extend Back to the Beginning, but Scientists Think Now That Globe Has Been Inhabited for 1,250,000 Years.

FIGURE 1. "How Long Has Mankind Lived on Surface of This Old Earth?" *San Antonio Express* (Texas), 8 Jan. 1928, 6D (© San Antonio Express-News/ZUMAPress.com), American Museum of Natural History Library, the Papers of Henry Fairfield Osborn (1857–1935) MSS.0835, Series IV: Books, Box 96, Folders 5–7: Post publication letters and reviews, "Man Rises to Parnassus," Folder 7

sion at Columbia University in 1891. The museum curatorship and presidency from 1908 enabled Osborn to carve out a place for the traditional disciplines of comparative anatomy and paleontology in university education besides the new experimental biology, and he recruited university graduates to set up these branches of research at the museum. Through this staff, the museum's department of vertebrate paleontology became an international hub in the exchange of experts, knowledge, and objects. The museum acquired an archive of earth history for the department through acquisition and exchange as well as the organization of expeditions. This made it possible for Osborn and a diverse team of experts to reconstruct the lost worlds in exhibitions for the new mass audiences. As figure 1 suggests, Osborn and his staff made the museum reach out beyond its walls also with regard to human evolutionary history.[1]

Figure 2 shows Osborn's library, located on the fifth floor of one of the turrets of the museum building. It functioned as his study and serves as a starting point to introduce my main concerns. It is like a microcosm of the global genealogies that Osborn helped establish and into which he inscribed his person and work. On the desk were photographs of his two sons. The walls were covered with engravings and inscribed photographs of pioneer scientists, including the paleontologist Georges Cuvier, evolutionists Georges-Louis Buffon, Charles Darwin, Alfred Russel Wallace, and Thomas Henry Huxley, the great geologist Archibald Geikie, and the hero discoverers of the supposedly oldest human-made tools: James Reid Moir and Ray Lankester. There were photographs of explorers such as Robert Peary, Fridtjof Nansen, Theodore Roosevelt, and Richard Byrd; cases filled with books and unbound pamphlets; and most of the remaining floor space was taken up by long tables that held fossil bones and teeth.[2]

Osborn stated in an address that his people came from a pure English stock that could be traced back to Scandinavia. He explained that his

FIGURE 2. Osborn's office, by A. E. Anderson 1905, American Museum of Natural History Library Photographic Collection, no. 333451

maternal surname *Sturges* was derived from the Scandinavian *Sturge*, meaning "strong." *Osborn* came from *Asbiörn*, Scandinavian for "divine bear": "Considering my roving propensities, combined with much old-fashioned religious sentiment, I like to think of the fusion of characteristics implied in the three words 'strong-divine-bear.' " On his mother's side, Osborn claimed among his closer ancestors a colonial warrior, "Indian fighter," and leader in political, military, and ecclesiastical affairs. His grandfather, Jonathan Sturges, rose to become a leading merchant of New York, a cofounder of the Illinois Central Railroad, and president of the Chamber of Commerce. Osborn presented his father, too, as a self-made man who engaged in the East Indian trade and became a railroad tycoon. With this heritage Osborn accounted for his roving propensity and strength. To explain his religious sentiment, he referred to his birthplace Fairfield (Connecticut), "with its stern convictions as to what is right and what is wrong, with its pioneer spirit of Christian education and civilization."[3]

These family morals come close to what Donna Haraway ([1989] 1992, ch. 3) has called "teddy bear patriarchy" in her analysis of the AMNH's African Hall. She views the exhibition that Carl Akeley developed under Osborn's presidency as a crystallization of the postbellum bourgeois values. The expression resonates with Osborn's self-designation as strong-divine-bear. However, it is meant as a pun on Theodore Roosevelt, after whom the toy animal was named. Roosevelt was chair of the museum board and the AMNH's most powerful patron. He became president of the United States in 1901, and in 1912 founded the Progressive Party, which stood for many reforms in education, housing, and labor. It was an alternative to both the traditional conservative stance toward social and economic issues and to various more radical streams of socialism and anarchism. It was associated with a patronizing philanthropy, also manifest in the museum's goals. Haraway in particular shows that the dioramas in the Akeley Hall expressed gender and race stereotypes. Finally, Roosevelt also embodied the cultural tropes of the new masculinity, the adventurous and strenuous life, and stood for the ideals of the outdoor and conservation movements (see, for example, Roosevelt [1905] 1991). As we will see, these moral concepts were conveyed by museum exhibitions as much as by Osborn's description of his forebears. And as the progressive politics were accompanied by conservative aims, the idolatry of the frontiersman, the adventurer, and the explorer took place within industrial capitalism. Indeed, the Jesups, Dodges, Morgans, and Roosevelts of politics, finance, and transportation constituted the trustees of the AMNH.[4]

This paradox of striving for social progress and a nostalgia for "the old ways" has often been ascribed to the transformations wrought upon society and landscape through the surge in industry, the increase in population and immigration, urbanization, money-driven business values, and such "isms" as corporate capitalism, commercialism, utilitarianism, materialism, and scientism. These transformations triggered a backward and inward orientation, a turn to the "original ways of being" most strongly connected to the loss of what was perceived as typically American wilderness and wildlife. The closing of the frontier in particular became a rationale for anxieties about the loss of manliness and adventure, of Darwinian struggle and individuality, and by inference about the degeneration of the individual, "the race," and the nation. These anxieties were at the heart of the cult of the primitive, of which the second decade of the twentieth century saw many expressions beyond the conservation efforts and the outdoor movement: the establishment of Boy Scouts and hunting clubs; the nature writing of Jack London, John Burroughs, and John Muir; as well as a more widespread aesthetics and religion of nature and neo-Romanticism. To preserve, restore, or re-create wilderness also meant to ensure the survival of the real American type, of chivalrous values, self-reliance, general fitness, and morality.[5]

Re-creation in the sense of the reconstruction of nature in urban settings and the regeneration of the modern citizen was also a goal of the museum (Mitman 1996). Rainger (1991) has elaborated on how Osborn's religious background and social position influenced his work in science and education. Osborn largely shared the moral outlook of the powerful and rich New York elite. They wanted the museum to be an instrument of modernization as well as a stronghold against the decay of traditional values. Like the universities, libraries, parks, and zoos that were being established, the museum was to function as a space of civic education. The racism and Nordic supremacism that lay beneath Osborn's genealogical self-identification as being of Scandinavian and pure English stock were the flip side of this progressive effort. In his intellectual biography of Osborn, Brian Regal (2002, ch. 5) engaged with the direct synergic relationship between Osborn's scientific and popular work and his involvement in the eugenics and anti-immigration movements, which he shared with friend and museum trustee Madison Grant. Osborn considered it necessary to prevent excessive immigration of south European and Asian types; to help the "multiracial" children in New York to improve themselves according to their potential; and to preserve the natural order of "races," classes, and sexes against the erosive forces of the blacks' and women's movements. "Preservation" was generally an important concept in Osborn's great project. He advocated not only the conservation of nature and animal species, but also the preservation of, above all, the "Nordic race" (also Rainger 1991, ch. 5).

Osborn's predecessor to the presidency, the banker Morris K. Jesup, had been involved in forest preservation from the 1880s onward (Adirondack Forest Preserve, 1885). Exhibits at the AMNH expressed concern for the preservation of American nature, animals, and "primitive peoples." These efforts of the 1880s and 1890s were developed into a specific museum policy of conservation under Osborn. He conceived of his scientific work and that of his colleagues at the museum as a grand effort in establishing an archive of the present and past for future generations: "'The reason why certain of our expeditions are being pressed so hard at the present time,' he said, '. . . is that the natural beauty and life of the world are vanishing with a rapidity that is unbelievable, both among the native races of men and of mammals on land and sea. Unless we secure the records of these races now we shall never secure them. . . . To get these things while they are procurable, so that other generations may know what the world's life has been, is a great labor and a great duty.'"[6]

Osborn thus also engaged in the Boone and Crockett Club, founded by Roosevelt in 1887 for animal study, protection, and hunting (another well-known paradox), as well as in the Association for the Protection of the Adirondacks, the American Bison Society, the Save the Redwoods League, the National Conservation Congress, the National Parks Association, the American Nature Association, and many more. That the effort of conservation for future generations encompassed cultural preservation is manifest in Osborn's participation in organizations such as the American Scenic and Historic Preservation Society and memorial associations and commissions. Regarding his family history, Osborn was a member of the Fairfield Historical Society and the New England Historical Genealogical Society. Both the preservation of American natural and cultural history, and of his stock were integral to the attempt to protect and restore "the race": an attempt to which his engagement in the Immigration Restriction League, the Galton Society, the American Eugenics Society, and the Aryan Society provide direct testimony (Osborn 1930b, 139–143).

However, for Osborn, the history and genealogy that needed to be preserved and reconstructed reached much further back in time. He succeeded in institutionalizing American leadership in vertebrate paleontology, and as president of the museum, he made the evolutionary history of vertebrates the main focus of research and exhibition. In doing so, he stood on the shoulders of Joseph Leidy, Nathaniel Marsh, and Edward Drinker Cope, and he could profit from the role fossils had played in nation building. As Keith Thomson (2008) and others have shown, paleontology had long since been providing historical narratives and American icons. This was true not only for the deep histories and animals such as the mastodon that were reconstructed from fossil remains, but also for the history of American paleontology and its pioneers. The traces of the American deep past were national treasures; they were also bones of contention between men and institutions devoted to paleontology, a natural history that was strongly associated with the westward movement and the resulting territorial conflicts. In "epic efforts" and public feuds, Marsh and Cope spearheaded the discovery of many dinosaur species in the 1870s. When Osborn later organized museum expeditions to the American western states and territories, this triggered the "second dinosaur rush," in which the Carnegie Museum of Natural History in Pittsburgh and all the other major museums followed with collecting expeditions.[7]

Concomitantly, dinosaurs conquered popular culture. A mastodon mount had been America's first reconstruction of a fossil vertebrate, and in the 1860s, Benjamin Waterhouse Hawkins mounted the skeleton of a Hadrosaurus. Such a Hadrosaurus mount was bought by the Princeton University Museum. Arnold Henry Guyot, one of Osborn's teachers at Princeton, hired Hawkins to paint murals of America's Late Cretaceous environments. No doubt, these early reconstructions influenced Osborn's later work at the AMNH. Generally speaking, even though there existed a tradition of mounting fossil vertebrates at museums, and even though the well-known Henry A. Ward of Rochester, New York, sold finished specimens, "it was not until Henry Fairfield Osborn pioneered the display of mounted skeletons at the American Museum of Natural History in 1891 that the modern era of fossil display began" (Thomson 2008, 312). With the mountings of *Diplodocus carnegiei* at the Carnegie Museum and the *Tyrannosaurus rex* discovered by Barnum Brown for Osborn at the beginning of the twentieth century, America had its new popular icons.[8]

Osborn deplored the fact that when his teams embarked on the job of reconstructing *Tyrannosaurus rex*, hardly anyone knew what a dinosaur was. He sent out invitations to important people, summoning them to the museum for a dinosaur tea—a *T. rex*. The event was covered by the newspapers, which triggered interest in the dinosaurs at the museum. In the end, Osborn would pride himself on having turned *dinosaur* into a household word: "More than that, it means something to them, a linking up of the present with the past."[9] Osborn took great pains to ensure that the general public made this link between present and past, to render the strange creatures and scenes from history meaningful to people. Fossil vertebrates and especially dinosaurs could well serve to teach the lessons of nature to museum visitors. But "our" own deeper history was most instructive. When groundbreaking finds were made, Osborn therefore turned to paleoanthropology. With the remains of *Pithecanthropus erectus* (today *Homo erectus*) brought to Europe from Trinil (Java) in 1895 by Dutch physician Eugène Dubois (1858–1940), there was fossil hominid evidence from outside Europe and beyond the age of the Neanderthals. In France, the murals at Les Eyzies proved the claim that the Cro-Magnons had engaged in cave painting, and the great paleoanthropologist Marcellin Boule described the nearly complete Neanderthal skeleton of La Chapelle-aux-Saints. Soon afterward "Piltdown Man" (*Eoanthropus dawsoni*) was "discovered" in Sussex, England—a combination of an apelike jaw and a rather modern-looking braincase that was only decades later definitively exposed as a forgery. Beginning with a tour of the major sites in France and Spain, Osborn made paleoanthropology his field of study and promotion. His friend and important collaborator, William King Gregory, took up primatology. Gregory had graduated under Osborn and began at the museum in 1911, eventually advancing to the position of head of the new Department of Comparative Anatomy in 1920 (Rainger 1991, 99–100).

It was especially through human bones that Osborn sought "our" evolutionary history and its meanings for present identities and future goals. As the apex of evolution, humans carried this history in their bodies. The human being was a "palimpsest—a new writing upon the almost obliterated traces of an old" (Osborn 1927a, 147). Bones were "full of proof of our past and present evolution and of signs of what the future man will be in body and mind" (ibid., 146). However, we will see that this encompassing picture of one evolutionary history within the body of all was in fact constantly undermined by the claim of particulate histories in bones. Indeed, during an interview given to a journalist in 1928, Osborn "revealed his uncanny ability to trace the racial ancestry and, to some extent, mental characteristics of people from the shape of the skull and head, a natural result of his profound study of human evolution. . . . He has often astonished acquaintances by telling them with great accuracy the racial strains from which they sprang, the country in which they were born, and often the exact locality of their natal or ancestral place."[10]

Thus, Osborn could read the evolutionary past from bones, and this past was seamlessly linked to a person's national origins and character in the present. But it was not only the human bones—fossil and recent—that allowed him to draw moral lessons from history and devise policies for the future. To begin with, Osborn wanted an integrated anthropology, one that synthesized the whole galaxy of inorganic and organic sciences (1927c, 481–482). In particular, he relied on archeology. In Osborn's museum study, there hung the photographs of the men who had advocated the acceptance of what they took to be the so-far oldest known tools. It was the progress from these humble beginnings to the most elaborate culture of the Cro-Magnons that seemed most instructive to Osborn. The ways of life of the extinct humans appeared as a place to look for the educational ideal. The AMNH was to be the antitoxin to the modern diseases: "Our greatest idea in the Museum of Natural History is to counteract the bad influences that exist to-day."[11] This was to be achieved not only by substituting the experience of nature "for those who cannot travel" (Osborn 1927a, 240; see Rainger 1991, 119–120), but by substituting the experience of "our" deeper history for those who cannot travel *in time*.

Around the turn of the century, the museum approached the half-million visitors mark; in 1920 it counted around 1,750,000 visitors; six years later, the number exceeded two million (AMNH 1927, 3). This was a considerable crowd to teach the lessons of evolutionary history to, but the museum instructed not only through exhibitions. At the end of the nineteenth century, the New York school system had been centralized, schooling was mandatory, and schools were desegregated. In the *World's Work*, Osborn boasted that the museum system reached the entire city, extending to five hundred schools, thereby seeking to develop the moral qualities of the youth. Everyone needed to learn that evolution was at work always and everywhere and that "unless we heed its laws we can expect only disaster."[12] In fact, during Osborn's presidency of the museum, the natural history lessons for teachers from the state of New York, for whom a special reading room had been built, were supplemented with lectures for school classes and the lending of learning materials such as small collections of objects or slides of exhibits. Quite literally, the trucks in figure 3 connected the museum to the network of schools. The museum's Department of Public Instruction (later Education) coordinated these outreach activities, which took the lessons of nature into the local public school classes that, in 1905, consisted of 70 percent first-generation immigrants. In 1916, Osborn spoke of hundreds, on some days even thousands, of children who visited the museum and for whom the exhibits had to be self-explanatory and illuminating. In the 1920s, he bragged that up to six million pupils and students received occasional instruction (Osborn 1927a, 244, 258; Rainger 1991, ch. 3).[13]

Finally, Osborn reached much wider audiences with magazine and newspaper contributions: "It proved impossible . . . to remain indifferent to the demand for the diffusion and dissemination of knowledge among both the adult and the juvenile population of our great city. This kind of education is carried on in the exhibition halls of palaeontology, which

FIGURE 3. The five trucks of the Department of Public Education for use in delivering slides and nature study collections to schools, by H. S. Rice 1917, American Museum of Natural History Library Photographic Collection, no. 311774

have made the hitherto unknown animals and men of the past familiar in every household, in magazine and newspaper articles that have broadcast interest in the subject throughout the nation and the world" (Osborn 1930b, 60). Time traveling was also facilitated through talks accompanied by lantern slides and well-illustrated publications. Through this, Osborn sought to awaken "latent predispositions and tastes" in his audiences, desiring that the young and old of all races be restored or developed to the best of their type. Osborn claimed that this was no theory but the record of actual experiences from museum visitors (Osborn 1927a, 259, quote from 260; see also Osborn 1925).

How exactly did Osborn think his verbal and visual reconstructions from the deep and Paleolithic past could reactivate latent predispositions and tastes? In order to answer this question, I have to engage with his notions of a visual memory, a racial soul, and a race plasm, which I do in chapter 1. From this vantage point, I situate Osborn's reconstructions of evolutionary history in the contemporary historical culture: did Osborn's strategies succeed with diverse audiences? His successful popular books *Men of the Old Stone Age* (1915) and *Man Rises to Parnassus* (1927) were very much expressions of the concerns of their time, so that prehistory was made living in the present. Through their engaging narratives and images, the books allowed "the men of the Stone Age" to travel into academic disciplines and households. To be able to survive in new and foreign fields, however, "the ancestors" were adapted to the needs of various knowledge cultures and to existing ideas about history. They also met with resistance and contempt (chs. 2 and 4).

Although *Men of the Old Stone Age* was Osborn's first major contribution to human evolution, he wanted more than paper ancestors. In a decade's effort, his team built the Hall of the Age of Man. In chapter 3 I show how they acquired casts of hominid fossils and artifacts from other institutions. This was hard work in and of itself, because a network of exchange had to be established at a time when Europe was at war. Finally, Osborn also wanted more than casts. He organized the Central Asiatic Expeditions in search of "fossil man." The expeditions were very successful and popular with the American public, but they were also locally contested and a failure in terms of finding human ancestors. However, if Osborn did not have his own fossil ancestor to put on show in the hall, he more than compensated with murals of prehistoric scenes. A close look at the way in which these were produced by the painter Charles Knight will bring to light the transformative aims Osborn had with the halls of the evolutionary past as well as dire conflict about the authority over visions of "our" history.

Osborn was so concerned about bringing the right narratives and images of "our" evolutionary past to people that he tried to control the reception of his books and the way his exhibitions were understood. In chapter 5 we will see that he even interfered with the productions of others. He had direct influence on authors of popular and children's books concerning evolution, and he cooperated with fiction writers. However, all kinds of fossil men came alive. The rise of paleoanthropology coincided with the transformation of the American centers into knowledge and expert cultures, in which general education and the mass media opened up the public and private spheres (Oleson and Voss 1979). In the mushrooming popular culture, there was an expression of yearning for the primitive, the original, and the past. Finally, Osborn's narratives and images from "our" evolutionary history were not only challenged by lay writers and in prehistoric science fiction: with the advent of genetics and the rise to prominence of the experimental sciences, he increasingly found his natural history under attack. His understanding of the evolutionary processes was jeopardized by the new Mendelian genetics and neo-Darwinism. His science from bones—and his politics from bones, too—were questioned even within the museum walls. Nonetheless, it was only after his death that a synthesis of paleontology and genetics would be reached that rendered Osborn's race plasm and racial soul obsolete.

CHAPTER ONE

From Visual Memory to "Racial Soul"

During his education and later as professor at Princeton, Osborn engaged in embryology, comparative anatomy, neuroanatomy, geology, and increasingly paleontology. He came under the influence of James McCosh—president of the university and among those religious leaders who publicly endorsed Darwinism, reconciling it with the Presbyterian faith. Osborn was also among his students in psychology, including aspects of mental imagery and visualization. On the matter of psychology, Osborn further studied and conversed with the British polymath Francis Galton. Osborn stated that "nearly a year devoted to research in visualization and imagery, as American colleague of Sir Francis Galton, cousin of Charles Darwin, was of inestimable aid in the art of picturing past conditions and forms of life, as well as in the art of reconstruction and restoration of mammalian forms" (Osborn 1930b, 55–56). This year of research promises insights into Osborn's practice of reconstructing evolutionary history in image, text, and exhibition. It also offers valuable clues to his understanding of how these reconstructions would affect the observer and reader.

In 1884, Osborn published three papers on visual psychology, for one of which he collaborated with McCosh in "A Study of the Mind's Chamber of Imagery" (1884). McCosh defined memory as the recognition of an object or event perceived in the past. All senses could imprint themselves on the mind through the stimulus of touch, sound, and image, with the most vivid memories stimulated from the sense of sight. Well-defined shapes "leave a photograph of themselves on our souls" (Osborn and McCosh 1884, 52), producing a store of images or photographs of objects in the mind's chamber of imagery. The more dramatic the event experienced, the more vivid would be the memory of it, including the thoughts and feelings it evoked. Besides the state of the brain (conceived of as a certain disposition of the molecules in the gray matter), the attention given to an object or event played a role for the strength of a memory and the power of its recollection.

To elaborate on the factors that impacted on the ability of forming images in the mind's eye, Osborn and McCosh carried out a survey of Princeton and Vassar College students, through the distribution of Galton's "Questions on Visualising and Other Allied Faculties" in 1881. The circa sixty questionnaires also found their way into Galton's *Inquiries into the Human Faculty* (1883). Osborn subsequently changed the questionnaire to better suit his purposes and distributed it among Princeton and Harvard Medical School students with the help of the psychologist and educator G. Stanley Hall. One of the results was that direct experience was often impossible to distinguish from mediated experience. Even written accounts, such as "a word-painting of the scene and of the man or woman" (Osborn and McCosh 1884, 55) given by travelers, biographers, and historians, might be turned into personal remembrance once their true source was forgotten. It seemed of paramount importance that the mind was imprinted with the right kind of images—be they direct perceptions or reproductions of such in speech, text, or image: "Nothing tends more to degrade the mind and sink it in the mire than low and sensual images rolled as a sweet morsel under the tongue. On the other hand, images of duty, of self-sacrifice, of courage, of honor, of beauty, of love, elevate and ennoble the soul" (ibid., 56).[1]

Despite their description of memory in terms of photographic impressions, the knowledge of the physiology of vision was advanced enough for Osborn to know that it was not an image, but a signal that traveled from the retina through nerve fibers to the visual area of the brain. Thus, in order for a person to recall an image, the signal had to be repeated. Nonetheless, in the paper he wrote on visual memory, Osborn used the metaphor of the tablet of wax or of steel to clarify the different degrees to which visual impressions imprinted themselves on human minds. Osborn agreed with Galton that the faculty of visualization was inherited, with a keener expression in women than in men, in children than in adults, and in "savages" than in civilized people. This apparent presence of onto- and phylogenetic gradations strongly supported the claim for the visual memory's naturalness, but did not preclude an aspect of nurture. On the contrary, the capacity for visual remembrance was seen to be subject to cultivation and improvement as well as to decay under disuse or excessive abstract thinking (Osborn 1884a).

The insights that the visual memory was stronger in "savages" and could weaken through abstract thinking seemed to confirm new tenets in psychology and education such as those held by Osborn's collaborator Hall. It seemed that while the development of abstract thinking was a particular achievement of civilized men, this process, and the concomitant weakening of the visual memory, had proceeded too far. According to Hall and a common notion of the time, education was a public good, but too much education and civilization in general caused neurasthenia, a state of "racial decadence," individual degeneration, and effeminacy. Both Osborn and Hall felt the lack of struggle in education and women as educators had a particularly weakening effect on the growing male. Hall's remedy consisted of an accentuated revival of the stages of human evolution from savagery to civilization in developing and educating the American boy into the American man (Bederman 1995, ch. 3). Certainly, the education of children in evolutionary history through objects and images in and from the AMNH could contribute to that aim. But more knowledge was needed about "the influences of heredity, of race, of cultivation or neglect" (Osborn 1884a, 450) on the visual memory.

This hope for insights into the connection between memory and heredity was expressed at a time when heredity-memory analogies guided many evolutionary theories. As the underlying mechanisms of heredity were unknown, some "Lamarckians" thought of the acquisition of a new character in this light. Cope, Osborn's paragon in paleontology, represented the American "neo-Lamarckian school." The memory analogy supported the assumption of the inheritance of acquired characteristics as well as recapitulation theory (Cope 1896). Characters might be inherited proportional to the intensity and frequency of the producing stimuli, just as in the process of learning, something is retained in memory through repetition over time. Morphology, as it unfolded in the process of ontogeny, therefore amounted to something like the organism's memory of its evolutionary history. Also where behavior was concerned, inferences were drawn from the observation that actions that were initially triggered by conscious thought eventually became automatic with frequent repetition—as in piano playing. Similarly, instincts were understood as the unconscious remembrance of intensely learned behavior and experience. Such behaviors and experiences were stamped on a species' germplasm or germ cells, some speculated by vibrations and wave motions, others by electrical potentials, or chemical changes. The German physiologist Ewald Hering argued that the vibrations of an external stimulus were transferred to the nervous system and from there to all organs. The nervous system could in time reproduce series of vibrations that in the beginning had involved the participation of consciousness. Because the nervous system reached the developing gametes, the vibrations were imprinted on the hereditary material (Hering 1870; Butler, e.g., 1880; Semon 1904; see Gould 1977, 85–100).

Osborn was aware of such theories that rendered it feasible to speculate about the possibility of remembering experiences made by ancestors that had somehow been imprinted in their germplasm. This would suggest that the repetition of stimuli habitually made in the evolutionary past could trigger similar ideas, feelings, and actions in the present human being as they had done in his or her phylogenetic ancestors. Osborn was also acquainted with developmental models of the mind that conceived of it as a hierarchical series of functional levels, with the voluntary functions overlaying and suppressing the more involuntary ones. The higher levels were only acquired later in evolution, so that according to recapitulation theory, children went through, and the insane represented, a stage on the evolutionary pathway that humanity as a whole had once traveled. However, the lower levels of the mind were also still present in healthy adults and were temporarily unleashed during dreaming. Such perceptions opened up the possibility that current humans somehow had access to primeval mental states and even memories.[2]

Osborn was clearly influenced by this knowledge in his understanding of how reconstructions of evolutionary history might work on the recipient's mind. However, he denied a phylogenetic memory of this kind. Recollections were triggered by a perception resembling earlier impressions or through the revival of an association of ideas. But these recollections were restricted to a retrieval of impressions or ideas from earliest childhood. He maintained, nonetheless, that in cases where a partial recollection came with an indefinite sense of when and how it originated, past real and imaginary images might become confused. An incomplete recollection may turn an imaginary image of the past into the vague remembrance of an actual perception of the past (Osborn 1884b). In sum, out of Osborn's research on memory arose the following conclusions for his later reconstruction and exhibition work at the museum: It was important to provide the right kind of strong and vivid impressions—

through word and sound, but most of all through images and visual displays. This would prevent degeneration of the capacity of the visual memory through excessive abstract thinking especially in modern men. It would build up a store of "good" images—of duty, of self-sacrifice, of courage, of honor, of beauty—that were favorable to the development of the character; if museum visitors were to appropriate the reconstructions as impressions of their own past, they had to be as true to nature as possible.

Moreover, despite the fact that it was the capacity for remembering that was inherited rather than the memory itself, Osborn thought that our minds were strongly phylogenetically shaped. The combined study of heredity, of prehistoric humans, and of the character of present men had brought him "the increasing conviction that our intellectual, moral, and spiritual reactions are extremely ancient and that they have been built up not in hundreds but in thousands—perhaps hundreds of thousands—of years" (Osborn [1913] 1924, xvii). Osborn regarded the lives of his idols, the men who in his judgment had achieved self-fulfillment, as a particularly illuminating source for the analysis of the human mind. They provided him with insights into the nature of the "racial soul, mind, temperament, and intellect." The lives of the naturalists, explorers, and nature poets Osborn adored became examples for "the best of these racial characteristics." Building on his knowledge of heredity and race, Osborn felt that his work on the biographies of the Scottish-American philosopher, author, naturalist, and conservationist John Muir, of the naturalist and nature writer John Burroughs, of Roosevelt, and others rendered him able to "separate the Scotch from the English and both from the Irish" (ibid., xvi). In Osborn's biographies we thus stumble on sentences such as: "In feature and in spirit of the Nordic stock, with a dash of Celtic temperament, Burroughs was true to his heredity" (ibid., 188).

In fact, Osborn thought that his writings on Burroughs and William Wordsworth were an original contribution not only to anthropology but also to literary theory. He boasted to the Columbia historian William M. Sloane, whom he had already known at Princeton, that "so far as I know, no one before has brought out this idea that the spirit of the poet reflects the past history of the race to which he belongs. Now that we have positive chronology of John Burroughs' people back for no less than twelve thousand years, this generalization is firmly based, not in speculation, but in knowledge."[3] Osborn's notion of the "racial soul" (Osborn [1913] 1924) came very close to a "racial memory." He claimed that

Wordsworth's "Ode on the Intimations of Immortality from Recollections of Early Childhood" made him realize that he himself had come to the same conclusion by different means, "namely, that the human soul is full of reminiscences and that it responds to conditions and experiences long bygone" (ibid., 183). The racial soul, in Osborn's view, was an adaptation to a particular natural environment that in the course of evolution resulted in race-specific predispositions of morals, of intellect, and of spirit. In "modern man," the soul responded to conditions and experiences that resembled those found and made in the formative phase of human evolution. In the 1880s, Osborn would still have been able to explain this racial soul on the basis of the inheritance of acquired characteristics in terms of a phylogenetic memory such as proposed by Hering. However, when he wrote the above words, the situation had changed.

Osborn's notion of a racial soul rested on the assumption of a racially specific evolutionary history and genetic makeup that was still heir to the American neo-Lamarckian school, for which use-inheritance, recapitulation, and also orthogenesis were core concepts (Bowler 1986, ch. 2). However, Osborn counted among his friends Edward Poulton at Oxford University, who was a strong advocate for Darwinian natural selection and who had translated the work of the German biologist August Weismann into English. Weismann had demonstrated the independence of the germplasm from the soma. As a result, Poulton had not given Osborn's "neo-Lamarckian school more than three years more of life" as early as 1888.[4] Indeed, Osborn began to search for ways to account for the fossil record without a direct influence of the environment on the hereditary material. As if to fulfill Poulton's prophecy, he reacted to Weismann's theory of heredity in *Nature* and *Science* in 1890. He defended the American neo-Lamarckian tradition, starting out from the age-old critique of Darwin's theory as unable to explain "the rise of useful structures from their minute embryonic, apparently useless, condition" (Osborn 1890, 110). He discussed traits such as the molars and feet of horses that manifested an evolution with definite direction. He granted that such progressive trends might not be due to the mechanism of use-inheritance. At the same time, he could not follow Weismann and Poulton in their seemingly exclusive reliance on natural selection. Evolutionary trends could not be accounted for by Darwinian randomness. One therefore had to assume some as-yet-unknown evolutionary factor to explain directed variation.

With the rise of genetics at the beginning of the new century, Osborn

(1907) sided with the neo-Darwinians such as Weismann and Poulton especially regarding the gradual nature of change. According to the Dutch botanist and geneticist of the first hour Hugo de Vries's influential mutation theory, evolution took place by large mutations that produced reproductive isolation from the parent stock. Selection was a purely negative agent, weeding out those new mutations that were not adaptive. Other influential Mendelians, such as the British geneticist William Bateson and Thomas Hunt Morgan in the United States, too, believed in evolution by mutations that amounted to saltations in evolution through the production of new phenotypic characters. Osborn repeatedly answered this claim by referring to the paleontological record that clearly evidenced gradual evolution. He emphasized that the paleontologists "have the best perspective, for they see the evolution of characters through long ages, without regard to particular organisms or species in which, for the time, they are perhaps being manifested."[5]

In the midst of the controversies about heredity and evolution, Osborn came up with a theory that could account for the trends he recognized in the paleontological record, particularly in the Titanotheres (extinct large-hoofed mammals, often horned) and Proboscidea (truncated mammals). The theory made some use of natural selection without completely surrendering to chance. The solution was a particular brand of orthogenesis, or what Osborn referred to as rectigradation. It stated that the potentials for the expression of evolutionary trends in characters lay latent in the germplasm and were activated by behavioral characteristics that were maintained over long periods of time. Changes in environment and habit would cause changes in ontogeny that kindled the hereditary potential and set in motion gradual evolutionary change along determinate lines. Natural selection thus selected among the end products of evolutionary trends instead of being causal in their formation. While such trends were often adaptive, trends that led to overspecializations such as oversize horns could lead to the extinction of the species (Osborn 1908, 1922b).

Osborn transferred the observation of evolutionary trends driven by the germplasm to the human realm. Also in the case of hominids, habitual behavior that represented an adaptation to a particular environment could trigger evolutionary trends lying dormant in the germplasm or race plasm ("race plasma," Osborn 1891). Once set in motion, a gradual unfolding of characters, similar to individual development, took over. As Rainger (1991, 124–130) has explained, this theory was not vitalistic or deterministic. Rather, in accordance with Osborn's philosophy of the self-made man, the individual as well as the environment were ascribed an active part. Humans had been "the architects of their own destiny" (Osborn 1927c, 483). Osbornian evolution was nonetheless progressive and potentially purposeful. It was a compromise between hard heredity (that however predetermined evolutionary trends) and the notion of progress through effort (or "Lamarckian" evolution), since long-term behavior could influence just what part of the race plasm would be activated.

Translated into a phylogenetic diagram, Osborn's orthogenetic theory corresponded to a structure with different hominid genera, species, and races on parallel branches. Once separated, the lines of the "modern races" showed no vertical connections, because Osborn considered miscegenation a relatively recent phenomenon. Because of the differential potentials of their "race plasms," and the different environments and habitual behaviors that had characterized their evolutionary histories, the "living human races" differed anatomically, but also in intellect, temperament, and spirit. They had different "racial souls." In the course of his career, Osborn accentuated the racial distinctions. In *Men of the Old Stone Age* (1915), hominids and *Homo sapiens* were still relatively young appearances. This had changed by the time of *Man Rises to Parnassus* (1927b). The so-called dawn man theory of the 1920s and 1930s turned relatively modern humans into creatures of the Oligocene, who had sprung in parallel with the ape family from a neutral stock. Just as the taxonomic and phylogenetic gap between apes and humans was thereby considerably enlarged, Osborn argued that the types of living humans (*europaeus*, *asiaticus*, *afer*) would be classified as different species or genera in zoology, with, for example, *Homo europaeus* ("Caucasians") containing what are in fact the species *nordicus*, *alpinus*, and *mediterraneus* (Osborn 1927b, 169).

Osborn had developed this kind of parallelism in the study of mammalian fossils. In *The Age of Mammals* (1910, 29–34), he explained what he called *adaptive radiation* and *polyphyletic law*. The concept of adaptive radiation was an adaptation of the Darwinian principle of divergence to explain the geographical distribution of species within his orthogenetic framework. By reacting to different environmental conditions, organisms triggered evolutionary development in diverging directions (*law of polyphyleticism*). This differentiation was limited by the fact that similar environments were found in different geographic regions and continents. The law of analogous adaptations (*parallelism* or *homoplasy*) said that because of a limited number of possibilities for adaptation, related or unrelated types of animals produced evolutionary changes running in parallel in similar environments in different geographic regions. Similarity of anatomy therefore had to be carefully distinguished from similarity of ancestry. Through this process of adaptive radiation, Osborn showed, many families of mammals had already differentiated in the early Eocene. By the end of that era, they had further subdivided into forest- and plain-loving types, which was followed in the Oligocene by a forest- versus plateau-loving divide. Osborn now postulated that this must have happened in primate evolution: Simiidae and Hominidae split in the Eocene, and the Neanderthaloids and the line leading to modern humans (pro-man stem) diverged in the Oligocene through adaptation to a forest region and plateau region, respectively. The law of irreversibility seemed to make it impossible that a hand, once adapted to the anthropoid tree life, could revert back to something closer to that of the "four-footed" primates. The human line had thus never gone through such an anthropoid and arboreal phase, but split off from a much more primitive primate form. This pro-man stem that led to modern humans was purely hypothetical—there were no fossils to support it. Osborn referred to the hypothetical ancestors as *dawn men* (Osborn 1930a).

This phylogeny reflected Osborn's view of biogeography that once again was developed in decades of study of fossil mammals. In Osborn's biogeographic scenario of evolution, there were centers of development and dispersal. Asia, and more particularly the Central Asiatic Plateau (Chinese Turkestan, Tibet, Mongolia), functioned as an important laboratory, where different types of mammals and hominids had evolved in the struggle for survival. The Central Asian Plateau's climate had increasingly worsened and the environment changed from lush vegetation to arid conditions. Of the Neanderthaloid stock, the most primitive hominid form, *Pithecanthropus erectus*, had survived in the lusher nature south of this laboratory, while the rest migrated over wide areas of Asia, Europe, and Africa (*Homo heidelbergensis*, *Homo rhodesiensis*, *Homo neanderthalensis*).[6] In contrast, Osborn thought that the harsher circumstances of the northern regions of Eurasia would have been favorable to the evolution of the higher types: the dawn men. From them evolved all so-called modern human races. And from these northern regions, the types Osborn called Mediterranean, Alpine, Cro-Magnon, and Nordic eventually radiated out and "subdued the entire Central Eurasiatic empire of the Neanderthals" (Osborn 1927c, 488). Osborn speculated, however, that what he termed *Negroid stock* would have arisen in Africa in regions more similar to the homeland of the Neanderthals of central and southern Asia (ibid., 1928a).[7] Clearly, these different hominid and human types would thus have very different "racial souls."

In both *Men of the Old Stone Age* (1915) and *Man Rises to Parnassus* (1927b), the Cro-Magnons of the "Caucasian stock" represented the apex in the succession of human types that invaded Europe from the east. Osborn described Cro-Magnon art as superior to the Hellenic. It seems that the Cro-Magnons had occupied just that perfect place between nature and culture, where life was still inseparable from the forces of nature and the struggle against it, while the human body and mind had already reached the highest refinement. Osborn was in awe of their "racial soul, mind, temperament, and intellect." And with their natural instincts intact, the Cro-Magnons had known better than to spare the females of the local Neanderthals when invading Europe, let alone accept them as mates. However, with the Cro-Magnons there also began a process of degeneration. A decline in artistry suggested by the archeological record was interpreted by Osborn as the effect of the Cro-Magnons' eventual estrangement from the natural way of life through the abandonment of nomadism. This negative trend was only exacerbated over time, so that however great the strength and courage of the later "Nordic race," who were contemporaries of the "lesser Mediterranean and Alpine races" in the Neolithic period, they had not had the Cro-Magnons' artistic mind. Subsequently, miscegenation contributed further to the degenerative development of the "races" (Osborn [1915] 1916, 272, 300; 1927b, 155–187).[8]

It has become clear what Osborn wanted to kindle in the readers' "racial souls" with his "word-painting of the scene and of the man or woman" (Osborn and McCosh 1884, 55) and his beneficial images of "conditions and experiences long bygone" (Osborn [1913] 1924, 183): the instinct against miscegenation, the love of a life in touch with nature and its laws, and the insight into the importance of struggle for individual and "racial" progress. It had been the lack of struggle in the lush jungle environments that had kept back the apes; it had been the fertile plains and abundant game that had limited the Neanderthals' improvement; and it was the present artificial and overcivilized way of life that threatened the stamina of the American. The Cro-Magnons and to a certain extent "the Nordics" could serve as good examples. They made clear that struggle was not enough. "The highest type of racial soul" possessed a sense of beauty and a reverence of nature, so admirably demonstrated in the nature poets such as Wordsworth and Muir whom Osborn cited in his scientific as well as biographical work. Against this background, in the following chapters, I engage with Osborn's books on human evolution and the Hall of the Age of Man in the order of their publication or opening. In the next chapter, I ask how *Men of the Old Stone Age* (1915) fared in the historical culture of Osborn's time.

CHAPTER TWO

Paper Ancestors? or "A Word-Painting of the Scene and of the Man or Woman"

As the photographs of Cuvier, Darwin, Thomas Henry Huxley, and others in his study illustrate, Osborn positioned his paleontology within the natural history tradition of the nineteenth century. In fact, he intended *Men of the Old Stone Age* (1915, hereafter *MOSA*) to be released on November 24, the day of publication of Darwin's *On the Origin of Species* (1859). More specifically, when looking at how the objects and the knowledge that went into *MOSA* were collected and produced, we see that Osborn wanted the book to make a significant contribution to European anthropology and prehistory. In a letter to his publisher, Charles Scribner, he appeared confident in this regard:

> I believe it will take rank as a great work because the treatment is entirely original and it lays the foundation for all future research as well as for future popular writing on this great subject. A distinctive feature of the book is that for the first time, as stated in the preface, the men are treated with their implements and with their geographic and climatic surroundings in a truly historic manner. . . . McGregor's remarkable reproductions, which are absolutely unique and published here for the first time, will greatly strengthen the work, in the same way that Knight's reproductions have in the "The Age of Mammals."[1]

The "historic manner" referred to the fact that this was no account as dry as bone, but a narrative about fleshed-out human beings in their ways of life. It was the images in particular that achieved this resurrection most dramatically. The illustrator Charles Knight had already advanced to the position of Osborn's right hand in the process of visualizing dinosaurs, mammals, and hominids from the evolutionary past in their environment. Small images by Knight could at that time be seen beside the mounts of fossils in the halls of vertebrate paleontology at the AMNH. In chapter 3 we will look into the gigantic project that Osborn and Knight had already taken up in the decoration of the exhibitions with murals when Knight began to work on the illustrations for *MOSA*.

The other artist mentioned in the quote, J. Howard McGregor, was one of Osborn's former Columbia students who had been brought to the museum. He specialized in the production of full busts from fossil hominid remains on what Osborn called "the strictly scientific basis of your mode of restoration."[2] Osborn made McGregor detail what seems to have been a very laborious and careful making of a man from bones in the museum's bulletin, *Natural History*. With the example of the Neanderthal, the stages the busts went through were explained and reproduced as photographs: the cast of the skull; the skull with the missing parts restored; the skull with muscles modeled out of plasteline, restored cartilages, and plaster eyeballs; and the hairless skull with all the soft-tissue applied (McGregor 1926). One of the photographs showed the restored cast superimposed on the model to visually emphasize the match. However, the final skull with sculpted facial expression and hair would indeed be first published in *MOSA*. By the time of *MOSA*, McGregor had made such "racial portraits" (ibid., 289) of *Pithecanthropus*, Piltdown, Neanderthal, and Cro-Magnon. They were clearly among the greatest fascinations of the book.

However, the possibly most laborious process in reconstructing the hominid types was not detailed in McGregor's account. Prior to reconstruction in busts, images, or words, Osborn's team had to get access to the fossils as well as artifacts and sites. Osborn was a vertebrate paleontologist at an American institution while the fossils, artifacts, and sites from which knowledge about human prehistory was produced were in the Old World. He therefore had to establish connections with European institutions and anthropologists. In fact, he also used the book to write himself into that community of experts. Osborn drew them more closely together by paying homage to their work and by establishing a network of friendly exchange. In the end, he relied on French but also British, German, and American expertise to the extent that he referred to *MOSA* as a collaborative work.

When Osborn began to dictate *MOSA*, he wrote to the French authority in the field of prehistoric cultures and art, Henri Breuil: "I am accepting absolutely your chronology both of the culture stages and of the art. . . . Needless to say I have vastly enjoyed your own contributions and shall make very extensive use of all your writings. I believe I have already written to you that I am proposing to dedicate the volume to Professor Cartailhac, yourself and Professor Obermaier."[3] In fact, Osborn regarded *MOSA* as "in a measure a tribute to the brilliancy and devotion of the French anthropologists as well as to the prehistory of France itself."[4] Osborn was greatly indebted to his "dear friend Abbe Breuil," who had shown him the great caves of the Dordogne.[5] As one service in return, he permitted Breuil to reproduce some of the images that went into *MOSA*. The two men would continue their mutual support, and Osborn made Breuil's work especially on the cave paintings more widely known through the facilitation of some English texts as well as through his own writing; Breuil was eventually awarded the Daniel Giraud Elliot Medal of the American National Academy of Sciences for outstanding work in zoology and paleontology (1924). The other great prehistorian whose support was invoked by the dedication, Émile Cartailhac, had worked with Breuil on the cave art in Altamira (Cantabria) and caves in the Pyrenees, whereto he led Osborn. The third figure in the triumvirate of French archeology was the German-born Hugo Obermaier, who was a friend and collaborator of Breuil and Cartailhac. He, too, had shown Osborn some of the Spanish sites.[6]

All three prehistorians were at least for a time at the Institut de Paléontologie Humaine in Paris, the first institution dedicated solely to paleoanthropology and prehistory. Breuil and Cartailhac had founded it with yet another French authority on whom Osborn greatly relied: the paleoanthropologist Marcellin Boule. Boule was among the colleagues Osborn addressed in the hope for a review of *MOSA*, in this case for *L'Anthropologie*. Despite occasional difficulties with Boule, Osborn could expect a positive reaction to his treatise. In the main he followed the interpretations of the French experts, among the most spectacular of which was Boule's expulsion of the Neanderthals from human ancestry. It was in this respect that Osborn diverged from the Czech-born anthropologist Ales Hrdlicka, who was acting head curator of the Department of Anthropology at the Smithsonian Institution. Hrdlicka thanked Osborn for including his view that the Neanderthals were the ancestors of modern humans, but complained that Osborn had represented his classification wrongly. To the Smithsonian, the Neanderthals were also of the same species.[7] Indeed, part of the aim in distributing *MOSA* so generously among colleagues was the correction of errors, which went beyond a difference in opinion.

The British paleontologist Arthur Smith Woodward and the anatomist Grafton Elliot Smith were not entirely happy with Osborn's cautious stance with regard to "their" Piltdown remains (that were only definitively revealed as a forged combination of a modern human skull with an orangutan jaw several decades later). Osborn thought it unlikely that *Eoanthropus dawsoni* (the formal name, meaning Dawson's dawn man) was of early Pleistocene age and ancestral either to Neanderthals or any "surviving race." Hominids of such modern brain anatomy were creatures of the later Pleistocene; even in late Pliocene times, the primate branch would not have evolved beyond the primitive type of *Pithecanthropus erectus* (today *H. erectus*). Correspondingly, Osborn only included a quick note on the controversy surrounding the supposed tools from European Tertiary deposits. He again followed the opinion of the French experts that these were unworked stones (Osborn 1915, 9, 85–86).

Besides communication, visits to collections and sites, as well as scientific publications, Osborn could draw on more popular accounts that testified to a life of "our ancestors" beyond scientific communities. At the time of writing, there existed a number of important treatises that helped to solidify the field of paleoanthropology/Paleolithic archeology. *MOSA* was part of the first wave of book-length accounts that attempted a synthesis of the new knowledge from archeology, paleoanthropology, paleontology, and geology, and often drew on comparative ethnology. Before and simultaneously with *MOSA*, the British geologist William Sollas published *Ancient Hunters and Their Modern Representatives* (1911, 1915); the British anatomist Arthur Keith brought out *Ancient Types of Man* (1911) and *The Antiquity of Man* (1915); and Obermaier wrote *Der Mensch der Vorzeit* (1912). Further testifying to an interested reading public, before the end of the year, Scribner informed Osborn that the 1,500 copies of the first edition had nearly all been sold. The publisher had already ordered another thousand, and a British edition was discussed with George Bell and Son. *MOSA* was a favorite gift book, and as the reviews began to appear, Scribner commented that they could not be better.[8] At this point, Osborn started a scrapbook, in which he collected reader reactions and other letters, news coverage, and reviews. It was an album for the album of the men of the Old Stone Age.

Indeed, profiting from the second phase in the communications revolution (Lightman 2007, ch. 1), *MOSA* received an enormous amount of attention. This was particularly the case in the American West, possibly because the book was based on lectures given at the University of California. It was greatly praised, and used in libraries and in classes at all levels. It was discussed in faculty clubs, graduate meetings, and literary societies, as well as in the private circle of friends and family; it was even the stuff of conversation between total strangers. It appealed to scholars, scientists, and professionals from a wide range of fields.

Osborn catalyzed this interest across knowledge cultures by sending out copies to people of different backgrounds asking them to review *MOSA*. In doing so, its meaning for history and as such its lesson for the present were closest to his heart. As we will see, he certainly tried to control not only the book's circulation but also its political messages and the morals it held for the perception of "self" and "other." However, the paper ancestors only became alive, and the meaning of evolutionary history was only generated in the process of diverse readings. So what did the papers write about *MOSA* and whom did the book reach? How did the word paintings and pictures of people and scenes long gone come alive in different discourses? How were they transformed in the process? Did Osborn's intended messages come through?

First of all, *MOSA* related to current historical consciousness. It was taken as a rendering of deep human history that shed light on the concerns about progress, imperialism, and "racial relations." Prehistory seemed to reveal that the course of history and the present developments were inevitable outcomes of much longer processes. In dominant readings, *MOSA* was brought up in support of "racial" hierarchies and xenophobia. The reception of the book also testified to a widespread interest in a human pedigree by far transcending the historical period, and to the allure of fossil bones as carriers of this past. The *Los Angeles Times* spoke of "relics, and 'mute, they chronicle an ancient tale,'" and the *Union* wrote that "the man that hunted and was hunted by the saber-toothed tiger in Southwestern Europe . . . left memorials of himself that have survived to the present, though the time that has elapsed makes that covered by all written history shrink to a pinpoint in comparison."[9]

One of the most influential reviews was Roosevelt's in the *National Geographic Magazine*, which at the time reached "the largest number of scientific and general readers of any magazine in the world," so that Osborn could not "imagine any more effective means of making this volume widely known."[10] In fact, Osborn not only solicited the review; he also instructed Roosevelt on what to write. Roosevelt stuck quite closely to the suggestions of his "Dear Fair,"[11] and praised *MOSA* as the most important book since Darwin's *Descent of Man* (1871). Beginning with the age of reptiles and the age of mammals, Roosevelt told of the origins in Asia of *Pithecanthropus*, Heidelberg, Piltdown, Neanderthal, Cro-Magnon, and Neolithic man, as well as of "the races still in existence." From there, these types successively dispersed and replaced the more primitive forms they met. A most crucial moment in human history had been the encounter between Neanderthals and Cro-Magnons, not without inference for the present:

> [The Neanderthals] were not our ancestors. With our present knowledge, it seems probable that they were exterminated as completely from Europe as in our own day the Tasmanians were exterminated from Tasmania. The most profound change in the whole racial (not cultural) history of western Europe was the sudden and total supplanting of these savages, lower than any existing human type, by the tall, finely built Crô-Magnon race of hunters, who in intelligence evidently ranked high as compared with all but the very foremost modern peoples, and who belonged to the same species of man that we do—*Homo sapiens*. (Roosevelt 1916, 125)

To provide this turning point in "our" history with the appropriate visual dramaturgy, the article juxtaposed Knight's picture from *MOSA* of the clubbed and stooped Neanderthals with that of the classically rendered Cro-Magnon figures engaged in cave painting.

Roosevelt reported that the Neanderthals were not the only ones the Cro-Magnon hunters had competed with. There had been another race in southern Europe "akin to the negro pygmies of present-day Africa. But these small negroids soon vanished, and the tall hunter-artists remained the sole masters of western Europe for what, judged by all historic standards, was an immense period of time" (ibid., 126). However, the Cro-Magnons had also disappeared nearly entirely and five human types entered from Asia that formed the basis of the existing, very mixed populations of Europe. Roosevelt found many of his convictions supported by this account of "our" history. He was an imperialist who had argued for the permanent acquisition of the Philippines, and he coined his expansionism in the rhetoric of the American westward pioneering. The American colonies in the Pacific were the new frontier and the Filipinos the new Apaches. For Roosevelt, *MOSA* was the story of the adventurous white male and his strenuous life, and of the competition for territory between "races" driven by a spirit of travel.

In general, Osborn's generation had lived through the "Gilded Age", during which the United States added nine states and staked imperialist claims not only in the Philippines but also in Samoa, Panama, Nicaragua, Hawaii, Cuba, Guam, and Puerto Rico (Sandage 2008). Roosevelt was not alone in his understanding that *MOSA* showed that imperialistic practices had always been a means of progress. *MOSA* documented that "higher and lower races" had always coexisted, and that the displacement of one by the other constituted the propelling momentum of history.[12] Osborn had not invented this model. Imperialist rhetoric was widespread in anthropological literature, and Osborn drew heavily on Sollas. In *Ancient Hunters and Their Modern Representatives* (1911), the geologist-anthropologist at Oxford University had described the successive invasions of Paleolithic races into Europe from the east. From comparative analyses he inferred that the races who had been driven out of Europe to the peripheries of the earth were now represented by indigenous peoples such as the Tasmanians, the Australian Aborigines, the "Bushmen," and the "Eskimo"—"Stone Age" peoples who in the present were the victims of imperialism and colonialism once again (Sommer 2005a). One newspaper praised Osborn's *MOSA* as "revivifying the past from remains of men and animals unearthed from layer after layer of deposit in the caves and river drift and finding indisputable evidence of wave after wave of ascending humanity that swept westward from unknown Eurasiatic origin, each lower race and type being in its turn dispossessed or exterminated by its higher successor, until finally there arrives the Cro-Magnon race, of a grade of culture that ranks it high in capacity and intelligence, even judged by modern standards."[13]

Of all these Paleolithic "dispossessions," the Neanderthals' displacement by the Cro-Magnons was often singled out. Boule had reconstructed the Neanderthals as a truly primitive type. At the same time, he had elevated the Cro-Magnons to their beautiful, artistic, and intelligent contemporaries—a move that was widely discussed in the French press (Sommer 2006). Osborn followed this move, while taking a more sympathetic, or rather paternalistic, stance toward the "poor Neanderthal." Nonetheless, with these new interpretations, the pair could function as the perfect foil for self and other. The papers loved the story of the encounter between the lowly, somewhat degenerate, ugly Neanderthals and the well-built, spirited, and culturally advanced Cro-Magnons. And when the historical subjugation of peoples and their territories was referred to in light of this prehistoric precursor, so were the "racial" hierarchies. *The New York Times* and the *Enquirer* zoomed in on the Neanderthals in "direct conflict with the Caucasian newcomers. This swift replacement of a lower race by a higher [was] the most profound change that ever occurred in the racial history of Western Europe." It brought the reign of the Cro-Magnons—"the heroes of Prof. Osborn's story."[14] The *Sun* specified that "the type of the race [of Cro-Magnon] is entirely Asiatic, Caucasian, and not African."[15] Even in today's measures, the Cro-Magnons were an excellent type compared to "existing races of much lower brain capacity, such as the Eskimo or Fuegian."[16]

As in the case of "racial" hierarchies, journalists who were fascinated with the imperialistic model usually expanded on resonances with particular current issues, such as the claim for "Nordic superiority," the fear of "racial degeneration," (im)migration or miscegenation. *MOSA* not only resonated with foreign affairs. It touched very much on issues at home. The United States had witnessed the defeat and dispersal of Native Americans, the mass movement of African Americans to the north and into the cities in the wake of emancipation, and in the last quarter of the nineteenth century, it admitted twenty million immigrants with immigration rising to a peak in the early 1900s (Sandage 2008). One moral lesson that was drawn from the Cro-Magnon-Neanderthal encounter was that the Cro-Magnons' instincts were at this point still healthy enough to resist the lure of miscegenation that later generations would succumb to: "In the replacements of savage as well as of historical peoples the men are often killed and the women spared to be taken into the families of the warriors." In contrast, "no evidence has thus far been found that the Neanderthal women were spared or allowed to remain in the country."[17] The unnaturalness of a coexistence or even alliance between the two types was driven home by yet another visual juxtaposition, this time of McGregor's busts (see figure 4).

Osborn's friend, the New York lawyer and naturalist Madison Grant, drove the same message home. In his discussion of *MOSA*, he emphasized how in the past, migration movements had gone along with degeneration through interbreeding. He was delighted about the moral authority that Osborn's prehistory seemed to lend to his eugenic politics and Nordic supremacism: "The bearing of such an astounding discovery on the questions of the present time, especially those relating to the

MYSTERIOUS APPEARANCE OF BEAUTY IN THE PREHISTORIC EUROPEAN COUNTENANCE

DURING the long interval from Aurignacian to Magdalenian times, that is, about twenty-five thousand years ago, a striking progress took place (from the standpoint of beauty) in the configuration of the European countenance. This is exemplified in the head of the so-called Cro-Magnon type of knowing man—*Homo sapiens*, as the authorities say. Neanderthal man, whose coarse features have been made so familiar in the literature of this subject, seems to have succumbed to the Cro-Magnon type. The latter was armed for offense, apparently, with a bow and arrow, against which the hapless Neanderthals had no means of contending. The subjugation of Neanderthal man and his extinction seem to have promoted the evolution of a lovelier human countenance. The Cro-Magnon type of face was common in Europe, comparatively, about twenty-five thousand years ago. Before that, say forty thousand years ago, faces

NOT HUMAN ENOUGH

The Neanderthal man of La Chapelle-aux-Saints, inhabiting the Dordogne region of central France in Mousterian times. Antiquity estimated as between 40,000 and 25,000 years. After the restoration modeled by J. H. McGregor.

were conspicuously Neanderthal in salient lines. Whether the Neanderthals were exterminated entirely or whether they were driven out is not known. The encounter was certainly between a very superior people, both physically and mentally, and a very inferior and somewhat degenerate people that had already been reduced physically by the severe climatic conditions of the fourth glaciation in the old world.* The Neanderthals were dispossessed of all their dwelling places by this new and more beautiful race.

In the replacements of savage as well as of historical peoples the men are often killed and the women spared to be taken into the families of the warriors. No evidence has thus far been found that the Neanderthal women were spared or allowed to remain in the country. In none of the burials of Aurignacian times, as this particular prehistoric era is styled, is there any evidence of the crossing or admixture of the Cro-Magnons and the Neanderthals. The human countenance gained from the circumstance. The chief source of the change, says the distinguished Doctor Henry Fairfield Osborn, in his new work, lay in the brain power of the Cro-Magnons, as seen not only on the large brain as a whole but principally in the almost modern forehead and forebrain. It was a race which had evolved in Asia and which was in no way connected by any ancestral links with the Neanderthals. It was a race with a brain capable of ideas as well as of reasoning, of imagination, and more highly endowed with artistic sense and ability than any other uncivilized race which has ever been discovered.

"In many characteristics the Neanderthal skull is shown to be nearer to that of the anthropoid apes than to that of *Homo sapiens*. This conclusion arrived at by Schwalbe, in 1901, has been more than confirmed by Boule's masterly study of the very complete skull of La Chapelle. After his detailed review, he concludes: As to the unity of the Neanderthal head form, these features are not peculiar to the skull of La Chapelle; in every case they are also found in the skulls of Neanderthal, Gibraltar, Spy, Krapina, La Ferrassie, which witness to the homogeneity of that human fossil type called Neanderthal. These features show a structural affinity between the fossil men of the Mousterian period and the anthropoid apes. It must be noted that many of these features may be found also in recent human skulls of the inferior races, but that they are very rare, very scattered, very isolated, and occur only as aberrations. It is the accumulation of all these features in every skull of a whole series which constitutes an assemblage entirely new and of great importance. In the skull, as in other parts of the anatomy of the Neanderthals, we should not expect to find every character intermediate between the anthropoids and recent man. The long Neanderthal face is somewhat similar to that of the Eskimo and is in contrast with the very short face of the existing Australians and Tasmanians. The depression at the root of the nose, just below the glabella, is very marked in all Neanderthals; there is less of the nose bridge than in any recent races, except those of the male Australians, yet the nose is not flattened but somewhat arched or aquiline. This feature is not characteristic of all the anthropoid apes, and in this respect the Neanderthals, Australians and Tasmanians are more different from the anthropoid apes than are some of the white races; thus the Neanderthal nose, far from resembling that of the anthropoids, differs from it more than does that of some recent human types. Many anatomists, following Huxley, have described the Australian and Tasmanian skulls as more or less Neanderthaloid, and some authors have gone so far as to regard these races as surviving Neanderthals."

Study of the Neanderthaloid forehead and eyebrow ridges, of the great depth of the face and of the peculiarly high, square form of the eye sockets, prepares us for a profile view of the skull found at La Chapelle, in contrast with that of the most highly developed and intellectual European type—the profile of the distinguished American paleontologist, the late Professor Edward D. Cope. He bequeathed his skull and skeleton for purposes of study and comparison. In the La Chapelle specimen we at once notice the flattening of the skull cap, the retreating forehead, the great prominence of the eyebrow ridges resembling that of the an-

VERY HUMAN

The head of the Cro-Magnon type of *Homo sapiens*, a race inhabiting southwestern Europe from Aurignacian to Magdalenian times. Antiquity in western Europe estimated as at least 25,000 years. After the restoration modeled by J. H. McGregor.

thropoid apes, the lengthening of the face as compared with the flattening of the cranium, the great prognathism or prominence of the face as a whole, and the special prominence of the rows of cutting teeth, as compared with the vertical or indrawn line and the recession of the tooth row in the Cope profile. This comparison also brings out the striking contrast between the high chin prominence of *Homo sapiens* and the deeply receding chin of the Neanderthals. The question is from which ancestral prototype of anthropoid ape are derived the physiognomical characteristics of *Homo sapiens*:

"It is possible that within the next decade one or more of the Tertiary ancestors of man may be discovered in northern India among the foot-hills

* MEN OF THE OLD STONE AGE. By Henry Fairfield Osborn. Charles Scribner's Sons.

FIGURE 4. Cro-Magnon versus Neanderthal—Howard J. McGregor's busts, American Museum of Natural History Library, the Papers of Henry Fairfield Osborn (1857–1935) MSS.0835, Series IV: Books, Box 99, Folders 6–10: *Men of the Old Stone Age* Scrapbook, Folder 7: "The Ape, the Ape-Man, the Adonis: Mysterious Appearance of Beauty in the Prehistoric European Countenance," *Current Opinion*, Apr. 1916, 265

migration of races, which we now call immigration, is clear, and the importance of maintaining the breed and stock of the finer races in unimpaired purity, must in consequence be sooner or later recognized."[18] For Grant, the white Anglo-Saxon Protestant male was the apex of human evolution. He believed that only Anglo-Saxons, a designation he used interchangeably with *Nordics*, could contribute anything useful to cultural development. Shortly after *MOSA*'s appearance, Grant expressed these views in his internationally influential *The Passing of the Great Race* (1916), where he, too, explained that the "Nordic race" originated in Central Asia and was now in danger, especially in America, of degenerating through intermixture with lesser breeds. Like other eugenicists, Grant recommended a check on immigration, "racial" segregation, and sterilization of "the unfit." Grant and Osborn had cooperated in the establishment of the New York Zoological Society and park, developing a deep friendship, so that Osborn had put Grant on the board of trustees of the AMNH. They were involved in the Immigration Restriction League, and in 1918, they would collaborate in the establishment of the Galton Society.[19]

Grant's discussion of *MOSA* was a warning against taking progress for granted, and indeed, the Cro-Magnons' rise to perfection, followed by a decline due to the loss of nomadism and racial interbreeding, could also support a cyclic view of history, according to which the achievement of high culture and civilization was inevitably followed by degeneration. The many species that had died out and civilizations that had fallen to pieces "suggest[ed] the query whether the life of a race as of the individual, does not in countless generations exhaust its vitality, and collapse—*sans everything*."[20] Furthermore, because *MOSA* was read as the nearly complete history of humankind leading up to the present situation, it could also seem that "the war of to-day is the resumed effort to force the north door of western Europe, where ingress has never been fully successful since the Romans came."[21] However, some readers chose not to emphasize the mechanism of advancement or decline through imperialism and war but an overall gradual progress of humanity, "a wonderful vista of the upward evolution of the human family from the speechless animal state into reasoning and speaking man."[22] The zoologist Theodore D. A. Cockerell was among those who highlighted this aspect that characterized Osborn's orthogenetic theory: "Though each successive generation has died, the stream of life has been continuous, and we of to-day represent the fruition of many hundreds of centuries of ceaseless endeavor. . . . Looking back on the story as a whole, it is hard not to see what the biologists call orthogenesis,—evolution having a definite trend and a definite fructification, seeming as though designed. This is the ideal of progress, the most characteristic peculiarity of our species as it exists to-day. Individually and racially, we believe that we are going somewhere."[23]

MOSA was thus clearly a product of, and acquired a life in, the historical culture of the time. Regardless of the kind of mechanism of historical development *MOSA* seemed to suggest, and regardless of whether the emphasis lay more on the hopeful view of past progress or on a warning against decline, the book was deemed important for historians. The *Newark Evening News* read Osborn as making "his opinion clear . . . that more recent European developments can be based directly upon what took place in those prehistoric ages," and the *North American* added that "all questions of philosophy, religion, sociology and speculative metaphysics depend for solution, in the last resort, upon such data as Professor Osborn has assembled and collated in 'Men of the Old Stone Age.' To know the characters and lives of the race's earliest ancestors is a prime requisite of knowledge of humanity's advancement . . . [and] from his final disclosures the way of the historian runs clear and unobstructed."[24] *MOSA* and other popular books complemented the drama of human history by adding most of its acts and thereby made history appear as the logical outcome of prehistory. They suited the taste of the time for complete and straightforward historical accounts. Osborn himself wrote to Breuil about the wide influence that "our work of popularization" has gained and pointed to the recent publications on human history by James H. Breasted and H. G. Wells that devoted the introductory chapters to prehistory.[25] Indeed, Osborn thought that the first ten chapters of Wells's *The Outline of History* (1920) were mainly based on his own work on the early history of the earth and the prehistory of man, and that the borrowing exceeded Wells's references.[26]

However, as hinted at in Osborn's letter to Breuil, *MOSA* also entered the scholarly discourses on history, and the education arena. Historians such as Henry Morse Stephens of the University of California (Berkeley) and Breasted of the University of Chicago were interested in the book for their own studies. Stephens also assigned the book for collateral reading in the course on general history. It was used in courses on anthropology and elementary paleontology as well as in further advanced courses: the number of students that came into contact with *MOSA* at this university alone was reported to exceed 1,300. R. S. Gortner of the Department of Agriculture at the University of Minnesota wrote to Osborn, saying that in their Biological Club (intended to function as an inspiration for graduate students), they had studied "prehistoric man" and in particular *MOSA*. Osborn was informed about the book's use in university teaching in various countries, including Japan. The topic of *MOSA* was also considered adequate for younger children, and the editor of *School* intended to write a series of articles in that journal about Osborn's book. Osborn himself thought it of value for educational theory and had set his sights on a review in the *Journal of Education*. *MOSA* confirmed his impression that since he had given up teaching and changed to a research professorship at Columbia in 1910, he could reach a much larger number of students than he had through lecturing.[27]

Beyond history and the historical sciences of life, *MOSA* informed the history and subject of a wide range of disciplines and allowed existing knowledge to be naturalized. The book seems to have resonated with a more widespread desire to found aspects of human life "biohistorically." Franklin H. Giddings of the Faculty of Political Science at Columbia put *MOSA* on his required reading list. Indeed, Henry Jones Ford, professor of politics at Princeton, regarded *MOSA* as containing material of great value to economists, and he agreed to review the book for the *American Political Science Review* (organ of the Political Science Association). Owing to its *longue durée* view of human development, he deemed *MOSA* relevant for the issues treated in his recently published *The Natural History of the State* (1915). Osborn's book could settle such fundamental questions for political scientists as whether the individual or the group had been more important in human evolution. *MOSA* indicated that the progress from animalism to savagery, to barbarism, and at last to civilization was not a matter of the individual but of social entities. Indeed, Ford gained the insight from *MOSA* that the evolution of the individual human being had been the outcome of a struggle between different societies with their characteristic institutions. Osborn's notion of progress through racial succession was thus adapted to the social scientist's view of the more recent past, and in the reading of another expert, *MOSA* formed a contribution to the theory and evolution of law.[28]

We have seen that Osborn had engaged in psychology and tried to integrate his understanding of contemporary individual and "racial minds" with speculations about the mental makeup of men in the Stone Age, an aspect of *MOSA* that triggered further interest. The scholar of the psychology of the primitive, Isador H. Coriat, for example, gleaned from the book that the phallic, sexual, and other symbolism present in the dreams of modern man was "merely a fragment of the mental life of our remote ancestors," of the symbolism evident in Paleolithic art and culture.[29] Coriat felt that Osborn's *MOSA* and *Totem and Taboo* (1913), in which Sigmund Freud founded his concept of the Oedipus complex in a phylogenetic fable, made psychology and anthropology very important for each other. While Coriat thought that our remote ancestors had a rich mental life, a Pittsburgh-based physician inferred from *MOSA* and William James's phylogenetic view of the brain functions (*Psychology* 1899) that primeval man had great sensory perception due to natural selection. This seemed to be supported by the fact that the Cro-Magnons could see enough in the dark caves to paint. The sensory faculties had become less important with the increase in intelligence, and among contemporary humans only atavisms showed extraordinary perception.[30]

Besides providing evolutionary depth to the knowledge of diverse sciences, *MOSA* could provide insights into the origins of science, technology, medicine, and art as human endeavors. The New York physician James J. Walsh had spent nearly two years in the Berlin lab of Rudolf Virchow, who was skeptical toward the theory of evolution. Walsh disagreed with much of what Osborn wrote, but was pleased that there nonetheless was something to learn from the book about the beginnings of medicine and surgery. Industrialists could warm to the topic and use prehistory to present their technological innovations in the light of a long-term progress. General Electric wanted to begin its commercial film on the history of lighting with a Cro-Magnon cave scene, to show how stone lamps had allowed the appearance of the first artists. From there, the film would follow the line of progress all the way to Thomas Edison's invention of the incandescent lightbulb and to the nitrogen lamps of the present day. Osborn was very willing to give them advice on the correct depiction of the Stone Age artists' technique.[31] Students of art, too, were fascinated by Osborn's introduction into prehistoric painting and sculpting, and thus to the origins of their métier. Osborn's book had "the unique distinction of being illustrated in part by 'the Upper Palaeolithic artists of the now extinct Crô-Magnon race.'"[32]

This misrepresentation of Breuil's drawings as those of the Cro-Magnons obviously rendered the book particularly authentic and added to the confusion of time levels. It inspired Cockerell to imagine that the Stone Age men might have practiced cave art for the very purpose of being remembered for thousands of years. In his drama about prehistoric cave life, the male mammoth hunt, and male cave art that he hoped would eventually be acted out, he thus mused:

> Bring chisel, paints, and on the cavern wall
> We'll paint the scene, that in the days to be
> As bright fire rays upon the picture fall

Our sons and their sons' sons may know that we
The mammoth slew, and thus began
A path of progress for the feet of man.
Treading this path, he can but upward go,
His life no longer simple ebb and flow.[33]

On a more humorous note, the Cro-Magnon painters of *MOSA* were the subject of a "A Stone Age Cabaret."[34] Even those who were appalled by Osborn's book took up the bard's pen. Theodore F. MacManus, LL.D., American poet and self-appointed authority on the "hideous moral and social aspect of Darwinian materialism," warned in verse:

If I was sprung from an anthropoid
This is the song I'll sing;
A Marseillaise to my brother-apes
For a bloody reckoning!
I pray you, therefore, wise savants,
Take counsel how and when
You choose to let us apelings know,
We are no longer men.[35]

MacManus feared that the implication of the evolution of humans from animals would be that there was no soul, no afterlife, and no God, and that this would instigate a hedonistic, libertine's way of life. Osborn received personal responses from people whose religious feelings were hurt and who feared the consequences of knowledge about prehistory for Christian morals.[36] *MOSA* could obviously be read as a threat to traditional values, and there were therefore also those who wished to confine the appropriation of Osborn's prehistory.

Osborn's answers support the conclusion arrived at by Regal (2002) and others: for Osborn, evolution was not at odds with religion. Indeed, "the ascent of man [seemed to him] to be one of the strongest proofs of the existence of God."[37] Catholic readers often had no quarrels with Osborn's description of the Cro-Magnons and the developments following their existence, but they had problems swallowing his rendering of a deeper past. Theistic evolutionists found ways to adapt *MOSA* to their creed: "There is nothing in [these volumes'] facts presented to interfere with a well-considered doctrine of design running through all nature leading up to the crowning act of creation which involved man's sudden appearance on the scene."[38] The progressive narrative presented in *MOSA*, Osborn's denial of any fossil evidence of a direct ancestor to modern humans, and his emphasis on spiritual and artistic evolution certainly facilitated a reconciliation with religious views.

Creationists could also seize Osborn's humanization of the fossil men in anatomy and mind. The apes, and with them humankind's connection to the animal kingdom, seemed to recede into a deeper and safe past. Although Osborn did not accept the age of 300,000 to 400,000 years given to the Piltdown man, he explained that human antiquity could be referred further back than previously assumed. The writer for the *Congregationalist* (Boston) judged that the Cro-Magnon bust that was reproduced in the article showed "a strong and striking face" that put "the ape-man, if there ever was one, back so far that it need not worry us."[39] Even the Paleolithic evidence for the coexistence of several types of hominids could play into the hands of Bible readers. It could be read as a confirmation of polygenist interpretations.[40] However, there was also resistance against what was perceived as scientific trespassing on traditionally religious territory. And some of the "evidence" in Osborn's book was quite an easy target for those who wished to discredit it. The naming of *Eoanthropus dawsoni* (Piltdown) as a new genus was seen as proof of the fact that "we map out our ignorance in long Greek names."[41] The dawn man was denounced as a forgery of anatomists who were overanxious to link their names with the discovery of the missing link. These were the charlatans of science who wanted to produce a sensation besides making money or a reputation.

Such attempts at undermining authority could be accompanied by the observation that Osborn was not an anthropologist but a zoologist by training. In a pamphlet, George Barry O'Toole was quite sophisticated in his effort to relegate *MOSA* and similar treatises to the realm of pseudoscience. Exploiting the fact that in *MOSA* Osborn would often relate controversial opinions on a certain find, O'Toole stressed that there was no agreement on any point of prehistory among scientists. No missing link had been found, and all fossils either belonged to the species *Homo sapiens* and differed from each other like "living races," or were monkeys or apes. Nonetheless, in line with Osborn's orthogenesis, O'Toole believed that the creator had endowed original stocks (classes) with certain potentialities that remained latent until called for by the necessities of the environment during phylogenesis. In stark contrast to such discussions, *MOSA* was also seized on by those who were on a crusade against religion. The *Truth Seeker*, for instance, elevated it to the new Bible that replaced the authority of a very bad book that had held and continued to hold back the march of civilization with the promise of a future reward in the skies.[42]

Obviously, *MOSA* was neither taken up nor rejected as a whole but partially accepted and transformed to fit people's existing understanding of their own and others' origins, histories, and identities. The latest word from science was not simply taken at face value and at times, it was even picked to pieces. The responses certainly indicate that *MOSA* could serve opposing political and moral purposes, even if the dominant reading seems to have been along Osborn's lines of the struggle of the individual and "the race" in the service of evolutionary progress. We have also seen how *MOSA* made its way into education and academic disciplines beyond history, where it resonated with the trend to historicize and also naturalize aspects of psychology, economy, and politics. Last but not least, the new knowledge of human prehistory inspired art: it is in such creative engagement that people often expressed their more personal experiences of ancient memories rekindled, as in the case of a woman who rhymed that the Niagara Falls

> Awake in me strange pre-historic joys,
> And glad, old memory that the soul enthralls;
> I hear the voice of primitive heart calls
> For love that animates, and rules, and buoys,
> Renewing strength and elemental poise,
> That beats down barriers like cardboard walls.[43]

While a poet might feel her heart and spirit opened through a particular sensation to unleash age-old memories, most readers inscribed themselves into prehistory via ancestors. *MOSA* was often perceived as "an authoritative account of what is known of the character and life of our earliest direct ancestors,"[44] even though Osborn did not think of the men of the Old Stone Age as direct ancestors of modern humans. What he did offer was the conjecture that the line of the Cro-Magnons had survived into the present day—an offer that was accepted in order to connect to this noble line. *MOSA* provided the opportunity to replace stereotypes of the apish missing link with images of a more worthy forebear:

> Now, therefore, let us stop thinking about the hideous figure which has been conjured up as our ancestor . . . , with protruding muzzle and retreating forehead, not yet fairly on its hind legs, progressing with a shamble, and not making much advance even at that. Let us rather concentrate our gaze on the man of the Cro-Magnon type, physically our equal, alert intellectually, artist as well as technician, religiously inclined, already looking up to the stars. And by way of lesson let us ask ourselves how much we, 25,000 years later, have profited by his example, and with a million-fold more opportunity, some of it contributed by him, have made ourselves worthy of such an ancestor by improving upon him in our thoughts, our words and our deeds.[45]

This journalist had certainly swallowed Osborn's message, and the noble Cro-Magnon was made into a figure of redemption. Commentators were thus inclined to claim the type as ancestor of all humanity, not merely of a chosen few such as may be found in the Dordogne. It seemed more likely "that their blood still flows in the majority of mankind."[46] The understanding of the Cro-Magnon as "Our Oldest Great Grandfather" imbued the moral lessons that the knowledge about their life seemed to entail with a particular authority, and these lessons could be strongly gendered. In line with Gail Bederman's (1995) diagnosis of a turn from the Victorian ideal of manliness to a more physical and aggressive masculinity, *MOSA* in some men evoked fantasies of a lost prowess, power, and freedom. In contrast to the refined, aesthetically inclined, and religious ancestors above, such readers transformed the Cro-Magnons into the Tarzans of the Paleolithic: "The man was tall. . . . The shoulders were broad and the chest massive."[47] Such male readers received Osborn's signal that especially "the men of the race" were in need not merely of conservation but of regeneration.

Human evolutionary progress was often placed entirely into male hands: it had been the men who fashioned tools and weapons, carved stone, ivory, and bone, and decorated the walls of their caves. But for such readers, it was the particularly masculine aspects that were taken on. It does not come as a surprise that prehistoric male hunting kindled desires at a time when the good old hunting days were coming to an end because of the loss of American "wilderness" and "wildlife." In these readings of *MOSA*, the yearning for the savage within was most strongly expressed. The conservative British *Morning Post*, for example, claimed that "there are traces of the aboriginal man in the most polished product of modern civilization, and that is the reason why 'good hunting' (in Mr. Kipling's sense of the phrase) so deeply stirs the souls of those who are able to come by it." Men should well remember their heritage and remind themselves that "woman, as he [the aboriginal man] draws or sculptures her, is a shapeless and hideous mass of flesh and rolls of fat—no more than a mere breeding-creature, no doubt, who sat at home in the tribal cavern and sucked at marrow-bones continually."[48] Indeed, for this journalist, figurines reproduced in *MOSA* proved that the Stone Age men had not cared much for women as people; they paid attention to the voluptuous body rather than the face. The male physique was set in contrast to Paleolithic woman who was described as smaller and of a lower brain capacity. Transformed into stories all about "him," interspersed with nasty comments about "her," *MOSA* could serve to remind the new woman of her natural place in the order of things. Still, in stark opposition to such mysogynism, some women could find the hope in Osborn's portrait of "our" male ancestors that man at heart was an artist and not a warmonger.[49]

On the public stage of the mass media, the Cro-Magnons were thus made into very different kinds of ancestors to serve people's diverse needs for identification. Beyond the pages of newspapers and magazines, the men of the Old Stone Age entered the private sphere and became part of everyday life. A Nebraskan farmer, for example, was so pleased with *MOSA* that he bought extra copies and mailed them to friends and near strangers, men whom he had met in traveling, and who were interested in Osborn's men of the Old Stone Age. Another man from El Paso, Texas, took *MOSA* to Faywood Springs, New Mexico, where he stayed ten days "and where the loneliness was such as to make any kind of men welcome, and 'men of the old Stone Age' double so." He, too, told of a lonely-looking ranchman who had driven by in an automobile and stopped for a moment to chat. The man was an amateur archeologist. When he showed him *MOSA*, "he literally grabbed it and . . . 'dived' into it out in the hot blazing sun for a long time and then surrendered it only because he had to be going after having carefully noted down the publisher's name."[50]

Merely hearing about *MOSA* could excite a craving to get to know the ancestors, and one librarian reported that "it was in as great demand as any work of fiction."[51] However, for those far away from the centers, the book was hard to get hold of, and Osborn would respond to such complaints by sending a copy to a private address.[52] Unfortunately, not even a nearby library guaranteed access to the book, for reasons beyond the fact that people who were not in a business or profession might need the signature of "a responsible citizen."[53] The editor of the *Brooklyn Daily Eagle* complained that he had placed a library reservation as long as three months ago and still had not received *MOSA*. Like many others, he set out on an odyssey to locate the book in libraries. He failed to procure it, which made him suspect that the public libraries were controlled by conservative politics, or that "they are being run by the men of the old stone age."[54] Regardless of whether there is some truth to this conspiracy theory, even the New York Public Library could not afford more than five copies, which did not satisfy the demand. To make matters worse, it could also prove difficult to buy *MOSA* at bookstores as it sold so rapidly.[55]

Osborn himself did his share to kindle the ancestor cravings. He clearly regarded the men of the Old Stone Age as family, if not in the direct line. He saw his task in tracing the genealogy of today's people beyond the historical past, and people responded to that: "I shall be very glad if you will, as you so kindly suggest, send me an account of my ancestors in Europe, with your autograph inscription. I have never been able to trace my ancestry more than seven generations back, which runs into the Thirty Years War; but the assurance you give me that your book will trace me back to the 'stone age' whets my antiquarian appetite."[56] In the reading of *MOSA*, people brought the ancestors to life: "I have spent a memorable evening with *Men of the Old Stone Age*—and I look forward to many séances with them in the next weeks."[57]

Of course, for the private reader as well, *MOSA* was not merely about the past. It helped to answer the question "what is man?"[58] Quite contrary to many treatments in the news, in some individuals *MOSA* kindled the hope for a more peaceful human coexistence. It could be read as inspiring mutual sympathy and aid, discouraging the exploitation of one nation by another, and even as supporting the peace movement. One reader thought that *MOSA* could serve the various peace organizations in counteracting the incorrect understanding of Darwinism that influenced many statesmen. A general public discussion of its findings might give rise to a new consciousness of "man" and change the attitude toward fellow beings.[59] Here as elsewhere, *MOSA* was not only adapted to lend support to certain realities; it was mobilized for a demand for change. It had not only conservative but also reactive and subversive potential. Indeed, for some readers, *MOSA* offered views into the depth of human history that changed the way they perceived the world: "It seems to conduct one *forward*, as well as back; upward, as well as down into the depths of the earth, and I find myself looking at the stars, even, with a new interest, because of the wonders revealed about the history of man."[60]

Readers often wanted to be part of the project of reconstructing the transformative history of human becoming. They sent Osborn stones they believed to be prehistoric implements, their publications on prehistory, information on interesting caves or other sites, and even a report about surviving Stone Age men in the woods of Flanders, most likely referring to a group of Romanies or Sinti. Some of these informants were disappointed that Osborn did not find the evidence convincing with regard to an occupation of America in the Stone Age, and they hoped that they might help to gather such evidence.[61] In these cases, the *our* in *our history* was as narrow as the nation. In general, private and public reactions to *MOSA* suggest that people were in need of ancestors who could serve as models and inspiration and who could remind them of a history within. They felt themselves led on time travels and were motivated to take part in the project of searching for traces of "our" history.

The fact that *MOSA* was so intimately moving for many readers owed a great deal to the reproductions of Knight's images and of McGregor's busts. Quite along the lines of Osborn's own thinking, the West Coast paleontologist John C. Merriam observed that, however skillful, "a word picture" could never achieve as much as a visualization. It was the imagery in particular that "contributed to make the perusal of this volume the nearest approach to a journey through the land of Men of the Old Stone Age."[62] McGregor's busts were portraits of humanlike and human beings in the style of the statues of great men in history. This drawing on visual culture clearly aided Osborn's attempt to humanize the Stone Age men and thus to bring them to life. McGregor rendered even Piltdown man so human that Osborn's rejection of Piltdown as ancestor could be disconcerting to those "who would be glad to claim as a remote ancestor one with such high-bred cranial contours, and who see in Piltdown a very suitable stump to which to attach a family tree."[63] Those for whom "whatever pertains to the origin of man has the deepest interest" valued the scientists' application of the newest technologies to build a complete head out of a skull. Combined with other information contained in *MOSA*, the images allowed the layperson to "devise something in the nature of a picture of the daily life of the men of the old stone age."[64] In the end, it was the interplay of the visualizations with Osborn's prose that made "the volume so fascinating, so dramatic, so absorbing" for the reader that it would demand "some sacrifice of his rest, for he will certainly sit up late at night until he finishes the volume."[65]

Osborn seems to have mastered the art of "giv[ing] us a word-painting of the scene and of the man or woman" (Osborn and McCosh 1884, 55) that he believed could turn a mediated account into a personal remembrance: "The ancient races of men are made to live again—they cease being fossils and become realities."[66] In order to make his history living, Osborn brought the past into the present and vice versa, evidencing "the author's rare faculty for the restoration of remote periods in terms of present-day thought and feeling. This genius for representing fossils as living beings [was] again evident in 'Men of the Old Stone Age.' "[67] In order to achieve this, Osborn drew on well-established genres. *MOSA* was a "fascinating romance" of "epic grandeur." At the same time, although Osborn might have had to bridge gaps in knowledge, it seemed that "for the most part he stands aside so that the carvings, drawings, skulls and stones may tell their own story."[68] Obviously, for some, Osborn achieved the impossible: to unite authoritative and objective scientific presentation—as if the objects spoke for themselves—with lifelike verbal and visual reconstructions that turned the hominids into the heroes and losers in the grand epic of human history.

However, the above reactions also indicate that Osborn's literary and visual technologies might have been too apparent. In other words, they might have been even more successful in the suspension of disbelief had the readers been less aware of the processes through which Osborn translated natural objects into scientific objects, and finally into living men and their stories. Thus, it was exactly his prose and the illustrations that offered points of attack to religious commentators. Walsh retorted that despite the fact that the missing link was still true to its name, Osborn claimed that the evidence was there and that the ape-men "can be actually presented in pictures which make their value as evidence indisputable." Walsh even claimed the agreement of William Diller Matthew—the vertebrate paleontologist who worked under Osborn from the mid-1890s to 1927—when he observed that the combination of the ape jaw and the human skull in the Piltdown bust was "little short of a deliberate imposition on the public." Walsh was well aware that the demonstration of the scientific nature of the method of restoration was a technique to realize and humanize fragments of bone. And so was the artistic genre and tradition that was invoked by placing "the head thus constructed (not reconstructed) on a pedestal such as has always served hitherto for human busts."[69] Indeed, in the 1916 edition of *MOSA*, Osborn followed the American anatomist Gerrit Smith Miller in calling into question the association of the modern human braincase with the ape-like jaw.[70] Osborn's identification of the Piltdown jaw as that of an extinct species of chimpanzee and of the skull as *Homo* rather than *Eoanthropus* (dawn man) rendered McGregor's bust of Piltdown on the basis of both bones awkward at best.

As a consequence, O'Toole situated Osborn's "constructions" in the context of forgeries such as allegedly committed by Ernst Haeckel, whom he accused of falsifying certain embryological drawings and of making up an ape-man.[71] The average reader had to take the diagrams and reconstructions in popular manuals and textbooks of evolution with the proverbial grain of salt. The alleged reconstructions or restorations in *MOSA* were in fact "pictorial fiction." The lively imagination of the scientific illustrator could turn any fossil fragments into "imposing confirmation in the shape of simian or pithecanthropic skulls overlaid with a veneering of human features." O'Toole's arguments against the "pseudo-science" that served "to gratify that natural human craving for a 'finished picture'"[72] were not easily dismissed, and similar concerns were heard from within the scientific arena. A reviewer in *Nature* cautioned that in view of the curiosity of the general public, the scientist was tempted to turn tentative hypotheses into established facts. He feared that Osborn had failed to resist this temptation in *MOSA*, which had reached a second edition in the United States within six months of its original publication.[73]

Paradoxically, this Fleckian argument that hypotheses become facts when they travel from the esoteric circle of experts to the exoteric mass of laypeople might partly have been provoked by Osborn's habit not only to inform readers on the practices of science and reconstruction, but at times to even prompt readers to understand the images as working hypotheses. He did not hide uncertainties or smooth controversies into agreement. Apparently unnoticed by any reader, because of the changes his interpretations underwent, his prose could even flatly contradict the visual reconstructions, the adaptation of which would have been costly and laborious. *MOSA* was in fact everything but an unequivocal presentation of *the* truth. Everyone who has read *MOSA* also understands that the readers' reactions detailed in this chapter testify to a great effort on their part. Despite Osborn's claims to the contrary, the book in itself certainly does not offer stories in the usual sense of the word, and its prose remains a rather dry discussion of data. Thus, the men of the Old Stone Age only really came alive in the process of reading and rewriting in the public, popular, and private spheres.

The much more serious critique was, however, that the missing link was still missing. In fact, when *MOSA* first appeared in print, Osborn already fostered the hope that the political situation in China would soon allow him to organize an expedition to search for the remains of our direct ancestor.[74] In the next chapter, I engage with this search in the context of the construction of the Hall of the Age of Man.

CHAPTER THREE

The Hall of the Age of Man

The Politics of Building a Site of Phylogenetic Remembrance

When Osborn visited the European metropolitan natural history museums that had been founded since the 1880s, he understood that to provide vivid impressions of the evolutionary past—impressions that appealed and added to the visitor's visual memory—he needed novel technologies of reconstruction and exhibition. In a visual culture marked by the gigantism in capitalism, architecture, and technology, and by the spectacle of Barnum and Bailey's Circus and the Ringling Brothers and Barnum and Bailey's Greatest Show on Earth, museum exhibits had to become more dynamic and captivating. The circus traveled across the country by railroad with its enormous equipment and used the methods of Frederick Winslow Taylor to efficiently rebuild the tents again and again, with every worker carrying out specialized duties that had been defined for each part of the labor process. Among the shows were reenactments of the recent Indian Wars, the Spanish-American War, the Boxer Rebellion, the building of the Panama Canal, and—as at the world fairs—exhibitions of "strange and savage tribes" from around the world.[1] The other great shaper of the new visual culture, film, also often focused on narratives of nation, history, and belonging (Davis 2008; McGovern 2008). Institutions of serious instruction like the AMNH tried to stay detached from mass entertainment, but it was not only the railroad circus's and the Wild West show's cultivation of spectacle and efficiency that Osborn shared; he also shared the nostalgia for a mythical past, spiced up with a celebration of scientific and technological advance.

For his very own mythical past, Osborn organized expeditions that brought back fossils from the western states and territories on train wagons and ships often for reduced or no fares due his connections to the magnates. He hired a team of scientific experts, technicians, and artists who, under his direction, revolutionized the mounting of fossil bones to fill the Hall of the Age of Reptiles (1905), the Hall of the Age of Mammals (1895), and the Hall of the Age of Man (1924) with lifelike fossil mounts, so that the visitors could walk through them in reenactment of the epochs in the history of life.[2] Osborn's pet project, which also proved one of the greatest challenges owing to the scarcity and value of hominid fossils, was the Hall of the Age of Man, constructed between 1915 and 1924. In the first part of this chapter, I engage with the acquisition of the hominid casts and paraphernalia, the search for the bones of "our true ancestors," and the setup of the exhibits. Constance Areson Clark (2008, 107) has pointed out that the Hall of the Age of Man was really a hall full of "elephants." And indeed, the hominid remains were to be in the company of mounts of contemporary fauna. Elephant evolution was a particular hobbyhorse of Osborn's, and one of the murals that Knight produced for the hall was an impressive statement of the importance of the rough environment of the Pleistocene for the evolution of these magnificent creatures (Sommer 2010c). It was one of several huge scenes decorating the Hall of the Age of Man, the production of which is explored in the second part of this chapter.[3]

If Osborn wanted the American public to meet the men of the evolutionary past and to learn about their ways of life in order to be reminded of a more natural state, he needed to get access to the precious resource. As we have seen in terms of the knowledge that went into *MOSA*, Osborn made similarly good use of his increasingly well-established international network where the acquisition of casts was concerned, and he sent his staff on tours to European museums. While the exchange of material was a sine qua non for Osborn's Hall of the Age of Man, it was also a means to stabilize the interpretation of objects. There was tension at work between the claim of ownership and priority in the description of a find, and the need to have it circulate through the network in order to achieve consensus on a particular interpretation. Finally, although the exhibits in the Hall of the Age of Man were kept up-to-date with regard to new discoveries and understandings, they nonetheless froze the knowledge about the past to a certain degree. As we have seen for *MOSA*, with the entrance into the public sphere, hypotheses could appear as facts and contentious fragments of bone as a particular hominid (see also Sommer 2010b).

The British Museum was a very important source for the "elephant" mounts in the hall (Proboscidea, trunked mammals). Osborn returned the favor by providing Smith Woodward, keeper of geology and Osborn's friend, with literature and reconstructions, for example, prepared from McGregor's models of the Piltdown brain and head.[4] For Osborn such presents were precious, as he held McGregor's work in the highest regard. In Osborn's eyes, McGregor combined an unusual artistic skill with a strong imagination in comparative anatomy. In the traffic of scientists, publications, images, and letters, as well as casts and molds, between the AMNH and other institutions, McGregor's reconstructions were indeed among the most valued arrivals. Osborn's gifts were particularly welcomed by Europeans during the war, when anthropologists and archeologists joined the army, lost colleagues and students, and were engaged in war work at home, when some of their institutions were turned into hospitals or places for medical instruction, or were used for other purposes related to the war, and when resources were scarce and travel difficult.

The war obviously curtailed this traffic, but there were other obstacles to circulation. In contrast to Smith Woodward's willingness to share the Piltdown "remains," Boule did not allow his assistants to show Osborn the Neanderthal skull from La Chapelle-aux-Saints—or its cast—when he was in Paris. It had been this nearly complete skeleton that had served Boule (1908) as the basis for the claim that the Neanderthals were a separate species and not ancestral to modern humans. This move set in motion a general trend toward hominid phylogenies marked by parallel evolution of the diverse "human types and races." Osborn's mind was set on acquiring a copy of the precious specimen, and he came into possession of an excellent cast of the skull "by a round-about route," through a Dr. Williams, whom he thought must have purchased it at the Institut de Paléontologie's lab.[5] Boule immediately pronounced it a forgery of the Germans, since the war degraded "*Homo sapiens* var. *germanicus* to the level of primitive *Homo ferus*" (my trans.).[6] National rivalry certainly hindered the circulation and influenced the interpretation of bones (Sommer 2005a), but Boule was known to be guarding his fossils particularly jealously. He now claimed that he had had two copies of the spectacular Neanderthal skull and jaw as well as the brain cast prepared for Osborn but that somehow something Matthew did had stopped that transaction. It was only when Boule received a cast of a *Tyrannosaurus* skull from the AMNH that he offered the said skeletal casts in exchange. Thus, in 1920, Osborn informed Breuil that "our Age of Man Hall promises to be extremely impressive."[7] To Smith Woodward he enthused that the hall was nearing completion, and that "in the central cases I am trying to collect casts of every specimen that has been found bearing on the question of human ancestry."[8]

It seems that the "trouble" with Boule continued, however, and when McGregor was sent to secure a typical archeological collection from France for the AMNH, Osborn wrote a letter to the Prince of Monaco, who had endowed the paleontological institute in Paris. The prince promised his support, but Osborn warned McGregor: "You will find it hard to dislodge the traditions so strong in certain museums that natural specimens belong to the curators and not to the world of science."[9] Osborn now also used his skull casts of the La Chapelle-aux-Saints and the Gibraltar Neanderthals as presents to open doors for McGregor, for example, at the University of Liège and at the Royal College of Surgeons, where the "oracle of British anatomy," the powerful Keith, resided.[10] McGregor wanted to obtain enough data to complete his Neanderthal and *Pithecanthropus erectus* restorations—full-body restorations, it seems, that were never realized. And indeed, at University College London, Elliot Smith showed him all his *Pithecanthropus* casts and let him use his laboratory to make an intracranial model. In British institutions, McGregor could study original fossils and take stereophotographs, and he received casts. Among these were such remains as the Galley Hill specimen. Like the modern-looking skull of the Piltdown "remains," in the aftermath of Boule's expulsion of the Neanderthals from modern human ancestry, relatively modern specimens of supposedly great age like the Galley Hill were instrumentalized by Keith and other anthropologists for the claim of a great antiquity of the modern human type in general and of the "human races" and the British in particular (Sommer 2007a, 200–205).

In Liège, McGregor found the geologist Max Lohest and the paleontologist Charles Fraipont very helpful. He took photos of the Spy Neanderthal material and was given casts of their entire series. McGregor's luck did not leave him in Paris, where he managed to get access to the La Chapelle-aux-Saints remains and where Boule showed him the unpublished La Ferrassie Neanderthal fossils. McGregor was also a favorite with Cartailhac, Pierre Teilhard de Chardin, and Breuil, with whom he went over his collection. He obtained good casts of the Cro-Magnon skulls from Les Eyzies and of Pleistocene artworks from the Museum at Saint-Germain-en-Laye—including the Brassempouy sculptures and the Venus of Willendorf. From France, McGregor planned to go to the Heidelberg museum, to Berlin to see the Le Moustier Neanderthal remains, and then to Prague. It was difficult to travel in Europe, and he felt he was losing a lot of time owing to visas, but also because of the Europeans' willingness to discuss their fossils. McGregor keenly felt that he was dealing with different knowledge cultures. While at times this was a trial, in the end he thoroughly enjoyed the more personal atmosphere in the French institutions and in particular the original fossils and artifacts. The journey was a resounding success, and he reported to his boss: "I trust the sum total of 'loot' in the form of casts may not seem insignificant."[11]

However, because new and important finds were made during the long work on the Hall of the Age of Man, the staff went through cycles of redoing the cases. There were also developments in scientific understanding that directly influenced the hominid exhibits. Most dramatically, Osborn changed his opinion toward the supposed tools from English Tertiary deposits and the Piltdown "fossils." Since writing *MOSA*, he had become aware of a quasi-modern horse in the Pliocene. As he liked to draw inferences from the evolution of horses, this find alerted him to the possibility of a greater antiquity of modern human anatomy. Furthermore, bone fragments of a second Piltdown man had been discovered that swayed general opinion in favor of acceptance. Osborn went to examine these in the British Museum in 1921. In the summer of the same year, he visited the British archeological sites in East Anglia, and the flints discovered at Foxhall (near Ipswich) convinced him of the existence of a Tertiary toolmaker in Europe.

Nonetheless, the Tertiary tools were not yet stabilized completely as traces of "our" past, and Osborn continued to closely monitor what the international community had to say on the matter of various *eoliths*, as the supposed tools from Tertiary deposits were called. He also employed the International Press-Cutting Bureau to send him newspaper articles from abroad on *Pliocene man*, *Piltdown man*, and similar keywords relating to Tertiary hominids. Furthermore, in this, as in all questions of archeology, Osborn turned to Nels Nelson, then associate curator of North American archeology at the AMNH (from 1923 associate curator of archeology, and from 1928 curator of prehistoric archeology).

One of Nelson's tasks was to summarize and appraise new archeological literature for Osborn, even letters containing archeological information. With regard to the eoliths matter, Nelson could reassure Osborn that they were gaining acceptance among prominent scientists in Britain. Early supporters of the discoveries of the English amateur archeologist James Reid Moir were Smith Woodward, Ray Lankester, and Keith. Now Nelson gave special mention to M. C. Burkitt, a young Cambridge graduate, instructor under the anthropologist Alfred Haddon, and, essentially, a student of Breuil's and Obermaier's. This well-connected expert accepted the supposed tools that came from below the presumably Upper Pliocene marine deposit referred to as Red Crag, found in Suffolk, Norfolk, and northeast Essex.[12]

Soon thereafter, the human workmanship of some of these flints was confirmed by Breuil and another leader in the archeological community, Louis Capitan.[13] With the sanction of the great French savants, some of those who had vehemently questioned the authenticity of the tools, like Sollas, changed their opinion. And it was only now that Osborn went public with the "man of sufficient intelligence to fashion flints and to build a fire, before the close of Pliocene time" (Osborn 1921a; also Osborn and Reeds 1922). The eoliths were now stable enough as traces of the human past to be allowed to enter the Hall of the Age of Man, and Osborn ordered a series from Reid Moir. In fact, he was so enthralled with the vision of human phylogeny the eoliths held that he financially supported Reid Moir's research in East Anglia.[14] It was this vision of the great age of modern humans that in the shape of his "dawn man theory" eventually brought him to launch a crusade against those who still clung "fondly to the ape ancestry theory."[15] Osborn helped turn the East Anglian eoliths sites into British national monuments, into the most famous spots in the early history of Great Britain (Osborn 1921a). Furthermore, after his trip to England, he considered it possible that at least the East Anglian "Tertiary man" was represented by the human "remains" from Piltdown.[16] When he returned from his visit to the eoliths sites with Reid Moir, Osborn was thus anxious to make *Eoanthropus* the star of his exhibition. He had the cases rearranged, devoting one to the history of the Piltdown finds, and he tried to press Smith Woodward to provide him with casts, including reproductions of the flint types and the supposed hide dresser. In order to be able to exhibit the flints from both sides, Osborn asked for duplicates. In 1922, he received the new Piltdown specimens and placed them in the case of honor of *Eoanthropus*.[17]

Osborn was now waiting for casts of the newly discovered Rhodesian man, also from the British Museum. Like others he considered the fossil pro-Neanderthaloid. He read this skull from Broken Hill (Northern Rhodesia/Zambia) from his new perspective on human evolution. He thought that it corroborated his view that humans never went through an arboreal stage, that the ancestors of man had walked erect for a very long period of time—perhaps as far back as the Oligocene. *Homo rhodesiensis* was therefore important, and Osborn had not shied away from instructing Smith Woodward to use a very fine thread saw to cut a horizontal section of the brain cavity to make a perfect intracranial cast. In placing the cranium, he should adopt the Frankfurt line, which Osborn had used in all the profile illustrations in *MOSA*, allowing for an exact comparison of the cranial and facial proportions.[18] To enable communication between scientists, the precious fossils were often turned into inscriptions, but in order for a paper proxy to be of any use, it would have to retain certain information such as the proportions. Thus, anthropologists had to standardize their processes of translating a three-dimensional object onto a two-dimensional surface—a difficult and never fully successful endeavor (Sommer 2007a, 155–159). It was in this shape that Osborn had his first access to the new find. However, owing to the considerable loss of information in a drawing, a stereophotograph, and even a cast, the study of originals was considered indispensable.

This problem also existed for the original *Pithecanthropus* fossils, which could not be studied by the museum staff. It took all of Osborn's cunning and another detour to get access to this particular treasure. When Dubois proved to be secretive, Osborn took the matter up with the Dutch minister in Washington, which resulted in a cable from Dubois inviting the Osborn team to see all his material. McGregor carried out valuable studies, but publication had to wait until Dubois's own paper would appear. Such things tended to render Osborn impatient; the scheme of an entire case about the Trinil discovery in the Hall of the Age of Man awaited its realization.[19] Most nagging was of course that Osborn considered none of these fossils to be the remains of a direct ancestor. *Pithecanthropus erectus* was generally interpreted as a primitive type that had survived into relatively recent times when much more advanced forms were already roaming the earth: "*Pithecanthropus* is another instance of the survival of a very primitive type of mammal in a primitive forested environment where food was plenty, there was little need of clothing, and safety was assured by concealment or flight rather than by combat with weapons" (Osborn 1929b, 216).[20]

Faced with the lack of a direct ancestor, Osborn, like other anthropologists, considered every possibility. He was intrigued by the story of the Foxhall jaw, a bone that had once been described by the American physician R. H. Collyer but was then lost (Collyer 1867). Osborn set his wife on the trail of the mystery. It was a romance to his liking, because from the common physician the jaw went through the hands of the most prominent geologists and anatomists in France and Britain, including Richard Owen and Thomas Henry Huxley. Most of all, the mysterious jaw had a prominent chin. In other words, it was of a more modern anatomy than the Piltdown "fossils," and its carrier would thus have served Osborn better as the shaper of the Foxhall eoliths.[21] However, even greater hopes were invested in a tooth found in America. Osborn interpreted *Hesperopithecus*—the Pliocene Western ape from Nebraska—as an intermediate between human and anthropoid, with a greater resemblance to the human (Osborn 1922a). Having finally in their hands the original remains of what they thought was a primate meant a great deal to the AMNH crowd; the sparse specimen was a very precious relic indeed. This preoccupation with a single tooth was mocked in the *Omaha World-Herald* in a cartoon that at the same time elevated the specimen to the status of missing link and compared it to the worth of oil and the spectacle of great sports events (see figure 5).

The value of the tooth for the scientists becomes evident from an incident. Gregory and Milo Hellman, an orthodontist and physical anthropologist at the museum, paid a visit to one of the best radiographers in the city. For comparative analysis, they desired radiographs of the *Hesperopithecus* tooth as well as of the teeth of a man and a chimpanzee. Unfortunately, the radiographer dropped the *Hesperopithecus* tooth on the floor and a large piece broke off, "which naturally made Doctor Hellman and me feel as if a great misfortune had fallen upon us."[22] Osborn replied to Gregory with long instructions on how such priceless material should be treated: a specimen should never be held except over a white linen sheet; a mold and cast should be made immediately; a special little box with a glass cover top should be prepared and fully labeled; and in it, the specimen should be placed in a fireproof safe. The Piltdown specimens, for example, were thus preserved in a safe in the British Museum, of which the curator Smith Woodward had the combination. At least the X-rays confirmed that the tooth of *Hesperopithecus* was closer

FIGURE 5. A tooth from the missing link, American Museum of Natural History Library, the Papers of Henry Fairfield Osborn (1857–1935) MSS.0835, Series III: Academic Papers, Essays, Lectures and Notes (1877–1892), Research Subjects, Box 86: Evolution, Folders 14–15: Newspaper articles 1922–1923: *Omaha World-Herald*, 29 Apr. 1922

to that of "the Indian" than that of the chimp. Alas, in the aftermath, the precious relic underwent degradation from anthropoid to peccary, and Gregory eventually published a brief note in *Science* recanting the interpretation of *Hesperopithecus* as representing a new genus of ape.[23]

However, even without this transformation, the tooth would not have satisfied Osborn. Although the media and even some of Osborn's colleagues at times stylized *Hesperopithecus* as an ancestor, Osborn never thought of it as a candidate. His hopes for a relic that could dethrone Piltdown from its most special place in the Hall of the Age of Man were engaged elsewhere. In fact, he had already sent his men to Asia to find evidence of "a high-browed race with surprisingly large brain capacity, ancestral to the Cro-Magnon type" in Tertiary deposits.[24] These so-called Central Asiatic Expeditions marked the peak of the museum's period of great explorations; they were the largest and most publicly discussed project. The leader of the expeditions, Roy Chapman Andrews, became an adventurer cult figure through the constant and able involvement of the media and the American and worldwide publics in the events (e.g., Andrews 1921, 1926). Andrews was a former student of Osborn's and now assistant curator involved in the reconstruction of the ancient worlds for the public. As Regal (2002, 136–138) has shown in his discussion of the expeditions, it was "Professor Osborn's prophecy as to the Asiatic origin of mammalian life" (Andrews 1926, 3) that initially motivated the endeavor. Preliminary expeditions had been made in 1916–1917, followed by the Central Asiatic Expedition in 1919 (in the interim, Andrews returned to the Far East as a spy for the navy [Gallenkamp 2001, 71–73]). By the 1920s and into the 1930s, the idea that the great Asian plateau had been a center of hominid evolution and dispersal was widespread among paleoanthropologists (Sommer 2007a, 193–194). Thus, the Third Asiatic Expeditions that began fieldwork in 1922 and, with interruptions, carried on until 1930 were understood as the search for the Tertiary ancestor of man in the high plateau region.[25]

However, beyond the Hall of the Age of Man, one of the aims of the Central Asiatic Expeditions was to collect large mammals for exhibition in Osborn's new hall of Asiatic life. While Walter Granger (associate curator of fossil mammals) and Andrews went with their team to China and Mongolia, Barnum Brown (the AMNH's fossil hunter "Mr. Bones")—also as part of the Third Asiatic Expedition—would go to India to work with the Geological Survey under C. E. Pilgrim in the Siwalik beds. Brown's collection and knowledge from fieldwork in India was to serve Osborn's research on the trunked mammals (Proboscidea) as well as these animals' exhibits in the Hall of the Age of Man. Nonetheless, Osborn mobilized money by fueling the craving for ancestors and wrote to Brown: "Your work in India will represent another region which, as you know, has yielded all the primates hitherto found in Asia excepting the *Pithecanthropus* of the Island of Java. It has been impossible to finance the Third Asiatic Expedition without dwelling on the possibilities of finding the ancestors of man."[26]

It does therefore not come as a surprise that with regard to finding the "missing link," there were rivalries between different Western research groups. When Andrews at one point intended to try his luck at the site where the French Jesuit anthropologist Teilhard de Chardin had discovered hominid specimens, Boule once again reacted with animosity. It was the American team's dearest wish to outdo the French in the race for ancestors, and the expeditions were seen as a great American endeavor.[27] Despite this desire, when the finished Hall of the Age of Man was presented to the public in 1924, no ancestor had as yet turned up, and Osborn had to content himself with Piltdown as a model for the human ancestors of an earlier time. He was delighted that McGregor visited the Piltdown site, "including the exact historic spot where Osborn recanted!"[28] Indeed, Osborn had contributed five pounds toward a monument to Piltdown man, and McGregor felt accordingly when standing at the site of remembrance: "I experienced a feeling of reverent awe such as possesses me on viewing the site of incineration of some illustrious martyr."[29] In the absence of the bones of the ancestor of the noble Cro-Magnons and living humans, the special exhibit for Piltdown at the AMNH was meant to induce the same reverence in the museum visitors.

However, the story by this time embodied in Piltdown and other hominids was not easily conveyed to the museum visitor. McGregor's busts of *Pithecanthropus*, Piltdown, Heidelberg, Neanderthal, and Cro-Magnon man had been arranged in successive cases that also contained the fossil and tool casts as well as explanatory material. Although Osborn thought that no ancestor was on show, the positioning suggested a direct line of human ascent. More generally speaking, visualizations of evolution that circulated in Osborn's context could be equivocal. Clark (2001) has shown that the ways in which images were understood by nonspecialists might well have differed from at least the messages consciously intended by their expert producers. Even if biologists tried to convey a more complex idea of the evolutionary process, the public was already so accustomed to the linear hierarchies from ape to man that evolutionary diagrams tended to be read as indicating a goal-oriented and hierarchical development.[30] In accordance with Rainger's (1991, 180) intuition, Clark demonstrates that exactly that kind of misunderstanding also pertained to Osborn's Hall of the Age of Man, where visitors felt a simple progress was being communicated in a *scala-naturae* fashion.

The messages from the Hall of the Age of Man were certainly ambiguous, partly because it was constructed over such a long time. The guide only added to the confusion: it combined photographs of the McGregor busts and restored skulls in what seems to be a straight line of ascent with strong visual and textual arguments for a branching hominid phylogeny. The exhibit's intermediality was further enhanced—and with it the inconsistency of the message—by the fact that visitors were invited to use the reading tables in the hall, where they were provided with Osborn's *MOSA* as well as with "Tour of the Stone Age in 1921" (reprinted from *Natural History*) that was representative of the changes Osborn's understanding of human evolution had undergone. However, a branching phylogeny and a linear line of progress were not incompatible. While the bones of the "true ancestors" had not yet been found, the fossils on the parallel branches could well serve as models for our ancestors of an earlier period, and the series of known hominids was understood in terms of an overall advance in anatomy and culture. This understanding structured the family tree in the showcase on man's place among the primates in the Hall of the Age of Man (Osborn [1921] 1923, 5). The genealogy of the primates was rendered in a branching fashion, and the line leading from the last common ancestor of all lines, *Propliopithecus* (in the tree signified by a jaw), to living humans was empty. There were no fossils to document our history and evolution from Oligocene times. At the same time, the horizontal series of skulls at the end of the branches visually reaffirmed the ladder from "the modern white skull" (ibid., image caption), via Cro-Magnon, Neanderthal, Piltdown, *Pithecanthropus*, gorilla, chimpanzee, and orangutan, down to the gibbon (Sommer 2010d).

Osborn also tried to guide visitor interpretations and reactions by explaining his theories in the press. In the year that the "finished" Hall of the Age of Man could be presented to the public, he took the step of declaring in the popular magazine *ASIA* that the arboreal theory of human origins had been abandoned. At this point, evolution was attacked by William Jennings Bryan and like-minded thinkers in an upsurge of religious fundamentalism. In a comment on the *ASIA* article in *Nature*, Osborn's theory was seen as an attempt "to counteract the influence of Mr. W. J. Bryan, who has lately published many attacks on the doctrine of organic evolution, and seems to have a large following in the United States of America."[31] Clark (2008) has read Osborn's scientific and public writings through the lens of the antievolution campaign of the American 1920s, revealing Osborn's conspicuous public involvement in the debates surrounding the Scopes Trial of 1925. Both Osborn and the Hall of the Age of Man were prominent targets of antievolutionists, exactly because of their power in shaping public consciousness. Since the bulk of the spite was directed at the "ape theory" of human origins, Osborn needed not just any Tertiary human bones, but a particular kind of Tertiary man—one less tainted by a relationship to the apes. In combination with Osborn's own religious background, the dawn man theory has thus been interpreted in this context as a strategy to soften protest and to gratify his desire to reconcile religion and evolution.[32] However, as Clark (ibid., 125–131) elaborates, this provoked criticism not only from his colleagues at the AMNH but also from other scientists and lay observers, and it was used by some antievolutionists to discredit science.

Even if not always successful, Osborn was certainly inclined to prevent controversy. He liked to replace terms that had become controversial with new ones that would render his prehistories and politics more palatable to groups of diverse backgrounds. *Dawn man* sounded better than *ape-man* or *missing link*. Similarly, Osborn suggested to the author and reviewer Charles Johnston that he use the term *retrogression* instead of *degeneration*, and *progressing* for *developing*. Osborn spoke of *birth selection* instead of *birth control* in his eugenic writings, which made the technology sound less like an intrusion into privacy and a curtailing of people's freedom. He recommended that his friend Julian Huxley, who is the focus of part 2, do the same in order to "avoid a great deal of prejudice." He also sent Huxley a copy of the *ASIA* article and explained that "the necessity for religion in connection with morals is becoming very acute in this country. There is a strong movement, headed by a number of scientific men but not all, in this direction. There is a fearful wave of crime, which we attribute to the total lack of moral training in the public schools. Jews and Christians are united for the necessity of such training."[33] However, if the dawn man theory was thus clearly meant to allow for Christian moral lessons to lower the crime rate and other offenses against society, Osborn took issue with the attempt to read the new label as a cover for creationism. Only "persons uninformed in the technical branch of science known as phylogeny" could interpret it as a denial of the theory of descent. To the contrary, it referred to "a long line of ancestors which for the time may be designated as 'Dawn Men.'"[34]

In fact, Osborn did not only continue to acquire casts for the Hall of the Age of Man after its opening;[35] he also held on to the hope of presenting to the public the bones of dawn men. Indeed, "Andrews returned with the most brilliant results, after discovering true upper Palaeolithic and Mousterian Palaeolithic in Mongolia."[36] However, just when hopes were high for finding the much older traces of dawn men, explorations were brought to a halt in 1926 by civil war and Chang Tso-Lin's dictatorship of northern China.[37] This distressed Osborn all the more because his dawn man theory was in the open and disputed by students and colleagues: "Gregory and I are having a wonderful discussion on whether or not man descended from the apes, which resounds as far east as Japan."[38] They had to "get through to Mongolia and settle this question of questions."[39]

Both Osborn and Andrews were therefore most relieved when, in 1928, they could get back to work. However, under the Nationalist Party that obtained control of China, that season's collection was held at Kalgan by the Society for the Preservation of Cultural Objects ("Cultural Society"). To Andrews it seemed that the Nationalists were "the devil's chosen swine."[40] He felt that Chang Tso-Lin had at least known how to keep the Chinese in order, which could only be done by cutting off heads. Luckily for the Americans, the archeological specimens—which might be interpreted as antiquities by the Chinese—were already out of the country. But the meaning of fossils, too, was an issue. For the Chinese, they were cultural objects and ancient objects of value, while the Americans now suddenly tried to pass the precious traces of past lives off as natural objects. It did not help that at one point Andrews had auctioned a dinosaur egg and sold it for a fictive price to catalyze support for the expeditions. The Americans were accused not only of stealing China's treasures but also of searching for oil and minerals, and of collaborating with the militarists.[41]

Although Andrews managed to get the collection released, for the 1929 season, the Chinese demanded that half of the expedition staff be Chinese and that all the new fossil types be studied at the AMNH by Chinese and then returned to their country. Andrews was getting more restless and intended to go public, but Osborn warned him to keep quiet and make it appear as if he had nothing of great value. Osborn had already set in motion his network, writing letters to members of the Department of State, and was planning to meet President Hoover himself. He also tried to exert financial pressure through his uncle Jack Pierpont Morgan, who could make further loans to China conditional on the continuation of the Third Central Asiatic Expedition. When the situation looked similarly hopeless for the 1930 season, Osborn accused the Chinese Cultural Society of "base ingratitude" and blindness toward "the real scientific interest of China."[42] However, following a revolt led by two generals against the "Nanking" Nationalists, Andrews finally managed to get permits from both warring factions so that regardless of the outcome of the conflict in China, the expedition would be able to return. After ensuring that this prolonged stay away from the museum would not be disadvantageous for his career once Osborn retired as president (which he eventually did in 1933), Andrews and the AMNH team carried out another season of research in China.[43]

It was Osborn's last hope for a find that could stand as proof for the dawn man theory. The remains from Zhoukoudian (near Beijing) to which Davidson Black gave the name *Sinanthropus pekinensis* (today *Homo erectus*) in 1927 were naturally of much interest to him. However, he came to the conclusion that it was not a dawn man, but Neanderthaloid. Nonetheless, he ordered casts with the intention of "a splendid showing" and "a place of honor in the Hall of the Age of Man." At the same time, Osborn wanted the greatest discovery for himself, and he was pleased that with his dawn man theory, he was "making the biggest sensation of anything I have ever done."[44] He thus made sure that the Central Asiatic Expeditions continued to be well publicized.[45] However, the Mongolian project finally came to an end because Andrews's application to the Chinese Commission for the Preservation of Antiquities was denied. Osborn took his revenge in *Science*. He depicted the Chinese as bad and backward, refusing to allow the "good Americans" to bring progress and enlightenment to their country (Osborn 1931a). In 1929, Andrews's popular book *Ends of the Earth* (1929) appeared, and when he followed Osborn's advice in preparing the *Natural History of Asia*, volume 1, he made clear that he felt he had won by giving it the title *The New Conquest of Central Asia* (1932). The American public seemed to think so, too, and Andrews—who had already received the Kane Medal from the Geographical Society of Philadelphia and an honorary doctorate in science from Brown University for his work in Asia—was awarded the National Geographic Society's Hubbard Gold Medal (1931).[46]

Indeed, with respect to finding big animals of the past, the Central Asiatic Expeditions were an immense success. Osborn boasted that the fossil beds were among the richest in the world and largely represented a period of as-yet-unknown land fauna.[47] The vertebrate collections allowed a reconstruction of the Central Asiatic plateau as "the chief theatre of evolution, not only of the land *Mammalia*, but of the giant land *Reptilia* of the world" (Osborn 1926a, 202). Osborn intended to make the most of what the expeditions had brought out of the "barbarous region," and he put Granger and other museum workers immediately and exclusively to work on the preparation, measurement, description, photography, exhibition, and cataloguing of the entire reptilian and mammalian collection. Special rooms, artists, photographers, and secretaries should be devoted exclusively to that work.[48]

While Osborn felt that the Chinese had thwarted his chance to present the world with the original bones of the dawn men, the cultural remains the expedition had brought back at least allowed him to infer that the central Asiatic plateau had been of similar importance in human evolution. Andrews described the human beings who had lived there some 20,000 years ago as "a hardy people of considerable strength and endurance. No weaklings could have lived in such open country, under the semidesert conditions and severe climate" (Andrews 1932, 413). Indeed, with the Central Asiatic Expeditions, it had exactly been Osborn's hope to "rout for all time the 'warm, tropical jungle theory' as the birthplace of man. Nothing great ever came out of a warm tropical jungle where life is easy, food plentiful and trees form the first line of defence." Andrews was sent to the place "where our ancestors struggled for their existence."[49] Thus the Dune People, as the men of the Central Asian Stone Age were called, could carry the lesson of the importance of a strenuous life for progress.

All in all, even without any dawn men on exhibition, in the Hall of the Age of Man, "reproductions"—even "facsimiles"—of the objects from all over the world were put together by conscientious experts "to demonstrate the slow upward ascent and struggle of man from the lower to the higher stages, physically, morally, intellectually, and spiritually" (Osborn 1927a, 253). The hominid and archeological casts alone could not carry that message, however. In order to understand how Osborn made the hall express especially the struggle that had been necessary for evolutionary progress, we have to move from the fossils to the murals. The cases might have conveyed progress, a noble genealogy, and the true nature of the hominid types. But it was through the murals of the animals in their natural surroundings during the age of man that Osborn imparted the message of the importance of struggle. It was through them that he worked for the transformation of the museum visitor, especially to counteract the degenerative trend he recognized in the urban population.

The AMNH was very much shaped by its location in New York City, where many of the transformations the country underwent were particularly pronounced. In 1898, five boroughs were combined to create Greater New York. The city's first subway line was completed in 1904 and helped unite a city increasingly marked by entertainment and consumer culture. The rapid urban and industrial development pooled money and power in the hands of magnates and caused exploitation and destitution as expressed by fin de siècle novels such as *Maggie: A Girl of the Streets* (Crane 1893).[50] Osborn feared for the constitution of the city's populace, exposed as it was to poverty, vice, and the unnatural cityscape. In the production of the murals, he thus followed the maxim that "the best training for citizenship, in the highly artificial environment which surrounds the mind and spirit of the boy and girl in our times, is to show Nature in all her beauty and attractiveness, in all her moral lessons and inspirations, as well as in her stern moods of command" (Osborn 1927a, 252). In the AMNH as a "nursery for good citizenship" (ibid., 257), the murals may mediate nature's male capacity as educator and commander, as judge of the living world according to its own laws, through its attractiveness.

Osborn's purpose for the murals was guided by an ideal of truth to nature that Lorraine Daston and Peter Galison (2007) have described for the natural history illustrations of the seventeenth to the middle of the nineteenth century. They see it as characterized by the attempt to capture the type that lies behind the variety found in nature. The role of the natural historian consisted of arriving at the general through careful study of the particular in all its diversity. The epistemic subject was thus conceived as creatively engaged in the process of knowledge production and reproduction. A trained perception granted the natural historian access to the knowledge that lay behind the surfaces of things. For the visualization of this truth, he depended on the illustrator who would render it with the necessary beauty. Although Osborn was working several decades later, we have seen how he styled himself as part of the nineteenth-century naturalist tradition. He would tell the story to everyone who would listen of how—as a young man in training in England—he shook Darwin's hand in Thomas Henry Huxley's laboratory (Osborn, for example, [1913] 1924, 57). The centerpiece of that tradition was in Osborn's view the emphatic and direct study of nature, and through reconstructions true to nature, the museum could substitute this healthy nature experience for the visitor.[51]

Osborn shared the notion of psychologists that pictures could function as substitutes for the real thing and evoke similar experiences

(V. E. M. Cain 2010, 289), and as we have seen, his own research suggested that realistic renderings might even evoke experiences habitually made in the prehistoric past of "the race" because "the racial soul" was full of reminiscences from a long time ago. Finally, strong visual impressions, even if mediated, could be internalized by the viewer as his own: "And the best way to learn one of these laws [of nature] is to see it in operation; this is far better than to read about it, for what is seen becomes part of oneself" (Osborn 1927a, 269).[52] If Osborn's purpose was "to present visually the laws of nature and of art in such a way as both to educate and to create a strong impression on the mind of the visitor" (ibid., 235), he needed the help of the artist (ibid., 236–237). Knight's reconstruction paintings that transported scientific knowledge via an emphatic-aesthetic sensibility were a centerpiece in Osborn's visual education in the laws of nature, and as visual historian Victoria E. M. Cain (2010, 290) has shown, Knight developed perspectives to engage the viewer more directly in the actions he portrayed. However, even more so than at the time described by Daston and Galison, in the first decades of the twentieth century, the notion that the artist was to be the executing hand of the naturalist's synthetic vision was the source of never-ending conflict.

Osborn left nothing to chance but set up a system of tight control over Knight's work to achieve the desired effect. In the project, which Osborn and Knight began with the decoration of the Hall of the Age of Man, the production of murals followed a predetermined series of steps that involved other museum experts besides Osborn and Knight: 1. The landscape illustrating a geological period was chosen and the most characteristic animals were selected; 2. The appearance of the animals (form, color, typical attitudes) were discussed in detail, also taking into account the scale in the context of the entire hall; 3. The artistic composition and color scheme were negotiated on the basis of a series of charcoal sketches showing possible arrangements of the animal groups; 4. A color painting was produced that would serve as model for the mural. This step took up four-fifths of the entire time required for the restoration, because Knight had to do research on the geology, fauna and flora, and because he had to come up with a satisfactory color scheme. It was characterized by many abortive attempts; 5. The larger panel was copied from the original color painting (Knight 1922, 279).[53]

A mural project was also discussed as part of its larger context, which could be facilitated by sketches and even a papier-mâché model indicating the walls, windows, and mural spaces in the hall.[54] Knight would receive briefs that contained a list of animals Osborn wanted him to include, the scene to be represented, and the literature, experts, and possibly additional material to be consulted.[55] Knight arrived at an image of a fossil animal in the way he drew living ones, for which he would carry out studies at the Bronx Zoo and in the taxidermy section of the AMNH.[56] For extinct organisms, he studied the anatomy on the basis of fossils and expertise, drew the skeleton, and applied the muscles. Furthermore, in order to be able to draw an animal from all sides and perspectives, Knight often made small sculptures (Knight 2005, 78). This also enabled him to place the three-dimensional figure outside, in order to observe the casting of shadows. The animals were then integrated into their environment on a small painting that was "copied" to mural size. Finally, the large painting was placed on a panel and put in place in the hall, so that Osborn could examine it again for final touches.[57] The paintings were photographed in color before permanent installation in the halls, to be copyrighted and distributed pending Osborn's permission of reproduction. In general, Osborn made sure that the murals were described in the museum's bulletin, *Natural History*, before being made accessible for other educational purposes.[58]

In order to efficiently channel the workflow, and to ensure sufficient control over the process to make the murals convey his particular vision, Osborn roughly defined the series of work steps in contracts that would become more refined over time. The terms and conditions not only included the final size, the price of the finished work (between $4,000 and $7,000), and the mode of payment, but also specified that the subject matter and composition were to be determined in consultation with Osborn. Knight would be paid in monthly installments and receive a certain part of the total sum (for example, $3,000) once Osborn approved a sketch. A part of the total sum (for example, $400) could also be withheld until the finished work had been approved. Furthermore, questions of ownership and rights of reproduction were central if Osborn wanted the paintings to become an effective part of the museum's function in distributing the moral messages from prehistory. From the 1920s, contracts stated that the museum had the right to buy the sketch on which a panel was based (for example, for $600), with copyright for the sketch and final work remaining with the museum.[59] However, despite these provisions, each step in the production of a mural was accompanied by time-consuming, nerve-wracking discussions of the practical aspects of the work, scientific and artistic issues, timing, payment, and copyright.

Of most interest in this context, the actual negotiations over exemplary paintings bring to light Osborn's insistence on the murals' role in allowing visitors to experience nature's laws in action.

The first mural project for the Hall of the Age of Man was the "Mammoth and Reindeer" panel. Knight made it very clear how he saw the balance of science and art in his restoration: "This decoration should of course, be primarily a work of art, in this way alone will it be of any value and interest. The scientific part should be unquestionably kept subservient." To the contrary, Osborn insisted on truth to nature when he retorted: "Mammoths twelve to thirteen feet high and Reindeer as large as the wapiti elk—may look beautifully in a state capitol building, but in a *Museum our murals must also conform to the truth of nature*."[60] It was not so much that Osborn did not agree with Knight on it being a work of art, but he did not think the mural conveyed the impression of the Pleistocene he wanted to make: "The general effect does not seem to me to produce the robust character of these mammoths and the severe conditions of life characteristic of the Pleistocene. I quite agree with you that the primary purpose of these paintings is that they shall be works of art, but they should also express the great broad truths of Pleistocene life and environment and should convey to the visitors to this great hall this supreme impression."[61] Only through being true to nature would his larger aim be achieved: to evoke the struggle for survival that had driven the history of the species and races and that was essential for future progress.

When the first mural was finally on the wall in September 1916, Osborn praised it in one of the many letters to the person who paid for the work: J. Pierpont Morgan. It was a magnum opus, both artistically and scientifically. It was "the best thing of the kind which is to be found in any Museum in the world." Osborn described it as perfectly mural in spirit and in harmony with the room and the fossils. Even though Knight still had to add a few atmospheric touches once the canvas was permanently fixed to the wall, Osborn was sincere when he said: "I believe that these murals, with the careful personal supervision which I shall give them and with Knight's undoubted talent, are destined to endure and to exert a great influence on all our visitors, young and old, for all time."[62]

The kind of great influence Osborn wanted to exert on the visitors, and the great effort this cost, can be illustrated with the reconstruction of the Rancho La Brea site. It was particularly close to Osborn's heart because of the potential it held for a transformative narrative.[63] In order to visualize "The Death-Trap of Rancho La Brea," Osborn began a clever exchange with the Museum of History, Science, and Art in Los Angeles, in January 1920. He wanted to study the museum's material from the Rancho La Brea tar pools and have Knight create a mural on the subject. The policy of the Los Angeles museum—as in fact of museums in general—was to describe the animal remains before sending out any duplicates. So Osborn wrote to director Frank S. Daggett, announcing that he intended to visit the museum on his way to Hawaii. He wanted to bring Knight with him to study the collection and visit the asphalt pool site to investigate the landscape. In exchange, Osborn suggested that Knight might copy some of his art for the California museum for a recompense that would partly finance his trip. As an appetizer, Osborn sent them examples of Knight's work. Osborn also communicated on the matter with John C. Merriam (professor of paleontology at the University of California, Berkeley), who had excavated and described the bones of saber-toothed tigers, great lions, wolfs, mammoths, mastodons, bison, horses, camels, ground sloths, and other mammals and birds from the site. In fact, Osborn informed all his connections in the area about the pending journey. He even sent the six oil paintings from which Knight copied the larger murals to the LA museum, offering an exhibit. After the financial questions had been settled with the Knights, the journey could begin, and although Daggett died very unexpectedly, the exhibition was realized and Osborn received his Rancho La Brea casts.[64]

The Rancho La Brea was a particularly large project, consisting of four murals for the southeast side of the Hall of the Age of Man. Two were to show the tar pool scene (panels A and B), one was to show a giant condor and birds (panel C), and another camels and *Felis atrox* (panel D). The panels were intended to run around the corner, and their preparation demanded more space. Rather than following Osborn's advice to work at the museum to be close to the objects, however, Knight secured the top room in the building of the Parkway Commission in Bronxville.[65] When the sketch for the first mural was done, Osborn did not appreciate Knight's dark tone. Knight cautioned against too much realism in order to stay true to the style of the hall, which he kept in mind when painting over the sketch. He assured Osborn that he would do whatever was required to make the sketch match Osborn's ideas as closely as possible. At the same time, he emphasized "how difficult (and with most artists wholly impossible) it is to paint a picture according to some one's else [*sic*] ideas and to be able to visualize at once what is in their minds."[66] The "idea" Osborn wanted to convey through the Rancho La Brea murals was similar to many of the word paintings of scenes long gone that he and Andrews distributed. Only this time, the dying took place on a much larger scale. It certainly imparted the struggle for existence to spectators:

> It is still far too somber. . . . Southern California in the dry season is a land of cloudless skies, of most brilliant sunshine, and of very dry heat. The light is dazzling, arising from the straw-colored or yellowish-brown vegetation. The air is pure and free from dust, except in the cities and along the roadways, unknown to the Rancho La Brea mammals. There are no dust or loess deposits, because during the rainy season the earth is covered with very rich vegetation. The theory that in times like this the Rancho La Brea tar-pools were most fatal is supported by Darwin's observations in South America (see "Voyage of the Beagle") that during the dry season animals congregated around the water-pools and were mired, forgetting all precautions in the madness of thirst. I recall that from the very first I have dwelt on this theory, and on the idea that the painting must give some explanation for the terrific loss of life. It was only during a certain period that this loss occurred, namely, the early Pleistocene, because no modern animals are entombed. Director Lucas examined the picture with me to-day and we both agreed that the key was too somber to represent conditions of brilliant sunshine and drought. The composition is admirable.[67]

By October 1921, the Rancho La Brea panel A hung in the Hall of the Age of Man. There remained panels B–D of the Rancho La Brea series, and the difficult negotiations about the reconstructions of prehistory continued between various parties. Knight was instructed to substitute the ground sloth with another animal in his sketch for the first. Osborn would have liked a bison—this icon of American identity—while Knight protested that the big animal would demand a completely new composition and that Osborn should choose a smaller animal.[68] Knight suggested that the museum pay him for a couple of weeks to study the mounted specimens and background at the Los Angeles museum; he tried to get $7,500 (instead of $6,000) for the three-part panel and ended up with $7,000. Osborn was displeased because it had taken nearly half a year to arrive at a settlement, which slowed down the work on the hall. In a memo to the executive secretary George H. Sherwood that was intended for Knight's eyes, he threatened to employ another artist. Knight

FIGURE 6. "The deathtrap of Rancho La Brea," photograph of Knight's scene painted for the American Museum of Natural History, Osborn (1921) 1923, 26, American Museum of Natural History Library Photographic Collection, no. 39442

made good progress after that. The complete Rancho La Brea series was hung on the wall by November 1922, and by May of the following year, it was finished (see figure 6). Knight implored Osborn to frame the panels in clear gold but he was overruled by the museum's art advisers.[69]

Despite a process of reconstruction that was fraught with conflict, the mural of the Rancho La Brea tar pits is clearly one of Knight's masterpieces. It is a true scene of horror. In fact, the painting did give "an explanation for the terrific loss of life"[70]: driven by thirst in the dry season, the animals were "mired in the treacherous pit" (caption figure 6). The painting was widely reproduced, and in 1925, another opportunity to disseminate this scene presented itself when Knight was asked to produce a mural on the Rancho La Brea finds for the Los Angeles museum. The new director, William Alanson Bryan, suggested that Knight spend the summer in California and work in cooperation with the local experts to finish the painting in time for the opening of the museum's La Brea hall in November. In contrast, Knight and Osborn thought that the composition of the larger Los Angeles mural (ten by fifty feet) should be copied from the New York one and then be sent to the West Coast where Knight would finish it. Bryan initially insisted that the museum's La Brea scene should be original, including "all of the larger and more spectacular animals" to render it "the most interesting and spectacular painting in any museum in the country."[71] However, the cross-continent negotiations took so long that Knight's proposal was finally accepted. Furthermore, although the Los Angeles museum denied Knight the right to engage in any publicity relating to the mural prior to the new hall's opening, Osborn suggested an article on it to the *Los Angeles Times*. Knight distanced himself from this move and even worked on some modifications to the New York mural, on which he invited Bryan to comment. Osborn was delighted with the finished panel, and so was Bryan, even though he observed that Knight had not integrated his comments.[72]

As far as the AMNH was concerned, Knight talked about bringing his work for the Hall of the Age of Man to a conclusion in 1923, and by the beginning of the following year, Osborn was concentrating all forces on the completion of the hall. Gregory, Matthew, and Nelson were working toward its opening. As Osborn emphasized, Knight's murals were central to the impression of the Hall of the Age of Man, and he instructed that "so far as possible skeletons [were] to be placed beneath the corresponding restorations by Knight."[73] Small paintings were also shown alongside fossil mounts, and additional material was often integrated to document the discoveries. The murals communicated with these complex pieces of exhibition as well as with each other. In sum, the exhibition was intended to evoke in the visitor's "racial soul" "conditions and experiences long bygone" (Osborn [1913] 1924, 183) that were conducive to a progressive expression of "the race plasm."

Furthermore, with the placement of the three notorious murals of Neanderthals, Cro-Magnons, and Nordic people, one of which can be spied in figure 7, Osborn could force visitors to embody his "race succession model" of human evolution. The murals were not exhibited as a series—quite at odds with the way they have since been so prodigiously reproduced. Rather, the human murals were placed above the entrances on adjacent walls. Instead of suggesting a linear line of descent, at times when there were a lot of visitors, streams of people would have entered the hall from different directions and walked beneath the three "racial portraits," as though reenacting Osborn's vision of the invasion of different human types into Europe during the Stone Age.[74]

However, although Knight was supposed to decorate all three paleontological halls, this project was never put into practice for the Hall of the Age of Reptiles: Osborn and Knight clashed seriously over matters of copyright, ownership, and the question of the priority of art or science in the production of the "Titanothere" mural. In this crisis, Knight took up a commission at the Field Museum in Chicago for murals in a Dinosaur Hall.[75] This could mean that the spread of the scenes from deep time

FIGURE 7. Photograph taken in the Hall of the Age of Man, with mammoth skeleton and part of the "Mammoth and Reindeer" mural as well as of the Cro-Magnon painting and showcase, by E. M. Fulda, American Museum of Natural History Library Photographic Collection, no. 39130

was considerably increased, but also that Osborn would lose his control over the messages they carried. After nearly twenty-eight years of painting at the AMNH, Knight had a worldwide reputation, and Osborn had supervised the work to such a degree that he referred to it as "Osborn-Knight restoration." Osborn believed that Knight was unable to be true to nature without his guidance, and eventually made the criticism that Knight's murals for the Field Museum contained many mistakes.[76]

That the disagreement was about control can also be inferred from the fact that Osborn's reaction to Knight's Chicago work stood in contrast to his general support. He and his staff—particularly Nelson and Gregory—were very supportive when Henry Field began at the Chicago museum of his great-uncle, Marshall Field. Field had received his scientific education primarily from Oxford under Julian Huxley, with an emphasis on comparative anatomy, paleontology, and anthropology. From 1922 onward, when the "Hall of Prehistoric Man" was a project, to its opening and to Field's presidency of the museum, Osborn lent his help and gave access to his publications and the AMNH collections. After all, the moral from prehistory could not be spread widely enough—particularly if Osborn had a say in it.[77] However, despite Osborn's skepticism of Knight's work under what he considered foreign or less expert guidance, Knight had sufficiently internalized the schemata developed under Osborn to reproduce them long after both he and Osborn had left the AMNH. The struggle for existence remained a motif at other institutions such as the Field Museum but also in Knight's own publications (e.g. 1935, 1946). Knight's images had a huge impact on how people imagined their deeper past, and they influenced paleontologists and paleoartists for generations to come (Sommer 2010c). They were an important ingredient in Osborn's profusely illustrated journal, newspaper, and magazine articles, as well as his books, one of which recounted the struggle of our ancestors to scale Mount Parnassus.

CHAPTER FOUR

Creative Evolution, or Man's Struggle up Mount Parnassus

Man Rises to Parnassus: Critical Epochs in the Prehistory of Man (1927b) was Osborn's second book-length engagement with human evolution. He wanted his new conception of evolutionary history to replace other ways of generating an understanding of self and other, and to enrich the historical consciousness with the meaning of a much longer and in many ways more formative prehistory. As a contemporary remarked with regard to Osborn's efforts, "If there were no other benefit from all his learning and work, it would be memorable to have eased the minds of anxious folks on the subject of their own genealogy." However, we will also see how this new "visible story of our world and the creatures that have peopled it through the ages" was negotiated on the platform of the media as well as in personal communication.[1] Like *MOSA*, *Parnassus* was adapted to private needs and strongly resonated with public debates, but the 1920s were a different time from the 1910s. There was the "noble experiment" of prohibition, in which the consumption of alcohol was associated with Catholics and immigrants, but which also increased the thrill of drinking and partying as well as organized crime. There was an economic boom that saw the development of the New York City skyscrapers. In this period, New York became the most populous urbanized area in the world. There was great social friction. The largest urban African American population initiated the Harlem Renaissance—a movement for civil rights as much as a strong expression of African American culture and art—and women entered new social spheres and took on new roles in larger numbers. The women's movement led to the Nineteenth Amendment, which in 1920 enfranchised women. It was the time of the new woman, the new Negro, but also of rampant individualism as well as socialism.[2]

In the production of *Parnassus*, Osborn continued to rely heavily on the help of museum staff and the expertise of his international network. This did not end with publication. Osborn would ask colleagues to write corrections directly into the book, so that he could integrate these as well as their new publications into the next edition.[3] At the beginning of the work on *Parnassus*, he had been taken on Paleolithic and Neolithic tours—as he termed them—to all the important European sites. In recounting these travels in time to the reader, he used the analogy of the anthropologist's and explorer's ventures into unknown regions to study foreign primitives, and he claimed to be guided by a similarly benevolent ethos vis-à-vis the people he met. In contrast to missions in anthropology, the paleoanthropologist could not improve the peoples he studied. But he could improve the reader's image of them: "All human races, fossil or living, demand our sympathetic understanding" (Osborn 1927b, ix). During these time travels, Osborn might have rested on the "Pleistocene Walking Stick" Brown gave him for his seventieth birthday. The cane was made of 50,000-year-old wood found in association with extinct animals such as ground sloths, and it was spotted with carbonate of lime crystals.[4] After all, Osborn had not only grown old; he now also had to travel much further back in time than for *MOSA* to witness "proofs of the existence of intelligent man and his flint culture over 1,250,000 years ago" (Osborn 1927b, blurb).

This was one reason why Osborn chose a new title for a book that was initially planned as another edition of *MOSA*. The new title also alluded to the fact that, at this advanced stage of his life, he was mainly interested in the spiritual, intellectual, and moral aspects of human evolution (ibid., ix). In contrast to the sciences that fed into technological progress, his science, rather like literature and poetry, offered transformative experiences. This spiritual quest was motivated by the moral decay he saw at work in his society. While Osborn argued for the reintroduction of the teaching of traditional religion in public schools as one means against this decay, the Old Testament contrasted with his scientific knowledge and personal life. In 1926, the famous author Upton Sinclair reminded him that he had supported the late war, that he owned stocks and bonds, and most of all that his books and his museum were full "of graven images of these primitive men." Sinclair, too, was anxious that morals should be taught and agreed that the survival of civilization depended on it. However, there was a need for a moral code that had a real relationship to the actual facts of the present world, and the actual needs of the people living in it. In this regard, he appealed to Osborn "as a scientific leader . . . charged with the responsibility of working out such a code."[5]

No less was demanded of Osborn by Will Durant, who interrogated the great minds of his time on the pressing problem of redefining the meaning of life. Osborn was in "good company": Durant, who had been at the Department of Philosophy at Columbia University, sent his questionnaire to presidents Herbert Hoover and Tomáš Masaryk, to David Lloyd George, Winston Churchill, Aristide Briand, Benito Mussolini, Marie Curie, Dean (William Ralph) Inge, Earl Bertrand Russell, Joseph Stalin, Igor Stravinsky, Leon Trotsky, Mahatma (Mohandas Karamchand) Gandhi, Richard Strauss, Albert Einstein, Gerhart Hauptmann, Thomas Mann, Sigmund Freud, George Bernard Shaw, H. G. Wells, Thomas Edison, and Henry Ford, among others.[6] Durant shared with Osborn and Sinclair a concern about the effect of the growth in scientific knowledge on human self-understanding. He feared that it had caused "a disillusionment which has almost broken the spirit of our race." No telescope or microscope had provided a glimpse of God. The importance of human existence had dwindled through insights from astronomy and geology, and "biologists have told us that all life is war, a struggle for existence among individuals, groups, nations, alliances, and species." Historians claimed that progress was a delusion. Psychologists had reduced the soul to brain functions and described the will and self as the helpless instruments of heredity and environment. Human life had been robbed of its meaning, and the changes wrought to society through the industrial revolution, with the help of contraceptives, were destroying the home, the family, and traditional morality, "and perhaps (through the sterility of the intelligent) the race."[7]

For Osborn, both Sinclair's moral code and Durant's meaning of life could of course be found in our ancestors. Through *Parnassus*, the meanings and morals associated with his dawn man theory should reach as many people as possible. In the book, Osborn explicitly called this theory post-Darwinian in the sense that it did not contain the notion of an ape-man ancestor in Africa that was halfway between the later hominid and anthropoid families and most likely arboreal. He gave his readership a more worthy ancestor to remember. But even if he distanced humans from extinct hominids, anthropoids, and animals in general, the higher social mammals could still provide insights into the moral equipment of early man. This mammalian heritage would have induced hominids to behaviors such as "the safeguarding of the family, protection and careful upbringing of the young, protection of the chastity of women, inculcation of absolute integrity both in word and deed, communal and tribal cooperation for the general welfare, reverence for higher supernatural powers, love of decoration, of beauty and of art" (Osborn 1927b, viii–ix). If this moral code could be found so far down our family tree, in mammals, including the first hominids, how could it not be part of the makeup of modern humans? Modern humans simply needed reminding.

Osborn responded to Durant by reassuring him that "lines of research which I am now carrying on convince me that we must restore the word 'creative' to the word 'evolution,' as distinguished from the old word 'created.' My intensive researches and explorations of the last fifty-five years make this research into the principles of the creative evolution and the motive forces behind it the most absorbing interest of my life."[8] It was exactly the understanding of evolution that Durant presented as commonly held that Osborn was set against. Indeed, progress was possible through a struggle for existence among individuals, groups, nations, alliances, and species, but for Osborn life had always been about more than competition and hardship. It was about sociality, beauty, and art. Far from being mere instruments of heredity and environment, humans—even in the Stone Age—were the makers of their destiny. Instead of God's creation there was the creative power of the evolutionary process and of man the tool- and self-maker.

In *Parnassus*, Osborn built his argument for ancient morals and human-induced mental and physical progress on the evolution of tools. Five hundred thousand years ago, "the first educator was one of our eolithic ancestors, sitting over the fire teaching his boys how to fashion flints, while his mate was teaching the girls how to prepare skins for clothing" (Osborn 1927a, 257–258). This scene not only serves as another imaginary origin of conservative family values and gender roles; it also hints at the importance of tool making. Osborn's team did not succeed in procuring the bones of these Eolithic ancestors, the dawn men; but the "Eolithic tools and their evolution" gave evidence of the dawn men and their ascent. Thus, when Osborn began to work on *Parnassus*, he pushed Nelson to come up with a phylogeny or genetic relationship of the known types of flint implements from the Eolithic up to the Neolithic. This would obviously support the status of eoliths as actual tools

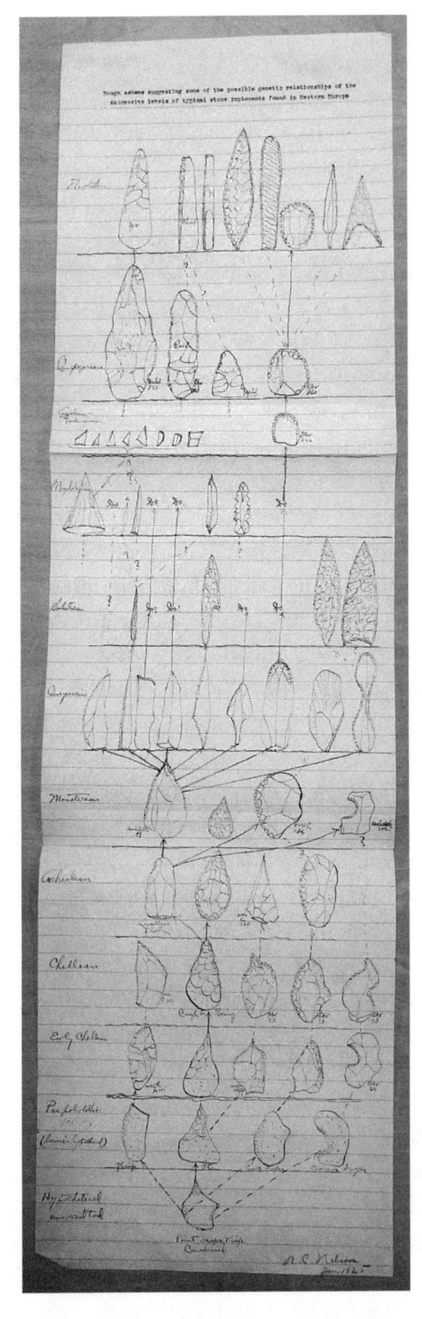

FIGURE 8. Genetic relationships of tool types, by Nels Nelson, American Museum of Natural History Library, the Papers of Henry Fairfield Osborn (1857–1935) MSS0.835, Series I: Correspondence 1884–1935, Box 16, Folder 11: Nelson, Nels C. (1875–1964)

and thereby human evolution from a toolmaker in the European Tertiary. Nelson was very uneasy about the task. He called into question the accuracy of the series of stages that Reid Moir had established in experimental flaking and which he claimed were similar to the evolutionary series of tool types from eoliths to neoliths. Nelson saw no genealogical relationship between the core and the flake tool types, or an evolution from the flake forms to Neolithic tools. He nonetheless submitted "a rough sketch plan," "done in great haste," that is shown in figure 8.[9] At the same time, Nelson warned Osborn to steer clear of the eolith matter in a general treatise with which he wished to avoid controversy. Nelson strongly felt that an interpretation of the eoliths should first be published in a technical journal and thoroughly thrashed out before being incorporated in a volume such as *Parnassus*.[10]

Osborn did not heed this advice; instead, he mobilized support for his claim that the human mind had evolved through hand-brain interaction. He drew on the publications and correspondence of contemporary neurologists such as Frederick Tilney. Tilney, in turn, was motivated by Osborn's work and letters to include a chapter on the prehistoric brain in his *The Brain from Ape to Man* (1928), which he dedicated to Osborn. In fact, Tilney visualized "man rising to Parnassus" on Osborn's terms. In figure 9, tool-less *Pithecanthropus erectus* has hardly even begun the ascent, and *Homo heidelbergensis* with his primitive club never made it beyond midway, actually looking backward. The dynamic figures in the struggle up the holy mountain are Piltdown man, Neanderthal man, and finally Cro-Magnon man, the last two of which face each other in competition. Of course, it is the Cro-Magnons who win the race and guard the top with advanced weapons, so that some of their members may celebrate their victory and express their well-developed minds in fine art.[11]

Osborn actually traced the belief in the importance of the trained hand in mental evolution back to Anaxagoras, an idea that had been given to him by Durant's *The Story of Philosophy* (1924).[12] He chose Aeschylus's description of the steady and parallel development of the practical arts and sciences, reason, and language in "Prometheus Bound"—Aeschylus's account of man's gradual rise to Parnassus—as a motto (Osborn 1927b, ch. 1). The structure of *Parnassus* was modeled on the Greek drama, including prologue and epilogue, so that content harmonized with form. The rising of man toward the top of Parnassus took place in front of the reader's eyes and was driven by demigods like Prometheus, by the pioneers and innovators of humankind. For good rea-

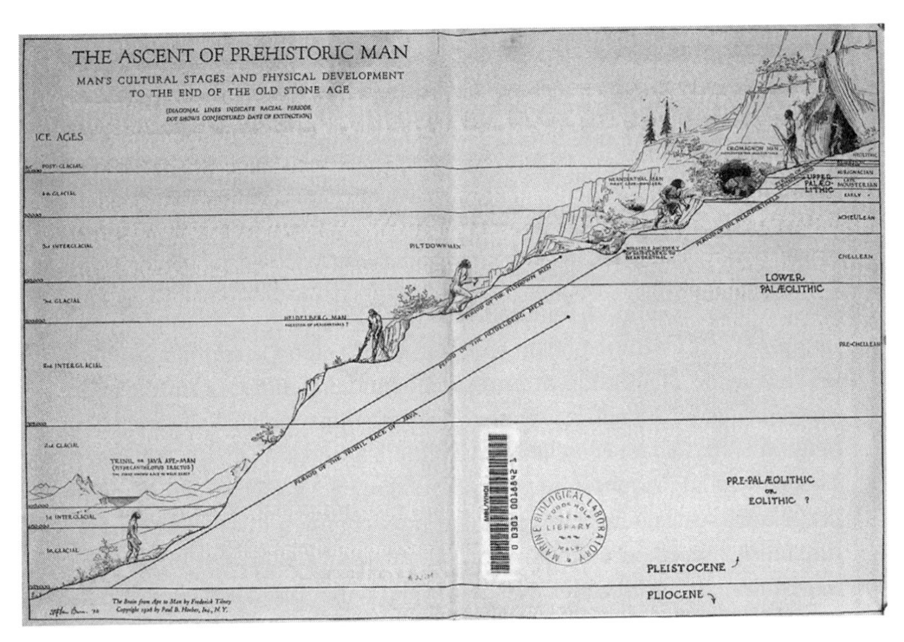

FIGURE 9. "The ascent of prehistoric man," Tilney 1928, dust cover

son, the book was not titled *Man's Rise to Parnassus*, but *Man Rises to Parnassus*.[13] Osborn rendered human evolution so creative, or processual, that it was theatrical. In this sense, Regal (2002, 168–169) has described *MOSA* and *Parnassus* as novels based on a true story. In fact, Osborn also performed man's rise to Parnassus in front of audiences, combining his dramatic narrative with a show of lantern slides.[14] While the truthful rendering of human achievement throughout history should resonate with the reader's "racial soul," the borrowing of the narrative frame from the cultural memory might facilitate the assimilation of that grand story.

However, despite all classic, neurological, and archeological support, the great drama was somewhat weak on evidence, particularly where its main protagonists, the dawn men, were concerned. As we have seen, Osborn had to content himself with Piltdown man as the "fossil" coming closest to the undocumented heroes. On the basis of its brain capacity, *Eoanthropus dawsoni* was described as having equal intelligence as some of "the living races." The long spiritual quest for the bones of the perfect, large-brained dawn man to fill the void of our direct ancestry was never achieved; but there was hope in his shadow:

> On Sunday morning, July 24 [1921], after attending a most memorable service in Westminster Abbey, the author repaired to the British Museum to see the fossil remains of the now thoroughly vindicated Dawn Man of Great Britain [i.e., Piltdown Man]. The few precious fragments of one of the original Britons, which had been preserved in a steel fireproof safe from the bombs thrown by German aviators and which will probably be thus guarded from thieves for all future time, were taken out and placed on the table by Smith Woodward, so that full and free opportunity was given for the closest comparison and study. (Osborn 1927b, 52–53)

This scene of worship at both the religious and the scientific altar from *Parnassus* at once makes clear that religion and science were not at odds; indeed, that a new reverence could be found in "our" deep history. This new religion would prevail over fundamentalism, enemies of civilization such as the Germans, and certainly common thieves. The scientific fetish presented to Osborn on the museum's altar, this relic of the large-brained Tertiary man, suggested the noble history also of the direct human line: man's steady rise to Parnassus. The house in which it was worshipped was a house of science that stood for open exchange in a common search for knowledge. *Parnassus* was a promise of redemption, but in order to enlist readers for the new religion, Osborn also wrote the threats of "racial degeneration" and moral, spiritual, and intellectual decline that were raised by Sinclair and Durant into his story from the Stone Age.

With his sense for timing and dramatics, Osborn made sure many friends would find the book under the Christmas tree. Sales were very satisfactory, with a second printing on its way a year after publication. Osborn's status and power certainly influenced the attention *Parnassus* received in the international press. The reviewer for *Dixie Magazine* called Osborn a giant and was impressed that "from left to right his titles read 'LL.D. Trin., Princ., Colum; Hon. D. Sc., Canb. [*sic*], Yale; president of the American Museum of Natural History; Research Professor of Zoology, Columbia University, Senior Geologist, U.S. Geol. Survey.'"[15] Osborn had been reelected year after year as president of one of the world's most important educational institutions, and his writings had brought him the reputation as one of the most high-ranking scientists globally. In the *New York Sun*, the entomologist, evolutionary biologist, and science administrator Vernon Kellogg called Osborn a distinguished zoologist, paleontologist, anthropologist, evolutionist, as well as natural philosopher, who had produced dozens of books and classics on evolution. The leading American physical anthropologist of the time, Earnest Hooton, wrote of Osborn as "perhaps, the best known American official natural scientist." Hooton, who himself was enamored with the myth of Nordic superiority, was particularly interested to learn that Osborn's name derived from a Norse one that had "been Anglicized into something signifying 'godlike warrior.' This afford[ed] new evidence of the improvement of the zoological status of the Nordics in their English home."[16] As Hooton's take on the name *Osborn* already illustrates, in *Parnassus* genealogy was invoked on several levels.

Osborn wanted to know whether his parable worked, if his epic shook the "racial soul" of his readers awake. He therefore followed and also interfered with the circulation and uptake of *Parnassus*. He engaged several newspaper-clipping agencies to look into the press mirror. However, Osborn's relationship with the media was ambiguous. By that time, newspaper articles had become such central tools to follow "public opinion" on certain issues that American clipping agencies found themselves in competition (te Heesen 2006). Osborn was concerned about the growing importance of "news" and the increasing number of tabloids.[17] At the same time, he thought that the press had a responsibility for education; it should function as something like the extended arm of *Parnassus*, in that it distributed its message far beyond the book's direct reach. Indeed, Osborn, as well as his work, was given ample space, and as in the case of *MOSA*, the newspaper reports on *Parnassus*, together with reviews and letters, were glued into an album. It allowed Osborn a more or less coherent look at the travel of his "ancestors" through the cultures of his time.

To begin with, readers reacted differently to the new outlook that the dawn men presented on their evolution. Reginald Smith of the Department of British and Mediaeval Antiquities in the British Museum was hopeful that Osborn's "acceptance of Pliocene man will influence a large number of readers hitherto dubious."[18] Others emphasized that his theories were well founded and archeologically based. Hooton judged the new data important and Osborn's new views, such as the acceptance of Piltdown man and the change of its date from Pleisto- to Pliocene, justified. Some commentators even presented the dawn man theory as scientific orthodoxy, while yet others agreed but completely misunderstood Osborn's claims.[19] Finally, papers like the *Portland Evening News* situated the dawn man theory in the anthropological landscape. It was observed that the underlying idea of parallel evolution, or orthogenesis, had been put forward by the late German physician and comparative anatomist Hermann Klaatsch, who had attributed peoples of the Mongolian type to an orangutan ancestry and African peoples to another anthropoid lineage. Also drawing on parallelism, the British anthropologist Frederic Wood Jones had postulated the alternative view that man and the anthropoid apes had arisen and evolved independently from a basal primate stock (Jones 1919, 1929; also Cope 1893). Osborn's claim of a human origin independent from the apes and an almost predestined separate evolution of the diverse "races of modern humanity" seemed to enjoy a considerable amount of professional support. However, the author was not convinced. After all, it was much easier to derive human anatomy from that of a primitive anthropoid than from an apelike tarsioid.[20]

Indeed, Osborn's move toward pushing the general trend of an interpretation of the fossil record in terms of parallelism and orthogenesis to the extreme of relegating the anthropoids to a mere side branch did not find many followers within the scientific community.[21] And at times, it was satirized in the press. The commentator in the London *Times Literary Supplement* quipped that any transatlantic millionaire would be jealous of Osborn who had huge assets at his disposal. He wondered how many more hundreds of millennia would be required when Professor Osborn would finally be able to date "the pre-crepuscular Adam of Central Asia, who presumably took a certain time to look round before he started off in the direction of Norfolk," where, supposedly, Eolithic tools had been found.[22]

However, there were many very favorable reactions, and even those who rejected the dawn man theory were not necessarily all condemning. The general picture of progress driven by human initiative resonated widely in the media. *Parnassus* was advertised as providing "incontrovertible evidence of the gradual rise and triumph in man of the spirit of discovery, of art, of science, as foreshadowed by Aeschylus."[23] The editor of *Nature* felt that the book "should be of decided service in stimulating interest in man's development and rise to higher planes," and the *Courier* hailed that it "triumphantly verifies the constant upward urge of man and his mind."[24] This was not only about the past but also about the key to a progressive future. The *Times Literary Supplement* commented that "the moral of the book, then, is that by interpreting human history naturalistically—that is, as the outcome of conditions which by taking thought we can govern to our increasing advantage—we can take our place among the pioneers of the race instead of succumbing to the decay liable to overwhelm all who identify Providence with a policy of drift."[25] In other words, if humans came to understand the mechanisms of evolutionary history, they would be able to consciously steer the course of it.

The *New York World* distributed to its readers Osborn's definition of the racial soul that was central to this understanding. It was the " 'spiritual, intellectual and moral reaction to environment and to daily experience.' . . . 'This racial soul [was] the product of thousands or hundreds of thousands of years of past experience and reaction—it [was] the essence or distillation of the spiritual and moral life of the race.' "[26] Thus, through that heritage in spirit, intellect, and morale, prehistoric experience was still present in "modern man," and humans could act on it. But if this soul was the product of hundreds of thousands of years of interaction with past environments, the question suggested itself of whether it could be at ease in the modern world. Were current crises due to the unnatural mode of life to which modern humans subjected their "racial soul"? Should they return to more healthy conditions that better agreed with their evolutionary heritage?

However, the concept of the racial soul also suggested that there was no such unit as "modern humans": "In Europe, for example, the soul of each of the three great races—the Alpine, the Mediterranean and the Nordic—[was] individualized, it [was] the product and summation of its own racial experience in the long past of its development."[27] Thus, all "races" had to perfect their own characteristics. The independent development of the "races" needed to be carried into the future. Osborn was represented as saying that "purity of race" had to "be preserved at all costs so that promiscuous mingling of higher and lower strains may not result in the production of an inferior type of man." It was understood that not all "races" were creative equally, or indeed created equally—Osborn's differentiation of the human types was at times read as polygenism, that is, as amounting to the claim that the "races" did not even share a common origin.[28]

The *New York Herald Tribune* reproduced Osborn's warning that civilization interfered with "the natural order of things." Instead of the laws ruling in the animal kingdom, there was now occupational specialization and racial interbreeding—nature had lost control. Civilized man was "upsetting the divine order of human origin and progress."[29] Terms like *progressive* and *creative* were used in this context in the sense of restorative, or re-creative. It was about a recovery of the "true racial type." Readers understood *Parnassus* as an appeal to follow "the moral of Evolution," as an appeal to the natural way of life and "racial consciousness."[30] An admirer from New Zealand sent Osborn a quote from a press notice that warned that "the only way to reverse the downward trend is to develop the racial soul. This means complete subordination of the individual to the good of the race."[31] Indeed, the *Lyttelton Times* of that country wrote that "with the soul's awakening, man may climb to the heights of Parnassus."[32]

However, the racial aspect, in particular the Nordic supremacism with which it was associated in *Parnassus*, was also met with resistance. The commentator in the *New York World* took up Osborn's claim that racial purity was now only found in the Scandinavians, while the Nordics had suffered dreadfully in Germany, and their original pioneer stock was dying out in America. This reader refused to assimilate the political message of the racial souls and in contrast spoke out in favor of "racial" mixing, denying that there had ever been pure stocks.[33] The writer for the *Times Literary Supplement*, too, was most perturbed by Osborn's expression of Nordic supremacism. Osborn's narratives were "presumably intended for Nordics only," which might not be too bad after all, because *Parnassus* was not a serious contribution to contemporary science. For this discussant, the evolutionary past held no authority over the present and future; *Parnassus* was unmasked as an offer of forged memories to "the Nordic stock."[34] One commentator even remarked that Osborn's thinly veiled politics amounted to potboiling.

At the other end of the spectrum, Grant defended *Parnassus* exactly because he valued it as a reminder of the Nordics' great past. He was therefore disappointed that despite similar views, Osborn did not refer to his *The Passing of the Great Race* (1916).[35] It seemed obvious to Grant that those who could not boast a noble heritage would resist such ways of writing history. Indeed, "the thought of Nordic supremacy [was] intensely offensive to those races and classes which [could not] claim relationship with it." With regard to the right to this noble ancestry, Grant took issue with the fact that in Europe the term *Nordics* was often used to denote the peoples around the Baltic, whereas it also included many other peoples such as the "Celts, Oscans, Hellenes, Persians, and the Aryan invaders of Hindustan," peoples without any connection to Scandinavia. In a review of *Parnassus*, he traced the coveted Nordic identity to "a branch of this generalized Nordic race in western Asia [that] pushed northwestward from southern Russia into Scandinavia [some 10,000 years ago], where its racial characters became greatly intensified and specialized." These characters were tallness, muscularity, blond hair and blue eyes, and an oval face with classic features. Their carriers were the ancestors of tribes known to the Romans as Germans. They spread from the Baltic Sea and differentiated into closely related tribes or nations all across the continent, "forming everywhere the upper classes and in some cases, as in Holland, England, Frisia, and Scotland, the bulk of the population."[36] Grant thus used *Parnassus* to unite a potpourri of different tribes and nations, of the upper classes, as well as the physically "strong and beautiful" in Nordic descent.

Parnassus was certainly understood as a statement in favor of the care of especially the "higher races"; it thus seemed to raise the question of which characteristics were of particular importance in "racial appraisal." While Grant mentioned physical and cultural traits, the *Times Literary Supplement* pointed to Osborn's insistence on temperament. Osborn claimed that humans of the Western and the Northern Hemispheres were descendants of Greeks, and they were restive, eager for new truth, and progressive, while "Orientals" were docile and stationary. He supported his characterization of these "racial souls" with the contrast between the Western "Prometheus Bound" and the Eastern "Book of Job"—the contrast between the passive resignation of a man who was content to regard himself as the creature of God and the adventurous man who believed his fate to lie largely in his own hands: "man as a self-maker."[37] Thus, while "racial isolation" was advertised as beneficial for every "racial soul," the active and strenuous life appeared as a technology of progress particularly suited for the sons of Prometheus.

Indeed, Osborn argued for the confinement of the ongoing movements of some human types. In the context of his presidency of the Second International Congress on Eugenics in 1921, he had been heard to declare that "the Great Nordic Stocks Must Be Maintained in the United States or the American Ideals Will Be Lost Forever—Increased Birth-Rate and Restricted Immigration Needed."[38] By the time *Parnassus* was published, laws that restricted the immigration of what were perceived as lower types from eastern and southern Europe as well as from Asia had been passed, and sterilization of those declared unfit could be enforced. Osborn was associated with leading members of the eugenics movement, and he amply voiced his eugenic ideas in the 1920s (Regal 2002, 122–123). Readers were able to read *Parnassus* in this context, and the book was in fact understood as an appeal to heed the teachings of eugenics.[39] Osborn maintained the belief in eugenic means until his death, and in his paper to the Third International Congress on Eugenics, he lamented the multiplication of "the incompetent and the criminal class." While this could be counteracted through contraception, contraception was also a threat, because it allowed the intelligentsia to have fewer offspring. Since highly educated women had fewer children, women should no longer be educated in a male way but made fit for matrimony.[40]

In general, Osborn's moral lessons from anthropology and evolution were not politics of race, nation, and class only, but also of gender, and they were motivated by private as well as public "crises." In 1928, Osborn complained that seven of his colleagues had been left by their wives because of other men.[41] As a symptom of the transformation of female identities, his friend Andrews was getting a divorce, and so was Osborn's son Perry. The changes forced on society by the new woman seemed to leave scars everywhere, and Osborn wrote to the chauvinist Andrews: "All is going finely at home and at the Museum with the usual terrible and dramatic difficulties that seem to surround life now-a-days while we are on the crest of the wave of feminism!" Again, also in deciding on the rightful gender roles, one had to take recourse to nature: "As old Dr. Peacock said in 'Nightmare Abbey' (buy a copy), 'Nature, i.e., science, is a great consolation in one's advancing sea of troubles.'"[42] Osborn's own wife, too, was a consolation and a wonderful companion to him, and she knew her womanly place until her death in 1930.

Nonetheless, where racial issues are concerned, Regal (2002, 182) has shown that Osborn was not as hardline as his friend Grant. It seems deliberate that Osborn did not refer to Grant's *The Passing of the Great Race* (1916) in *Parnassus*. Indeed, in the preface Osborn had written for Grant's book, he urged Grant to more sympathy. At least some comments on and reactions to *Parnassus* treated above confirm the impression that although Osborn agreed with Grant that the key to history lay in biology, "race," and heredity, he believed that favorable moral and intellectual traits were distributed throughout all the "races," if unevenly. Appeals for sympathy in *Parnassus* were taken on by more moderate readers. Some even still perceived Osborn as doing "his bit to quelch race prejudice."[43] However, other private readers of *Parnassus* were more in line with Grant's threatening picture of the passing of "the one great race." A Mr. Nelson felt pressed to opt for quitting any public assistance for malformed infants and public health and welfare for both the mal- and well-formed. But he doubted that such ideas would be carried by politicians, and he therefore had little confidence in Osborn's rise to Parnassus.[44]

With Osborn's use of metaphors such as "higher members of creation"[45] and the concept of a racial soul, it does not come as a surprise that similarly contradictory readings were still possible with regard to the meaning of his evolutionary view for the science-religion question. Peter Bowler (2007) has emphasized that, while there is a tendency in the secondary literature on the 1920s to paint an evolutionists-versus-fundamentalists picture, the religious landscape of the decade was very diverse, including liberal versions of Christianity. Correspondingly, *Parnassus* provoked a range of reactions. Its religious solemnity was often welcomed. A member of Saint Bartholomew's Church in New York enthused: "I took it to bed with me that night and read into the wee hours with it. It is a great achievement, and I congratulate you upon another noble contribution to the proper study of God which is man."[46] A teacher at the Union Theological Seminary in New York was positive that his students would enjoy the book, because "I suppose everyone of the four hundred and thirty-two students registered in this Seminary believes in the evolution of man and should welcome the appearance of this book of yours."[47] Indeed, people were at times inspired to worship Osborn: "Now, on the upward slope of Parnassus, near the end of the trail, I praise you and thank you for your service to the truth of God and the life of man."[48]

Someone from South Carolina considered the epilogue too somber because he could make out the beginning of the greatest reformation the Old World ever saw: "Jesus can still lead man on—the real Road to Parnassus."[49] To the contrary, a reviewer for *America—a Catholic Review of the Week* gave a negative answer to Osborn's final questions of whether the youth had gained in reverence and in faith, and women in modesty and love of family. As already apparent from the reactions to *MOSA* in the 1910s, and much more publicly voiced in the context of the antievolution movement of the 1920s, evolutionary theory, even if expressed as spiritually as in *Parnassus*, raised fears: "If we have [lost in spirit], we can, in great part, thank the evolutionists for teaching our young folk that they have animal ancestors on whom to lay the blame of their escapades."[50] While such readers held Osborn responsible for the very crises he himself wanted to counteract, others thought that *Parnassus* indeed presented an alternative way to preserve traditional morals without subscribing to religious fundamentalism: "Had this book made its appearance previous to the Bryan-Fundamentalist controversy of 1925, Mr. Bryan would have found himself confronted with Professor Osborn's convincing and irrefutable reply to the questions he raised."[51]

While some readers thus understood Osborn to oppose religious fundamentalists, others felt that *Parnassus* was playing into their hands. An article in the *Argonaut* carried the title "Osborn Bows to Fundamentalism." Its author, the advocate of science Maynard Shipley, applauded Osborn's sequel to *MOSA* but found a fly in the soup. By this he meant Osborn's introduction of "the Dawn Man of the long pre-Stone Age" (quoted from Osborn 1927b, 74) with the aim to remove the supposed myth of ape descent from the stage and the movies, from caricatures of our pedigree, and from scientific discourse. However, if the ape-ancestry hypothesis was becoming out of date, Shipley observed, it was thanks to the fundamentalists and not through comparative anatomy and zoology. Shipley cited Gregory (1927a) to attack Osborn's "demonking theory" and to defend Darwinism with the evidence that sciences such as comparative anatomy and comparative physiology gave to the kinship of man and ape.[52] Shipley's interference seems to have frightened Osborn. He was alarmed that the message of *Parnassus* was getting out of hand and sent Shipley his papers on the relation of man and the apes in the hope that he would drop the charge of fundamentalism. Somewhat ironically, in his comment, Shipley, too, reprimanded the producers of popular culture for portraying the human ancestor like a monkey, and he believed that the *Hesperopithecus* tooth about four million years ago had belonged to an American toolmaker. He therefore had little difficulty with Osborn's generous allotment of time to human ascent.

Finally, Regal (2002, 14–17, 173–175) has shown with regard to Asia as the cradle of humankind that Osborn's dawn man theory resonated with ideas dominant in the occult movement. Not entirely to his liking, notions such as the racial soul seem to have appealed to spiritualists like Basil Crump. Crump believed in an Atlante-Aryan race who, after the sinking of Atlantis in the Miocene, colonized other parts of the world and built the pyramids. Osborn's *Parnassus*, as well as the theory of Wood Jones, was understood to come closer than other anthropologists' theories to the secret-doctrine teaching (Blavatsky 1888). In Crump's view, apes represented a degenerate offshoot from the human line from a time when humans still resembled giant apes, although they were already thinking and speaking beings. Osborn's ideas on the antiquity and importance of the human mind were adapted by Crump to the notion of a thought-produced universe. Even Osborn's protestation that he himself had " 'little of the mystic' " could not deter Crump from counting him in as one of them.[53]

Parnassus was thus instrumentalized in support of a wide range of belief systems: spiritual philosophy, different forms of religion, antimaterialism, chauvinism, masculinism, progressivism, eugenics, racialism and racism, human brotherhood, and so forth. Nearly all of the central concepts in Osborn's reconstruction of human evolutionary history were controversial: racial soul, racial purity, racial mixing, racial superiority, orthogenesis, progress, and degeneration. However, most readers, it seems, still engaged in the cult of ancestry, individual as well as "racial": "I am something of an ancestor-worshiper since my visit to Japan, so I will approach the book reverentially."[54] The commentator in the *Seattle Times* also read *Parnassus* as being about "our great greats, down to the hundredth and thousandth and millionth generation."[55] *Parnassus* once again allowed readers to bring these ancestors to life: "We spared it for the afternoon hours here where we have quiet hours between 2 and 4 and look out from our little salon window at the snow-covered mountains and can talk . . . without interruptions. Eleanor lies on her couch and Louisa[?] reads aloud and I look at the cloud mantled heights and in imagination see the skin-clad men with their stone-age weapons tracking the bears."[56] Some joined Osborn in his glory that "while these anthropoid apes were luxuriating in the forested lowlands of Asia and Europe the 'dawn men' were rising in the invigorating atmosphere of the relatively dry plateaus of central Asia."[57] And they were following in his journey to "befriend[] the dawn man, remove[] from his reputation the bar sinister of ape descent, and credit[] our Stone Age ancestors with moral traits found amongst primitive peoples in these days."[58] A Japanese reader lauded the fact that "the sympathetic attitude toward the ancestors of man is admirably kept up so that one feels that one is reading really about his ancestors."[59]

In fact, readers again demonstrated that they not only integrated the fossil men into their family history but also made their own contributions to a deeper historical culture. A New Zealand correspondent informed Osborn about his popular book on evolution. Elisabeth Fulda, formerly an artist at the AMNH, wrote to Osborn that she was in the process of writing a children's book on prehistory. A teacher of a class at the Children's Museum in Boston asked for Osborn's permission to directly quote from *Parnassus* to the children and for some of Knight's illustrations. Finally, when life changed again drastically with the Great Depression, *Parnassus* could still bring consolation. A woman who had lost her job was inspired by "the most thrilling story" to pursue her goal of becoming an archeologist. She "sat up nights to finish it, and begrudged not one moment of lost sleep!"[60] The would-be archeologist was again not the only one to refer to Osborn's prose. Others, too, experienced *Parnassus* as a great read: "I began it this morning and I read and read and read until I had finished it. How can he make a scientific subject, heated with the fullness of scientific knowledge, fascinating to an ignorant 'laywoman'?"[61] As with *MOSA*, Osborn seems to have succeeded in providing a word painting "of the living man and his mode of thought."[62]

Furthermore, Osborn's use of the classics seems to have served the purpose of fusing prehistory, history, and the look into the future. To begin with, it was "such pleasure to have the discoveries of man's progress so delightfully related to the great literature that has marked the historic period."[63] Although in general the book was much more prosaic, this feature could make Osborn's version of the ascent of man look epic. The *New York Times* reported that "the branch known as Hominidae evolved slowly through unnumbered generations from its quadrupedal primate ancestry into the modern members of the human cast [*sic*], capable of building vast empires and unique civilizations, able to compose majestic symphonies and to develop into more powerful masters of their environment." In a "momentous drama" man had evolved into "Nature's aristocrat"—and he had better make sure he remained in that class, because human evolution was "not the musty romance of the past, but a drama which [was] still going on."[64] Indeed, Henry E. Christman of Alfred University regarded the facts of prehistory even as more relevant than those of history, and its story "as interesting as the best of our novels; and certainly more elevated and worth while reading."[65]

Some members of the press ranked Osborn as "a master novelist," comparing him to Muir and John Burroughs.[66] Even the *New York Times Book Review* praised the "true romance of ancient days," in which Osborn unfolded a "fascinating story with the swing and vigor of a Norse saga," and which surpassed such books on human evolution as published by Boule and Sollas.[67] Clearly, form, narrative, and language were recognized as instrumental to the success in kindling "racial souls." Such praise must have been music to Osborn's ears; as we have seen, to him, literature, poetry, and natural history were closely related approaches to a deeper understanding of nature, and he considered poetry as particularly able to both give expression and speak to the "racial soul," to evoke memories of times long past. If Osborn's dawn man theory had fictional elements and was rendered in the style of the romance, this was in line with his belief that fiction, too, could do its share in the distribution of the right kind of verbal paintings of our deeper history.

However, we will see in the next chapter that this was only the case if fiction was in essence true to nature. A particular concern here was evolutionary time. Osborn had attacked director Frederic A. Lucas for jumbling together animals of different periods in the halls of the AMNH. No wonder that the press and science fiction writers thought dinosaurs had coexisted with humans (Hellman 1968, 149–150). Osborn wanted to correct such errors and interfered in the production of not only educational literature but also fiction. He was in correspondence with many authors, from those of poetry to children's books, and these exchanges reveal still more places where "our evolutionary heritage" was negotiated and where different desires were articulated. Finally, while Osborn was increasingly irritated by the proliferation of popular histories and images of "our" evolutionary past, his own more scientific accounts came in for criticism from expert communities.

CHAPTER FIVE

History Within between Science and Fiction

One of the avid readers of Osborn's prehistory was the naturalist and explorer Ernest Ingersoll. Ingersoll had written juvenile novels and books, including some on paleontological subjects such as *Knocking Round the Rockies* (1883; about the Hayden Geological Survey, 1874–1877) and *Nature's Calendar* (1900). Osborn set Ingersoll on the track of writing a study of "the dragon," a mythical creature Osborn considered to be the cultural progenitor of his dinosaurs. After all, the Chinese had dubbed Andrews and him "Men of the Dragon Bones," an expression Osborn (1924) had borrowed for an article title to which he now referred Ingersoll. When Ingersoll's dragon project was on its way, Osborn received installments of the book for comment. Once it was ready, Ingersoll wanted an introduction to authorize his work. This gave Osborn the chance to present the Chinese as living in a prescientific, mythical past in which dinosaurs were still dragons, with the inference, I presume, that this was what came of impeding the work of scientists, who were now the real authorities on the truth about history. Be that as it may, Osborn promoted Ingersoll's book *Dragons and Dragon Lore* (1928) and added it as another treasure to his author's library, a place reserved for the trophies he procured in the lands outside science and of fiction.[1]

Another author who sought Osborn's cooperation was Lunette E. Lamprey (pen name Louise Lamprey), who wrote juvenile books for Little, Brown and Company under the series title Children of the Ancient World. They were intended for children between seven and ten, and for supplementary reading in schools, as well as information for teachers of history. The whole idea for the series—which sold fairly well in England

and the United States—had come to her during a children's camp. And this project, too, was about correcting popular notions of history and origins. Lamprey observed that grade-school children were "naturally intensely interested in primitive life," but she regretted that they "should get their impressions wholly from Indian tribal life and not know anything about their own prehistoric ancestors."[2] The first book in the series was therefore titled *Long-Ago People: How They Lived in Britain before History Began* (1921) and went back into the prehistoric times of cavemen. The following volumes dealt with Greece, Rome, and Egypt, and the next was to be called "Children of Ancient Gaul" (published 1927). She sent Osborn the general sketch, explaining that "the idea is to make the life of the primitive race real and interesting enough to give children increased interest in museum exhibits and in any prehistoric dwellings they may visit when travelling."[3]

This living history was a project to Osborn's taste, even more so because Lamprey drew on *MOSA* for her story. She assured him that in her work, she used "only what seemed authentic material";[4] the illustrations by Margaret Freeman were similarly intended for educational purposes and had to be historically correct. Lamprey liked to intersperse her stories of healthy, inquisitive, and self-reliant Stone Age children with comparisons to the children of the present day, who in similar situations on their own in a demanding but also wonderful environment might have had to go hungry or cold. The prehistoric parents were portrayed as good teachers of hunting techniques, toolmaking, and housekeeping, as well as good models of alertness, courage, and morals. Lamprey occasionally accentuated this by introducing an "effeminate" man into the Stone Age scene (Lamprey 1921, 17–28).

For *Children of Ancient Gaul*, Osborn was asked by the publisher for a "sentence or two that we might quote with your name in the nature of endorsement of the book on the score of historical accuracy."[5] In appreciation of his expert support, Lamprey sent Osborn a copy of the school edition (there was also a trade edition). Osborn was delighted by the use she made of his writings. The book deserved to be placed into the hands of his grandchildren at Christmas, which was always a mark of high approval.[6] He sent Lamprey an inscribed *Man Rises to Parnassus*, also recommending she read his *Creative Education in School, College, University, and Museum* (1927a), in which he developed his grand ideas on transformative education. However, not everyone liked the kind of self-understanding Lamprey wanted to implant into children's minds. Referring to the antievolution laws passed in some states during the 1920s antievolution campaign, someone protested: "No wonder the South is waking up to this dirty stuff and passing legislation to stop the use of such books in the schools."[7] Osborn sympathized with Lamprey, and they kept up their cooperation in the project of putting the right kind of identity-forming (pre)history into children's heads for years to come.[8]

Osborn's prehistory became associated with another delicate issue when the writer for a juvenile audience, Emma Gillmore, was inspired by his work for her *How and Why of Life* (1932) that explained the evolution of sexual reproduction and gave a bird's eye view of mammalian evolution. Gillmore had seen mounted specimens from the AMNH in European museums and she felt inspired to contribute something to this international effort to promote the evolutionary view.[9] She sought Osborn's approval when sending him an advanced copy with the request for a review. She was treading dangerous ground, but Osborn thought she handled the delicate topic of reproduction well enough, the material proof of which was his presenting a number of copies to some of his younger friends. After all, Osborn saw no harm in conveying a moral code of gender and reproduction through the lens of evolutionary history. He also approved of her use of AMNH exhibits and felt the need to emphasize that it had taken the museum "forty-two years [of] collecting, mounting and restoring these animals and assigning them their appropriate Greek names,"[10] so that topics from dragons to sex might be illuminated by them.

A particularly Osbornian history was given in *The Earth for Sam: The Story of Mountains, Rivers, Dinosaurs and Men* (1930), the first in a series of illustrated science books for children by W. Maxwell Reed. For Reed's information on archeology, Nelson arranged a collection of tools in chronological sequence. Reed was generously supplied with books from the museum, and this time it was the librarian of Osborn's library of paleontology who checked the text for scientific accuracy. Osborn granted ample reproduction of Knight's and McGregor's restorations, and many other illustrations stem from the museum. The book is a treasury of prehistoric imagery, and especially nice are Karl Moseley's cartoons. In these, one might glean support of the tarsioid theory—or a mockery of it (see figure 10)?

Reed took an agnostic position with regard to eoliths and Tertiary man. However, following Osborn's staging of evolution, in *The Earth for Sam*, primate groups are successively expelled from their homestead in

FIGURE 10. Meet the tarsioid ancestor?, by Karl Moseley, Reed 1930, 266

the Gobi, because they can no longer cope with the roughening living conditions: "The next group that could not stand the intellectual strain of remaining a member of the ancient but progressive family, traveled far to the west and eventually came to Africa. We call them Old World monkeys" (Reed 1930, 286). These "conservatives" that had been unable to keep up with the "progressive" branch were at the time of writing still lingering in Africa (ibid., 308, 311). To connect the past even more strongly to the present, the successive waves of ever higher primates that traveled from Asia to other continents were described in terms of "waves of English, Irish, or Italian immigrants" entering America (ibid., 356). Indeed, present "primitive savages" were said to be much like the Neanderthals, "as any one knows who has seen and smelled an Eskimo in his tent" (ibid., 362). In contrast, throughout the book, but most strongly in the final chapter, Reed used "the white primates" for *humankind*.

Reed ended his racist history with a reference to the geographer Ellsworth Huntington's *The Character of Races: As Influenced by Physical Environment, Natural Selection and Historical Development* (1924), which justified Teutonic supremacism on the basis of environmental determinism. Reed recommended Huntington's book as a sequel to his. It naturalistically portrayed the continuation in historic times of the prehistoric struggle between races for supremacy over territory that alone could explain why "some tribes of human beings are so bright and powerful" (Reed 1930, 375). Reed brought into the children's rooms the Osbornian hierarchy of races in past and present and the associated "racial succession model." However, in contrast to Osborn's emphasis on the role of the spirit, of beauty, and art in human evolutionary history, to Reed, this process was the result of "the relentless law of the elimination of the unfit" (ibid., 329). Finally, Reed reproduced Amédée Forestier's Cro-Magnon scene from the *Illustrated London News*. This rendering would have been much too erotic for Osborn's taste. Most importantly, Osborn must have disliked the "Negroid" look of what he considered to have been Paleolithic supermen. Reed "tainted the blood" of Osborn's prehistoric Greeks by granting that they might have had a few African ancestors.

Clearly, Osborn tried to ensure that popular and educational books rendered the deeper past in line with his own science and politics. While this may not astonish, the fact that he demanded the same standard of fiction comes as more of a surprise. Indeed, Osborn compared his success in turning the AMNH into an efficient center of calculation, where the objects and knowledge gained in the field were collected, analyzed, exhibited, and immediately translated into narratives and images to enter the historical cultures internationally, to the speed of travel in Jules Verne's *Around the World in Eighty Days* (1873). He bragged for one particular trophy from Central Asia, the *Baluchitherium*, or the giant rhinoceros of the Miocene, that within nine months of the discovery of its remains in Central Asia, he had made the creature known to millions of people (Osborn 1926a, 193–195).[11] He envied the ease with which writers like Verne in his *Journey to the Center of the Earth* (French original, *Voyage au centre de la terre*, 1864) could produce and distribute lost worlds. Osborn was therefore intrigued by the genre of prehistoric fiction that entered its golden age with H. G. Wells's *The Time Machine* (1895). When the tutor of Arthur Conan Doyle's children was introduced to him, Osborn promptly responded by inviting the teacher and his famous employer to visit the museum.[12] No doubt, the creator of *The*

Lost World (1912) would be "interested to see some of the real bones of the animals which [he had] treated so delightfully in imagination."[13] Although Osborn could not get hold of the famous author at once, Doyle's sons visited his museum and "were very interested in the Reptiles."[14] The success of Osborn's dinosaurs finally secured him a visit from Doyle himself, further traces of which are unfortunately lost.[15]

However, we do have evidence of the exchange between Osborn and another prehistoric fiction writer: George Langford. Langford liked to produce bas-reliefs of Pleistocene animals and Paleolithic humans, even fossil mounts, as well as poetry. Osborn was very much in favor of these endeavors and promised Langford that his mammoth relief and his poem "The Dawn of Art" would be included in the Cro-Magnon case in the Hall of the Age of Man. Osborn also wanted to publish the verses in *Natural History* together with Knight's Cro-Magnon mural.[16] The following year, Langford announced the publication of *Pic, the Weapon Maker* (1920), a prehistoric story for the young that he himself had illustrated with ink drawings. It was greatly inspired by *MOSA*: "It is a novel and although fictional, is an endeavor to adhere as closely as possible to the facts as known or reasonable deductions from them."[17] The story may remind the current reader of the *Ice Age* film plots: it revolves around a youth of the Mousterian period who associates himself with a mammoth and a woolly rhinoceros. Langford situated it in the Valley of the Vézère near the archeological site of the Le Moustier rock shelter. Pic discovers the lost art of Acheulian flint retouching, which leads the three friends to such prehistoric sites as found in the Seine and Somme regions, the Thames valley, and Belgium.

The story was meant to "interest people in prehistoric things,"[18] and Langford therefore asked Osborn for an introduction. Osborn agreed, and he also encouraged Langford to do his own archeological exploration and sent him publications on promising spots. However, before Osborn would write the preface, the museum was again to check the text for scientific accuracy. After Matthew had read the *Pic* manuscript, he assured Osborn that "I regard Langford's work as showing an unusual insight and comprehension of the subject."[19] Langford himself asserted that he tried to stick to the true atmosphere of the time, and that the book was free from objectionable matter that could corrupt women or children. Further to Osborn's liking, Langford idealized his human character and kept him clean from any "simian affinities."[20] Langford shared Osborn's belief that the characters of an evolutionary story should be rendered sympathetic and humanlike in order to allow the reader to identify with them. He also followed Osborn's theory in building the story on the importance of tool production in human evolution.

Finally convinced of Langford's true vision of things, Osborn "hastily dictated the *Preface*."[21] Of course, this went through the hands of one of his colleagues before going out—this time Gregory's. The problem with the preface was that knowing Langford's and his own views were in line, and thinking of a Paleolithic hero, Osborn had the Cro-Magnons in mind, while Langford's wonderful story was about a Neanderthal boy. Gregory agreed that "this fact makes Professor Osborn's *Preface* almost entirely inapplicable to your book, as he is evidently referring exclusively to the Cro-Magnon people, for whom he has the greatest admiration. These are the 'great people' . . . 'intellectually endowed quite as richly as ourselves' of his *Preface*."[22] Langford revised the preface, making as few changes as possible and referring to the Neanderthals as a side branch rather than as ancestors of today's people, so that Osborn might accept the editing.[23] Reading Osborn's published introduction, one notices that he was pleased with the generally sympathetic stance Langford took on so-called prehistoric man. He emphasized that even among "the most primitive races of mankind," even among the Malays and Polynesians of today, there was a "sprinkling of fine characteristics." However, for Osborn, true refinement was only found in the Cro-Magnons. Therefore, now aware of the fact that the book dealt with Neanderthals, he made sure to specify them as "a very ancient and primitive branch of the human race" (Osborn 1920, xi). The reader should keep in mind that the author "may idealize these primitive men" (ibid., xii).

There were further issues to contend with; the publisher had a cover image made that included a "man who was a sort of composite of Joe Chamberlain face; Neanderthal body and legs; and gorilla arms and hands."[24] Langford succeeded in keeping it off the book itself, although it went on the paper jacket. When Osborn finally held the product of tedious negotiations over "our ancestors" in his hands, he liked it, even warming to the thought that "there were certainly some bright and decent people among the Neanderthals."[25] However, some readers of the book disapproved of the fact that Langford's mammoth and rhino spoke; someone even accused him of "nature-faking." In harmony with Osborn's philosophy, Langford retorted, "Those who shudder at the bare mention of 'missing link' and 'simian ancestors,' might view their shady past with equanimity if the matter were presented to them in palatable form."[26] In fact, some readers appreciated exactly Langford's brushing-up of their ancestors and his animation of their past that stood in stark contrast to "a considerable flight of books on the life, love and labors of the original man-monkey, and their collateral idiocies of the Tarzan type."[27]

Because Langford was producing a book series called The Long Ages Ago, the process of negotiation between "science" and "fiction" continued. Osborn was content enough with *Kutnar—Son of Pic* (1921) to order a number of copies for some of his young friends as Christmas presents, injecting the story into his own family's understanding of their past. He considered the animal drawings improved and hoped that the men could be further corrected as soon as McGregor completed his life-size model of the Neanderthal man.[28] While this demonstrated Osborn's patronizing side, as Osborn meddled with Langford's fiction and art, Langford did not shy away from interfering with Osborn's science either. When Osborn renounced his earlier suspicion with regard to the human workmanship of some of the East Anglian eoliths, Langford wrote to him about his trip to England in 1905, when he went fossilizing, and visited the Ipswich area and the local museum. Langford had picked up many fractured flints from the Red Crag, and he highly doubted that they had been fashioned by humans. However, in the end Langford gave in to Osborn's expert opinion, allowing that "it sounds as though you have really unearthed Pliocene Man."[29] While Langford maintained his fascination with the Neanderthals, "The *Mousterian* Flint-worker" eventually became a hominid of a "bent-kneed misshapen kind / With protruding maw, massive chinless jaw."[30] Langford thus complied, and when he was working on an account of elephant evolution from a layman's point of view, Osborn and Matthew provided him with scientific information. Osborn strongly recommended that Langford wait for his monograph on the subject.[31]

Unlike Langford's strange fascination with Neanderthals, the French fiction writer Victor Forbin's imagination was kindled by the Cro-Magnons. Forbin dedicated his *Les fiancées du soleil* (1923) to Osborn and described his story along the lines of Osborn's prehistory, as an account of the triumph of the intellect over brute force, of humans gaining mastery over the demanding environment at the close of the glacial period thanks to their spirit of invention and reason. Indeed, Forbin's genuine interest in science was noted by the museum team. The book was translated into English, and Forbin estimated that a letter by Osborn, his "cher Maître," would be of great assistance in finding a good publisher.[32] Osborn's statements had already been used in promoting the French original, which sold well, and they seem to have helped in securing Henry Holt for the production of the annotated edition of *Les fiancées du soleil* (1925) for school and college use.[33] Osborn was very pleased with the American edition of Forbin's book. He secured it for his author's library in the museum and ordered copies to give to his friends.

In general, this was a science-art cooperation to Osborn's liking. Forbin truly revered Osborn and drew inspiration for his next novel, *Le secret de la vie* (1925), from the scientist's *The Origin and Evolution of Life* (1917).[34] Osborn therefore enjoyed working with Forbin on an article for *L'Illustration*, for which Osborn provided the photographs. Expressing his curiosity about Forbin's work in progress, "La fée des neiges," Osborn once again showed his esteem of literature as a means to translate the scientific knowledge about evolutionary history into a transformative story: "You will be interested to see in my address in the forthcoming number of *The Scientific Monthly* my acknowledgment of Balzac's tribute to palaeontology in 'La Peau de Chagrin.' I am glad your writings are so successful; you certainly deserve it all."[35]

Osborn was satisfied with the prehistoric fiction of these authors, but the popular culture of his time was peopled by a wide range of missing links (Clark 2008). There indeed was "a considerable flight of books on the life, love and labors of the original man-monkey, and their collateral idiocies of the Tarzan type."[36] And there was no one as successful in their production and distribution as Edgar Rice Burroughs, creator of Tarzan. In his prehistoric fiction, Burroughs heavily relied on the paleontological halls of the AMNH and the writings of its members such as Andrews's *On the Trail of Ancient Man* (1926).[37] In imitation of the scientific travelogue, the reader of Burroughs's prehistoric fiction is again and again treated with naturalist descriptions of the plants and beasts encountered on the journeys through the scenes remote in time. In imperialist fashion, Burroughs's explorers of prehistoric worlds are able to survive in and partially dominate their new surroundings because they have appropriated the knowledge that was created at home through traffic in the opposite direction, through exactly the kind of work Osborn and his team carried out in transporting the objects faraway in geography and time to the United States. The Burroughsian naturalist-adventurer even manages to settle scientific controversies through eyewitness experiences. Owing to this advantage over the scientific experts in the real world, he can correct their mistakes. At least in the case of the controversy around *Diplodocus*, the Burroughsian explorer verifies Osborn's way of reconstructing the creature (Burroughs 1918b, ch. 3).[38] Furthermore, in the setup of his lost worlds, Burroughs used structural devices reminiscent of Osborn's biogeography and orthogenesis. His lost worlds contain hot spots of evolution where differentially advanced human types compete. One also encounters hierarchical series of hominid types spread out in geographical sequence that simultaneously represent stages in individual development. Finally, the way in which Burroughs brought the individuals and "races" of the Stone Age alive gives evidence of his acquaintance with current paleoanthropology, especially as presented in *MOSA* (Burroughs 1918a, 1918b, 1918c, 1930; Sommer 2007b).

Burroughs's fiction, like Osborn's science, delivered quite explicit comments on sociopolitical issues. Both claimed to reveal natural social and racial hierarchies, and the true essence of men, women, blacks, or whites, in what Burroughs more honestly referred to as the "interesting experiment in the mental laboratory which we call imagination" (letter from Burroughs to the *Daily Maroon* in 1927, quoted in Porges 1975, 223). Famously, Burroughs had developed this kind of experiment in "Tarzan of the Apes: A Romance of the Jungle" (1912) ("The Tarzan Theme," *Writer's Digest*, June 1932, reproduced in Taliaferro 1999, 14). But the experiment could also be set up in the primordial and by inference natural setting of the lost world, into which competitors, removed from their familiar cultural environment, could be placed, exposing them to the Darwinian struggle for survival. Who would take which place in the hierarchy of human sexes, types, "races," nations, and individuals? In *Tarzan at the Earth's Core* (Burroughs 1930), Tarzan himself was actually placed in such a primordial setting, and lo and behold, because of his good biological heritage and unharmed by overcivilization, Tarzan proves himself the undisputed hero of the Paleolithic, but close on his heels follow other, especially American adventurers.

Reminiscent of Andrews's travelogues (e.g., 1926, 7, 45, 108) and Osborn's messages from the Stone Age, in the Burroughsian primordial settings, English and American heroes tend to undergo a transformation that looks like regression, but is, in fact, an advance. In the struggle against nature, each other, and against forgetting in the timeless places, the civilized among them fall back on their primeval instincts and reanimate their essential qualities. The change in looks from pampering

American apparel to the bare necessities of the primitive warrior is accompanied by an inner metamorphosis. Burroughs's heroes find the way back to their true type: "force, energy, initiative and good judgment combined and personified" (Burroughs 1918b, ch. 1). Although their clothes, weapons, and ways of thinking and acting begin to approach those of "the savage," Burroughs's heroes maintain their moral and intellectual superiority. They shed of their breeding only that which is cumbersome in a life close to nature. The thin layer of degenerate cultivation peels off when the effects of an effeminizing kind of education that emphasizes the indoor study of books over direct nature experience, and of an overprotective and artificial environment, slacken. The outcome is the fittest type in the struggle for survival among fossil and living "races": "With one squad of a home-guard company I could have conquered [the lost world of] Caspak" (ibid., ch. 7). To emphasize this point, Burroughs places his American hero on the back of a beautiful thoroughbred, itself a product of an evolution after the principle of racial purity. Thus, in line with Osborn's hope in the transformative power of prehistory, Burroughs performed true acts of remembering in his protagonists who are "full of reminiscences" and respond "to conditions and experiences long bygone" (Osborn [1913] 1924, 183).

Burroughs also brought eugenic concerns into the Stone Age (Burroughs 1918c), and at times his politics through fiction indeed seem a vulgarization of Osborn's more scientized histories. In his unpublished essay, "I See a New Race," from the 1930s, Burroughs slurred that "every one knew that there was something quite wrong with the way in which man utilized the powers that evolution had given him. He was not far from perfect, but he did not appear to be improving as the centuries unrolled. There were many, in the 20th Century, who believed that the masses were less intelligent than the Cro-Magnon race of Paleolithic times. But, even worse, it was apparent that as the stupid multiplied without restriction the whole world was constantly growing stupider" (quoted in Taliaferro 1999, 266).[39] If Burroughs was among those who received the message of the Cro-Magnon, Osborn must have been appalled by his lack of decorum. He certainly did not approve of some of the freedoms of imagination Burroughs took in his fiction. For dramaturgy's sake, Burroughs not only placed prehistoric human races in environments populated by ravenous dinosaurs; he also created ape-men and some fantastic human-animal hybrids.

However, Osborn was powerless against Burroughs's typewriter. As

John Taliaferro (1999) shows, Burroughs's career is a case study in the mushrooming popular culture. His stories were the product of a one-man writing factory. Serialized on the cheap paper of the pulp fiction magazines, they flooded the newsstands. Subsequently, Burroughs's romances were republished as first editions and fifty-cent popular books and translated into over thirty languages, purportedly selling up to sixty million copies during his lifetime. Burroughs distributed Tarzan through all available channels from print media to radio and film. He was smart enough to trademark his most famous hero and had him fight in the comic strips of daily newspapers, which were a mainstay since the turn of the century. Tarzan as merchandize greeted children and adults from the ice cream cup to pajamas. Burroughs was also one of the first writers to incorporate when he became tired of negotiating with publishers. From 1923 onward, he published all his stories through his own Edgar Rice Burroughs Inc. And like Osborn, Burroughs always had his audience in mind and stared into the press mirror. His production process could even be participatory, as when he asked "Tarzan's friends" which kind of ending they would prefer for the movie.[40]

If Osborn was on a crusade to infuse people's minds with the right kind of evolutionary narratives, in trying to destroy those dinosaurs that roamed the lost worlds at the same time as humans, he was up against Burroughs's own efficiency and business cunning. In fact, Osborn must have detested the pulp-fiction culture as much as the blatant departure from truth to nature that Burroughs and other contributors allowed themselves. Still, their creatures were all over the place. Osborn's history within was of enormous influence, but he ultimately had as little control over the "word-painting of the scene and of the man or woman" from prehistory that was implanted in readers' minds as Burroughs himself had over Tarzan (Krüger, Mayer, and Sommer 2008). The same holds true for the images that were added to people's visual memories. Although some of the prehistoric films that began to appear on screen were inspired by Knight's pictures, they diverged widely from Osborn's prehistory; the first feature-length prehistoric fiction movie, *The Lost World* (1925), once again rendered humans the contemporaries of dinosaurs (Milner 2012, 174–177).

Osborn not only had to witness how his ancestors were toyed with in the tides of popular culture; he also had to see his evolutionary history increasingly criticized from different scientific corners, not least from within the AMNH and Columbia University. Before ending this first part of the book, I have to address these changes that will allow me to point to discontinuities but also continuities with part 2. At stake were Osborn's orthogenesis, his concepts of the race plasm and the racial soul, as well as hominid phylogeny and classification—in other words, the science as well as politics of evolution. Let me begin on a positive note, however, with how a visitor may still have experienced and applauded the museum in 1930:

> The thrills which ran up and down my back in the Hall of Dinosaurs, the enthusiastic admiration for the results of the Central Asiatic excavations, and the pleasant sensation that nowhere a rigid pattern holds sway, but that, for example, one can take an animal, as in the case of the horse, and connect its past and present, a thing which . . . has been so successfully done in the presentation of man—all this makes the visit a pleasure. . . . The intellectual life of the Museum goes far beyond the exhibitions. The School Service Division, the cooperation of the Staff in animal conservation, in the establishment of conservation parks, etc., are examples of the activities which show that here the exhibits are but a nucleus around which everything is arranged which joins mankind with nature and maintains the connection. I do not like to mention names, but that of Henry Fairfield Osborn must be mentioned. For in him the great investigator is combined with the solicitous financier and the kindly man with a genuine love for those who cannot see the whole, as nature gives it to them, but to whom the substituted Museum must drive home the magnitude of creation.[41]

When Osborn retired from the museum presidency in 1933, the spirit of the museum had already begun to change. In the context of the Third International Congress of Eugenics of 1932, the new Hall of the Natural History of Man opened, and Gregory proudly reported to Osborn on an exhibit of a series of twelve dissections. It allowed the visitor to compare four successive layers of the muscles of the foot of the bear, the chimpanzee, the gorilla, and man. It demonstrated the close affinity between man and the anthropoids. The new exhibit powerfully forced on the visitor evidence of the derivation of the human foot from that of an animal adapted to an arboreal life. That this was intended as a malicious joke on Osborn's self-identification as divine bear is improbable, but the aim of conveying in the halls a spiritual atmosphere and noble hominid genealogy was certainly under attack. Osborn not only disagreed with the knowledge conveyed; he also disliked the way it was presented. He strongly felt that man was being degraded to an engine at work. He found the human models repellent and suggested placing a full-size statue of an athlete at the center of the entrance. However, Gregory disagreed with Osborn's notion that the exhibit should elevate the visitor through contemplation of the beauty of the human body: "Thousands of artists and poets have celebrated the beauty of the human body but I do not feel that we should enter that crowded field. The human body as a living engine and as a product of evolution is the sole subject of the exhibit in Section I."[42] Art was kept at a distance from science; aesthetics and knowledge were severed. Gregory tried at least to console Osborn with the information that the other half of the exhibit would be devoted partly to "the races of man." A series of beautiful figures were intended as the centerpiece.

As the Hall of the Natural History of Man illustrates, the anthropoids moved into the spotlight. Gregory informed Osborn that at the American Association for the Advancement of Science meeting, where the AMNH had an exhibit at the popular position of honor in the auditorium, "our fine gorilla bust, made by Raven in West Africa makes a fine decorative center piece."[43] With *Man's Place among the Anthropoids* (1934), Gregory finally rebutted theories such as those held by Osborn and Wood Jones (Rainger 1991, ch. 9). Incredibly enough, Osborn still found room for negotiation: "I observe in Figure 3 that our views are slowly converging and I hope in time they will be altogether uniform."[44] It has to be admitted that the family tree of the primates is yet another very ambiguous image. It does seem to suggest that the main stem leading to "the modern human races" was clearly distinguishable throughout primate evolution, melting at the bottom into the early offshoots of the lemuroids and tarsioids. However, the theory of Central Asian dawn men was not only seriously challenged by the contemporary anthropoids; the australopithecine fossils were also gaining in status (a skull was first discovered at Taung in the northern Cape Province of South Africa in 1924). The primate tree from Gregory's book that Osborn referred to was a wall painting in the AMNH exhibitions, and it featured "the southern ape" at the missing-link position. An *Australopithecus* cast had been installed in the Hall of the Age of Man in 1930,[45] and Gregory eventually changed the guide to the hall, presenting southern Africa as the most likely place of origin of humankind and *Australopithecus* as the most humanlike of apes (Osborn [1925] 1947; already 7th ed., revised to 1938).

Osborn knew that his history within was losing ground, and he was also aware that his evolutionary politics were under attack. The most central figure of the antiracism movement in the United States was Franz Boas, who was in the anthropological department of the AMNH (1896–1900) and taught anthropology at Columbia (1896–1934). Boas not only deconstructed the racial types of physical anthropology in his scientific studies; he also engaged in public campaigns (Barkan 1992b, 76–90). Boas and his Columbia students such as Ruth Benedict and Margaret Mead brought cultural relativism to the wider American audience and they spoke out against Osborn. Osborn was alarmed and he warned Gregory to tone down his writings on the subject of race (Rainger 1991, 178). The former museum ethnologist Robert H. Lowie even refused to sign the tribute for Osborn's seventieth birthday because of his racial prejudices (Hellman 1968, 197). However, the most aggressive critique was voiced after Osborn's death:

> From any strict scientific viewpoint, Henry Fairfield Osborn, sterile as a creative worker, holder of unscientific and inhumanitarian racial views, and finally a betrayer of science in his compromise with the church, deserves no honorable place in the annals of science. . . . Much more sinister were his Nordic-superiority race theory and his notion of selective birth control which, had it been carried out, would have meant a systematic extermination of the races displeasing to him. But his theory was also his practice. Under his administration, the Museum of Natural History could have been transported to Nazi Germany and been at home. Its personnel was kept pure of Semitic coloring, though there are many Jewish scientists eminent in the fields of research that it covers. Appropriately enough, one of the last of his honorary degrees was awarded him by a Nazi University.[46]

Although this sweeping condemnation of Osborn goes too far, for many, the time had come to remember the past differently and to build new genealogies. Furthermore, during Osborn's lifetime, his understanding of the workings of evolution had been attacked beyond anthropology. He had felt increasingly threatened by the new experimental sciences, in particular genetics. With Thomas Hunt Morgan's research group at Columbia, Osborn was at the source of world-leading laboratory research in the field. Morgan's group studied drosophila mutants and confirmed the Mendelian laws through further insights into the mechanisms of heredity. They could identify the gene as a segment on a chromosome and understood that one gene influenced more than one character and vice versa. The interaction of genes and their linkage on chromosomes explained how discrete inheritance could have continuous phenotypic effects. Mutations as well as chromosomal recombination and crossing-over resulted in new and novel combinations of genes (Morgan 1916, 1919, 1923). The results of Morgan's and his students' work made it possible to view these processes as bringing about the organismal variation on which natural selection works (Allen 1978, 1980; Kohler 1994).

For Morgan, the hereditary processes were ruled by chance. He therefore had little patience with Osborn's idiosyncratic way of accommodating his orthogenetic theory to genetics.[47] Osborn on his part adhered to his main tenets of the role of the interplay of behavior and environment in triggering evolutionary trends (Osborn 1926b, 1926c, 1930c, 1931b, 1933, 1934a, 1934b). He could make out in the writings of Morgan and others the revival of Darwinian theory and the demise of Lamarckism. Still, two years before his death, he maintained the possibility of "an internal perfecting principle" (Osborn 1933, 200). The trick was to attribute Darwin's *chance* to the ignorance of the Victorian era; Osborn claimed that Darwin himself had allowed for the possibility that chance would turn out to have been a placeholder for a principle subsequently discovered (ibid.; Rainger 1991, 132–145; Regal 2002, 65–74, 77–79). By the time Osborn wrote this, the Mendelian laws of character transmission as specified by the Morgan group and others in cytological experiments had been brought together with Darwinian natural selection in population genetics in such works as R. A. Fisher's *The Genetical Theory of Natural Selection* (1930), Sewall Wright's "Evolution in Mendelian Populations" (1931), and J. B. S. Haldane's *The Causes of Evolution* (1932b). These authors developed mathematical-statistical models for the change in allele frequencies in populations under the influence of factors like selection, migration, and drift. Evolution was described as a process determined by mechanism and chance, and such notions as Osborn's creative evolution were labeled mystical (e.g., Wright 1931, 154).

However, Osborn was not alone with his variations on the orthogenetic theme. Orthogenesis, but also "neo-Lamarckian" and vitalist interpretations survived into the 1920s and 1930s. Such mechanisms could not only explain trends in paleontology, but were also in harmony with recapitulation theory in embryology, and the adaptive geographic variation of organisms encountered by biogeographers and systematists (Burkhardt 1980). As Bowler (1986) and others have shown, paleontologists, and in particular paleoanthropologists, continued to rely on parallel evolution that was either explained by use-inheritance or directed genetic evolution as late as the 1940s (see also Sommer 2007a, part 2). There were also disciplinary rivalries involved. Osborn and some of his colleagues acutely felt that the laboratory and experimental sciences were gaining in prestige, while geneticists were increasingly alarmed by the special status attributed to chemistry and physics.[48] Morgan regarded genetics as the basis of development and evolution and put Osborn's science down as speculative. On the contrary, throughout his engagement with genetics, Osborn (e.g., 1931b) emphasized that only the fossil record could provide reliable insights into the history and mechanisms of evolution.

Indeed, a full integration of this record into the new Darwinian framework was only achieved after Osborn's death, by yet another scientist at the AMNH. George Gaylord Simpson had worked under Osborn after he came to the museum in 1927 as Matthew's successor at the Department of Vertebrate Paleontology. When studying at Yale University during the 1920s, Simpson was already acquainted with Mendelian genetics and the principle of adaptation through mutation and natural selection; and he was influenced by Matthew's biogeography and neo-Darwinism. Simpson liked Osborn and gave him credit for his early attempt at synthesizing knowledge on evolution and for establishing some fundamental principles such as adaptive radiation (Hellman 1968, 205).[49] Nonetheless, at the AMNH, he began to oppose Osborn's germinal predestination (Rainger 1991, ch. 8). It was only after Osborn's death, however, that he argued against orthogenesis and parallelism by using genetics, addressing the implausibility of genes remaining dormant for very long times in different evolutionary lines. In the 1940s, Simpson began to speak out against orthogenesis as well as Lamarckian and vitalist interpretations of evolution more frequently (Sommer 2010a).

Simpson's paleontology was part of the so-called evolutionary synthesis that integrated the insights from theoretical population genetics with knowledge from further biological disciplines, in particular field sciences. In his *Tempo and Mode in Evolution* (1944), he reinterpreted the paleontological record in the light of the population genetic principles and at the same time showed paleontology's unique contribution to the study of evolution. Within this framework, there was no need for orthogenesis. Where such trends were indeed visible in the fossil record, they were the result of directing selection over prolonged periods of time. In contrast to Osborn's emphasis on determined development, Simpson stressed the role of adaptation and genetic drift, especially in the evolution of higher taxonomic units (Gould 1980; Swetlitz 1993, ch. 6). In this rethinking of his field, Simpson was also influenced by Theodosius Dobzhansky, whose *Genetics and the Origin of Species* (1937) reoriented the outlook of many biologists. Dobzhansky had joined the Morgan lab in 1927 and tested Darwinism in lab experiments and the mathematical models from population genetics in studies on the genetic variation in natural drosophila populations. He thereby also translated the mathematically challenging models into something understandable to field biologists (Adams 1994; Provine 1994). A third "architect of the synthesis" was Ernst Mayr—also at the AMNH from 1932 to 1953. As associate and later full curator of birds, Mayr worked on the immense bird collections to clarify systematics and biogeography, and he undertook fieldwork. His Columbia lectures formed the basis of *Systematics and the Origin of Species* (1942), which was a keystone of the reformed systematics and species concept within the synthesis. Against the earlier typological approach, he defined species as actually or potentially interbreeding populations that are reproductively isolated from other populations, emphasizing the role of geographical isolation in the formation of new species (Bock 2005).

To characterize these scientists' approaches to biology, the adjective *organismic* was used. It served to distinguish them from the genecentrism of theoretical population genetics. For the naturalists, organismic variation in populations stood at the center of interest. The selective value of a gene had to be considered in the context of other genes and the ontogeny and behavior of the organism in a polytypic population (Dobzhansky 1964, 451; Beurton 1999). Simpson emphasized that while the individual organism may seem unimportant in view of the fact that genetic mutations occur at random, the fate of mutations, their combination, segregation, spread, survival or extinction, as well as equilibrium point within a population depended not only on the number of organisms, as Wright, Fisher, and Haldane had demonstrated, but also on their behavior. This particular view from the organism served Simpson also to distinguish "the old from the new paleontology": "The most limited and formerly the most common sort of work in these fields consisted of the examination and comparison of a supposedly representative, dead individual of any given species. . . . Now both zoologists and paleontologists, each in the ways permitted by their materials, are more likely to study animals as functional, active organisms and not only as static morphological exhibits. This shift in emphasis has given rise to a new concept of the individual and to a new orientation of the individual in the scheme of things" (Simpson 1941, 9). There is a certain irony in the fact that the paleontologist at the AMNH talked of exhibits as static. But Simpson rightly pointed out that in this view, the individual organism was far more important than in Osborn's orthogenesis, where, to begin with, the direction of a taxon's evolution was predetermined by the germplasm once a trend had been triggered.

In his important revision of horse evolution, Simpson would try to persuade his general reader that "[fossils] are animals, just as full of life as you are, even though they occur at different points in the endless stream of time. Within their own segments of this stream, they breathe, eat, drink, breed, fight, and lead their own lives" (Simpson 1961, xxxiv–xxxv; see also 1951). Even the paleontologist had to think from entire organisms that had a dimension in time as well as in space. Through such an understanding of the individual organism the paleontologist would arrive at a new concept of species. Species could no longer be represented by an archetype, with individual variation as a simple deviation from the main theme. Rather, a group consisted in the sum of individual variation and its changes through time and distribution in space; the species concept had become statistical. While a synthesis of research on all levels on which phenomena of life manifested themselves was the ultimate aim, Simpson stressed that the individual organism was the focal point:

> The group is not an entity in the sense that the individual is an entity. A group achieves adaptation and progresses only in the sense that the individuals composing it do so. Satisfaction is an individual compulsion and not a group achievement. Evolution is not a thread on which individuals are strung, but a structure composed of individuals. A species is not a model to which individuals are referred as more or less perfect reproductions, but a defined field of varying individuals. A phylum is not a supermodel that abstracts the immutable features of a group of specific models, but a flowing river of ever-changing individuals. (Simpson 1941, 14)

This poetry of opposites that drew a line as sharp as possible between the old and the new biology was supported by drawings that juxtaposed visualizations of the concepts of orthogenesis and random variation and directing selection; the three-dimensional and four-dimensional organism; the archetypal and statistic species concept; and the phylum as a string of beads and as a structure made up of individuals. Although addressed to the Academy of Sciences, the images seem to be made for children. The lesson needed to be understood.

The turn to the integrated organism—integrated in individual and evolutionary time, in functional and geographical space, and in population structure—was associated with a rhetoric of revolution and novelty. However, the picture of a sudden rupture in the history of paleontology suggested by Simpson has been problematized. Simpson's paleontology did not appear from nowhere (Sepkoski 2012, ch. 1), and the rhetoric of the second Darwinian revolution also served ideological and institutional aims (Smocovitis 1994, 1996). This is particularly conspicuous in the case of Julian Huxley, who will be at the center of part 2. Huxley was another prime mover of the evolutionary synthesis. He had in fact worked on Osborn's orthogenetic paleontological series before Simpson. Huxley contacted Osborn in the early 1910s, and Osborn was more than usually pleased with his role as counselor, because the man in his mid-twenties was of the best possible stock. He was the grandson of Thomas Henry Huxley, the very British naturalist Osborn so cherished. Osborn had known not only the grandfather but also the father, Leonard Huxley, and after a first personal encounter, he assured Julian Huxley that he would always be welcome at the Osborns'.[50] Huxley was interested in paleontology and busily studied Osborn's books, beginning with *The Age of Mammals* (1910) and *Men of the Old Stone Age* (1915). He was looking for descriptions of large fossil groups, including the respective species and their characters. Like Osborn, Huxley was most interested in the larger trends in evolution. Throughout his venture into paleontology, the young man of "heredity charm"[51] could count on Osborn's assistance. However, Huxley—like Osborn's protégés in general—would undermine the evolutionary theory of the doyen of paleontology.

During the 1910s and 1920s, Huxley was working on heterogenic growth in a wide variety of organisms. He sent Osborn the paper in *Nature* that came out of his research on the fiddler crab at the Marine Biological Laboratory at Woods Hole. In this paper, he concluded by drawing inferences for correlative growth beyond the animal just studied: "The apparent orthogenesis of horns, etc., in mammals, especially urged for Titanotheria by Osborn (1918, etc.), is no true orthogenesis at all, in the sense that it is not caused by determinate variation of the hereditary constitution" (Huxley 1924, 896; see also 1926b). Heavily relying on Osborn's help and work, Huxley came to the conclusion that paleontological trends and parallel evolution were much better accounted for by natural selection favoring general increase in body size that would result in correlated increase, for example of horns, by means of a hereditary growth mechanism common to the related lines. The horn changes would thus be automatic results of general size changes. "That is, orthogenesis would mean nothing else but directional evolution, and had better be dropped in favour of that term, because its past use has often implied directional evolution which is also determinate owing to internal causes" (Huxley 1924, 896).[52] Huxley again presented his solution in direct opposition to Osborn's work on evolutionary trends in horn development in the book-length treatment in *Problems of Relative Growth* (1932, 218).

However, Huxley—as well as Simpson—actually embraced what Marc Swetlitz (1993, ch. 3) has called the prevalent paleontological consensus that was at the core of Osborn's understanding: new higher taxa (phyla, classes, orders, families) evolved from the more generalized representatives of ancestral taxa; and evolution within a higher taxon proceeded from a more generalized common ancestor to differently specialized descendants, with specialization eventually resulting in extinction, degeneration, or stagnation. Also like Osborn, Huxley thought that the first process lay at the basis of evolutionary progress. And again like Osborn, he was most concerned about the meaning of evolutionary progress for humankind. Already in the textbook *The Science of Life* (Wells, Huxley, and Wells 1929/1930/1931/1934, chs. 6.3–6.5), he brought his understanding of the paleontological record to bear on the human realm. Huxley, like Osborn, extrapolated from evolutionary history a plan for human betterment. He believed that knowledge of the evolutionary past held the key to controlled future progress, if only humans embraced its lessons.

As exponent of the new biology and evolutionary synthesis, the challenge lay in conceptualizing progress in accordance with the neo-Darwinian mechanisms and from the perspective of the organism. Thus, although Osborn called *The Science of Life* a masterpiece in popularization, he strongly disagreed with Huxley's account of human evolution, somewhat desperately referring him to the Switzerland-based anthropologist Adolph Schultz, who agreed with his own dawn man theory.[53] Osborn's time had come to an end, but he understood that Huxley, too, aimed at bringing the history within alive in the world. Huxley had "inherited in a rare degree the art of popular exposition of scientific facts which was one of the chief characteristics of his distinguished grandfather."[54] And like Osborn, he worked toward the aim of reminding people of their history within not only through a great effort in popular writing and public appearance, but also through powerful organizations.

PART II

History in Organisms

Julian Sorell Huxley (1887–1975) at the London Zoo and Other Institutions

London Zoo is a space where many layers of time converge. Under the heading *Heritage*, panels explain its architectural history and the changes in animal keeping, from the attempt to render their compounds naturalistic to the functionalism of the modernist architecture of the early twentieth century. An emblem of the latter is the Penguin Pool by Berthold Lubetkin that was finished the year before Julian Huxley took over the directorship of the zoo in 1935. It manifests the understanding that "culture" is not at odds with "nature"—an understanding that was shared by Huxley, who promoted Lubetkin's work at the garden. The building expresses the notion that while an animal's needs have to be taken care of, this does not require an imitation of its natural surroundings; this can be achieved just as well by providing a naked reinforced concrete oval, with nest boxes, a swimming pool, stairs, and slides. The Penguin Pool may thus stand for Huxley's belief in general that humans can and even should interfere with nature to make it a place as conducive as possible for human as well as animal well-being. In fact, this was more about humans than animals, also in the case of the Penguin Pool, since such enclosures should in the first place render the animals observable and entertaining to visitors. Today, the Penguin Pool is no longer thought of as a suitable hosting space for the animals; it has become a Grade One listed English Heritage building. Thus, London Zoo, like other zoological gardens, is a site of remembrance that entangles the past with the present and documents layers of history (Shapland and Van Reybrouck 2008).

For Huxley, that was true in yet another sense. While I might think of the animals in a zoo as more or less closely related to humans, for Huxley, they embodied the course of our evolutionary history. At the opening of the Bronx Zoo, Osborn had rhapsodized that "the Mastodon is beyond recall, but before long his collateral descendant, the elephant, will be here; and this afternoon, as you wander through the ranges, you will see restored to their old haunts all the other noble aborigines of Manhattan."[1] To Osborn, the living elephant recalled the mastodon and the animals that once roamed the range of Manhattan drew attention to the changes civilization had wrought and that the zoo, to a certain extent, was to reverse by bringing people into contact with a reconstruction of the past. The zoologist Huxley, too, intended his zoo to facilitate transformative experiences in the visitors. As we will see in chapter 7, Huxley believed that animal bodies could make people aware of their own history within. He wanted visitors to understand the animals as archives of evolutionary history and their own bodies as the manifestation of the most progressive trend of that history. In fact, Huxley had come to think of humans as the only species that still had the potential for further progress—if only evolutionary history and evolutionary mechanisms were rightly understood and the conscious planning of the future carried out accordingly.

While Osborn had been surrounded by fossils in his museum study and library, Huxley actually lived in close proximity to the animals. From his lodgings, he could hear their sounds, and similar to Osborn's study, the zoo rooms embodied Huxley's genealogical outlook: "As a link between new and old, almost as a symbol of continuity, there hung above the new Secretary's mantelpiece one of the famous portraits of his grandfather. 'What an agreeable destiny,' wrote Wyndham Lewis, who saw it shortly after Huxley had moved in [at his rooms in Regent's Park], '–to inherit, as it were, the animal kingdom.'" (Clark 1968, 256) Indeed, Huxley thought that entire humankind had inherited the animal kingdom, and that the main trustees, such as evolutionary biologists like his grandfather Thomas Henry Huxley and himself, had a duty to convey this self-understanding. Thus, family genealogy mattered, too; in particular his descent from the great evolutionist was of central importance to him and in his career.

This is visualized in figure 11, which shows him below the portrait

FIGURE 11. Julian Huxley under portrait of grandfather Thomas Henry (and beside photo of brother Aldous), ca. 1941, Julian Sorell Huxley—Papers, 1899–1980, MS 50, Series XIII: Memorabilia, Box 150: 1940–1949, Folder 1: 1940s: " 'Any Questions?'," *Life*, 17 May 1943, 31–34, on 34, courtesy Woodson Research Center, Fondren Library, Rice University

of his grandfather and beside a photograph of his brother Aldous Huxley. It is also meaningful that figure 11 is an image published in the papers that somebody clipped and sent him to protest about the fact that a man of his renown would pose with a cigarette in his hand. The photograph therefore indicates that Julian followed in Thomas Henry's footsteps with respect to public standing as well. However, the family legacy also troubled Julian Huxley, especially because in contrast to his grandfather, he wanted to imbue evolutionary history with moral authority over the present and future of humankind. His extensive interference with public concerns provoked the spite of the Council of the Zoological Society. Furthermore, in contrast to Osborn's aims, which had largely been in harmony with those of the museum trustees, Huxley's changes to the zoo met with animosity. In negotiations over the space, "the new and the old biology" clashed.

Like Osborn's enormous effort to spread his messages beyond the museum, Huxley established what he referred to as an extended zoo. In chapter 8 we will see how Huxley, with such friends as Lancelot Hogben and J. B. S. Haldane, engaged in a crusade to plant the right understanding of the hereditary process and evolutionary history in the mind of the public. The belief held by Osborn and many of his generation that current "racial," social, and gender orders represented the natural order had to be rooted out. The old ideas had to be replaced with the insight that the history within was in harmony only with societies in which there was equality of opportunity. In the writings of Huxley and his friends, the novel understanding of heredity and evolution became a political tool of a new kind. The significance of individual variation to Darwinian evolution was put to work as a weapon against a eugenics that aimed at a few "good" human types, as well as against a racial anthropology that perpetuated thinking in discrete human groups with the "Nordic" at the top of the hierarchy. Huxley and colleagues perceived this kind of science not only as outdated, but as in the service of nationalism with its potential for war, of the middle class with its privileges, and of laissez-faire politics and fascism.

Already Osborn had been troubled by the tension between the good of the individual and that of the group. But while looking into the deeper past from the perspective of the early twentieth century had made him emphasize the need to harmonize the self-determination of individuals with the goal of "racial" progress, Huxley increasingly read from evolutionary history the importance of integrating free individuals through a feeling of panhumanism. Huxley shared the concern of many cultural analysts about the possibility of developing a mature personality in consumer, mass, and media societies, because it was through the individual, if through the individual en masse, that the evolutionary potential of the species could be unleashed (Huxley 1926c, 22). Diversity and individuality rightly understood were valued positively as prerequisites to progressive evolution. For Huxley, "the well-developed well-patterned individual human being is, in a strictly scientific sense, the highest phenomenon of which we have any knowledge; and the variety of individual personalities is the world's highest richness" (Huxley [1964] 1992, 84). This was the case because each individual person could realize "an important quantum of evolutionary possibility" (ibid., 85). In the human phase of evolution, this possibility referred to cultural evolution, which Huxley conceptualized in analogy to biological evolution. Cultural progress had to be catalyzed by providing the individual with "a psychosocial technology," which included "ideological machine-tools like concepts and beliefs for the better processing of experience" (ibid., 113–114).[2] Not every kind of concept or belief would have positive effects. Key ideas needed to be selected for or against depending on whether they were still advantageous in a certain cultural environment. In general, in the human—or psychosocial—stage of evolution, progress consisted of the gradual development of society in accordance with the increasing knowledge about the natural world.

While there should be diversity of ideas to ensure the continuation of development, shared key ideas could integrate humanity. Toward the end of his zoo years, Huxley mobilized his utopia of establishing a democratic world society along the evolutionary processes of differentiation and integration. He positioned this utopia internationally as the possibility of a common future beyond the war. As will be of concern in chapter 9, however, it was as first director-general of UNESCO (1946–48) that he hoped to finally be in a position of power to work toward that aim. At the same time, Huxley never got rid of an overhanging sense of crisis. The experience of two world wars intensified apprehensions, and the threat posed by weapons of mass destruction, the chasm between the haves and have-nots, but also overpopulation, the exhaustion of natural resources, and the disappearance of wildlife became ever more alarming. The latter were particularly threatening in Huxley's view, because nature in the sense of our evolutionary heritage was one of the main sources of human fulfillment through knowledge, but also through aesthetic contemplation and adventure. Just as for Osborn, for Huxley the evolutionary approach to nature held the key to a new spirituality. Huxley had already been involved in conservation during his time at London Zoo, but this endeavor took center stage in his work for UNESCO, followed by his engagement in the foundation of the International Union for the Protection of Nature and the World Wildlife Fund.

Huxley referred to his evolutionary utopia as scientific or evolutionary humanism, and by the time he took up leadership of UNESCO, this worldview had become decidedly ecological: humankind needed to recognize the entangled nature of an evolved world, take stock of and preserve it, and scientifically manage and analyze it to apply the knowledge thus gained to further personal and collective advancement. It was more urgent than ever for humankind to understand itself as evolution become conscious, and to take on its responsibility as evolution's sole trustee. To this purpose, UNESCO and other organizations should ensure the establishment not only of natural parks but also of museums, libraries, and other institutions in the service of trusteeship of the entire human cultural as well as natural heritage. Huxley's involvement in educational policies in Africa under the Colonial Office and for UNESCO nourished the belief that this continent held a particular potential for further evolution. It was a reservoir of evolutionary history as well as potential; its people seemed to have retained a greater degree of freedom for development than "the West."

However, within UNESCO just as at the zoo, Huxley's by now totalizing project met with resistance. As we will see in chapter 10, by the 1960s, his ideas had become out of sync with significant political, social, cultural, and ultimately scientific developments. Although his offerings of the morals of evolutionary history in popular writings, radio talks, films, and exhibition had enjoyed considerable success, by that time, his liberal synthesis, his planetary vision, and his elitism and anthropocentrism more frequently attracted criticism. And while his views on eugenics had been progressive in the interwar years, they could now provoke outcries of disgust. Finally, some scholars resented more strongly the notion that everything and thus all knowledge ultimately rested on evolution and evolutionary biology. There had also been groundbreaking developments within the life sciences: with the new molecular technologies, the conviction grew that the ultimate place to search for phylogeny and evolutionary history was in the gene—a notion that went along with what has been called a reductionist view of life. This gene-centrism was a thorn in the side of Huxley and others who stood for the organismic, synthetic, and humanist perspective. In order to understand that contrast, and in fact the gist of what follows, we first have to turn to the development of Huxley's ideas on evolution and its larger meanings.

CHAPTER SIX

If I Were Dictator

The Modern Synthesis, Evolutionary Humanism, and a Superhuman Memory

Although it was during his directorship of London Zoo that Huxley focused on evolution and systematics, the development of his philosophy was influenced by prior research interests up to the early 1930s, especially in ethology and experimental physiology. He engaged in experimental embryology during a stay at Naples (1909–1910) and as demonstrator (1910) and later senior demonstrator (1919) in the Department of Zoology and Comparative Anatomy at Oxford University. The relative security of this post made it feasible to marry Juliette Baillot, and the work that he continued as professor of zoology (1925) and honorary lecturer in experimental zoology (1927) at King's College, London University, culminated in *The Elements of Experimental Embryology* (Huxley and de Beer 1934). Huxley interfered with the developmental process in organisms from sponge to chicken to study the mechanisms of dedifferentiation and differentiation, and he investigated the hormonal control of metamorphosis. He also studied the relative growth of organs during development in organisms from crab to deer. This alerted him to the fact that species could not be defined on the basis of size relations of their body parts, and we have already seen how his research on relative growth enabled the insight that developmental mechanisms could help to explain supposedly orthogenetic trends in the fossil record (Huxley 1932).

Furthermore, his work on rate genes with his former Oxford student Edmund Brisco Ford provided alternative explanations for the phenomena paleontologists had often attributed to recapitulation. If the rate of a process during development was correlated with the time of its onset and final equilibrium level, a mutation that increased the rate would make the process start earlier and continue for longer. The new process would repeat all the steps of the original one, but in accelerated and abbreviated fashion, and add novel steps to its end (Huxley 1936b, 93). With their rate-gene research Ford and Huxley contributed to the understanding of gene-environment and gene-gene interaction. Gene expression might depend on circumstances such as temperature, and Ford and Huxley described how genes timed and influenced the degree of activity of other genes. Like the work of the Morgan group and others, these studies in embryology provided insight into the complexities of heredity and development that would strongly impact Huxley's perspective on the history within and the meaning of human diversity (e.g., Ford and Huxley 1927, 1929; Baker 1976; Dronamraju 1993, 11–30; Olby 1992).[1]

In contrast to Osborn's struggle with the new genetics, a visit to Morgan's Columbia University fly group in 1912 had inspired Huxley to regard the research on drosophila mutants as providing very important links between Mendelism and Darwinism and he argued that point with Morgan. That year Huxley took on a professorship at the Rice Institute in Houston, Texas, to build up a biology department, and he was able to attract Hermann Muller from Morgan's lab. Once back in England after engagement in the war effort, Huxley wrote to Osborn that he felt "very lucky in having got hold of young Haldane, son of the physiologist, to lecture in genetics in the department" at Oxford.[2] Like Osborn, Huxley observed that "the geneticists are coming back to a view-point very similar to that held by Darwin himself."[3] But in contrast to Osborn, Huxley had no problem accommodating this emerging synthesis of the two opposed camps. In a report for the British Council, Huxley would hail the successes in synthesizing aspects of biology in British interwar science as a renaissance. These were achieved in C. D. Darlington's cytological approach to heredity, the mathematical approach to genetics that reconciled biometrics with Mendelism, as practiced by R. A. Fisher and Haldane, and in experimental embryology and developmental physiology that culminated in the synthesis of biochemistry and *Entwicklungsmechanik* in the work of Joseph Needham's and C. H. Waddington's school at Cambridge. There was also Charles Elton's ecological work on fluctuations in wild populations, and Huxley's own on the relative growth of body parts and on Richard Goldschmidt's concept of the rate gene with Ford. The man Huxley endearingly called *Henry* was instrumental in the synthesis of mathematical population genetics and the study of natural populations through his cooperation with Fisher. And as Huxley noted in his outline of British science between the wars, this process preceded the integration of many other fields.[4]

Indeed, in a paper of the period, Huxley described biology as embarking on "a phase of synthesis" to the British Association for the Advancement of Science (Huxley 1936b, 81). He discussed various fields of biology in the light of the new Darwinism and how their recent observations were in accord with the theory of evolution through genetic variation, natural selection, and adaptation. The paper was welcomed, and Huxley was asked to expand on it. This time, the name he gave to the phase in biology would stick: *Evolution: The Modern Synthesis* (Huxley 1942a). He emphasized that the new insights from genetics, embryology, ecology, biogeography, paleontology, and systematics contributed to a unified picture of evolution through variation and selection. Huxley's influential book was intended to lay the basis for an integrative biology from the molecule to human culture on the principle of Darwinian evolution. He drew heavily on the above-mentioned advances in knowledge, and as with many of his writings, Ford went through the manuscript for him—a favor Huxley returned. Huxley also drew on his own work, not only in the laboratory, but also in ornithology. Like many British conservationists, Huxley was a devout bird-watcher from childhood onward and later called for organized action to preserve Britain's wildfowl.[5] He was among those ornithologists who in the early twentieth century carried out in-depth life-history studies of birds, and he thought about the role of sexual and natural selection in courtship.[6]

The move from a typological species concept (à la Osborn) to a populational one has been identified as particularly important to the paradigmatic change of which Huxley was part, and organismic variation interpreted from a Darwinian perspective was a topic of special interest to Huxley. He published several papers on clines. He described the different patterns of the gradations in measurable characters within species and through space and time. He coined the terms *geocline* for gradations in geographical space, *ecocline* for such that were due to differences in the habitat, *ontocline* for clines in individual development, and *chronocline* for Osborn's paleontological trends (Huxley 1938a). In the introduction to *The New Systematics* (1940b), he also discussed possible mechanisms underlying the observed patterns of variation. This constituted a comparably multidimensional conceptualization of the species as undertaken by Simpson. At its center stood the organism as the level on which selection worked, but the synthesists thought that for a full understanding of evolution, all levels from alleles to genotypes, phenotypes, and polymorphic populations in their natural environment needed to be integrated.[7]

However, there has never been consensus on the meaning of the modern synthesis. Thomas Junker and Eve-Marie Engels, for example, have an encompassing understanding: "By Evolutionary Synthesis we mean the historical attempt of the 1930s and 1940s to develop a materialist, gradualist, and selectionist theory of evolution that explains as completely as possible the evolution of organisms—the transformation of species as well as the splitting of species, micro- as well as macroevolution—and thereby integrates the findings of as many biological subdisciplines as possible. This was most successful for genetics, mathematical and ecological population genetics, systematics, and paleontology" (my trans.).[8] Although similar definitions are widespread, the literature on the modern synthesis is enormous and there is no agreement.[9] Indeed, the smallest common denominator among historians and philosophers of science appears to be that it came to a reduction in the number of acceptable evolutionary mechanisms. Explanations for evolutionary phenomena had to be based on the variation-selection theory; vitalist, "Lamarckian," orthogenetic, or saltationist alternatives were no longer acceptable.[10] As Stephen Jay Gould (1980) has observed, toward the end of the 1940s, the synthesis "hardened" further in its prioritizing of natural selection (also at the expense of mechanisms such as genetic drift), and William Provine (1988, 61) has even described the evolutionary synthesis as an "evolutionary constriction."[11]

In fact, as protagonists of the synthesis, Huxley, Dobzhansky, Mayr, and Simpson emphasized this constriction as one of their greatest achievements.[12] For Huxley, the expansion of the synthesis to traditionally neo-Lamarckian and/or orthogenetically inclined fields such as paleontology, ecology, taxonomy, and biogeography ended "The eclipse of Darwinism" (Huxley 1942a, chapter title 1.3) that had kept biology back since the late nineteenth century. The synthesis undermined the understandings of evolution supported by these mechanisms. However, as historians and philosophers of biology have shown, Huxley, Dobzhansky, Simpson, and Mayr were far from abandoning the notion of evolutionary progress. Their own notions of progress are usually understood as absolute as well as relative. The synthesists recognized in the evolutionary process a relative kind of progress in the sense of degrees of adaptation to changing environments. But they also established criteria that allowed for more generalized meanings of progress, and at times even for an absolute kind of advancement that culminated in humans. The challenge was to develop conceptions of evolutionary progress that remained within the "evolutionary constriction."[13]

The problem of progress was a topic of great concern throughout Huxley's career. In his early work *The Individual in the Animal Kingdom* (1912), he developed a notion of progress as a rise of the level of organization that consisted in an increase in the complexity of the division of labor between organs and body parts through differentiation and specialization with concomitant increase in their integration. During his time at today's Rice University, he also presented a series of lectures called "Biology and Man" in which he discussed the subject. In a long process of thinking about trends that were overall progressive in evolutionary history, progress came to refer to a growing independence from the environment (mechanisms for stabilizing the inner milieu, efficient reproduction and brood care) and to an augmenting control over the environment (differentiation of the nervous system and thus improvement of cognitive faculties). In contributions such as "Progress, Biological and Other" (revised in Huxley 1923) and "Progress Shown in Evolution" (Huxley 1928), Huxley emphasized that this kind of progress could be explained without vital principles, solely on the basis of organismic variation and Darwinian selection that led to adaptation.

Huxley continued to reason that direction and progress could be made out retrospectively in the phenomenon that new dominant groups of organisms had repeatedly developed from more generalized types. Such new groups subsequently adapted to different niches by way of differentiation (cladogenesis), a process that reduced their adaptive flexibility. Thus, the successively dominant groups reached a degree of specialization that rendered further advancement in organismic organization impossible (stasigenesis) (Huxley 1955). For Huxley, specialization therefore meant either an improvement of the basic organization of an organism, or the adaptation to a particular environment as a result of differentiation. The prime example of a positive specialization, or progress, was humans, who not only lacked narrowing specialization, but who, with the evolution of the brain, had gained the capacity to simulate the specializations of other organisms (such as flight or swimming).[14]

In fact, by the 1930s Huxley believed that all nonhuman evolutionary lines were overspecialized and had reached their end points. The human being was the sole "trustee, spearhead, or effective agent of any further evolutionary progress" (Huxley 1950, 20).[15] Huxley made it clear that although humans were the last stage in an evolutionary succession of dominant types, they like any other organism were a product of random and directional, but not goal-directed processes.

However, the contingent emergence of humans nonetheless signified something new that not only turned them into very special animals but also altered the nature of evolution: "By means of tradition, man at last overcomes, if but partially, Nature's veto on the inheritance of acquired characters."[16] The development of a new level of consciousness made possible the transgenerational communication of experience and knowledge, to which Huxley also referred as a process of heredity (Huxley 1941a, 1–33 ["The Uniqueness of Man," first published 1931], 4). Although in analogy to purely biological evolution, it was possible to distinguish more somatic (maintenance) from more genetic elements (transmission) in evolution by tradition, as, for example, between material production and the education system, these were not hermetically separated. In psychosocial evolution, as Huxley called the human stage, the processes of maintenance, transmission, and transformation were short-circuited. Huxley declared that these were humankind's main mechanisms for change: "The resultant shareable, transmissible, and progressively transformable tradition gave rise to the new type of entity or organization technically called cultural, and evolution in the psycho-social phase has been essentially cultural, not biological or genetic" (Huxley 1955, 7).

Thus, while Huxley could no longer allow for the inheritance of acquired characters within biological evolution, on the cultural level, individual achievement could be handed on to the next generation: "In this new phase of evolution there is no longer any distinction between soma and germ-plasm as there is in the biological phase: the whole psychosocial system evolves and is itself transmissible; it is both formative and formed" (Huxley 1964, 50).[17] In other words, just like prominent thinkers such as Ernst Cassirer ([1942] 2007, 484–485), Huxley thought that with the development of symbolic forms humans had solved a problem that organic nature as such could not. As Huxley tried to communicate through the zoo and other institutions, this opened up the possibility of teleology, of planned cultural evolution along progressive lines.

At the same time, although Huxley conceived of evolution in the psychosocial phase as mainly cultural, to a lesser degree (and less consistently) than Dobzhansky and Simpson, he maintained the possibility of a goal-directed biological evolution, because the human-made environment was now the substrate against which natural selection "measured" fitness (Sommer 2010a, 55–58).[18]

For Huxley, the question of how progress in the psychosocial stage might be achieved could be answered by looking at history from the perspective of evolution. In analogy to cladogenesis, anagenesis, and stasigenesis in the biological phase of evolution, Huxley observed that cultural evolution showed short-term optimizations through adaptation (one-sided specialization through differentiation), long-term optimizations through progress (general specialization), and nonadaptive moments (limitation). These processes were producing differences between cultures, progress in cultural entities, and the survival of obsolete units. However, in psychosocial evolution, cladogenesis was counteracted by a high degree of convergence through the exchange of ideas and techniques between individuals, communities, religions, and cultures that produced a strong unity across the variability: "But finally, that cultural differs importantly from biological evolution in respect . . . of the presence of diffusion and a consequent tendency to convergence as against divergence, of the immensely increased importance of mind and mentifacts, notably the accumulation and better organization of knowledge, and in many other ways" (Huxley 1955, 24).

Before Richard Dawkins's *memes* (1976, ch. 11), Huxley defined *mentifacts*—also called *memoids*—as components of human cultures that were not primarily of material (*artifacts*) or social (*socifacts*), but of mental function. They were materialized ideas that had a social life; they might comprise elements as diverse as machines, mass communication, scientific, legal, economic, and political systems, works of art, philosophy, social hierarchies, and styles of cuisine.[19] Against the background of his thoughts about clado-, ana-, and stasigenesis, Huxley's point was that the selection of mentifacts had to become a conscious process. Their survival should depend on their fitness for adapting a particular culture to the increasing knowledge from the sciences. In general, adaptation in humans meant the adjustment of belief systems (rather than biological systems) to the steadily improving knowledge about the natural world (rather than to the natural world itself) through psychosocial selection. This was Huxley's new categorical imperative; the human being was "the necessary agent of the cosmos in understanding more of itself" (Huxley 1950, 20). As early as the 1920s, Huxley forced his audience to recall that humans were the embodiments of the evolutionary processes and as their apex the movers of evolutionary progress: "Remember that now in the fullness of time, the cosmic forces through whose agency we have been evolved, have made us the trustees of progress, and entrusted to our conscious free-will the future course of evolution" (Huxley 1926c, 24).

This was the core of what Huxley called *scientific humanism*. It was an idiosyncratic integration of the synthetic evolutionary science with a new humanism—a humanism that came out of the context of the institutionalization of the history of science. A central figure here was the chemist and mathematician George Sarton. The history of science was at the center of Sarton's philosophical and historical system, and he attempted to institutionalize it two years before the family emigrated from Belgium to the United States through the launch of the journal *Isis* (1913). As a home to the journal, the History of Science Society was founded in 1924. *Isis* should be the instrument of discipline building from Sarton's positions at Harvard University and the Carnegie Institute; but *Isis* should also stand for something else: for the lessons of tolerance and wisdom that history presented. Sarton's conception of history and the role of science therein were influenced by nineteenth-century thinkers such as Auguste Comte and Herbert Spencer, as well as utopian and socialist ideas. Sarton's grand aim was a progressivist universal history that was founded in positive science and that worked toward the brotherhood of man. If Sarton paid lip service to what at times appears as a strikingly contemporary conception of science in society, it was this liberal faith that guided most of his work (Thackray and Merton 1972).

Sarton presented his program under the title "The New Humanism" (Sarton 1924), a French version of which was published as early as 1918. History of science should bring together the classical humanist and the humanized scientist. He defined science as the common thought of the whole world, as the organized body of all the facts and theories from which almost all arbitrariness had been excluded, and on which enlightened people were unanimously agreed. Because positive knowledge was the common patrimony of all humankind, the domain of science was internationalism. Moreover, science constituted the central axis of human advance and provided the fundamental method of social organization. While no logic might be found in the course of political events, wars, and other human catastrophes, there certainly was a very rigorous logic at work in science. Sarton defined humankind's purpose as the creation of immaterial values, such as beauty, justice, and above all, truth. True relations were necessarily beautiful, and they could not be unjust. Against all odds, humankind's essential purpose was thus not the struggle for survival and supremacy, nor competition for the world's resources, but the creation and diffusion of spiritual values, which was made increasingly easy as communication technologies shrank the globe. Although it was not only scientists but also artists who were the true creators and guardians of these ideals—the stewards of the future of humankind—Sarton *did* perceive a natural order of knowledge, with mathematics as the foundation, followed by physics and biology. Scientists were pursuing a sacred task. It was thanks to them that the advance of mankind could go on indefinitely.

In "The Faith of a Humanist" (1920), Sarton revealed an organicist notion of society in which scientific activity and knowledge functioned somewhat analogously to the hereditary material. While literature, the arts, and religions were wonderfully beautiful human creations, they were not slowly evolving as they were handed down from generation to generation—this status was reserved for positive science. The arts might be the phenotype of each generation, but the sciences were the genotype. In this model, the history of science was the skeleton of the history of humankind; it was the stuff that carried and organized the growth of civilizations. The history of science was not cumulative and progressive in the same way as science, but it was closer to the hereditary material than other tissues. It was the connection between the geno- and the phenotype, because it made evident the progressive nature of science and thus of humanity. By 1931, Sarton had developed this notion of the history of science as a world-transforming technology in the book-length *The History of Science and the New Humanism*. Because the history of science demonstrated that only the natural sciences were cumulative and progressive, they took the central position in the humanist frame. At the same time, the natural sciences were inherently human because they were a human desire, a human activity, but also a reflection of nature in the human mind. Every increase of knowledge about the natural world would touch on the knowledge about humankind; both natural and human history were about the human. The new humanism was a program to understand and at the same time to increase our part in cosmic evolution, and it was shared by a generation of historians of science who were internationally integrated around institutions such as *Isis* and the History of Science Society.

Another key figure in this context was Sarton's British friend Charles Singer, who had studied medicine and then focused on the history of his field. He would also be the driving force in the foundation of the British Society for the History of Science in 1947. Relying heavily on Sarton's new humanism, Singer defined a *scientific humanism* in his opening of the first issue of *The Rationalist: A Journal of Scientific Humanism* in 1929 (Singer 1929).[20] It was essentially an attack against historians who neglected the true *movens* in the history of humankind—science—and against the educational system that did not convey the transformative power of science and the history of science. For Singer, acquaintance with the craft of science would help a student to live his life, but the knowledge of the history of science would show the student why his life was worth living. It would acquaint him with the purpose of human existence.

Singer taught history of biology and medicine at Oxford and University College London, and formed a network around him that connected the three main centers for the teaching of the history of science and medicine, which included Cambridge (Cantor 1997). Among Singer's acquaintances, friends, and collaborators was Huxley, who, for example, partook in a summer school on science and civilization that Singer co-organized in 1922. Like Sarton's and Singer's, Huxley's thinking was imbued with nineteenth-century values. His grandfather Thomas Henry Huxley had been a strong believer in the progress of civilization through the advancement of science. However, in Julian's family, there were important exponents of British idealism as well as empiricism. The Huxleys and Arnolds brought together a scientific and literary elite that represented both, the cultivation of intellect and feeling, science and religion, truth and beauty (Divall 1992; Patten 1992; Waters 1992; Wiener 1992). From the early 1920s, Julian Huxley recognized in history a tool to integrate these opposites; history could discover what he called "the soul of science" (Huxley 1926a [originally published as "The Soul of Science," *Observer*, 27.11.1921, 4], 165; Mayer 2005).

Obviously, the label *scientific humanism* and what it stands for point to the central role of science in Huxley's philosophy: "Without the impersonal guidance and the efficient control provided by science, civilization will either stagnate or collapse, and human nature cannot make progress toward realizing its possible evolutionary destiny" (Huxley

1941a, ch. 13 ["Scientific Humanism"], 276). Huxley presented his scientific humanism as a purely naturalistic system, and scientific methods had to be applied to all human concerns:

> It [scientific humanism] insists that the same scientific procedure can be applied to human life as has been applied with such success to lifeless matter and to animals and plants—scientific survey, study, and analysis, followed by increasing practical control. It insists on human values as the norms for our aims, but insists equally that they cannot adjust themselves in right perspective and emphasis except as part of the picture of the world provided by science. It realizes that human desires and aspirations are the motive power of life, but insists that no long-range or comprehensive aim of humanity can ever be realized except with the aid of the pedestrian and dispassionate methods, the systematic planning, the experimental testing which can be provided by science alone. (ibid., 274–275)

Science had to be applied to the human realm, inform worldviews, and guide social planning, but it also had to take into account human values, desires, and aspirations.

In *Biology and Human Life* (1926c, 23), Huxley defined the biologically founded ideal of society as "a democracy of material opportunity freely surrendering itself to the guidance of an aristocracy of thought." This is one instance in which the military metaphor brings his totalizing vision dangerously close to a totalitarian vision, even though—as we will see—he attacked totalitarian systems. Also in *If I Were Dictator* (1934a), which was an early comprehensive expression of his ideas for world development under scientific humanism, Huxley suggested a central planning council constituted by experts as well as a central science council that would adjust the social and economic system to the future possibilities of science (rather than the other way around). As for Sarton, for Huxley some scientists were more equal than others. However, in Huxley's thinking, biology was more relevant to the project than the physical sciences, and the study of evolution was key. Since the progress of humankind (and, by inference, of the world) depended on the human understanding of the evolutionary processes in biology and culture, the evolutionist had a particular responsibility to formulate possible perspectives for the future. Huxley saw anagenic progress at work in history in a series of systems of ideas that determined societal organization. Thus, tribal societies structured by magic belief preceded the God-centered systems of the Middle Ages. Although these were already organized around the notion of human progress, progress was considered to be under supernatural control. Even societies that focused on science in the hope for progress by means of its mechanistic and reductionist approach did not bring about true progress. Only Darwin opened the door to an evolution-centered ideological organization that allowed progress in the sense of a holistic development under human control.[21]

Owing to the centrality of evolution, Huxley also made use of the term *evolutionary humanism*.[22] However, the special role of the evolutionist should not obscure the fact that it was an inclusive conception. In Huxley's view, walking through that door opened by Darwin depended on a synthetic understanding of the cosmos, and therefore on an integration of the natural sciences, anthropology, social sciences, and the humanities, but also of other knowledge cultures such as politics, the arts, and everyday life. An elite group with diverse expertise was needed, and Huxley invested time and energy in organizations and propaganda networks to campaign for syntheses of all kinds. He was, for example, chairman of the Idea Systems Group, in which he cooperated with journalists and writers, humanists, as well as social and natural scientists. Along the lines of his understanding of cultural evolution, this group sought to discover which idea systems and key ideas globally were still functional in their current social environments and which were not. How could existing ideas be adapted and new ones developed to meet the current problems of population growth and shortage in resources? Which ideas would fit the demands of societies structured by communication technologies? How could ideas serve industrialism and the spread of its products as well as further inter- and supranational organization? Huxley felt that guided adaptation of ideas and societies was becoming more and more feasible because societies were increasingly self-conscious: they recorded their development in social, economic, demographic, and natural surveys, monitoring public opinion, resources, and crime among other factors.[23]

In steering the evolution of idea systems on the basis of the knowledge thus gained, some world philosophy should be aimed at, but not uniformity. The goal was a cultural plurality grounded on common general beliefs. The study of idea systems and their components should eventually allow intervening in their further development toward "a new evolutionary view of man's relation to the cosmos at large and his destiny within it."[24] The ultimate dream was an institution that would function as a clearing house for all the world's knowledge, creative products, and expert advice, for knowledge management and distribution in every genre and media of communication, with the general aim of individual, populational, and human fulfillment. Throughout his career Huxley lobbied for such centers of calculation, and with the Idea Systems Group, he tried to raise funds for a "New Humanist Institute" that would promote evolutionary humanism on a global scale.[25]

However, to change ideas systems, communication of adaptive knowledge and thought was not enough. Walking through that door opened by Darwin depended also on Mr. and Mrs. Everyman. Huxley envisioned a fulfillment society that provided the opportunity for an open, complex, and holistic self-realization to those individuals who consciously strove for aesthetic, intellectual, and spiritual perfection. The largest possible number of individuals would be given the broadest spectrum of possibilities to unfold their potentialities through education, accomplishment, adventure, cooperation, and meditation. Beyond personal initiative, incentives would be needed for social thinking and acting, so that cooperation, altruism, sensitivity, and sympathetic enthusiasm could spread.[26]

From the beginning, one of Huxley's greatest aims had thus been to infuse "the ideas and interests of everyday life" with his view of history and the future (Huxley 1912, ix). One big project toward that aim was the massive and very widely distributed *Science of Life* (1929/1930/1931/1934), sometimes dubbed the first textbook of modern biology. It was a manifestation of Huxley's and the Wellses' evolutionary humanistic outlook. Following in the track of H. G. Wells's *The Outline of History* ([1920] 1929), it was intended as living knowledge in ordinary language that "reflects upon the conduct of our lives throughout" (Wells, Huxley, and Wells 1929/1930/1931/1934, 2). It was to contribute to "*one*, the training of all the individual faculties to as high a level as possible, speech, drawing, the full use of hands and body generally; *two*, the development of a persona and of the self-knowledge and the practical psychological commonsense necessary for happy personal conduct and the filling of a distinctive rôle in life," and most of all to "*three*, the establishment of a picture of the universe in accordance with reality, the realization of the great adventure of humanity and of a personal rôle in that drama; and *four*, the special technical training and experience needed for the due enactment of the individual rôle" (ibid., 1466). *The Science of Life* was an early and very successful example of Huxley's myriad attempts to transform people by making them realize that they were evolution personified and could thus work toward the perfection of self and group.

In the end, the individual was the entity through which collective human evolution—the coevolution of idea systems and social systems—would have to be achieved. At a time when scholars began to analyze and theorize the relation between individual and collective forms of memory, grappling with their constructedness and symbolic and media-molded nature,[27] Huxley and the Wellses wrote that "by means of books, pictures, museums and the like, the species builds up the apparatus of a super-human memory. Imaginatively the individual now links himself with and secures the use of this continually increasing and continually more systematic and accessible super-memory. The human mind . . . grows up, it takes to itself more or less completely the growing mental life of the race, adds a personal interpretation to it, gives it substance and application" (ibid., 1474). Through the internalization of "the knowledge of the race," individuals partook in a superhuman memory to which each personal life could contribute. Knowledge, built over generations, materialized, stored, and managed for general use, would attune individuals to common purposes. Humankind had to head toward "a collective human organism, whose knowledge and memory will be all science and all history"—a process, at the end of which stood "the promise of Man, consciously controlling his own destinies and the destinies of all life upon this planet" (ibid., 1475).

In the endless struggle for the realization of that promise of consciously directed psychosocial and biological evolution, Huxley's own mentifacts were far from free-floating. They were bound to paper, radio waves, or celluloid, and wherever possible to exhibits arranged in three-dimensional space. I therefore have to look at Huxley's cosmology in connection with actual practice. A couple of years after the completion of *The Science of Life*, Huxley took on the secretaryship of the Zoological Society of London. As we will see in the next chapter, he worked toward the implementation of his goal to transform humans by the means outlined above, albeit still on a relatively modest scale.

CHAPTER SEVEN

Evolution in Action

The Zoo as a Site of Phylogenetic Remembrance

Toward the end of 1941, as secretary of London Zoo, Huxley departed for the United States on behalf of the Ministry of Information to lecture on postwar reconstruction and social services. The Council of the Zoological Society of London had agreed to this unpaid leave. Nevertheless, Huxley received a telegram informing him that the council used his prolonged absence due to the impossibility to get home from the United States to suspend the post of secretary under the emergency powers because of the financial problems caused by the war. In his reply by cable, Huxley stated that the right procedure would be to renominate him without salary and ask him to find other work during the war. He had already been working for half his salary for about a year from October 1939. The ensuing controversy resulted in Huxley's resignation after less than seven years' tenure.

The controversy makes clear that the suspension of the post was not solely motivated by financial reasons. A considerable part of the elderly fellows accused Huxley among other things of having carried out research during his salaried working hours. Although the zoo had been opened in 1828 in part to allow fellows to study diverse animals, with Huxley's secretaryship the question arose of who should carry out research and what kind of research. Huxley belonged to the younger generation of biologists who were reforming what *biology* meant. Together with a cohort of like-minded scientists, he questioned the strong focus on comparative anatomy and taxonomy in the Victorian biological tradition and helped usher in the age of experimental research, most notably in genetics and developmental physiology. At the same time, the new biologists were interested in integrating what were becoming isolated biological specialties. They wanted to bring together the younger and the traditional branches like paleontology, ecology, and embryology under the framework of a Darwinism reformed by the insights into heredity. Drawing on all branches of biology, traditional taxonomy and systematics were revived by an approach Huxley called "evolution in action" (e.g., Huxley 1939, 82; also 1953)—a revival that was catalyzed through the foundation of the Association for the Study of Systematics in Relation of General Biology (1937), in which Huxley was involved.[1]

As Joe Cain (2010) has shown, as secretary of the Zoological Society, Huxley attempted to implement this general biology. His predecessor, Peter Chalmers Mitchell, had already brought about scientific reforms during his secretaryship, which lasted more than three decades. Among these was the Prosectorium, where under the society's pathologist, researchers engaged in parasitology, bacteriology, and anatomy.[2] The department also accommodated voluntary researchers. Huxley could, for example, bring to the lab a refugee from Nazi Germany, the former director of the Breslau zoo, to work on the digestion of the sloth. Indeed, Huxley spoke of the Zoological Society as "in the first place . . . a scientific society" (Huxley 1936c, 104). When he took over the secretaryship, this was more propaganda than reality however, as there were few professional zoologists among the fellows. To amend the situation, Huxley introduced a new kind of research fellowship that facilitated the participation of professional scientists, and young scientists in particular. He also improved the zoo's library and increased the number and quality of scientific papers presented at the monthly scientific meetings of the Zoological Society that took place at the zoo. Finally, he reformed the society's *Proceedings* by subdividing it into Series A, devoted to General and Experimental zoology, and Series B, on Systematic and Morphological zoology. He thus gave space and even priority to the new biology vis-à-vis the proceedings' previous focus on systematic work and anatomical and morphological studies. In the words of one commentator: "Now the scope of the Society's publications is much wider, and they are beginning to receive some part of the work on those branches of zoology, of more recent development which now attract the bulk of the more able young men."[3]

Where Huxley's own work was concerned, his time at the zoo was framed by his address to the British Association for the Advancement of Science on "Natural Selection and Evolutionary Progress" (1936b) and his codification of the new evolutionary outlook on biology in *Evolution: The Modern Synthesis* (1942a). This book—that was notably a collaborative achievement—"was to cover not only variation and natural selection, but every topic bearing on the subject, from the biochemical basis of heredity to the evolution of consciousness, with its effects on human cultural development, and the problem of defining evolutionary progress" (Huxley 1970, 236–237). On the same lines, the zoo should be a site of basic and applied research, bring scientists from various fields and scientific societies together, and not only illustrate evolutionary progress but function as a testing ground for the scientist as citizen as well as for the citizen as scientist. It was to contribute to the preservation of wildlife, further the understanding that humankind was its sole trustee, and help adjust the public understanding to the new scientific knowledge of nature in order to turn citizens into participants in the project of progressive evolution. After all, during his zoo years, Huxley also published a collection of his essays under the title *The Uniqueness of Man* (1941a), which was a particularly strong statement of his scientific humanism.

There was more at stake, therefore, in Huxley's quarrels with the council than biology or even science in the narrower sense. Generally speaking, some of the younger members of the scientific elite viewed science as part of society. Contrary to the entrenched disdain for applied and popular science, they felt that scientists were also citizens and that it was their duty to communicate knowledge and to engage with the pressing issues of their time. It gained Huxley's particular approval that these scientists had become conscious of their social obligation and of their responsibility in postwar reconstruction: "Haldane, Hogben, myself, Needham, Waddington, Darlington and others have devoted a good deal of energy to the popularization of science and to writing on various general questions from the angle of the scientist."[4] However, the council objected to Huxley's enormous effort to bring science to society and vice versa, which in their view could only debase true science. Most of all, Huxley did not merely seek to intensify communication between science and society: he treated science, politics, and society as tightly interlinked. Shortly before taking on the post of secretary, Huxley published *Scientific Research and Social Needs* (1934b), based on a BBC radio series. After a tour of British research laboratories and university departments, he believed more than ever that science had to be applied to the planning and organization of society, to social problems from education to the penal system, as well as to one's worldview. But "science too is a social activity, and itself demands scientific study" (ibid., x).[5] On the contrary, the majority of the fellows of the Zoological Society believed in science as something restricted to experts' work remote from worldly affairs.

It was obvious from the beginning of Huxley's tenure that this would not be the case. Even his election was broadly covered in the press, and the comments made clear that Huxley was regarded as public figure and scientist in one person. Huxley's upcoming election to the post of secretary was met with general enthusiasm, but there was also some anxiety due to his experimental past.[6] Although as director of the zoo he had to hand in his vivisection license, this was about his credibility as trustee of the animals. People were particularly reminded of his experiments on the role of the thyroid gland in the development of the axolotl that had provoked the press to make fantastic conjectures. Huxley thought he had brought back an animal from the evolutionary past, "recreating" a salamander-like land animal that he wrongly claimed for thousands of years had only existed in tadpole form (Huxley 1920). What was more, Huxley had publicly defended vivisection when he came to the aid of his friend Haldane, who had attracted criticism with *Possible Worlds* (1927), in which he showed the "average man" (ibid., xl) what was happening in laboratories. Huxley would not give up the fight with the council regarding the possibility that employees of the society could hold a license, but he had to reassure the public and to shift research emphasis to another of his hobbyhorses.[7]

Huxley had already gained the reputation of a pioneer of evolutionary ethology before his zoo secretaryship (Dunbar 1989). He now emphasized that "although no experiments involving a vivisection licence are carried out under [the society's] auspices, it conducts scientific studies on the diet of animals and on animal behaviour, and encourages visiting scientists to use the Gardens for study" (Huxley 1936c, 104). He had plans to improve facilities for behavioral studies and for bringing in more research students. There were close connections with the Avicultural Society, which had zoo rooms on loan,[8] and in 1936, the Association for the Study of Animal Behaviour was founded under Huxley's presidency (Durant 1986). Indeed, the zoos at Regent's Park and particularly at Whipsnade lent themselves for the observation of animals. Whipsnade had been initiated by Mitchell in 1931; the grounds covered a square mile on the boundaries of Bedfordshire and Hertfordshire and animals could be given much more freedom than in the crowded, thirty-five-acre city zoo. At Whipsnade, Huxley observed the mating behavior of birds and carried out quantitative analysis over the years. At Regent's Park, he, for example, documented a kind of hibernation state in hummingbirds.[9]

One of the most exciting branches of behavioral research Huxley promoted at the zoo was the study of animal sounds. The German refugee Ludwig Koch was working on two sound books with the ornithologist Max Nicholson. Nicholson was Huxley's friend, and Huxley helped in finding a publisher for *Songs of Wild Birds* and *More Songs of Wild Birds* (Nicholson and Koch 1936, 1937) that were great successes with the general public. The collecting of bird songs also stood in the context of a growing awareness about the loss of wild fowl, and an excited *Morning Post*, for example, reported that an extinct bird had sung to the Royal Society of Arts in London from a record.[10] Under Huxley's zoo directorship, Koch also recorded animal vocalizations at Regent's Park and Whipsnade. His recordings for *Country Life* at the London zoos resulted in the first mammal sound book, *Animal Language* of 1938, for which Huxley wrote the text (Huxley and Koch [1938] 1964). Koch had begun to record animal sounds in childhood and went on to record everything from famous figures like Otto von Bismarck to musicians and singers, factory and other city noises, as well as farm animals. He wanted to preserve the sound not only of people but also of things, such as the last piano on which Johann Sebastian Bach had played. He dreamt of a Sound Institute, where natural sounds, music, folklore, the human languages of the world, dialects, famous voices of the past, and so forth would be stored for future generations. This plan accorded with Huxley's interest in the establishment of cultural archives, but there would be no Sound Institute—partly because eventually "all these sounds were drowned by the shouting of Hitler" (Koch 1955, 68). However, early in the war, Huxley would introduce Koch to the BBC, and the British birdsong recordings with Nicholson actually provided the basis for the BBC Sound Recording Library. The BBC later acquired the Koch collection, and Koch was put in charge of its Wildlife Sound Unit. In this position, he could continue to develop the collection—and his own voice, by greatly exaggerating the German accent as a trademark for his popular broadcasts.[11]

At the zoo, Huxley integrated Koch's recordings into his research. They were played to the animals to study their responses, which ranged from great agitation, to reacting only to their own species or to the other sex, to no reaction at all (Huxley 1970, 236). The fact that an animal recognized the calls of members of its own species even though it had never lived among them was for example interpreted as indicating an innate basis for such responses (Huxley and Koch [1938] 1964, 12). In *Animal Language*, Huxley explained the biological functions of the diverse sounds, such as individual recognition, group coordination and cooperation, as well as mating, warning, threatening, and deflection, even false signaling or sound mimicry, such as when birds imitated the hissing of snakes. Huxley emphasized that there were proximate and ultimate causes to sound production. The layperson was certainly right to assume that birds sing out of pleasure. Vocalization was always affective. However, as a biologist, Huxley was especially interested in its evolutionary function. He envisioned a new branch of ethology that would experimentally investigate the meaning of the differences in vocalization between cognate species and determine the relation between developmental flexibility and hereditary determination. He also imagined practical applications—collectors, for example, might use the records to lure animals into close range (ibid., 14).

Most importantly, the study of animal sounds related to Huxley's great interest in evolutionary progress. He found a wide range of complexity at work, culminating in "sound-signalling in the highest animals acquiring almost the status of true speech" (ibid., 17). He followed the attempts to teach the great apes human languages and was interested in the larger issue of cognition associated with the language question. However, the qualification *almost* in the above quote was key, and he would inform the public that

> painstaking efforts have been made to teach chimpanzees to talk, but after over a year's training the animal had only two or three words at its command. We know that there are definite "speech centers" in the human brain: these particular bits of cerebral machinery must be very poorly developed in the chimpanzee, and without them he can no more achieve fluent speech than a man could weave without a loom. Human beings have thus reached a new stage in the development of language. We may call it the stage of speech, and define speech as the use of arbitrary symbols to convey information, including words for things. Man has been defined as the tool-making animal; but in a very real sense the most important tools that he possesses are words. . . . Vocal speech remains the basic method of human communication, and so of the essential human capacity, not only for transmitting experience from gen-

eration to generation but for accumulating it. It is interesting to reflect that sound production, which appeared late in evolution and exists only in a small minority of living creatures, has in man become the necessary foundation for any future progress that evolving life may be destined to achieve.[12]

Ever since Darwin's *On the Origin of Species*, the question of the evolution of animal, and especially primate, language had loomed large, and in the last decade of the nineteenth century, sound playback experiments had been carried out with primates (Radick 2007). The experiments of this kind that Huxley and Koch performed did not provide significant insights. However, the above quote indicates that the animal language work could be mobilized to support Huxley's claim that the emergence of true language and conceptual thought, and thus tradition and culture, indicated a transition from unconscious to potentially consciously guided evolution. Sound production was a rare quality even in the animal kingdom, where it evidenced a gradation in mechanism and function. The apes' complex natural communication indicated a high intelligence and advanced degree of sociability. However, it was only in humans that there was speech, and in the human phase of evolution, speech and written language had come to be *the* means to progress (Huxley and Koch [1938] 1964, 24–25).[13] In Huxley's three-phase model of inorganic, organic, and human evolution, nonhuman primates thus appeared as boundary animals. Huxley followed primatological studies such as those of his friend Solly Zuckerman, who had been research anatomist at the zoo and continued his primate studies there.[14] However, Huxley also made his own observations, and he believed he had witnessed the rudiments of art: "I was lucky enough to catch 'Meng,' the baby gorilla (alas, now dead), repeatedly tracing the outline of his own shadow on the wall with his finger."[15] Indeed, Huxley would not only hang a painting by a chimpanzee besides his Picassos to drive the evolutionary lesson home; many years later he partook in popular experiments to compare the artistic abilities of great apes with those of children.[16]

In fact, throughout his career, Huxley attempted to teach his evolutionary lessons to the public through animals. Mary M. Bartley (1995) has observed that he drew inferences for the human sphere from the concept of mutual sexual selection developed in his studies of the great crested grebe, which he thought was linked to the species' small sexual dimorphism (Huxley 1914). He demanded equality for women in human societies and presented "the virtuous 'marital life' of birds—monogamy, shared responsibility, caring for offspring—as examples of proper relations between the human sexes. In a very real sense, he was promoting a 'learn from the animals' moral that other Edwardian writers of animal stories had been advancing for decades" (Bartley 1995, 95). Through knowledge of animals and especially through contact with animals, it seemed possible to change people's attitude. Obviously, the greatest chance to do so came with the secretaryship of the zoo, where Huxley, among many other efforts, was involved in an exhibition that demonstrated Mendel's laws of heredity with living animals.

In this "learning from animals," the zoo publications were of central importance. Huxley produced a new zoo guide that provided a stark contrast to that published by Mitchell (1934), as well as a popular booklet that emphasized what the park had to teach on evolution. In the official guide (Huxley 1936c), Huxley suggested different routes through the zoo, informing on the evolution of many of the animals thus encountered. He explained how classification within an evolutionary framework functioned and specified the most important adaptations found at the zoo. The circa two million yearly visitors at Regent's Park were invited to observe how the animals' bodies were adapted to procure for themselves the right kinds of food, and that their colors permitted them to blend into the background, warned enemies of their poisons, or attracted mates. He explained how coloring—like sound and smell—carried meaning and was a form of communication, thus also allowing insights into the sense perceptions and the sociality of the animals. Huxley had organized a special exhibition on color change in animals that merged into their background in 1935, and he later introduced the Nocturnal Zoo to show the adaptations of animals that were active in the dark. When announcing the late night openings to the press, he promised a spectacle of big eyes and big ears, and indeed, the Nocturnal Zoo became a particular highlight.[17]

Huxley also used his popular booklet *At the Zoo* (1936a) to make sure the visitors would understand how such adaptations had come about. He emphasized that contrary to popular belief in the inheritance of acquired characters, the adaptations visitors could observe in the living animals were the result of natural selection over enormous periods of time. Huxley was in fact publicly fighting a member of the society council on this matter,[18] and in the booklet, he linked the new understanding of evolution to the central issue of evolutionary progress. We have seen that Huxley considered brood care as one of the progressive trends in evolution, and he instructed the visitors to pay attention to the increasing complexity in reproduction: from the budding of sea squirts to the wasteful egg laying of fish in the Aquarium, and from to the bisexual snails in the Insect House, to the sex-changing frogs in the Reptile House, to the protection and brooding of eggs in the Bird House, up to the gradation in infant care through the mammalian system that culminated in the intimate relationship of a chimpanzee mother with her precious and few young during their long childhood (see figure 12).

Furthermore, episodes of evolutionary history could sometimes be read from animals' bodies, and these, too, added up to a narrative about progress. Our remote ancestors' move from water to land was now reca-

FIGURE 12. Wasteful reproduction, "Sex Life of the Fish" by David Low, Huxley 1936c, 101

pitulated in the life cycle of the amphibian and was embodied in a particularly striking way in the lungfish on show at the Aquarium. The reptiles, including the birds, had evolved eggs that contained the necessary liquid to become full land animals. The anteaters in the Rodent House manifested the evolution from reptile to mammal in that they still laid eggs and were not perfectly warm-blooded. Furthermore, in the bodies of some animals, visitors could make out remnants from the evolutionary past, such as in the boas and pythons in the Reptile House that had retained a pair of claws from their lizard origin. Among the higher mammals, the coats of young lions were spotted, a relic from an evolutionary ancestor that had been a forest dweller.

Visitors were thus encouraged to observe the animals according to scientific criteria and from an evolutionary perspective. They were also invited to read their own bodies in this light—as the most highly developed product of the evolutionary process. Already in *The Science of Life*, Huxley and the Wellses had referred to the human body with its vestigial organs as a "Museum of Evolution," "recapitulating the past of the race during individual development" (Wells, Huxley, and Wells 1929/1930/1931/1934, ch. 5.3, quotes from 415 and 416). Humans had "the same family secrets in [their] embryonic cupboards" as other mammals (ibid., 416). In a nearly poetic, bitingly humorous passage, they had made use of the metaphor of remembrance. Even the antievolutionist William Jennings Bryan

> indulged in reminiscence of the sea-life led by his fishy forebears by constructing with his amnion a little "private pond" of fluid in which he might embryonically float, and by piercing his neck with gill-clefts, only to do away with them when he subsequently recapitulated his ancestor's greatest feat, the conquest of the land; recalled the furry, four-footed stage of his genealogy by his tail, all ready to be wagged, and his coat of flaxen down; and even, after birth, was unable to help recalling what he later regarded as a blot on his escutcheon—his simian past—by the active, semi-prehensile big toes on his babyish feet and his soon-lost ability . . . to support his own weight when hanging with his hands. (ibid., 418)

Huxley now instructed zoo visitors to think of their own body hair and the remains of a tail at the end of their backbones. He quipped that "the opponents of evolution are apt to forget their own embryonic tails" (Huxley 1936a, 72), but any individual "climbs up its own family tree" (ibid., 73). However, through the observation of evolutionary history and in the realization that this progressive history was incarnate in their own bodies, visitors should arrive at a larger philosophical insight. At the zoo, they could gain the understanding that the increase of intelligence and the efficient care of the young were two of the main trends of evolutionary progress that enhanced control over and independence from the environment up to the most powerful human form. The visitor thus "may reflect that if the blind forces of past evolution could effect so much, man's conscious powers could, if he set himself the task, achieve even more startling progress in the thousands of millions of years still before him on this planet" (ibid., 80).

In order to achieve further progress, humankind had to overcome its destructiveness. When studying the labels that Huxley had had attached to the cages in tandem with the booklet, visitors on the one hand acquainted themselves with the species' geographical distribution across the globe as a result of divergent evolution through adaptation. On the other hand, they should learn about the negative impact humans had had on wildlife. Conservation had long been an issue for Huxley, and while at the zoo, he established cooperation with the Society for the Preservation of the Wild Fauna of the Empire that had zoo rooms on loan, as well as with the British Section of the International Council for Bird Preservation that had the society's office as its registered address.[19] Huxley was aware of the importance of public support for the cause of conservation, and he emphasized the animals' perfect adaptations and long-term evolution. In a radio talk that was reprinted in the *Listener* under the heading "The Lesson of the Dodo," he warned: "Don't forget that if a species becomes extinct, something has gone out of the world for ever. Something unique, built up by the slow fashioning of evolution, has been irrevocably destroyed. We can't bring it back."[20] Huxley increasingly regarded the animal kingdom as a kind of threatened inheritance—not only his own, but that of humankind at large.

Not unlike Osborn, Huxley felt that humans were particularly susceptible to moral lessons during childhood. He therefore had plans for a better integration of schools into the zoo activities. In fact, he was particularly impressed by the New York Zoological Society's education scheme. In London, there already existed special facilities for teachers and their school classes. There were also reduced rates for teachers who wanted to study biology in the Aquarium with their classes. Mitchell, in cooperation with the London County Council, had established a series of expert lectures at the zoo that were designed for LCC teachers who intended to take school children on tours to the garden. Huxley not only reintroduced these, he also hoped that lectures for ordinary visitors could be arranged following the model of the British Museum. These should be on fascinating subjects that could be illustrated at the zoo, such as classification, distribution, and evolution. To this end, a lecture hall might be built. And he further envisioned a children's museum along the lines of the Children's Galleries at the Science Museum.

Thus Huxley also addressed the young on the matter of conservation. He told the University of London Animal Welfare Society that "by visiting the Zoo children become familiarised with many varieties of animals and their sympathy with the animal kingdom is increased in proportion to their greater knowledge, especially if they see the animals being carefully treated."[21] Children were frequently Huxley's audience when he lectured on rare animals and the disappearance of wildlife. To such events, he brought along zoo animals, from lion cub to alligator, to drive the lesson of preservation home by way of intimate encounters.[22] With similar didactic intent, and inspired by the Berlin animal kindergarten, Huxley immediately established a pets' corner and later a children's zoo in the zoo grounds, where kids could get into close contact with animals. The children's zoo was opened by Edward and Robert Kennedy (see figure 13); among other eminent guests were the princesses Elizabeth and Margaret—it was a great success. However, the pets' corner and other such popular projects were criticized by members of the Zoological Society who thought that "the study of animal life must be carried out in a quiet way. The openings at night and the children's playground [seemed] to introduce the first elements of unrest."[23]

Nonetheless, Huxley persevered, and his aim was once again more ambitious than instilling a passion for animals and their preservation. In the *Zoo* magazine, the society's popular periodical that Huxley introduced in 1936, he described the effect humans had had on the animals they domesticated, and how in turn, the animals had shaped human cultures. He also discussed how humans had changed animals in the wild by affecting their environment: "Man has affected animals in other ways; either intentionally or accidentally he has affected their geographical distribution and numbers to a very large extent, and sometimes by this indirect means has caused modifications in them by altering their environment and presenting them with new problems of adaptation."[24] Whether they wanted to or not, for better or worse, humans had been

ZOO GOATS RAISE A 'STORM'

THE goats were friendly enough, but it seemed as if this little visitor wanted the seat all to herself at the re-opening of the Children's Zoo at Regent's Park yesterday. Right, Teddy Kennedy, son of the United States Ambassador, who took part in the opening ceremony, received one of Dean's toy animals from Professor Julian Huxley as a souvenir.

[1935]

JULYSIA the Tonic

FIGURE 13. Ted Kennedy and Huxley with toy animal at the opening of the children's zoo, Julian Sorell Huxley—Papers, 1899–1980, MS 50, Series XI: Clippings, Box 136: 1909–1939, Folder 5: 1931–1939: "Zoo Goats Raise a 'Storm' " (the date on the clipping must be wrong; it is probably 1938), courtesy Woodson Research Center, Fondren Library, Rice University

and still were shaping the global environment and wildlife. Besides being more conscious about how they interfered with the environment of animals, humans also had to take better care of themselves. In a letter "To All Girls and Boys" published in the *Teachers World and Schoolmistress*, Huxley explained how the experiences encouraged in the pets' corner were meant to relate to this larger concern:

> An interest in animals is valuable in very many ways. It helps us to be sympathetic with other kinds of creatures and to avoid and hate cruelty to all

> dumb beasts. It adds a great interest to everyday life. . . . Then an interest in animals leads one to be interested in the study of life in general, or biology, as it is called, and this is very important. A knowledge of biology helps us to keep healthy and happy, and makes us understand more about the workings of our minds. The future of the race—whether it shall become degenerate or whether it shall become stronger and better in body and mind—depends on the understanding of biology.[25]

Thus, at the zoo, visitors were to be led from the petting of a goat to the realization that they were the shapers of global evolution and of the fate of humankind. Again, there were obstacles. A blatantly obvious degenerative force threatening human progress was the human tendency to wage war. On this point, too, the animal kingdom could be instructive. At the Dominion Theatre—under the auspices of the National Council of Education—Huxley lectured to an audience too large to be seated. In the talk, he demonstrated quite counterintuitively that ants, which might be observed on the zoo's "ant island," provided an argument against war: like humans, ants made war on their own species, stole, kept slaves, and exploited child labor; like humans, they were agriculturalists and domesticated animals. The moral lesson was implied in Huxley's verdict that ants had not experienced any progressive evolution for the last thirty million years. They were stuck in their barbarous condition. Humans, to the contrary, were not wholly instinct driven, but had intelligence and thus great degrees of freedom. It was a question of making the right use of it.[26]

Also with the intent to make the British population "stronger and better in body and mind," Huxley widely publicized the insights provided by the running of the parks, particularly regarding the knowledge of animal-environment interaction and nutrition. He campaigned for education about the right kind of diet and for a governmentally controlled basic nutrition standard. He was involved in the making of a film for which experts and working-class housewives were interviewed. In its advertisement, he emphasized that if the children of Britain were as well nourished as the gorillas at the zoo, the national physique would be much healthier. Newspapers spread his message that "if the Zoo can keep young apes from tropical forests healthy and vigorous throughout our English winters, it is demonstrating to the nation what a more adequate and scientifically balanced diet might do in assisting our children to grow up strong and disease-resistant."[27] The scientific insights into the role of vitamins and other accessory food factors were being applied at the zoo, improving the animals' health and reproductive success. The milk fed to the animals was certified tuberculosis free and irradiated with ultraviolet light to increase its content of vitamin D. Huxley therefore exclaimed that "I wish all the children of this country enjoyed as good food as their ape cousins in the Zoo!" (Huxley 1936a, 15). The zoo functioned as a testing ground and model. It made clear that considerable economic changes were needed; the lowest paid members of the community had to be provided with increased purchasing power or at least with cheap or free milk and other protective foods.[28]

The wider social goal of improving the British in body and mind was also associated with Huxley's architectural plans for the zoo. He was involved in the planning of the new Elephant House and the Rodent House with its nocturnal section at Regent's Park.[29] Many buildings at Regent's Park were up to a hundred years old and not built according to modern standards of health and exhibition. Huxley wished he had the money to tear down and rebuild the giraffe, parrot, lion, antelope, and cattle houses among others. He thought that, with the new buildings, Regent's Park should not try to give an illusion of nature, but to create stage settings to show the animals to best advantage. A notable success in this direction was in his view the controversial Penguin Pool by Berthold Lubetkin and Lindsay Drake of Tecton architects that absolutely repudiated any attempt at imitating the birds' natural habitat, but instead enhanced their amusing quality for the visitors' benefit. For Huxley, the modernist architecture at the zoo was another aspect of its model character; it should set the example for a scientifically planned society in which functional and hygienic buildings would be based on the study of human needs. Like the gorillas' diet, their house was an experiment in social engineering (Gruffudd 2000).[30]

With regard to nutrition, the possibility of turning the zoo into a model for scientific planning on a larger scale came with the war. Grassland was turned into agriculture to feed animals as well as to provide vegetables for the restaurants (catering departments). The wartime gardens attracted much interest, and Huxley wanted to extend the scheme into an example for imitation. He approached the Ministry of Agriculture with the idea of an exhibition of agricultural stock to show civilians how to grow more food. As part of the War Utility Exhibition, the Mappin Terrace Pavilion restaurant was turned into a museum on the use of by-products, and the ministry used the lecture room for demonstra-

FIGURE 14. The zoo in wartime, Julian Sorell Huxley—Papers, 1899–1980, MS 50, Series XI: Clippings, Box 137: 1940–1959, Folder 1: Biographical Materials, 1940–1949: "Bactrian Camel in London Zoo," W. Suschitzky-Pictures, "Man Out of Zoo," *Time*, 30 Mar. 1942, courtesy Woodson Research Center, Fondren Library, Rice University

tions. There were exhibitions on the keeping and breeding of goats, rabbits, and pigs, as well as beekeeping and a soil-less culture. The Reptile House became a row of rabbits and a small aquarium, while the Farnham Royal Poultry Society staged an egg show, and the Committee of the Red Cross Agriculture Fund created a poultry exhibition.[31]

The war of course brought many problems for Huxley (see figure 14). Apart from the financial difficulty that forced the society to drastically reduce the zoo staff, the most dangerous animals, such as poisonous snakes, spiders, and scorpions, had to be killed, and others like the large cats, the bears, and the full-grown apes moved to more secure locations.[32] And indeed, cages were hit in air raids and animals escaped. If these could not be coaxed back and showed any signs of being dangerous, they had to be shot by the trained riflemen who formed part of the zoo's Air Raid Precautions personnel.[33] However, in the midst of air raids and antiaircraft squads, Huxley recognized an opportunity to observe the animals' reactions to the bombing,[34] and once again, the public was involved in the project. Huxley published a questionnaire in the *News Chronicle* with the request that the public record their observations and send them to him: "How does your cat or dog behave when bombing begins?" (see figure 15). Not only did Huxley's own parrots frighten people by imitating alarms, it turned out that cows gave poor quality milk, hens might be put off laying, and some birds even experienced "mass hysteria."

Nevertheless, Huxley's conclusion was that animals in general reacted calmly to the turmoil of an attack: "A good proof that birds are not bothered by air raids in general is afforded by the starlings and black-headed gulls. These invade London in large numbers every autumn: this year they have come back as usual, just as if there wasn't a war on."[35] This

FIGURE 15. "Animal Behaviour in Air Raids," by Julian Huxley, *News Chronicle*, 4 Dec. 1940, Julian Sorell Huxley—Papers, 1899–1980, MS 50, Series VI: Publications by Julian Huxley, Box 98: 1936–1943, Folder 6: 1940, courtesy Woodson Research Center, Fondren Library, Rice University

was corroborated by his observations at Regent's Park zoo, which experienced several severe attacks. At one time, the Camel House, the only original building from 1830, was badly damaged, "yet the camels were sitting in their now open cage as if nothing untoward had happened!" (Huxley 1970, 255) In this respect, the animals and life at the zoo could set another good example for London and Britain at large. Like the agricultural projects, they demonstrated the moral-lifting lesson that "Britain can take it" (ibid., 254). By participating in Huxley's research, the public could acquire this lesson firsthand. Huxley also spread the word abroad: "When I was in the U.S.A. last winter I was often asked if this was really true [the damage to the zoo and the harm to the animals], and was told that the news had brought the war home to Americans more than anything else. A visit to Regent's Park at the present moment would, I think, convince a neutral observer, as much as any other single aspect of life, that London is carrying on with very reasonable efficiency in spite of the aerial *Blitzkrieg* and all its fury."[36]

What these examples make clear is that contrary to many fellows of the Zoological Society who criticized Huxley's reforms from the beginning, Huxley understood science to comprise the lab, the field, as well as its economic, social, and political application. Science involved the experts, the patrons, the officials as well as the average citizen. Science should inform every aspect of life. The zoo could well function as a model for such a science of and in life, especially in wartimes, when the synthesis of scientific research, scientific and civic education and action, and tension-releasing entertainment seemed particularly imperative. Indeed, in a short story about a zoo visit in the *Sketch*, the zoo appeared as a peaceful place in times of war and chaos. People could do good through the adoption of an animal—Punch Round Table (staff and guest meetings of *Punch* magazine around a big table), for example, had adopted the male lion—or learn how to imitate the zoo's wartime produce scheme. Most impressively, the animals simply stayed calm during the air raids and so would the ideal citizen. The zoo taught people how to stay sane, and how to improve one's physique and morals.[37]

However, in Huxley's scientific humanism, for the individual's fulfillment, active knowledge was not enough. Reminiscent of the way Osborn envisioned creative education, Huxley thought that for the optimal development of the individual's personality, the acquisition of knowledge about nature had to be coupled with experiences of natural beauty and art. The zoo also needed to be aesthetically pleasing: "It's true that the

Zoo is one of the few places where you can enjoy yourself and even get a meal after dark in the open air, that the fairy lights and the music are gay, and the floodlighting really lovely—I always think that the floodlit flamingoes on the Three Island Pond look like a ballet which has just stopped for a moment; and the curious unreal green of some of the trees is again like a stage decoration."[38] Huxley did not really differentiate between natural and artistic beauty: flamingos could evoke similar emotions as a ballet, and real trees could produce impressions just like a stage setting. The flamingo island, like the chimpanzee and gibbon islands at Whipsnade, were actually developed during Huxley's directorship. In contrast to the Regent's Park houses, these islands were planned in the newer barless style of exhibiting animals in order to enhance the visitor's experience. Huxley sought advice from the international zoo world, most importantly in this regard from Carl Hagenbeck Jr. from Stellingen. Huxley had in mind a master plan that would guide the development of the garden toward noble vistas and handsome buildings properly situated. Animals should be exhibited across a ditch or lake. He dreamed of a big open-air enclosure for apes or baboons, of lakes and streams, huge flying capes, and of more paddocks for more animals such as elks, roe deer, camels, buffaloes, and Barbary sheep. But such projects were inhibited by financial problems.

Beyond merely contemplating beauty, Huxley wanted visitors to engage in creative and artistic activities as well as scientific pursuits. He was acquainted with the New York zoo's studio for animal painters and sculptors, and his early projects included a zoo school of art. He established an art studio, in which an art master gave lessons with animals as models.[39] The art studio, designed by Tecton architects with the support of the London and Middlesex county councils and the blessing of the Royal College of Arts and the Council of Education, started out well enough. The studio opened in April 1937, and on average about one hundred students per week partook in the classes. However, the lightening and the transfer of the animals to the studio proved difficult, and there were few private students, so costs exceeded expectations. It was closed during World War II. The council had not been in favor of the project or in agreement with Huxley's preference for the modern architecture of Lubetkin—"a Russian and possibly a Bolshevik"—to begin with.[40] And Huxley's idea to install John Skeaping's larger-than-life horse sculpture at Whipsnade in 1937 met with resistance. The animal artist and sculptor helped at the studio and had produced a frieze for the restaurant at Whipsnade. With the support of H. G. Wells, George Bernard Shaw, and Kenneth Clark (director of the National Gallery), Huxley prevailed. But after his retirement, the horse was discarded. Similarly abortive was Huxley's attempt to place Eric Gill's statue *Humanity*—a nude female—in Whipsnade in 1938. All in all, the council was not in accord with Huxley's plan to turn Whipsnade into a "general cultural centre."[41]

However, even if Huxley could have implemented all his plans for the zoo grounds, the sites were not enough for his terrific vision. Like Osborn's museum, Huxley's zoo was to be a center of calculation from which the moral authority of nature emanated throughout the world: "This principle, of taking the Zoo to the people—shall we call it 'extension work'?—is capable of enormous expansion." On a primary level, extension work simply meant zoo publicity. The zoo often made the news in many papers, and the *Daily Mail* published regular reports on it. The BBC arranged for a weekly talk on the zoo, and this "Zoo Man's hour" was very popular. But extension work went beyond this. Huxley rhapsodized in the *Evening News* that "the Zoo might make and sponsor its own films, which would then go out to the ends of the earth. . . . Again, the Zoo might have its own cinema, where only animal and nature films were shown. . . . And it is bound to utilise ever more fully the technical achievements of modern invention in the shape of wireless, the cinema, printing, television, and so forth, for the purpose of bringing the Zoo to the people and extending the circle of its activities."[42] These were not all Huxley's ideas. He was strongly inspired by the zoos he visited in Canada, the United States, and Europe, for example, by the Frankfurt zoo's newspaper and cinema.[43]

Regarding the plan of a cinema in the London Zoo, Huxley actually had a commercial partner to build and run the movie house. The idea was that the cinema would show not only animal films but also Walt Disney cartoons (see figure 16). However, the treasurer opposed the scheme and the council feared it would lower the prestige of the learned society. Nonetheless, Huxley did set up a nature film production that carried the zoo lessons to a wide audience. In 1937, a subsidiary Zoological Film Productions was founded in agreement with Strand Films. Strand provided the cameraman and half the capital, and Huxley secured the rest through donors. They produced films on evolution and other topics for schools and commercial showing as well as footage on animal behavior and other material for scientific study.[44] In his capacity as supervisor of all biological films made by Gaumont-British Instructional, Huxley was

FIGURE 16. "Zoo education movement," Julian Sorell Huxley—Papers, 1899–1980, MS 50, Series XI: Clippings, Box 136: 1909–1939, Folder 5: Biographical Materials, 1931–1939: *Evening Standard*, 5 Jan. 1935, courtesy Woodson Research Center, Fondren Library, Rice University

further engaged in the production of nature pictures for schools. Finally, his own *The Private Life of the Gannets* (1934), that he had made for Alexander Korda, was a great success and shown at hundreds of cinemas in the country; it was even presented at the International Exhibition of Cinematographic Art in Venice.[45]

In film, as in radio and print, Huxley recognized a medium that could carry the zoo into the world, but also one that could bring the "wild parts of the earth to the centre of civilisation." He gave twelve radio talks on the making of nature films, explaining the difficulties of the technology and the shortcomings of current results that, for example, never made the audience laugh. One of the main problems was that nature films became dead once they were moved to cold storage. Parallel to the dream of a Sound Institute, Huxley suggested the establishment of national and international film libraries. These would have an annotated catalogue of films on different subjects, a collection of copies of as many as possible, and facilities for borrowing or hiring. He hoped that at least in the larger cities there would be built special cinemas of the newsreel type for showing particular kinds of films—one for nature films, another for educational subjects, and so on. Museums, too, should have films available. The Natural History Museum might have small peep-show projectors alongside exhibits of stuffed animals, showing the same animals in a state of nature: "After looking at an elephant group, for instance, or an enlarged model of the locust, you would repair to the adjacent projector, press the button, and see a short film of elephants charging through the African bush, or of a swarm of locusts darkening the sky."[46] For Huxley, the living organism was the most instructive by far, and film was the best available proxy.

In the end, however, both of Huxley's ambitions as secretary of the Zoological Society—reforming science and bringing it to society—were of limited success, partly because the fellows blocked many of his plans. After his resignation, the society even returned to the old organization of the *Proceedings*. Throughout Huxley's tenure, the council increasingly tried to hem him in. New plans had to be approved before anything was undertaken to implement them. Special committees were set up to control Huxley and his staff, such as the Permanent Estimates Committee, which would approve all expenditures, and the Emergency Committee, to supervise the acquisition of new animals. Huxley might have been a visionary, but to the society's chagrin, his interest in daily administrative business was very limited. In 1938, an assistant treasurer was installed to report to the treasurer any financial mismanagement; he had access to the staff, the books, and any correspondence concerning financial matters. In the same year architect W. G. Holford, professor of civic design at Liverpool University, was hired to prepare a plan for the zoo's development. Council members especially objected to Huxley's extension work and his frequent presence on public platforms. In 1940, a special committee was established to keep an eye on Huxley's "outside activities." In September that year, this special committee demanded his permanent presence at the zoo, and Mitchell even suggested that he submit a list of all the places he went to.

Nonetheless, despite many setbacks, Huxley's zoo was popular. In 1936, there were already around two million visitors to Regent's Park per year and 500,000 to Whipsnade. Visitor numbers rose to a new record before the war. This popularity was reflected in the press, and sometimes in personal letters. Overall, the media and the general public liked the zoo innovations and they liked Huxley. One avid zoo visitor wrote to him:

> I am writing simply as a Zoo-lover myself, and as the mother of three boys to whom "going to the Zoo" has become, not merely a pleasant habit, but part of the important business of living. The Children's Zoo will, I know, be a lasting influence on the life of my youngest son. We went there every week; we know all the animals as friends; we enjoyed all their fads and foibles. It became part of the fabric of his life. The Zoo magazine strengthened these ties, and was immensely popular. . . . We must hold on to the really good and true things for them [the children], and refuse to let them go. I believe the Zoo, and your lovely Zoo films, to be outstanding examples of these. The vision, the sense of beauty, and the humour you have given us in your running of the Zoo matter supremely.[47]

Clearly, with such visitors, Huxley's attempt to render the zoo a strong experience and an important part of people's lives was successful. Flanked by a myriad of medial distributions, the messages the animals were made to carry might settle in.

The popularity of Huxley and his secretaryship became particularly apparent when the press reacted stormily to the news of the council's suspension of the post (see figure 17). He also had allies within: younger and scientific members of the society such as Haldane, Fisher, Ford, and H. G. Wells organized strong support for Huxley. This culminated in his victory and reelection before the council. However, the fight of Huxley and the so-called Informal Committee had been associated with a plan for general reform, so that when none of his supporters were elected to the council, Huxley felt forced to resign.[48] Ironically, the controversy once again turned the zoo into a testing ground for larger social issues. As Joe Cain (2010, 359–363, 369–371) has observed, the press presented it as a matter of amateurism versus professional science, of old versus young, conservatism versus progressivism, and of aristocracy versus democracy.[49] This perception was enforced by *Evolution: The Modern Synthesis* (1942a), which was internationally hailed as Huxley's magnum opus and compared to Darwin's *On the Origin of Species*.[50] As a speck of hope in times of war, the reviewer for *Nature* saw in the book the potential to renew biology and to unite its diverse disciplines.[51] It was welcomed as the new standard exposition in English on modern evolutionary theory, and even though it was not considered an easy read, it

FIGURE 17. Monkey for Huxley in zoo affair, Julian Sorell Huxley—Papers, 1899–1980, MS 50, Series XI: Clippings, Box 137: 1940–1959, Folder 1: Biographical Materials, 1940–1949: Cartoon by Neb. (Ronald Niebour), courtesy Woodson Research Center, Fondren Library, Rice University

was perceived by the press as unsettling not only scientific orthodoxy but also religious belief, and even the popular myth of a teleological history. Harking back to the zoo controversy, one reviewer commented: "It will upset the complacent and static-minded reader as badly as its author did the static and complacent Zoo. The reader will either have to throw the book away, or think."[52]

Huxley's grip of the mathematical theory of natural selection and of chromosomal research, his knowledge of the latest findings in paleontology, and of the natural history of rare and obscure animals, dazzled reviewers. It was noted that "there are several books in the literature dealing with evolution—to mention only those by Haldane, Dobzhansky, Darlington, Goldschmidt, Hogben and Wright—to which we are greatly indebted for a clearer and wider horizon in regard to interpretation of the processes of evolution. Huxley's present book goes further; it gives us a deeper understanding, by showing how the manifold processes of evolution are interrelated; it puts before us the gigantic structure as a whole—a structure in which all the component parts have a rôle to play—and attempts to assign to each part its respective rôle."[53] Huxley developed this integrated system into an evolutionary cosmology, and it is clear that its relation to national and global politics (to which I turn in the next chapter) was a source of particularly poignant criticism from the fellows. However, despite the commentator's estimate, not only Huxley, but also some of the other authors mentioned in this review established a "gigantic structure as a whole" that explained the order of things from molecule to man. As I show in the following chapter for Haldane and Hogben, Huxley was not alone in drawing conclusions from evolutionary history and heredity for progress toward the right kind of societal organization.

CHAPTER EIGHT

Scientific Humanism in the Extended Zoo

History Within as the Basis of Democratic Reform

On the twenty-ninth of October 1940, Huxley sat in the zoo's air-raid shelter, writing the introduction to *The Uniqueness of Man* (Huxley 1941a) accompanied by the noise of guns outside and the sound of Sibelius's *Voces Intimae* inside. If *Evolution: The Modern Synthesis* was Huxley's magnum opus of the new biology, with *The Uniqueness of Man* he wanted to make a strong statement on humankind's place within the synthetic framework. The essays collected therein had appeared between 1927 and 1939 in a wide range of venues: the science magazines *Discovery* and *Scientific Monthly*, the *Strand Magazine* with its close to 500,000 readers, the *Nation*—the flagship of the American left, *John O'London's Weekly*—the leading literary magazine of the British empire, the *Atlantic Monthly*, the *Virginia Quarterly Review*, and the *Times*. Now, in the midst of anti-aircraft firing, Huxley tried to unite the essays under the hopeful perspective of his scientific humanism. To Huxley, the National Socialist and other fascist regimes represented the negation of any civilized order and were the irrational and dark threat to the democracies, but he also took issue with the laissez-faire brands of democracy with their mechanistic and economic ideals: "They were founded in freedom and promised prosperity and equality. But in place of freedom, men have found themselves enslaved to the impersonal machinery of the market; their purely political equality has been accompanied by gross economic and social inequality; and the promise of prosperity has been replaced by mass insecurity and frustration" (Huxley 1941a, viii). This disillusionment, Huxley observed, had set in motion a revolution and that revolution had to be taken to its end. Social services like subsidized housing, free milk, social security legislation, health insurance, and free education were only the first tentative steps in the direction of a new democratic world order. Amidst the tumult of a London under attack, Huxley appealed to human tolerance, solidarity, and altruism against a disproportionate individualism, and declared that "if civilization is to recreate itself after the war, it can only do so on the basis of what, for want of a better word, we must call a social outlook" (ibid., vii–viii).

Huxley—the biologist—did not consider himself outside his terrain when intervening in politics, since "biology [had] some relevance to the task" (ibid., ix) of postwar re-creation. After all, with regard to the scientist as citizen, he had the example of his grandfather. In fact, Julian appropriated the figure of Thomas Henry Huxley as a model scientist in a lovely BBC radio play he put together with the producer Douglas Cleverdon on the basis of letters and publications called *A New Judgment on the Great Victorian Scientist, T. H. Huxley.* The play highlighted Thomas Henry's contribution to the progress of biology, and especially his fight for the acceptance of evolutionary theory in science and in public. It showed that he had been interested in the meaning of evolution for other areas of knowledge such as anthropology, philosophy, theology, and ethics. Finally, Thomas Henry had been a connoisseur of literature and the arts: "His life reminds us that even in these modern days of specialization it is still possible to be a whole man, to cultivate every aspect of life: moral, emotional, intellectual, aesthetic, social, and political; still possible to be a whole man, not merely a professional specialist, however eminent."[1]

However, while Thomas Henry could be instrumentalized as exemplary scientific humanist, he presented a problem with regard to the question of the relevance of biological knowledge, and especially evolution, for the task of social reform. He had denied that any lessons for human society could be derived from evolution. It had taken Julian nerves and time to come to terms with the problem, but when he wrote the preface to *Uniqueness*, he felt ready to take up the issue with "Darwin's Bulldog," and he did so in an imaginary interview published in the *Listener.* Established in 1929 to print radio talks, the weekly magazine evidenced a boom in science topics, and the highbrow publication eventually reached a distribution of around 50,000 (Bowler 2009, 210–211).

On its pages, Julian explained to the ghost of Thomas Henry: "There [in *Evolution and Ethics*, 1893] you stated (I remember the passage vividly) that the ethical process of society depends not on imitating the cosmic process but in combating it, and by the cosmic process of course you meant mainly the ruthless struggle for existence. As an evolutionist, I never understood how man, himself a part of nature, could fulfil his destiny by fighting against that same process which gave him birth."[2]

Julian protested as an evolutionist against Thomas Henry's meaningless cosmos, but mostly he protested as a human being in a world riddled with crises. He lamented that "we have had nearly thirty years to adjust ourselves to the collapse of the world system that seemed so stable and so full of promise in your time . . . : first the war of 1914–18; then a period of cynical disillusionment; then the most spectacular economic collapse in history; then the rise of Fascist aggression." Julian ascribed Thomas Henry's denial of an evolutionary ethics to the different historical context the Victorian scientist had found himself in, but also to a different understanding of evolution. Thomas Henry's evolution had been red in tooth and claw, whereas by Julian's time intelligence and cooperation seemed to have played as important a part in evolution as brute force and competition. In fact, in some cases, competition within the species had even been harmful.[3] But there was more to be had from evolution than a moral argument against war and for cooperation. The entire future development toward a democratic world order had to be founded on this cosmic principle. In the essay "Scientific Humanism" (1931) included by Huxley in *The Uniqueness of Man* (1941a), he declared that "humanism, with the aid of the picture given by science, *can* achieve a framework strong enough for support. In the light of evolution, it can see an unlimited possibility of human betterment. And it can see that possibility as a continuation of the long process of biological betterment that went before the appearance of man" (ibid., 266–267; see also T. H. Huxley and J. S. Huxley 1947).[4]

We have to look outside the Huxley family traditions and several decades back in time to locate possible inspirations for this particular brand of scientific, and in essence evolutionary, humanism. One source can in fact be made out as early as the 1910s, when Julian Huxley embarked on the job of establishing a biology department at the Rice Institute in Texas. He did so with his usual enthusiasm and was assisted by eminent American biologists. In letters and during visits, the embryologist Edwin Grant Conklin of Princeton University, who had actually drawn Huxley's attention to the opening, advised him on where to buy which lab equipment and lab animals, how to deal with lab assistants, and how to found a library as cheaply as possible. During the years that were formative for Huxley, they also began to discuss the "Methods and Causes of Evolution."[5] Conklin held a lecture series on the meaning of evolutionary history for the present that he eventually published as *The Direction of Human Evolution* (1921). He drew inferences for human social development from the tenets of the paleontological consensus. He reasoned that while mere diversification and adaptation to particular conditions could well lead to an evolutionary dead end, humans could continue the progressive trend of increased control over and freedom from the environment. If the increasing division of labor were accompanied by intensified cooperation, the stifling effect of specialization would be averted and higher levels of general organization could be reached.[6]

For Conklin, such a progressive trend was also visible in the evolution of sociability from the one-celled organism, to the multicellular structure, the organs of a complex animal, and finally in increasingly socially integrated groups of animals. He concluded that "there is no doubt that evolution of human society has been in this direction, and the entire past history of living things indicates that further progress of society must be along this line" (Conklin 1921, 89). Most importantly for the context of this chapter, Conklin was more explicit about the nature of "further progress": "And all of these lines of social progress are correlated with the growth of democracy" (ibid., 97). Democracy was defined as "*a system which, ideally at least, attempts to equalize the opportunities and responsibilities of individuals in society*" (ibid., 100, emphasis in original). In order to achieve perfect specialization and cooperation, humankind had to balance the antagonistic demands of social and egotistical instincts, the interests of society and the interests of the individual. "Fanatical individualism or socialism," Conklin wrote, "find no foundation or counterpart in biology, for life and all of its activities consist in compromise, balance, adjustment between opposing principles" (ibid., 98–99). Democracy was therefore the only social system that harmonized with evolutionary history; its principles were the same as those found in nature. Indeed, Conklin talked of "the biological bases of democracy" (ibid., 100). A democracy had the best chance to achieve a perfect state of checks and balances. However, if it failed to grow into a highly integrated organism, a democratic nation might also become the breeding place of cancer cells.

In the early decades of the twentieth century, Conklin saw the challenges to his democratic vision in laissez-faire capitalism and in socialism—but also in traditional eugenics. Conklin was in the vanguard when he took issue with an understanding of eugenics according to which genius bred genius and "imbeciles" propagated "the feeble-minded," and in which greater intelligence was associated with the upper classes. Against this notion, Conklin claimed that the Mendelian laws of heredity were inherently democratic. Whereas the law of entail was aristocratic in that it ensured the social inheritance of property from father to son, no such thing was possible with regard to genes, because with each new generation, the hereditary traits were newly mixed. Therefore, Conklin concluded, human beings must be judged according to their own merits "and not by the merits of some ancestor whose good traits may have passed to a collateral line" (ibid., 132). Good and bad hereditary characters were distributed among all classes but also among all nations and "races." In fact, the phylogenetic diagram to represent humankind was not the family tree with clearly separated branches but "a net in which every individual is represented by a knot formed by the union of two lines which may be traced backward and forward to an ever-increasing number of knots and lines until all are united in this vast genealogical net of humanity" (ibid., 134). Conklin thus based the demand for equal opportunities in a social system on the workings of heredity, and he derived the right kind of societal development from evolutionary trends. He supplanted the human family tree with its neatly separated branches with a networked kinship structure that suggested a phylogenetic basis to human brotherhood. Thus, while he attacked the biological determinism inherent in much of eugenicist writings and emphasized the role of environment, and while he declared that class, race, and national antagonism were no biological necessities but fomented culturally and politically, he nonetheless argued for democracy by means of biological principles.[7]

Finally, Conklin emphasized that in the human sphere, the progressive natural trends were no longer automatically perpetuated. Indeed, nearly verbatim like Huxley, Conklin stated that "the past evolution of man has occurred almost entirely without conscious human guidance; but with the appearance of intellect and the capacity of profiting by experience a new and great opportunity and responsibility has been given man of directing rationally and ethically his future evolution" (ibid., 93). In fulfilling this purpose, the biologist would have to lead the way: "No one who has felt the force and sweep of the great doctrine of evolution, can doubt that biological principles underlie the physical, intellectual, and social evolution of man—that biology is a torch-bearer not merely into the dark backgrounds of human history, but also into the still more obscure regions of the future development of the race" (ibid., 110). The group of British experimental biologists to whom Huxley belonged took on this torch and preached the necessity of a conscious planning of human evolution. They shared Conklin's Wellsian vision that "in spite of narrow-minded and reactionary politicians, we or our descendants will yet see the whole human race brought together into a Society of Nations, a 'Federation of the World' " (ibid., 75).[8] And they, too, supported their political ideals by drawing on the evolutionary past.

The geneticist Darlington wrote to H. G. Wells the following about the three biologists I am referring to: "When I was very young, Galdane, Guxley, and Gogben (as the Russians called them), seemed to be the three Magi" (Darlington to Wells, 6.6.1976, quoted in Tabery 2008, 730). The initials of the three magi formed in reality a triple-H.[9] Oscillating around Cambridge, Oxford, and London, Haldane, Huxley, and Hogben cultivated their friendship during lunches, dinners, and weekends, as well as in letters and telephone conversations. They were bound together by common scientific interests and cooperated in research and organization, for example, in the establishment of the *British Journal of Experimental Biology* and the Society for Experimental Biology (1923) (Erlingsson 2006).[10] For the development of their status as magi, Hogben attributed great importance to Huxley's time at the Rice Institute, during which he had been updated on the progress in American genetics. Subsequently, Huxley publicized the advances in American biology and brought the great American geneticists (such as Muller and Alfred Henry Sturtevant) to Europe, thereby establishing knowledge exchange with British scientists like Haldane and Hogben. In fact, it was not only people and ideas, but also drosophila cultures in test tubes that moved from American to British labs. However, in accord with a holistic philosophy of life, Huxley and his friends also worked on more integrative projects; Huxley and Haldane, for example, collaborated on the highly successful textbook *Animal Biology* (1927).[11]

Beyond an interest in diverse branches of biology, the three magi shared a passion for politics and social planning. In his autobiography, Hogben reminisced about the members of the Tots and Quots club in the London of the early 1930s, to which the three belonged: "Our common bond of interest was the right use of science for human wellbeing" (Adrian Hogben and Anne Hogben 1998, 139). The politics of the triple-H were not identical. Hogben and Haldane were socialists, while Huxley might be called a liberal. To various degrees, they became radicalized in the 1930s, with Haldane admitting to Marxism toward the end of the decade and eventually joining the Communist Party in 1942. However, prior to World War II, the effects of the Great Depression and what they perceived as conservatism at home as well as the rise of Nazism brought them closer together. As Hogben put it: "When possible, those who profess minority creeds unite forces against traditional authority. It then seems to be a psychotherapeutic necessity for those who do so to delude themselves that they share a unique substratum of positive aspirations" (ibid., 45). Even if in retrospect it could seem like a delusion to Hogben, the triple-H had common social and political goals. Gary Werskey (1971b) has elaborated on the similarities in Huxley's and Haldane's conception of a scientifically planned society, and he has identified Haldane and Hogben as great revolutionary scientists of the 1920s and 1930s (Werskey 1978).

In fact, the triple-H were part of a wider movement in the Britain of the 1930s that was driven by an interest in the relations of science and society, the social responsibility of the scientist, the relevance of biology to human values and to the human present and future, the paradox of individuality in mass society, and problems of integration and progress.[12] Their approach to these issues was often subsumed under the label *scientific humanism*. In *Dangerous Thoughts* of 1939, Hogben wrote: "Scientific humanism is the creed I profess and the profession I try to practise" (13). It referred to the insight that the pursuit of science was intricately related to the responsibilities of citizenship because it transformed society.[13] As we have seen in chapter 6, the label of scientific humanism was not of Hogben's or Huxley's coining. It came out of the movement for the institutionalization of the history of science. In his abundant popular writings on the history of science, Huxley followed Singer in demonstrating a general advance in knowledge that changed people's worldview as well as the exterior world. However, for Huxley, as for the more socialist and Marxist historians of science, these processes were not necessarily for the good of people or nature.[14]

Of the triple-H, Hogben put most weight on the interdependence of science and society in his histories of science. The history of science was connected to social and economic history, for example, in the concept of the struggle for survival and laissez-faire capitalism, or in racial anthropology and the colonial system. The influence was mutual, with culture and history also impacting the way the sciences developed, which was, for example, obvious in the European expansion, when new animals and plants had been collected, or in the role of industrialization for the development of the earth sciences (Hogben 1938, 920–970, 1048–1075). For Hogben, one's view on the history of science was linked to one's stance in the nature-nurture debate that will concern us below. For if organisms, and especially humans, were a coproduct of nature and nurture no one could fully disentangle, then the scientist was already a cultural being. Or as Hogben put it, the genetic determinists "appear to hold that Newton would have written his *Principia* if he had been born in Tasmania" (ibid., 1057). The individual was not disinterested and detached in his or her research, but guided by upbringing, training, personality, and other factors. Some questions would be asked in a certain context, others not, and some scientists would try to defend their position in society or to advance themselves.

From Hogben's antielitist, anticlassist, and anti-imperialist perspective, science appeared as good science only if it was for the good of the people, if it answered to the common needs of entire humankind. It had to be concerned with moral as well as material advancement. Only in this sense could the contribution of scientific discovery to the progress of civilizations be measured, and the impetus science received from an expansion of opportunities for the satisfaction of the common social needs understood. Hogben shared with Huxley and Haldane the belief that progressive science could not thrive on its own. It depended on a favorable social context. While science was the key to a progressive development, that key needed a conscious effort to be turned. In order to engage in a science that was truly progressive, the scientist had to be a citizen and carry his or her science into the world (ibid.).[15] Indeed, especially Huxley and Haldane, but to a lesser extent also Hogben, brought their ideas to life for diverse audiences in radio talks, all kinds of popular text genres, and documentary film. In fact, the process of turning evolution and genetics—in the more restricted sense of a laboratory and field science—into a political weapon strongly relied on these media technologies. The early twentieth century saw important developments in the publishing industry, as publishers sensed a more general interest in science. Science, too, profited from mass circulation and photography. However, while the expansion of education and the movement for self-improvement among parts of the middle and working classes created a favorable environment for nonspecialist and popular science, the sciences had by then become professionalized. The gap between the scientific and the lay communities had widened.

Nonetheless, as Peter Bowler (2009) has shown, a significant portion of British scientists did from time to time communicate their science to larger audiences even if popular writing could hinder a scientist's career. Huxley experienced this not only in his quarrels with the Council of the Zoological Society but also when he was at length denied a fellowship in the Royal Society, partly because of his great public engagement. Huxley and Haldane were certainly outstanding in their efforts to disseminate their ideas on science and society. To the consternation of his colleagues, Huxley had resigned from his professorship at King's College in 1927 to collaborate with H. G. Wells and his brother on *The Science of Life* (Wells, Huxley, and Wells 1929/1930/1931/1934). Haldane was well aware that scientists were generally disgusted with "the vulgarizing tendency" of newspapers, which rather encouraged him to write for them. Hogben entered the public arena with a certain unease. He prefaced his best-selling self-education books with apologies, attributing his popularizations to long commuting train rides or illness. Huxley's, Haldane's, and Hogben's engagements in the communication of science can only be understood in the light of the strength of their ideological fervors. Besides common activities in clubs and societies, this engagement also proved to be a means of bonding between scientists who understood themselves as citizens and who intensely drew on each other's ideas.

Huxley was the first of the three magi to take on Conklin's torch. Building on his experience in writing regularly for quality magazines, where he had an agent place his articles, Huxley established himself as a contributor to daily papers and more popular magazines after the Great War. He was a welcome voice on the radio, and by the time of World War II, he was among the dominant public figures (Bowler 2009, 222–223). One of Huxley's most successful activities was the BBC's *Any Questions?* It was a Home Service general knowledge show for the troops, the experts on which were dubbed the "Brains Trust." It was so popular that it was eventually broadcast also on home wavelength. In 1942, about ten million people were following the humorous discussions of topics raised by the public such as missing links, phrenology, sea serpents, and the abilities of animals (Thomas 1942).[16] In contrast to the lightness of the Brains Trust, however, Huxley's public statements on the current situation of the world were deadly serious. Tailored to an all-powerful biology of evolution, scientific humanism provided a framework with which he could argue for social reform and against national and "racial" violence. In concert with his friends, he formulated a biology of social reform that was couched against laissez-faire ideas and the class system at home that was perceived as allied to classical eugenics, as well as against fascism with its foundation in race science. They reinterpreted our history within into an argument for social equality.

Central in this process was a reevaluation of variation. Mathematical population genetics had provided the knowledge that genetic variation was not lost but that gene ratios remained stable over time in ideal populations, and the study of natural populations had revealed an amazing genetic diversity. Haldane had worked on mathematical formulations of the effects of selection and other mechanisms on such genetic variability, and he explained: "As I come to the study of society from that of genetics, it is natural enough that I should be prejudiced in favour of human diversity and should hope that my country will not try to suppress it" (1932a, 40). Genetic variability in humans of course meant genetic inequality, but seemingly paradoxically, Haldane and his peers used the fact of genetic inequality to argue for political, economic, and social equality. Biological inequality was turned from a problem of conservative politics into a purpose in progressive evolution.

Under the title *The Inequality of Man*, Haldane reissued in 1932 essays that he had written for "several different audiences, varying from the readers of the daily Press to the members of the Royal Institution" (v). He emphasized that "I do not believe that a recognition of the inequality of man would be a blow to democracy (or rather to representative government based on universal suffrage)" (ibid., 24) and that "the recognition of innate inequality should lead not to less, but to greater, equality of opportunity" (ibid., 26). This was the case because "the ideal society would enable every man and woman to make the best of their inborn possibilities. Hence it must have two characteristics. First, liberty, which would allow people to develop along their individual lines, and not attempt to force all into one mould. . . . Second, equality of opportunity, which would mean that . . . every man and woman would be able to obtain the position in society for which they were best suited by nature" (ibid., 220). Diversity, not one or a few molds of people, was the basis for progressive societies.

The triple-H had different understandings of this ideal society, and for Haldane, democracy only constituted a rite of passage. However, at this point in societal development, democratic reform was the means against class society: "I do not know what would be the ideal form of government in a community where that equality had been achieved. Democracy appeals to me, not as an end in itself, but as the most hopeful route, at least for England, to a classless society" (ibid., 220). Haldane's attitude was inconsistent and changed over time. In 1932, he could not accept "American and Communist ideals," because they aimed at "the standardization of man" (ibid., 227). Later, in "Biology of Inequality" (Haldane 1938, 13–41), he criticized the notion in the Declaration of Independence that all men are created equal, because rather than highlight the necessity of economic equality, it supported a laissez-faire politics. Haldane now held that, to the contrary, socialism and communism were not based on the precondition of human equality, but demanded that each person should work to his or her ability and be rewarded according to that work, or to his or her needs respectively (ibid., 13–15).

However, the quintessence of such considerations remained the critique of systems—be they religious like Calvinism or political like fascism—that "idealize one particular type" (Haldane 1932a, 38) rather than further the fulfillment of each individual according to his or her potential. There was "something to be said for human diversity" (ibid., 40). This was not only about an efficient management of diversity rightly understood; this was also about the understanding that a progressive society needed all kinds of people, since "a man who can look after pigs or do any other steady work has a value to society" (Haldane 1938, 101). Huxley, too, argued for the "significance of human variability" (1941a, 34–84 ["Eugenics and Society," first published 1936], 75). It was "of the utmost importance for the material and spiritual progress of civilization" (ibid., 76). The understanding that the great human flexibility and variability were prerequisites for progress in the human phase of evolution lay at the heart of his scientific humanism (1941a, 1–33 ["The Uniqueness of Man," first published 1931], for example, 21–22). Political systems that cajoled or forced people into homogeneity lost their potential for further advance, and classical eugenics was perceived as in the service of such systems. The triple-H recognized in current eugenic ideals the method of professional stockbreeders, which is the striving for a particular kind of excellence.[17]

In their diatribes against "conservative eugenics," the reevaluation of diversity was connected to the nature-nurture question that began to trouble geneticists in the 1930s.[18] While traditional eugenics had assumed many complex characteristics such as intelligence to be expressions of Mendelian genes, the triple-H thought it impossible to separate heredity from nurture in the case of human beings. Haldane accused those biologists who pretended to do so—the conservative eugenicists and the fascists—to be as dogmatic as the Catholic Church. The most fundamental critique came from Hogben, however, who was strongly influenced by his lower-class background. On 23 October 1930, in his inaugural address as holder of the chair of social biology at the London School of Economics, "he delivered a somewhat sarcastic attack upon eugenics," accusing eugenicists of thinking that "reproduction and inheritance in human beings proceed on the same lines as in albino mice, fruit flies and the garden pea."[19] To the contrary, Hogben argued in scientific papers and popular writings that even for "simple" organisms, the relationship between environment and heredity was nonlinear. Uniformity of environment could be created in an infinite number of ways, bringing out genetic differences not measurable before and obscuring genetic variability previously visible. Some traits might be expressed in a wide range of environments, while others were more milieu specific. Hogben shared the developmental perspective with Haldane and Huxley as well as Conklin, and he pushed it to the conclusion that the genetics itself depended on the context (Hogben 1932; 1933; 1939, 1057–1064; Tabery 2008; see also, for example, Haldane 1938).[20]

Hogben's inaugural address caused quite a stir, but Huxley defended him as "on the verge of being a genius" and as "one of the very few of [his] colleagues for whom [he had] a deep personal affection as well as intellectual respect."[21] In fact, Huxley relied on his friend's science to the degree of reasoning that in order to allow every individual in a human society to take the place that best suited his or her genetic potential, the social conditions and economic resources had to be equalized—and equalized up (1941a, 34–84 ["Eugenics and Society," first published 1936], 45). In the 1920s, Huxley still thought that negative eugenics was scientifically possible and socially practicable by means of consultation and voluntary sterilization, at least in cases of single recessives such as deaf-mutism. Positive eugenics seemed on the verge of being scientifically cognizable and socially feasible through encouraging the particularly endowed to reproduce. However, at that time he was already skeptical about the possibility to improve the existing highest quality of the population by directed mating. To this end, one would need much more knowledge.[22]

Matters became more complex in Huxley's mind when it seemed increasingly impossible to keep the cultural and the natural apart and to define what was a positive and what was a negative trait outside the perfect social milieu. Differences in environment first had to be abolished to bring to light the genetic differences between individuals and stocks; until then, general conclusions could only be guesswork:

> In other words, a truly scientific eugenics can only begin when equality of social opportunity has been achieved. Furthermore, some types of social structure (like our own) will tend to produce effects the reverse of eugenic, while others will promote eugenic results. There is a vast field for the social reformer, the humanitarian, and the socialist in linking social improvement with the improvement of the race and an additional incentive to their efforts at planning human environment and human organisation if they can feel that these are destined not merely to improve the lot of individuals, but to help extension towards further progress.[23]

Therefore, eugenics was essentially a social science. Even though the triple-H continued to support voluntary sterilization, and in particular contraception, as a means to social justice, eugenics as a biological tool to steer evolutionary progress mostly vanished into the future (see also Haldane 1932a, 220). It was therefore not a matter of giving up eugenic aims and practices, but of adjusting these to "the biological as well as social realities." Thus, in the foreword to a pamphlet of the Eugenics Society, in which Huxley was engaged, he explained that "the inherent diversity and inequality of man is a basic biological fact; and Eugenics is the expression of a wish to utilize that fact in the best interests of future generations."[24]

Along similar lines, Haldane considered it ridiculous to judge the worth of people according to an unjust society rather than judge the worth of a society according to its degree of injustice: "Many of the 'unfit' are unfit for society as it is to-day, but that is often society's fault. The attempt to prevent them from breeding really involves the appalling assumption that society as at present constituted is perfect, and that our only task is to fit man to it. That is why eugenists are generally conservative in their political opinions" (1932a, 88). There were thus "undeveloped human resources" and the best hope for the future lay in giving everyone a chance (ibid., 223). Even with regard to the criminal, "we must take him as we find him, and attempt so to order society that he does not commit crime" (ibid., 42). Practiced in society as it stood and based on a wrong understanding of heredity, eugenics was "an instrument of class war" (Haldane [1932] 1937, 227). Humankind's biology should not be adjusted to an unfair societal system, but vice versa, society had to be brought in line with the facts about heredity. Like Huxley, Haldane hoped that under equalized social conditions, biological inequality would not produce a stratification like the British class structure or the Indian caste system both of which were based on descent, but occupational groupings according to ability. Notwithstanding the fact that the triple-H's own arguments for social reform were based on biology, they accused others of justifying social goals with biological knowledge; thus, Hogben was most infuriated by "the pastime of decking out the jackdaws of class prejudice in the peacock feathers of biological jargon" (1939, 53–54).[25]

Obviously, the triple-H addressed scientists as well as laypersons with their propaganda for a new biological outlook on social problems. Huxley's Galton Lecture to the Eugenics Society (Huxley 1936d), for example, provoked protest from both.[26] To estimate the degree of radicalism in the triple-H's views, one may consider the lecturer who followed Huxley. As the conservative *Morning Post* reported, this lecturer welcomed National Socialist Germany as a eugenic experiment of the greatest importance.[27] In fact, through the pens of the triple-H, the new biology, or the implications it had for the nature-nurture question, became equally effective against a British eugenics that was understood as a technology of the class system as against a fascist eugenics that was seen as a technology of racism. In *Science for the Citizen: A Self-Educator Based on the Social Background of Scientific Discovery* of 1938, Hogben attacked both in one breath: "Today the advance of human genetics is held back because the prosperous classes refuse to tolerate any challenge to their intellectual privileges. The natural mission of the middle class or of the Aryan race has now replaced the divine mission of the Church militant. There is no need to ransack Nazi publications for illustrations of a temper which exists [in] the official organ of the English Eugenics Society which . . . 'has always been especially interested' in 'that portion which is popularly called the upper and middle classes' " (1938, 1073).[28] As self-education literature, *Science for the Citizen* was the perfect tool to influence the consciousness of the time, and although being a demanding read of immense size, it was a great success and sold very well (i.e., 50,000 copies by 1940 according to Bowler 2009, 112).

Hogben thus criticized not only eugenics but also racial anthropology; the same year, Haldane, too, took issue with that branch besides eugenics in *Heredity and Politics* (1938).[29] In *Dangerous Thoughts* of 1939, Hogben reinforced the acid attack on British, German, and American eugenics and physical anthropology as directed against such scapegoats as the working classes, Jews, and colored people, as well as certain kinds of immigrants in the interest of the upper classes, "the Aryans," and "the Nordics," respectively. Finally, also Huxley used the new understanding of heredity and of the nature-culture relation to argue against existing notions of race in popular talks and articles such as "The Concept of Race in the Light of Modern Genetics." He attributed "racial," national, as well as class differences in IQ, aptitude, and character, and the claimed sexual differences mostly to natural, social, economic, and educational environments. Such stereotypes had a history: "An amusing example is the exclamation of the third-century Greek gossip-writer Athenaeus, 'Who ever heard of a woman cook?'"[30] This obviously did not mean that there were no genetically codetermined differences between humans, but they were unlikely to correlate with social groupings; as such, they had to be valued favorably.

Racial anthropology was most forcefully debunked as a false understanding of inequality or variability in *We Europeans: A Survey of "Racial" Problems* (Huxley and Haddon 1935). In its production that began in 1932, Huxley collaborated not only with the Cambridge anthropologist Alfred Haddon but also with Singer. Although Huxley had made racist statements about African Americans in the 1920s, by the 1930s, he was a leading antiracist. In *We Europeans*, which became one of the most influential antiracist books of the time, Huxley and Haddon showed that anthropology lagged behind other areas of biology that had become statistical. They explained that instead of establishing means for certain characters to produce ideal racial types, the new biology concentrated on actual spectra of variation within populations and species. With this method, it was impossible to describe discrete human groups, and the term *race* should be given up. This was of course not only due to the biological reconceptualization of variation, but also because "every time you use the word 'race' you are playing into Hitler's hands."[31]

In one of his digests of the arguments made in *We Europeans* for the *Yale Review*, Huxley was precise about the current shortcomings of racial anthropology. He explained that the classification of human groups on the basis of cultural elements did not reflect genetics or common ancestry: "Terms like 'Celtic,' 'Jewish,' 'Indian,' 'Arabic,' or 'Irish' may serve to denote a people or group of peoples bound together by tradition or history or language or religion or geographical contiguity, or united by cultural affinity or political usage (or misusage), even though the members of such a people are diverse in origin. As designations of ancestral types they erroneously apply a linguistic or a nationalistic terminology to a concept which should be defined only in terms of genetic transmission, or at least in terms of resemblance in physical characters."[32] Huxley cautioned that ethnologists might define their studies by reference to previously determined hypothetical types, which could stimulate generalization from inadequate data. If the collection of data such as on head size was to have any value for the biologist, it had to be accompanied by precise information on the nature of the group that had been measured: whether, for instance, it was a true random sample or whether there had been some selection, and if so, what the nature of the selection was. He warned that "not a few of the anomalies which are encountered by those who have endeavored to trace the 'racial affinities' of the varieties of man have arisen from an a-priori approach to the problem. We have attributed characters to a preconceived type, rather than determined the actual type by its characters."[33]

Furthermore, in *We Europeans*, Huxley and Haddon emphasized that humans were the exception in the animal kingdom: "In other animals, the term *sub-species* has been substituted for 'race.' In man, migration and crossing have produced such a fluid state of affairs that no such clear-cut term, as applied to existing conditions, is permissible" (Huxley and Haddon 1935, 107–108). They concluded that "the essential reality of the existing situation, however, is not the hypothetical sub-species or races, but the *mixed ethnic groups*, which can never be genetically purified into their original components, or purged of the variability which they owe to past crossing. Most anthropological writings of the past, and many of the present fail to take account of this fundamental fact" (ibid., 143). Human biological kinship did not amount to a tree structure, because "in man the convergence of evolutionary lines is quite as frequent as their divergence, and multiple ancestry is at least as important as common ancestry in considering the nature and origin of any group" (ibid., 111). "In man, the branches constantly meet and unite and produce new types of shoots," therefore, "the conventional ancestral tree may have some advantages for representing the descent of animal types; it is wholly unsuitable and misleading for man" (ibid., 266). In fact, *We*

Europeans was essentially a history of human migration and intermixture coextensive with the history of the human species. When a fraction of these processes was mapped onto Europe, the result was not a neat tree but a confusing net that nonetheless came short of representing the true complexity. This was about a redefinition of evolutionarily determined identities and kinships, and Huxley like Conklin rejected the icon of the family tree. On the basis of a true understanding of the history within, the phylogenetic diagram of humankind had to be conceptualized as a net, or alternatively, one might think of endlessly merging and diverging streams.[34]

In the words of one reviewer, *We Europeans* was "the first important attempt to bring together genetics and ethnology in rational relation. This [was] a real contribution to knowledge. It [made] science the handmaiden of tolerance."[35] The work received the John Anisfield Award given each year under the sponsorship of the *Saturday Review* to a sound and significant book published in the previous twelve months on the subject of racial relations in the contemporary world. It was generally received as "an opportune prophylactic against the spreading virus of racialism."[36] Indeed, the critique expressed in *We Europeans*, coauthored by the renowned Haddon who was of the older generation, rocked the heart of the British anthropological community.[37] Huxley would also continue to collaborate with his friend Singer in publicly denouncing racist statements and he supported rescue emigrations from Germany. These activities took up most of Singer's time, and until bombing began in the war, when Huxley had to stay at the zoo, he would visit Singer at his quiet cottage in Kilmarth (Cornwall).

Hogben, Haldane, and Huxley thus argued against a thinking of diversity along the lines of the old dichotomies and stereotypes. In the process, the diversity found among individual human beings as the result of genetics and environment could acquire quasi-religious quality. For Huxley, the notion of genetic variation as the result of the blind (and just) hand of Mendelian distribution and selection, combined with the ideal of cultural diversity as the result of individual creativity under conditions perfect for personal fulfillment, became a creed: "I believe in diversity. . . . We can try to forbid certain attitudes of mind. We could theoretically breed out much of human variety. But this would be a sacrifice. Diversity is not only the salt of life but the basis of collective achievement. And the complement of diversity is tolerance and understanding. . . . The only faith that is both concrete and comprehensive is in life, its abundance and its progress. My final belief is in life."[38] The belief in life, in evolution and its mechanisms as embodied in the human organism, would inspire humans to value diversity rightly understood and to build a more peaceful world (Sommer 2014).

Although the triple-H deconstructed the biological basis of elitism and racism, they wanted to retain the power of biological arguments. In other words, they had to show that there were wrong kinds and right kinds of science. They did so by reference not only to the truth about nature but also to the social effects on science. In the end, they made an entangled argument of science in a particular social system, science as a model for a particular social system, and the science or biology of a particular social system. In *We Europeans*, Huxley and Haddon made fun of the science underlying the National Socialist myth of a pure and superior Aryan race: "Let us make a composite picture of a typical Teuton from the most prominent of the exponents of this view. Let him be as blond as Hitler, as dolichocephalic as Rosenberg, as tall as Goebbels, as slender as Goering, and as manly as Streicher. How much would he resemble the German ideal?" (Huxley and Haddon 1935, 25–26) At the same time, the extrapolation from sound science to politics was defended. A similar strategy was used by Huxley against Johann von Leers's *History on a Racial Basis* (1936). For Leers "race" was everything; it was Ariadne's thread through history. However, there was only "one great and good race," to which Leers referred as *Aryan*, *Nordic*, or *German*, and to which—Huxley pointed out—Leers obviously counted himself. Drawing on Arthur de Gobineau and Madison Grant, Leers attributed all progress to this "race," while everything bad had come from the "Jewish race." "The purity of the Nordic race" thus appeared as the most valuable treasure, and Hitler as the "race regenerator."

Huxley asked: "How is it possible to get across such nonsense even to the most elementary or the most distraught of minds? For there can be no doubt that [Leers] and his colleagues do appeal to a wide audience with pseudo-scientific rubbish of this order. . . . In this country such a work would not be taken seriously, but in Germany it has attained a wide circulation. It is designed especially for use in schools."[39] To begin with, Leers was an active member of the Nazi Party and director of the Division of Foreign Policy and Foreign Relations at the German Institute for Politics. He had been trained as a lawyer and not as a physical anthropologist or ethnologist. Huxley could therefore relegate his book to German pseudoscience that told history falsely, and that undermined the sense of history of those who were fed with such rubbish. However, in harmony with his sense of the interrelation between science and society, a general readership could only buy into such rubbish if there was some kind of mass delusion going on. In stark contrast, true biology, as it was still possible in a less contaminated environment, retained its beneficial social potential: "The violent racialism to be found in Europe to-day is a symptom of Europe's exaggerated nationalism . . . Meanwhile, however, science and the scientific spirit can do something by pointing out the biological realities of the ethnic situation, and by refusing to lend her sanction to absurdities and the horrors perpetrated in her name. Racialism is a myth, and a dangerous myth at that" (Huxley and Haddon 1935, 287).

The argument for the interrelations of science and society became even more explicit in Huxley's Macmillan war pamphlet *Argument of Blood* (1941b), in which he discussed the disastrous state of German science under the Nazi regime. With the removal and killing of Jewish university staff, the infiltration and administrative control of universities by National Socialists, and the turning of science into a machinery of war (physics, chemistry, engineering) and racist ideology (anthropology), German science had ceased to be science and degenerated to pseudoscience. Huxley made clear his belief that true science was beneficial to the population, and that it not only depended on a democratic environment, but was itself democratically organized: "The freedom of learning and of research is an essential part of democratic freedom in general. Let us in Britain realise this and be prepared to defend it against any encroachment" (ibid., 35).

If the democratic potential of British science in contrast to German science could be carved out by reference to the corruption through National Socialism, the refutation of a particular kind of British science could go along with attributions of "young versus old" or "conservative versus progressive" in a society under reform. Haldane especially emphasized that those who held the key to the new biology were also the authorities with regard to the right social system: "The average degree of resemblance between father and son is too small to justify the waste of human potentialities which an hereditary aristocratic system entails. If human beings could be propagated by cutting, like apple trees, aristocracy would be biologically sound. England would presumably be governed by cuttings of Cromwell and Chatham; America . . . by cuttings of Washington and Lincoln." Because heredity did not work like cutting, but according to the Mendelian laws, Haldane like Conklin before claimed that equality of opportunity, rather than aristocracy, conformed to the natural processes. If the hereditary process had functioned like cutting, aristocracy would be justified. Haldane thus added: "But until the art of tissue culture has developed very considerably, such possibilities need not even be thought of" (1932a, 18).

Haldane's exuberant humor often makes it hard to grasp his stance, but one thing seems clear: there was nothing Haldane shied away from thinking. To the contrary, the potentialities of tissue culture research played an important part in his vision of progressive human evolution. It came with the hope of a future eugenics that would be truly biological and not mainly social. The question of diversity was also raised in relation to the theme of science-society interdependence when it came to the technologies of tissue culture. The scientists at the laboratory for tissue culture at the University of Cambridge in the 1920s and 1930s presented their research as a cure for all ills and alluded to the possibility of immortality. As a consequence, tissue culture research was discussed in the press not only in connection with babies in bottles, but also with an infinite growth of cell cultures (Wilson 2005). The new research therefore seemed to hold promises as well as threats. Like the modern environments that gave rise to the monotonous mass or mob, the potentially unlimited reproduction of the same cell raised the specter of homogenization.

This tension in the tissue-culture theme is at work in Haldane's prophetic essay *Daedalus, or Science and the Future* (1924), but also in Julian Huxley's and Aldous Huxley's science fiction works "The Tissue-Culture King" (1927) and *Brave New World* (1932). Both Huxleys painted a dark picture in which tissue-culture research led to the artificial breeding of human castes. In the anonymous poem "Still More Peers; or Tissue Culture Ltd." written by a University College London zoologist and in Julian's possession, Waddington's tissue culture experiments ended in his being able to produce brains that are bought en masse, especially by the upper classes and aristocracy.[40] Again, in the wrong social environment, science might be abused by those in power, and the substrate of evolutionary progress—diversity—might be impoverished. On the contrary, Haldane looked into the future and saw in laboratory reproduction a liberation of humankind and a potential increase rather than decrease in individual variation by means of a non-class-specific eugenics. In fact, Julian was also not altogether dismissive: after all, in "The Tissue-Culture King," the negative homogenizing effect of artificial breeding is due to the fact that the science in his fiction is part of a religiously based and despotic kingdom, rather than of a science-oriented and democratic society. In Huxley's view, science needed democracy just as the ideal society was only realizable on the principles and with knowledge of the sciences; and it was the mark of his ideal of democratic science as well as society that they estimated quality higher than quantity, and cultural plurality grounded on common general beliefs higher than uniformity.[41]

As we have seen, Hogben, too, regarded science and society as strongly interdependent. This meant that science needed a revolutionized society to work for everyone's good, and that there was no clear-cut distinction between basic and applied science. Hogben saw a role for genetics in the creation of new types of plants and animals in such a future (for example, by combining genes for resistance and high yield of fruit through selective breeding or crossing). And ultimately, the evolution of the human species itself would be brought under the control of "biotechnics" or "biotechnology" (1938, 1005). But this would have to go hand in hand with a new social contract: "Modern science offers us a NEW SOCIAL CONTRACT. The social contract of scientific humanism is the recognition that the sufficient basis for rational co-operation between citizens is scientific investigation of the common needs of mankind, a scientific inventory of resources available for satisfying them, and a realistic survey of how modern social institutions contribute to or militate against the use of such resources for the satisfaction of fundamental human needs. The new social contract demands a new orientation of educational values and new qualifications for civic responsibility" (ibid., 1089). Hogben was hopeful in view of the international organization of "scientific workers" in democratic countries and saw a general mobilization against not only fascism but also capitalism. Constructive democratic statesmanship might yet bring about enough state support for science and the insurance that its technical resources were commonly exploited (ibid., 1081, 1083).

Thus, Hogben's vision of biotechnology and Haldane's dream of tissue culture make clear what Diane Paul has succinctly stated: "Haldane and Hogben criticized many features of the contemporary eugenics movement but their goal, and that of their colleagues on the scientific Left, was the reformulation, not the defeat, of eugenic ideology" (1984, 572). Yet, because of the triple-H's awareness of the complex relations between science and society, the conditions under which a reformulated eugenic project would be acceptable were severe. At the outbreak of World War II, the triple-H were among the signatories of a statement published in *Nature* that expressed the hope that eugenic concerns would guide the reproductive choices of individuals in a future in which social conditions were improved and just, in which community concerns took center stage, and a federation of the world had come into reach, thus rendering it possible to make beneficial use of what would by then be a much better knowledge of heredity (Crew et al. 1939).[42]

It was toward such a federation of the world that Hogben increasingly worked. In "The Creed of a Scientific Humanist" of 1939, democracy in its present form seemed doomed, communism perverted, and even a certain brand of socialism insufficient (Hogben 1939, 13–24). Salvation lay in the scientific humanist program that opened up the possibility of a world government by federating nations with simultaneous increase in local self-organization with the help of expert knowledge (ibid., 21–24). In its realization, scientific and societal practices and goals would become synchronized. Societal advance would be modeled on scientific practice as the prototype of all common human action. Well into the war, Hogben made another contribution to this now pronouncedly global project of scientific humanism with his *Interglossa: A Draft of an Auxiliary for a Democratic World Order* (1943): "The writer believes that the alternative to barbarism is repudiation of national sovereignties in greater units of democratic co-operation, and that day-to-day co-operation of ordinary human beings on a planetary scale will not be possible unless educational authorities of different nations agree to adopt *one and the same* second language" (Hogben 1943, 11).

Huxley worked toward the same aim during the end of his zoo years. He propagated his most written-out plan for the future of democracy in a series of radio talks he gave when touring the United States in early 1940 in the service of American war intervention and collaborative postwar reconstruction. Harking back to Conklin's and his own thoughts of the 1920s,[43] Huxley elaborated on the basis of an analysis of history and the present situation his belief that democracy within a nation and ultimately a democracy of nations had to find the balances so natural to life itself between individual and community or state, between rights and responsibilities, between local organization and central planning, between layperson and expert, and between freedom and security. Because the expansion of social services, scientific planning, development policy, and international collaboration could also take place in totalitarian and fascist states, these natural balances were crucial for guaranteeing civil liberties. Contrary to political fanaticism and scientific dogmatism with their reinforcement of mental unity and biological homogeneity, the natural processes of balancing demands that were only seemingly antagonistic ensured the persistence of the diversity that was so essential for progress in social as well as natural evolution (Huxley 1941c).[44]

In fact, just as Huxley had made out progressive and limiting trends in the phase of organic evolution, he observed such trends in recent human history. He contrasted revolutions, mostly toward totalitarianism (Germany, Italy, Turkey, Spain, Portugal, Japan, China, Russia, and in "a pale sort" in Vichy France), with transformations that were evolutionary, as in Scandinavia, the British Dominions, the United Kingdom, and the United States under the New Deal, where measures of social security had been introduced. In such evolutionary change and in international convergence through the League of Nations, Huxley recognized certain progressive trends on a global scale: a trend away from laissez-faire toward planning and governmental control; a trend to take noneconomic motives and aims more seriously; an increasing concern with the material and human resources of developing regions, and a growing realization of the necessity for some strong international organization. Huxley condemned the developments in Japan and Germany, but he appreciated in general the effort to embark on the mission of a new world order. He hoped that the United States and other democratic nations would strive toward a new world order of another kind. Because these nations stood for a balance of the individualistic and communistic interests, and because they esteemed diversity, such an attempt would ultimately prove progressive rather than a shortcut to an evolutionary dead end (Huxley 1941c).[45]

Huxley actually believed that the environment of war favored thinking in terms of "all of us." In 1941, he observed a rapid permeation of British society with new ideas that were destined to become in a real sense collective, part of a group-thought and a group-will. The individual might resist the new modes of thinking, but then would try to grasp their meaning and their implications, and finally find intellectual satisfaction in their assimilation, a process that would initiate a sense of community of purpose. In this scheme, new ideas arose in a progressive minority—even if of multiple backgrounds—to whom Huxley obviously counted himself, and they were eventually backed by public opinion, albeit stirring resistance in conservative and reactionary circles. In the end, the generation of group-thought and -will would have to transcend the nation, uniting entire humanity on the basis of a shared evolutionary outlook.[46]

This was the year that Huxley's US radio talks were published as *Democracy Marches* (1941c) and *The Uniqueness of Man* (1941a) came out. And, indeed, the public reception of his scientific humanism was generally enthusiastic, both in Britain and America, where the latter appeared under the misleading title *Man Stands Alone*. It was a welcome message of hope from the evolutionary past: "The story of delicate adaptations by which man escaped the dull fate of a bird and the tiny brain of an insect, then developed eyes by tree-life, and next hands by standing erect on the ground, talked because he was gregarious, and perfected the only kind of body in which this kind of consciousness could exist, is a biological melodrama, beautifully told."[47] As we have seen for Osborn, Huxley's reconstruction of "our" evolutionary history was read as the melodrama of human becoming, and another reviewer, too, understood the role of narrative, when he called it "a tragi-comedy enlightened by hope."[48]

According to this story, humans had just barely escaped dull evolutionary fates by coincidentally acquiring the ability to think and to act on their thoughts in a coordinated fashion. In other words, humans were able to consciously define what the future would be, if only there were a concerted effort: "The kind of society to be aimed at is one which has self conscious social purposes, one in which every individual understands those purposes, his own potentialities and his own functions in relation to collective social life. It is to this end that all the agencies for moulding public opinion should be turned. Could there be a more splendid or inspiring purpose?"[49] This conscious social purpose was provided by Huxley's creed which—though it asked for similar educational, nutritional, economic, and other advantages for all people regardless of class, color, nationality, religion, political belief, or any other consideration—"is not socialism, nor communism, nor fascism, nor democracy as practiced; it is 'scientific humanism.'"[50]

In the end, *Uniqueness* was understood as an appeal to people to finally make good use of their evolutionary heritage: "Huxley has written an intensely fascinating and stimulating analysis of what he calls the uniqueness of man, those social values which are man's peculiar heritage and which enable him alone of all living things to attain intelligent social organization. If you believe that man can control his destiny if he will but make the effort, intelligently and sincerely, here is a guidepost

FIGURE 18. The scientist as citizen, Julian Sorell Huxley—Papers, 1899–1980, MS 50, Series XI: Clippings, Box 137: 1940–1959, Folder 2: Reviews, 1940–1949: Illustration by Christopher Stull Patri, "Essays of a Great British Scientist," *San Francisco Chronicle*, 25 May 1941, courtesy Woodson Research Center, Fondren Library, Rice University

that is brilliant and hopeful in a world in chaos. Beg, borrow or steal this book—but read it."[51] This general tenor was also well expressed in the *San Francisco Chronicle*, in which the reviewer enthused that "one finishes his book with the feeling that, though there be war, hunger, injustice, and all other evils in the world, we are still advancing and, if we use intelligence, can scarcely fail to advance, since nature is working with us, not against us, and we are working with nature." And also this paper considered it "a book which should be required reading for every intelligent adult in the world today—particularly today." What made Huxley's vision singularly important was that "he is able to support his arguments with observations drawn from the whole of modern science."[52] Correspondingly, in figure 18, we see a scientist in a lab, but a scientist dressed in the garb of the ordinary citizen.

However, some commentators felt less at ease with Huxley's atheism, or with his planning frenzy and what they took to be the demand to surrender human reproduction and social planning to the government and the scientist.[53] Such concerns were voiced especially in connection with Huxley's program as presented in *Democracy Marches* (1941c). Here, he seemed to stray even further from the biological terrain. The reviewer for the *Spectator*, for example, sensed Huxley's elitism when he wrote that one "might wonder if the picture of the working of our social services was not a little too rosy. It is a picture painted from the prosperous south. The depressed areas enter into it only as a place from which a worker, retrained in a new skill, escapes with ease."[54] In another review article, this picture of the planner from afar seems actually visualized. We see the giant scientist, far removed from the humble life in the valley. But his plans for the new world order overshadow the landscape and point away from the life we know. The iconography is threatening, and the gesture more that of the fascist than of the democratic fellow-citizen (see figure 19).[55]

Clearly, Huxley envisioned the evolution of democracy to be led by an elite of scientists like himself, and on a global scale by Britain and America.[56] At the same time, he knew that the goals of his scientific humanism depended on the active participation of many, and he transcended his top-down notion of sociocultural evolution in a circular model: "One of the ways in which a unified self-consciousness can develop is through a really living culture. The theatre, the cinema, painting, writing—these can be at once the antennae of society exploring new worlds of experience, and also a mirror reflecting the soul of a nation and making it visible."[57] The improvement of education, leisure, recreation, and culture was therefore paramount, as well as enhanced civic participation in politics and government: "Man won't come into his heritage by leaving everything to the experts: it's up to all of us" (Huxley et al. 1944, 96).

As the commentators on *The Uniqueness of Man* understood, and as Huxley explained in the context of the radio talks on "reshaping man's heritage," by *heritage* he meant the inner and outer human environment as they had been formed by biological and cultural evolution.[58] It was a legacy and a destiny at the same time: As evolution become conscious of itself, humans had inherited the world; they held the trusteeship over the future course of its development. It is this notion of heritage that will gain center stage in the next chapter, and in harmony with the emphasis on diversity we have noticed in this one, it was especially the diversity of that heritage that was understood to constitute the potential for fur-

FIGURE 19. The scientist as planner, Julian Sorell Huxley—Papers, 1899–1980, MS 50, Series XI: Clippings, Box 137: 1940–1959, Folder 2: Reviews, 1940–1949: Drawing by Edward Shenton, Ellsworth Huntington, "Huxley and Evolution," *Saturday Review*, 12 Apr. 1941, 5, courtesy Woodson Research Center, Fondren Library, Rice University

ther advance. Through his engagement with the Colonial Service and especially UNESCO, Huxley worked toward the management of this heritage on a global scale. The biological and cultural richness of the world had to be preserved and developed through analysis and the circulation and application of knowledge; it had to become the basis for panhuman integration and worldwide progress.

CHAPTER NINE

Evolutionary Humanism

Planned Ecology and World Heritage Management through the Colonial Office, UNESCO, IUCN, and WWF

In mid-November 1970, at the Second International Congress of the World Wildlife Fund in London, an eighty-three-year-old Sir Julian Huxley, fellow of the Royal Society, was awarded a gold medal in the category of conservation research and promotion of conservation.[1] In his acceptance speech, Huxley dated this engagement in conservation back to the early decades of the century, and he expressed his pride at having been instrumental in creating national parks and reserves "in this overcrowded island" and "in the miraculously preserved Pliocene lifescape—as Theodore Roosevelt called it—of East Africa." His "proudest achievement was at Unesco, where [he] succeeded, against considerable opposition, in setting up I.U.C.N. [the International Union for Conservation of Nature], which has . . . stimulated the formation of other bodies concerned with similar aims—like the World Wildlife Fund. And above all [it has] . . . succeeded in impressing on this bewildered [originally "overpopulated"] world the urgent need for the conservation of Wild Life and unspoilt scenery before they are wiped out by so-called progress."[2]

At the congress there was a message drafted, to be sent to all government leaders of the world, calling upon them to stabilize the size of their populations and "to establish effective trusteeship for man's heritage in the biosphere and for a long term improvement in the quality of human living, its arts and its amenities."[3] Huxley chaired the session in which the message was drafted, and it was not only the rhetoric but also the aims that carried his signature.[4] For Huxley, natural heritage had to be preserved not only owing to the fact that it was entrusted to humankind as the spearheads of evolution but also because it had an important part to play in further advance. Reminiscent of Osborn's view, for Huxley the "Pliocene life-scape" of Africa and other patches of remaining "wilderness" could function as reminders of ways of life more in harmony with human nature. In experiencing them, overcivilized humans may regenerate, and as we will see, those humans who still lived in such regions were understood as holding the greatest potential for a progressive future. It is also significant that the World Wildlife Fund medal was conferred for conservation research as well. Indeed, Huxley had enthusiastically embraced the new science of ecology from its beginning and he refined his scientific humanism along its lines.

The British pioneer of modern ecology, Charles Elton, had been a student of Huxley's at Oxford and his assistant on the Spitsbergen expedition of 1921. Huxley wrote an introduction to Elton's by now classic *Animal Ecology* ([1927] 2011), emphasizing the importance of applied ecology as well as of the academic field.[5] Thus, in *The Science of Life*, ecology was introduced to the general reader and student as "a fresh way of regarding life" (Wells, Huxley, and Wells 1929/1930/1931/1934, 961). In this groundbreaking textbook, ecology was brought to bear on the scientific humanist goal of consciously steered progress along the lines of evolutionary principles. The authors observed that in the past, human interference with the ecological web had mostly happened without sufficient in- and foresight, as when new organisms like pests were brought to colonized countries, soil was exhausted through monoculture, or finite resources were opened up. Such human interventions upset the natural balances; species had been exterminated and the environment polluted. In contrast, Huxley and the Wellses again called for applied ecology. In the future, a concerted effort by the sciences of life would be needed to develop ecological webs in a beneficial direction, by controlling pests and diseases, by genetically improving organisms, and by creating the desired ecological interdependencies. This was to be a human-centered ecology, aimed at adjusting environments to human interests. Indeed, as Peder Anker (2001) has shown, the British Ecological Society had urged for the inclusion of the human impact on environments and of human environments in ecological research almost from its foundation in 1913. But it was in the 1930s that concepts and approaches from human ecology really gained momentum through the work of some of the society's founders and members. Elton, Huxley, and H. G. Wells were among these, and so were others who play a part in this chapter, such as Nicholson and Edgar Worthington. As we will see, Nicholson—Huxley's ornithologist friend whom we have met in chapter 7—was a central figure in the British and international conservationist movement, and he promoted applied ecology in a humanist frame as late as the 1980s (Nicholson 1987).

Huxley's applied ecology was particularly close to what Hogben called *planned ecology*: "Man has it in his power to become an active and intelligent directive agent in the evolutionary process, using his knowledge of the diversity of living creatures to decide which are essential to his own welfare as objects of use or of aesthetic satisfaction, and using his knowledge of the properties of living matter to adjust the environment of the species he chooses as members of a rationally planned ecological system" (Hogben 1938, 971). Humans had long since begun to turn the world into their own ecological system, but the process now had to be subjected to conscious scientific planning. The species that were harmful to humans or competed with humans for resources would be killed, while those species that were beneficial to humankind and that humans used for food, shelter, ornament, or pleasure would be sustained or introduced. As shocking as this sounds, it does not seem to have contradicted Huxley's conservationist agenda, since destruction should affect mainly pests and parasites (through regulation of physical environments, segregation, specific poisons, and hyperparasitization) (ibid., 964–970, 971–1009).

In order to steer the global ecology, one needed knowledge of the workings of evolution and of the contemporary diversity of living organisms that was their result. If this knowledge were implemented in interdisciplinary efforts to engineer ecological systems worldwide and integrated into idea systems globally, humans would finally shoulder their responsibility. Scientists therefore had to survey the natural diversity and work toward its preservation, while making it accessible to everyone by means of efficient management and modern media technologies. Within a scientific humanist and human ecological framework, the same was true for cultural diversity. In the development and experimental implementation of this program, Huxley's experiences in Africa were central. In 1929, as a member of the Colonial Office Advisory Committee on Native Education, he was sent to East Africa to report on the role of biological science in education and on the value of nature conservation. British Africa offered important opportunities to consolidate ecology as an integrative approach to natural and human resource management; it was perceived as a living laboratory (Tilley 2011). In connection with the work of British scientists such as Huxley, Anker (2001) therefore speaks of an imperial ecology.

For sixteen weeks, Huxley traveled the four territories Uganda, Kenya, Tanganyika, and Zanzibar as well as the Anglo-Egyptian Sudan and the Belgian Congo. In retrospect, he thought of the event as "one of the most important turning-points in [his] life" (Huxley 1970, 179). For one thing, he felt it helped him to overcome some of his earlier racial prejudices. Disturbingly, he described how like the giraffe and the ostrich that looked merely silly in a zoo, while their observation in Africa brought to light their evolutionary rationale, "Africans" had looked odd to him in the South of the United States, while they shone in all their physical adaptations and cultural diversity in the variety of African landscapes. Huxley wrote that "on top of all this variety of nature and man there impinge Western civilization and Western industrialism. Will their impact level down the variety, insisting on large-scale production to suit the needs of Europe and Big Business, reducing the proud diversity of native tribes and races to a muddy mixture, their various cultures to a single inferior copy of our own? Or shall we be able to preserve the savour of difference, to fuse our culture and theirs into an autochthonous civilization, to use local difference as the basis for a natural diversity of development?" (Huxley 1931, 7).

Huxley thus feared for the human biological and cultural diversity and he feared for animal diversity, too. He made precise suggestions for the establishment of national parks. He envisioned a system of two different kinds of nature park, one where tourism would be encouraged for enjoyment and one in which nature should be left alone (with the exception of scientific access).[6] The two kinds—the national park and the national park sanctuary (the latter on the Belgian model)—might be adjacent to each other. Like the (legally uncertain) game reserves, the current forest reserves were insufficient for the purpose of protection because they were run by men whose responsibility it was to build up an efficient forestry business. Even though Huxley perceived the greatest threat to wildlife in the white man's hunting, he also recognized a conflict between nature and "tribes." He suggested that game reserves surround national parks as elastic buffers between the sanctuaries and the claims of sportsmen and cultivation, and that such reserves be established everywhere as breeding reservoirs for wildlife and especially game.

This recommendation accords with John M. MacKenzie's (1988, chs. 8–10) observation that legislation on game protection in British Africa followed on the heels of two international agreements, the Convention for the Preservation of Wild Animals, Birds and Fish in Africa (1900) and the Agreement for the Protection of the Fauna and Flora of Africa (1933). The interim period, he argues, marked the shift in emphasis from preservation to conservation that is evidenced in Huxley's suggestions. The reserve, like Huxley's sanctuary, was a locus of preservation, where game was protected and managed away from humans, with the possible exception of controlled white hunting. In fact, the establishment of such reserves had been strongly motivated by the attempt to maintain game for elite hunting and to curtail subsistence hunting. On the other hand, the approach of conservation with its scheme of the national park brought humans back in, but as observers, photographers, and, in Huxley's vision, certainly as scientists. National parks should counteract the ills of industrialization and urbanization, serve the maintenance of a healthy populace, and even civilize. This was about middle-class national and international tourists who had the money and leisure to travel by ship and plane, train and car, and who enjoyed the growing number of hotels and the general infrastructure in these rural areas.

From the beginning of the colonial period, regulations on land, game, fishing, and forestry curtailed the lifestyles of native African communities for whom the increased efforts for conservation contributed to the inability to use wildlife as they had done in the pre-colonial era. Although Huxley regarded the experience of wildlife as so beneficial for humans that he would not have appreciated its limitation to certain groups, this belief also affected his ideas concerning native land and hunting. In this regard, his recommendations went against the White Paper of 1923 that granted general priority to the native interest. In the case of the Kenyan Maasai reserve that was also a game reserve, Huxley proposed its partition into a Maasailand native reserve and a Maasailand national park, albeit with the natives' consent. Huxley saw a profit for humans as well as animals in this plan, as both reserve and park would be Maasai territory and they would receive the revenues from tourism. Tourism in turn would open up the beneficial effect of nature and wildlife to the entirety of humanity: "Humanity does not live by bread alone; in East African wilds a stream of men and women down the generations may find quickening, refreshment, inspiration" (Huxley 1931, 255). The Maasai had already ceded territories to the British in two "agreements" and been put in reserves to the advantage of colonial settlers. Now, because Huxley feared for the animals' future, he suggested that the Maasai were to yield to the benefit of animals—like people in the establishment of national parks globally and until today.[7]

The African megafauna gained particular value in Huxley's scheme because they represented an evolutionary world heritage of a special kind. The large mammals had died out and had been killed in the course of evolution in many parts of the world. Whereas in Africa, there were not only the layers of human prehistory Louis and Mary Leakey were bringing to light; there had also survived Pleistocene animal groups and Pliocene landscapes. While Huxley considered the mammalian living fossils as incapable of further progressive evolution, they might still catalyze transformative experiences in humans. However, it was mostly African "tribal" diversity, though also a heritage of the past, in which Huxley saw an asset for the future. The "tribes" were not simply to be preserved, but to be given the opportunity to express their potential for development: "It can never be our aim . . . merely to preserve a human zoo, an Anthropological Garden" (ibid., 137), because "there is no more deplorable spectacle than a community of human beings consciously 'being primitive' for the delectation of tourists' sentimental minds" (ibid., 245). At the same time, certain peoples such as "the pygmies" might still be preserved in a cultural stage reminiscent of prehistoric human ways of life. Years later, in a lecture given to a juvenile audience at the Royal Institution, Huxley posed "the question whether rare types of human beings should be preserved as well as rare animals."[8]

Huxley was thus ambivalent toward the then current notion that "blacks" belong to the land, that their "tribal" life is somehow natural and primitive in the sense of unspoiled. Like the land itself, the logic went, these "tribes" therefore had to be preserved through "white" interference (Dubow 1995, 170). Huxley did think he had found in East Africa a part of nature that was still widely unspoiled. He partook in the widespread antimodern sentiment associated with a fear of "degeneration," when he recognized in regions of Africa the salvation for a modern civilization marked by mass culture, cheap production, the horrors of war, of slums, of land loss, and so forth: "I see Africa chiefly and most thrillingly as the one part of the world of continental magnitude in which (without the destruction or degeneration of an old civilization) there could arise a new civilization, consciously planned or at least consciously guided from its beginnings" (Huxley 1931, 452). At the same time, this kind of reasoning ran counter to the belief also existent at the time that Africans are unsuitable for civilization (for example through Western education). Huxley saw in them a part of humankind who had not yet exploited to the full their potential for progress and might do so by profiting from Western experiences.

Africa was Huxley's chosen land for a laboratory of planned evolutionary progress, and he considered the current political system unsuitable for such development.[9] Despite differences between colony, protectorates, and mandated territory, the political structures Huxley encountered were similar. There was a governor or resident as the real executive who was directly under the Colonial Office, and the legislative council consisted of officials of departments of administration and nonofficials elected by nonnatives. More power needed to be vested in the local populations. Huxley mused that, as the first step toward self-government, one might reform councils so that the elected members would have a majority; but could *self-government* really apply to a government made up of a small immigrant minority? One also had to tackle the question of a vote for the Indians who had been brought in to build railroads. Finally, should the four territories be turned into a federation under one governor?

Huxley was uncertain about these points, and he emphasized that in order to frame a good native policy, much more knowledge was needed. He feared that the great majority of the English at home and many of those on the spot were ignorant on the subject. He suspected that most Europeans dismissed the Africans as mere savages, and classed them all together as "niggers." Such prejudices needed to be countered by information about the elaborate social organizations of all the peoples of Africa as well as about their biological diversity. Other stereotypes had to be removed: Europeans needed to be taught that Africans could make as good use of their lands as the whites, as recent studies had shown, and that they would profit from a more thorough education system.[10] Europeans would do well to remember that until the present century, the natives of Africa had met with little but greed, hostility, and exploitation on the part of the European and Asian invaders, and to this day, gross inequality prevailed (ibid., 25).

A few years before the *We Europeans* project, Huxley rejected racial typology and racism. And in the context of the growing controversy regarding the notion of white superiority as a justification of colonialism, Huxley, Hogben, and Haldane criticized studies that purportedly showed the innate backwardness of "the African race" as part of their general efforts detailed in chapter 8 (Tilley 2011, ch. 5). Huxley also objected to General Smuts's idea that British unemployment could be dealt with by opening up the unused lands of Africa. Smuts hoped for as perfect an institutional and territorial segregation of "black" and "white" as possible, with the "blacks" working for the "whites" who would drive development. Smuts's vision of a white backbone from Kenya to South Africa, with Africans segregated and developing along their own lines (Smuts 1930, 35–103), was opposed during the interwar years by those who wanted indirect rule for eastern Africa. Huxley agreed with Smuts that there could be no equality between "black" and "white" populations. However, for him, it was white settlement that should "be not equal but subordinate to native development and native production" (Huxley 1931, 449). For Huxley the solution lay in giving as much power as possible to local authorities, at the same time providing British guidance, including scientific expertise. Indirect rule as was the case in Nigeria and Tanganyika seemed the way to go. Instead of advocating that a form of British rule be imposed, Huxley suggested the principle of ruling through the native leaders with the aim of maintaining and supporting native rule within the defined limits. Development should be achieved by slowly molding existing institutions with the help of local traditions, pride, and initiative, and toward freedom and variety.[11] Obviously, this scheme harmonized with Huxley's general ideal of planning by combining local and central or global interest and control as well as lay and expert knowledge.[12]

A key aspect of this project was an education system that accorded with the evolutionary and ecological perspective. Biology and geography in the form of the study of life and its environment had to be the core of native academic education, and all other disciplines could be revived through evolutionary biology. Huxley wrote that "if we are to educate African children into being citizens of the world, we can only do it properly through making them better Africans" (ibid., 377), by educating them in all aspects of applied and pure biology in Africa. This kind of Western knowledge on parts of Africa had in itself a history, not least involving missionaries who named places and people as well as ordered and collected stones, plants, and animals; they engaged in cartography and natural history more broadly, exploiting, appropriating, and supplanting local knowledges in this process of making African lands amenable for European domination (for example, Harries 2007, chs. 4 and 5). Huxley emphasized that in Africa biology was more important than chemistry and physics because Africa's problems were about hygiene and agriculture. However, within the larger frame of scientific humanism, education had to go further. Children had to "realize that in books there exists the chief repository of the world's thought, and that they should be able if they so desire to draw upon this heritage of ideas and beauty by reading" (Huxley 1931, 324). Huxley therefore stressed in his report to the Colonial Office that Africans had to have access to the cultural as well as natural heritage, including the East African, to gain knowledge and develop their personalities and societies.[13]

To ensure that Africa would be developed toward a balanced human-animal habitat, propaganda was needed globally as well as locally. Huxley published articles on his journey in such Anglophone venues as the *Times*, *Cornhill Magazine*, the *Nineteenth Century*, the *Saturday Review*, the *Contemporary Review*, the *Atlantic Monthly*, and *Harper's Magazine*. He also produced a popular travelogue, *Africa View* (1931), which Sarton praised for its scientific humanism. Huxley's main purpose was "to interest people at home in this extraordinary continent and our share in the responsibility for its development" (ibid., 16). On the official front, the argument for conservation proved difficult with the colonial secretary, who associated it with privileges for white hunters. Nonetheless, Huxley saw his vision beginning to materialize, and "within a few years National Parks were established in all three East African territories" (Huxley 1970, 196).

As an immediate result of his report to the Colonial Office, Huxley was made a member of the committee advising Lord (Malcolm) Hailey's monumental African survey on science, administration, economics, and human problems that was highly influential with later colonial administrators (1933–1938; Hailey 1938). The survey that arose out of an initiative of Oxford University academics was intended to suggest how to scientifically manage African resources and human populations. It had a strong emphasis on science and on colonial reform. Huxley lobbied for ecology as an integrative approach and, after Elton had to decline the offer, installed Worthington as director of the scientific survey. Worthington had already been part of the growing scientific diaspora in British Africa that strongly relied on vernacular science. The colonial administration had hired him to study the East African lakes because the fishing industry was threatened by overfishing, and Worthington used the knowledge of local fishermen and anglers to develop food chains beneficial to British settlers.

Worthington's contribution to the survey, *Science in Africa* (1938), was an amazing collection of knowledge from a wide range of fields that presented Africa and knowledge on Africa as an ecological complex. In this endeavor, Worthington drew heavily on a large number of scientific experts—in geology, meteorology, soil science, forestry, botany, agriculture, zoology, entomology, animal industry, and human health and anthropology among others—as well as on colonial officials at home and on site. The volume was a plea for the production of more knowledge about Africa (especially surveys) and for the integration of native methods, the concern for sustained environmental yield, as well as the consideration of the great variety of environments and the interconnections between various problems into research, planning, and measures. The survey indeed partook in the shift from an emphasis on European to African agriculture, the growth of ecology, and the transition from the belief in colonial self-sufficiency and concomitant exploitation to a humanist ideal of bottom-up development. The survey eventually resulted in the establishment of the Colonial Research Fund (1940), of interterritorial research institutes in British East Africa, and of the Scientific Council for Africa south of the Sahara (1950). More generally speaking, Hailey's demand for increased welfare of Britain for Africa was instrumental for the Colonial Welfare and Development Act of 1940. In sum, despite its goal of more effective colonial control, the survey had a liberalizing effect on official policy (Tilley 2003; 2011, ch. 2; also Cell 1989; Anker 2001, 208–218).

Huxley's view on Africa also impacted conservation in Britain through his leadership of the Wild Life Conservation Special Committee, which included Nicholson. It was installed as advisory committee to the National Parks Committee that had been set up by the Ministry of Town and Country Planning. The work of the Wild Life Committee culminated in the foundation of Nature Conservancy under an act of Parliament codrafted by Nicholson in 1949. Its report outlined the foundations on which present British nature conservation policy is still based (Ministry of Town and Country Planning 1947). It suggested that reserves should safeguard the typical landscapes of Britain, functioning as reservoirs of samples of a particular fauna and flora. Even before his directorship of the London Zoo, Huxley had begun to develop a concept of the reserve as a kind of revitalized zoo that was suited to the needs of the new biology:

> In place of the encyclopaedic zoological gardens of yore, placed often in the proximity of great cities, with their smoky air and abundant contaminations, governments are now setting up reserves and small stations, here, there, and everywhere, in which particular species and groups of species may be observed less for structure than behaviour. This sort of observation passes insensibly into experiment—experiments in behaviour, experiments upon habit and intelligence, experiments in physiology. . . . There are now even bacteriological zoos, so to speak, where collections of living cultures of this, that, and the other infection are available for the experimentalist. (Wells, Huxley, and Wells 1929/1930/1931/1934, 23)[14]

Like the zoo, reserves had to serve diverse needs; besides being wildlife haven and laboratory, they should function as classroom, recreational space, and economic asset. When commenting on wildlife conservation in Ghana, Huxley would support Worthington's "suggestion of areas of what one might call Whipsnade type, intermediate between Zoos and National Parks,"[15] comparable to the institution set up near Nairobi. Through the Wild Life Conservation Special Committee, Huxley recommended that the management of reserves should include scientists, carry out research and surveys, be a central bureau and advisory service, and encourage educational measures. He even suggested that, in addition, local administrations define scientific areas, educational reserves, and geological monuments. Huxley thus argued for a system of reserves as the site to implement some of the measures that had so often been blocked at the London zoos and were stopped short by his premature dismissal. However, the suggestions in the committee's report were only implemented in a very rudimentary way (Evans 1992, 71, 75–78).

In the early 1940s, Huxley also made the larger issue of the development of the colonies one of his topics as a public persona, and in 1944, he was again sent to Africa by the Colonial Office as a member of the Commission on Higher Education in the British Colonies. He was deeply impressed by West African art and culture. In the newspapers, he voiced his support for indirect rule as a way to self-government. He envisioned a federalization of Europe in a European Council and the status of mandate for all colonies and their representation by a Mandates Commission in a reformed League of Nations. He demanded among other things large sums of money for the benefit of local inhabitants. After government would have been handed over step by step to former colonies, they could themselves unite to larger structures such as tropical federations to take their place beside Europe, the USSR, and pan-America.[16] As we have seen, at that time Huxley also lobbied for the even more daring vision of a federation of the world in the context of postwar reconstruction. In combination with his engagement in numerous organizations such as the League of Nations International Committee on Intellectual Cooperation (1922), Political and Economic Planning (1931), and the Next Five Years Group (1934) (Ritschel 1997, chs. 4 and 6), his public standing led to an involvement with the organization of the United Nations that was initially planned as educational and cultural (UNECO).[17] Huxley lent his support to his close friend Needham in the campaign to include the *S* in UNESCO and to establish an independent science division, which gained final momentum when Americans dropped an H-bomb on Hiroshima.[18] In 1945, Huxley was asked by the head of the Education Office if he wanted the post of full-time secretary of the Preparatory Commission with the possibility of becoming director-general of the organization once it was formally set up[19]—a possibility that materialized at the UNESCO conference in 1946 (20 Nov.–10 Dec.).

Earlier that year, Huxley had submitted a pamphlet entitled "UNESCO: Its Purpose and Its Philosophy" to the Preparatory Commission. He proclaimed that UNESCO—through education, the natural sciences, the social sciences, the humanities, the arts, and mass media—should aim at a single world culture, at a synthesis of East and West in scientific humanism, and at psychosocial progress on the basis of the knowledge gained from the science of evolution. The evolutionary approach "shows us man as now the sole trustee of further evolutionary progress, and gives us important guidance as to the courses he should avoid and those he should pursue if he is to achieve that progress. An evolutionary approach provides the link between natural science and human history. . . . It not only shows us the origin and biological roots of our human values, but gives us some basis and external standards for them among the apparently neutral mass of natural phenomena" (Huxley 1946, 8).

Through UNESCO, Huxley hoped to realize his scientific humanist project of a democratic world culture. The integrated planning of global development would follow the principles and ethics of evolution, and humans everywhere would develop their self-understanding from their biological roots. In order to achieve further evolutionary progress, variation remained key. In a chapter entitled "The Principle of Equality and the Fact of Inequality," Huxley once again stressed that "it is therefore of the greatest importance to preserve human variety; all attempts at reducing it, whether by attempting to obtain greater 'purity' and therefore uniformity within a so-called race or a national group, or by attempting to exterminate any of the broad racial groups which give our species its major variety, are scientifically incorrect and opposed to long-run human progress" (ibid., 19). As sole trustees of evolutionary progress, humans had to protect not only their own diversity but also that of their living and inanimate environments. The International Committee on Intellectual Cooperation, in which Huxley had been engaged, had already demanded that the preservation of the natural as well as cultural heritage should be part of the League of Nations' responsibility.[20] In his program for UNESCO, Huxley now broadened the understanding of heritage along these lines.

In "UNESCO: Its Purpose and Its Philosophy," Huxley proposed the means for preserving and cultivating the cultural and natural diversity. Knowledge had to be pooled and made universally accessible to allow the establishment of a world culture on the principle of diversity in unity. This goal should be catalyzed through international institutions such as a World Bibliographical and Library Centre, an International Clearing House for Publications, an International Home and Community Planning Institute, an International Theatre Institute, as well as through the production of internationally conceived films and radio programs. UNESCO had to "foster all methods which, like microfilm, make for easy storage, multiple reproduction, and rapid transmission of knowledge" (ibid., 56). Citizens should be involved more creatively and given the potential to actively acquire this knowledge and use it in their everyday lives. To this purpose, libraries, museums, and other cultural institutions where knowledge was managed and acquired had to be set up or reformed. Echoing his plans for the London Zoo, zoological and botanical gardens could be understood as "living museums," where instead of dead specimens there was "the presentation of living creatures or of nature in action" (ibid., 57). Techniques of exhibition and education had to be revolutionized, and gardens had to be integrated with school systems. Analogous to his notion of the extended zoo, UNESCO should "explore all the new means of projecting museums and their collections outside their walls—notably by films and television, as well as by abundant and improved reproduction." Zoos, botanical gardens, nature reserves, and national parks, as well as museums, libraries, reading rooms, art galleries, and national and historical monuments were part of the plan "of preserving the world's scientific and cultural heritage and of making it available" (Ibid., 56).

Huxley's demand that the world should be developed along the lines of his "evolutionary humanism" (ibid., 25) was controversial within UNESCO; some considered it propaganda for natural science and an atheistic doctrine. Indeed, Huxley had declared that "the advance of natural science, logic, and psychology has brought us to a stage at which God is no longer a useful hypothesis" (Huxley 1941a, 277–290 ["Religion as an Objective Problem"], 281). He had observed that in the course of history, people and societies had become secularized and the energies that had been put into the cultivation of religious identities had been diverted to new forms of identification on the basis of race or nation. Huxley had expressed the hope that in the future those energies would be invested more profitably in universal evolutionary humanism.[21] Obviously, Huxley's appeal to UNESCO to build universal brotherhood on the basis of common phylogenetic identification was an insult to members of other belief systems. As a result, the pamphlet was circulated not as an official document but as a representation of Huxley's personal views. Nonetheless, the programs for the natural, social, philosophical, and humanistic branches were developed along the lines of Huxley's suggestions, even if references to his evolutionary humanism were omitted. UNESCO should encourage global cooperation and surveys in these fields as well as the centralization, organization, distribution, and application of existing knowledge and the creation of new knowledge with the aim of world progress and rapprochement between human values and cultures.[22] However, just as the inclusion of science into UNESCO had not seemed self-evident, the need to include the preservation of the natural diversity in its mission was not obvious. Delegates at the Mexico Conference of 1947 did not immediately see why UNESCO should try to protect rhinoceroses or rare flowers; the safeguarding of unspoiled scenery seemed outside its purview. Nonetheless, Huxley and other nature lovers "persuaded the Conference that the enjoyment of nature was part of culture, and that the preservation of rare and interesting animals and plants was a scientific duty" (Huxley 1973, 51).

Huxley did not have much time. Because his nomination as director-general had been controversial, his tenure was limited to two years, so that during the General Conference of 1948 in Beirut, a successor was elected.[23] In his farewell speech to the conference, he refrained from referring to his scientific humanism and humanism in general (this was in fact left to the new officeholder, Jaime Torres Bodet). However, he demanded that the member states examine the achievements in their own countries against the aims of UNESCO: Were they scientifically analyzing political, economic, and social structures for planned development? Did they grant freedom to the press and make use of the media for education? Was education promoted in all social classes? Had the creative arts and architecture been encouraged? Were they investing in the natural sciences, educating in them, and applying them to maximum profit? Were they preserving their traditions—ballads, costumes, and craftsmanship? And last but not least, were they keeping alive their natural heritage—their landscapes and their richness in life forms—as well as their cultural heritage and diversity—their historical and cultural monuments—for social, aesthetic, and economic purposes? Had the countries moved toward internationalization and the international sharing of cultural and natural treasures?[24]

Still in office, Huxley left the Beirut conference to partake in a major step in the international coordination in nature conservation. Following the decision that UNESCO should work toward this aim, he traveled to Fontainebleau (France) to give the opening lecture at the foundation event of the International Union for the Protection of Nature (IUPN), in the formation of which he as well as Nicholson had been involved (Huxley 1973, 61–62). The IUPN was renamed in 1956 as the International Union for Conservation of Nature and Natural Resources (IUCN) to emphasize the ecological approach. It was the first global conservation organization—established at the initiative of the Swiss League for the Protection of Nature under the auspices of the French government and UNESCO as a network of government bodies, private organizations, and experts.[25] With this and following achievements, Huxley felt he and his collaborators had succeeded in guiding UNESCO in the right direction (ibid., 78).[26] He continued to try to influence its policy after his directorship, pushing for projects in biology (particularly ecology), conservation, and for population policy. He also suggested experts in these fields for posts.[27] Huxley aimed at scientizing, centralizing, and rationally and globally organizing conservation, and UNESCO, with the cooperation of the IUCN, was the organization through which this seemed possible.[28]

It was also through UNESCO that Huxley had the chance to return to Africa. In February 1960, the United Kingdom National Commission for UNESCO, of which Huxley was a member, had made a strong and successful case for education, particularly in Africa, at the General Conference of UNESCO. The plenary discussions had become more difficult owing to the diverse interests of the increased number of member states of East and West and ex-colonies.[29] The year also witnessed a change in British empire politics under Prime Minister Harold Macmillan, who demanded that the United Kingdom decolonize sub-Saharan African territories where national consciousness was rapidly growing and integrate the African parts into its Commonwealth. It was during this "Wind of Change" that Huxley undertook his journey to south, east, and central Africa to report on wildlife and natural habitat; it was one of his most influential missions for UNESCO in matters of conservation.[30] During a thirteen-week journey, Julian and his wife Juliette collected information on the now nearly forty national and "tribal" parks and some candidates. They also visited government departments, universities, research institutions, museums, and private organizations that were concerned with conservation.

In the preface to Huxley's report on eastern and central Africa the then director-general of UNESCO, Vittorino Veronese, continued the discourse that was by now pervasive: "African wild life is remarkable for its abundance and variety. It is famous throughout the world, and the Africans may well be proud of this scientific and cultural heritage" ("Preface," in Huxley 1961b, n.p.). Huxley feared that this heritage was more endangered than ever, not least through the new methods against the tsetse fly and diseases of livestock, technological and agricultural development, as well as the spread of money values. He observed that since his last report, organized poaching and population size had increased. These were clearly linked issues, because owing to the increase in animal and human populations in late-colonial Kenya, for example, inhabitants and their property were more often the victims of wild animals (without compensation). In view of the population growth, in the wake of World War II, the government had also moved reserve inhabitants to crown lands where wildlife was plenty. As Reuben Matheka (2008) has shown, direct African representation in the legislative council from 1957 enhanced the agitation against wildlife policies—that had been on the rise since the late 1950s—hand in hand with political resurgence. Kenyans demanded better animal control, compensation for damage, and a share in the financial gain from conservation. Rather than a wholesale condemnation, these were therefore demands for reform. However, because of unresolved difficulties and economic problems, on the road to independence (1963), game laws were increasingly defied and independence associated with the end of conservation.

Huxley recognized the threat to landscape and wildlife posed by the rapid emergence of African governments. However, he was more sanguine in his appraisal than other commentators of the time, because he trusted in the growing world interest in nature and wildlife and the rapid advance of ecology. In fact, Huxley's approach to the African problem and promise within evolutionary humanism was now fully garbed in ecological terminology. In a confidential interim report of his discussion circle, the Idea Systems Group, the term *ecology* had been proposed as a substitute for a concept of evolution that was still too tightly associated with the struggle for existence, the survival of the fittest, and the notion of a missing link. There was still a need in the eyes of the group to replace outdated methods and ideas with the quantitative approach and population thinking, and notions of competition and absolute values with reasoning along the lines of adaptation, equilibrium, and relativism. *Ecology* stressed interrelatedness, cooperation, conservation, and constructive development of resources; it implied careful surveys of all the elements in a given situation and their interdependencies. With the development of human ecology that focused on economics and sociology, it seemed possible to bridge the gap between the natural, the social, and the psychological sciences. In the ecological garb, the distinctive mark of Huxley's psychosocial phase of evolution consisted in the fact that the cultural kind of ecological climax was no fixed endpoint of development. Idea systems like habitats may replace each other successively, but there was no given final state to a system because new ideas or changes in outlook could always be introduced and a new equilibrium reached by planned development of natural and social environments.[31]

Huxley had also been discussing with Elton ways of conceptualizing grades of evolutionary development in ecological communities, and in the report on his mission in Africa, he drew on Elton's *The Ecology of Invasions by Animals and Plants* (1958).[32] Huxley emphasized the difficulty and danger of conscious human intervention in habitats by removing or adding foreign species. Nonetheless, it was "clear that as the African territories are developed, their wild habitats will cease to be strictly 'natural.' To put the matter in another way, if any habitat is to be conserved, it must be managed. It is further clear that habitat management must be scientific" (Huxley 1961b, 43). Echoing Alfred Lotka's attempt to describe the entire organic and inorganic world as a system of energy transformations (Lotka 1925), the ethics of conservation had two fundamental commandments: "Thou shalt maintain energy-flow; and Thou shalt not sacrifice the eternal or the continuing to the temporary or the expedient" (Huxley 1961b, 28). In an article for the *Endeavour* that was dedicated in particular to the ecological aspects of his African survey, Huxley tabled the biomass per unit area to show the richness of the East African habitats. He demonstrated that the management and cropping of wild animals yielded a higher rate of annual increase than the cultivation of domestic cattle.[33]

With the aid of scientific surveys and the type of study that the Conservation Foundation called *ecological reconnaissance*, which investigated the whole human and environmental ecology of an area, Huxley wanted to plan a network of national parks. These parks would provide pride, prestige, and protein as well as money to local "tribes." In 1961, the Maasai actually followed other communities in agreeing to the establishment of two game reserves under UN recognition on their territories, among other reasons to secure their lands in independence and to receive the profit from tourism (Matheka 2008, 126). For Huxley, properly managed national and "tribal" parks would preserve a representative sample of Africa's fauna, flora, and localities of geological, archeological, and historical interest for the enjoyment of future generations. They would preserve areas of outstanding natural beauty (in the phrase officially used by the National Parks Commission in Britain), and wilderness areas for solitude and adventurous sports (ibid., 60–61). Most of all, they would preserve Africa's large mammals, "the only climax community surviving from prehuman evolution" (ibid., 15). They would allow personal fulfillment to world citizens and function as a resource for the development of a uniquely African civilization, a process to which Huxley referred as "African humanism" in a lecture series at the University of Ghana that included a talk on what he now called *ecological fulfillment*.[34]

In order to effectively work toward African and global ecological fulfillment, national parks should have professionally trained staff and researchers as well as labs for local and visiting scientists; facilities for study courses and for visiting groups of school children, townspeople, chiefs and elders, professionals, and so forth; guides to accompany visitors and an information bureau with maps, guidebooks and short guide-pamphlets; facilities to make and show films; and a museum with relief models, explanatory diagrams and pictorial charts, study collections, exhibition collections of the main interesting animals and plants, fossils, geological specimens and diagrams, meteorological, historical, archeological and ethnological material, and a section on conservation; as well as game and bird watching facilities. In short, they should become living archives for the development of the individual and the collective in tune with the growing knowledge about nature. As such, they were conceived as local with a global effect: "If they are efficiently managed and intelligently developed, they could become chosen places of world pilgrimage, where people of every race, creed and colour will come to learn, to admire, and to enjoy" (Huxley 1961b, 84).

Ideally, the national-parks-system project would be supported by the UN through the financing of surveys, and the territories could then apply to the World Bank for a loan. In addition, the world public had to be induced to help through organizations and private initiative, and this is where the idea of yet another institution came in: Huxley believed that a central fund under the direction of an expert body should be established to campaign for "a kind of world-wide Green Belt" for our over-mechanized civilization (ibid., 91). In fact, when returning from Africa, he published a series of articles in the *Observer* that set in motion the events leading up to the foundation of the World Wildlife Fund (WWF) (Huxley 1960a. 1960b, 1960c). In these influential articles, Huxley described the observation of free-ranging animals as one of the most exciting and moving experiences, "comparable with the sight of a noble building or the hearing of a great symphony or mass" (Huxley 1960c, 23). Besides the notion of Africa as a resource for personal fulfillment, he repeated the importance of its diversity for evolutionary progress; Africa was at the same time a place of prehistoric heritage and of potential for future development.

Huxley's African symphony moved the public, including the British businessman Victor Stolan, whom Huxley introduced to Nicholson. Even though Stolan's romantic views were unacceptable to Nicholson, their meeting sparked the establishment of the WWF. Huxley was present at two of the nine meetings of the London planning group. In the course of 1961, the WWF enthusiasts met to discuss the organization and put together an international declaration on "Saving the World's Wild Life." They stated that the eleventh hour had struck, and "doubtless feelings of guilt and shame will follow, and will haunt our children, deprived of nature's rich inheritance by ignorance, greed and folly." However, they also observed that "hundreds of thousands of people have bought best-selling books and millions have watched films and television programmes about the world's endangered wild life. Many of these have felt: 'If only I could do something to help!'" The new organization was to channel such help and coordinate the existing conservation bodies worldwide. "Mankind's self-respect and mankind's inheritance on this earth" would be preserved through concerted action including the ordinary man and woman whom the message of our heritage had reached.[35]

Among the objectives of the organization were thus the collection, management, and dispersal of funds for the conservation of world fauna, flora, forests, landscape, water, soil, and other natural resources; the acquisition of lands; research and investigation; education at all levels; cooperation and coordination of conservation efforts; the design and circulation of material for exhibition, courses, and promotion; the exchange of students and specialists; and the promotion of conferences and meetings. The WWF would try to safeguard threatened places of high importance (such as Ngorongoro Crater in Tanganyika) and rare species (e.g., Arabian oryx and white and Indian rhinos), to counter new technical threats (such as toxic chemical spraying and oil pollution), and to deal with risks from political upheavals (e.g., for the former Belgian Congo national parks). Again, science, and in particular ecology, would play a crucial role, and the WWF was to represent a new scientific and strategic approach to conservation. Further along the lines of Huxley's planning ideal that combined the expert and the layperson, the central and the local, the official and the private, the WWF should have experts evaluate the situation in cases of emergency to wildlife, contact authorities, and raise public awareness. A permanent center of calculation should also be established. IUCN was to set up an operations group that should maintain a "war-room" with up-to-date world maps and diagrams showing the main current threats to wildlife and wilderness. It should identify projects and campaigns to avert the dangers and assess the available and required resources.[36] If in this picture Huxley's language of evolutionary ethics and phylogenetic brotherhood was replaced by the image of men fighting a war by all means, the efficiency and determination clearly expressed his own feeling of urgency.

The international board of the foundation was mainly drawn from the network of the IUCN that welcomed the WWF project at its conference in Morges of 1961. The conference focused on postcolonial Africa where, with the independence of the Republic of the Congo and civil war, many feared for the safety of the national parks. Switzerland therefore seemed a neutral place to found the WWF. Inspired by Huxley's UNESCO report, the IUCN started an Africa project that culminated in the international Arusha conference in September that year. The international community succeeded in making the new African governments allies in conservation, and the Arusha conference also witnessed the foundation of the African Wildlife Leadership Foundation and the union of eastern African societies in the East African Wildlife Society, further consolidating long-ongoing local efforts. An Arusha Manifesto was launched that according to the WWF "biographer" Alexis Schwarzenbach was "nothing but a summary of Julian Huxley's postcolonial nature conservation strategy" (my trans.).[37] Indeed, it stated that "in accepting the trusteeship of our wildlife we solemnly declare that we will do everything in our power to make sure that our children's grandchildren will be able to enjoy this rich and precious inheritance."[38] The event of the conference functioned as a stepping-stone for the WWF. Huxley's *Observer* articles were reprinted and distributed by the Fauna Preservation Society in the form of a brochure that included a foreword by Julius Nyerere, Tanganyika's first prime minister, and by Iain MacLeod, an architect of British decolonization.[39] "Save the World's Wild Life" became the WWF's first publication and contained the charter drafted by Nicholson. Within the month of the Arusha conference, again together with the *Observer* articles, it was presented at a press conference at the Royal Society of Arts where the foundation of the WWF was officially announced. Huxley gave a speech, and the event sparked a sensational surge of donations.[40] The appeal to people's phylogenetic conscience was crowned with success.[41] However, as Matheka (2008, 128) has shown, the focus on African conservation did not outlive the 1960s, and international efforts continued to ignore the concerns of the communities affected by wildlife conservation.

The cause of the large mammals had played a crucial role in gaining public support for the efforts to preserve the world's natural heritage and it would continue to do so (see figure 20).[42] In Huxley's own eyes, their cause was only equaled by that of birds. Through the *National Geographic* great-ape stories that began to pour out to millions in the years following the WWF foundation, it would increasingly be these particular mammals that moved the hearts of people, and as "close

FIGURE 20. Taking care of giraffe, elephant and co., Julian Sorell Huxley—Papers, 1899–1980, MS 50, Series IX: Organizational Materials, Box 120: University–Z, Folder 4: Organisations: World Wildlife Fund, 1962–1964, 1970, n.d., cartoon by David Langdon, "World Wildlife Fund, British National Appeal: Dinner at the Mansion House, Tuesday, 6th November 1962," courtesy Woodson Research Center, Fondren Library, Rice University

cousins" the apes made a special appeal to the phylogenetic conscience (Sommer 2000). Huxley was still interested in apes scientifically, and he befriended George Schaller, Robert Hinde, and Desmond Morris.[43] He also became acquainted with Jane Goodall. He visited her at Cambridge, traveled to Tanganyika, and kept her updated on the primatological research carried out by others while she was in the rainforest. He tried to bring her into contact with Konrad Lorenz and Nikolaas Tinbergen. Huxley and Goodall shared the desire to turn Gombe Stream Reserve into a national park for scientific research, and Huxley continued surveying and campaigning for conservation in Africa throughout the 1960s.[44] Huxley and Goodall also shared a passion for the dissemination of knowledge about human evolutionary history, of which the chimpanzees seemed living memories. In 1971, Goodall, who had named one of "her" chimps after Huxley, sent him her book *In the Shadow of Man* (1971), which in her view made a more important statement than her scientific papers.[45]

At the time Goodall's book reached Huxley, some four years before his death, the ominous shadow humans cast on the existence of their cospecies had been his longtime concern. Nonetheless, his view of conservation was anthropocentric. Conservation was important in human trusteeship of evolution. As they had been for Osborn, humans were the true center of Huxley's universe. He did his due to regain them their central position after they had been degraded to the status of an insignificant inhabitant of a small planet among seemingly innumerable stars. He gave them the role of "the rare spearheads, or torchbearers, or trustees—choose your metaphor according to taste!—of advance in the cosmic process of evolution" (Huxley 1963, 714).[46] Despite Huxley's understanding of human biological kinship, in fact of all relations in nature, in terms of networks, it thus seems apt that director-general René Maheu chose to plant a tree to commemorate Huxley's work for UNESCO at the institution's country house.[47] The act could symbolize the seeds Huxley planted in the fledgling institution, his efforts for conservation, or his view of the world through the lens of evolution that revealed to him a picture of humanity climbing up ever higher on the tree of life.[48]

If Huxley's motivation for conservation was anthropocentric, his vision of a world society remained indebted to the Enlightenment tradition and thus Eurocentric. Through his engagement with the Colonial Office and UNESCO, Africa took a special place in Huxley's utopia: Africa carried the hope for planned evolutionary success, if only it would learn from the mistakes that other regions had made in their development. This patronizing stance has been criticized by Glenda Sluga (2010) in her analysis of the concepts of "the world citizen" and "the one world" that were central in the early years of UNESCO. She identifies Huxley's influence as director-general in steering the organization toward an imperialist and liberal development stance vis-à-vis (former) colonies. Focusing mainly on the African policy Huxley laid out for UNESCO—rather than on his ideas in the late 1920s and early 1930s—Sluga classifies him as reactionary.[49] Huxley's earlier stance might be seen in the eighteenth-century tradition of what Mary Louise Pratt (1992, chs. 3 and 4) has identified as the narrative of anticonquest, in the sense of strategies of representation through which European bourgeois subjects promoted

European hegemony under the veil of goodwill. In the scientific anticonquest narrative, science appeared as innocent enterprise, but colonial practices and the expansion of European science "to the periphery" were at least tacitly embraced. The anticonquest was a narrative at times marked by humanism, egalitarianism, and critical relativism that in the end betrayed the ideal of a "civilizing mission." While Huxley's version of it might not have been reactionary in the spectrum of opinions in the late 1920s and early 1930s, he maintained his evolutionary humanist philosophy of common global development through decades, while UNESCO was founded and the institution and the world at large underwent significant changes. Certainly the situation had altered by 1960, the year in which seventeen independent African nations joined the UN, holding close to 20 percent of the votes in the General Conference.

Perrin Selcer (2012, S177) has further shown that UNESCO's interference with international science, which had been a program during Huxley's directorship, came to be viewed as totalitarian when the organization issued statements on race in the 1950s. Scientists objected to the fact that, with an obviously political agenda, the organization declared what would count as correct scientific findings on the subject.[50] In general, Huxley's evolutionary humanism was a totalizing project: everything from the individual personality, to science and technology, the mass media, natural and cultural resources, and national and international political, social, and economic systems had to be developed in small local to global organizations along the lines of a planned ecology. Huxley summarized it for the readership of the *New Scientist* in 1963 as concerning the three habitats that humans inhabit, the planetary, the social, and the psychological. Besides biological ecology, humans therefore also needed a social and psychological ecology. Psychological ecology referred to the exploration of the individual and collective mind, the realm of thought and feeling in interaction with the facts of experience. For all three kinds of habitat, he described special mechanisms of change and possible ways to conscious, directive, and progressive evolution.[51]

These concluding observations point toward the themes that will be explored further in the final chapter of this section. While Huxley's brand of humanism gained in momentum in the postwar years, by the 1960s, in the changing political atmosphere, his ideas began to appear conservative. And it was in particular the three aspects of anthropocentrism, Eurocentrism, and a totalizing project and understanding of history that were increasingly under attack. Huxley's human-centered and holistic outlook was also threatened from within science. In the course of the application of the powerful serological, biochemical, and molecular techniques to the reconstruction of phylogeny, evolution—including human evolution—was "reduced to" a matter of molecular change.

CHAPTER TEN

The Ascent of Man Defended

This preoccupation of his [Charles Darwin] with origins is revealed in the titles he chose for his two greatest works—the *Origin of Species* and *The Descent of Man*—though *The Evolution of Organisms* and *The Ascent of Man* would in fact have been more appropriate.
—*Huxley (1964) 1992, 27*

In the postwar years, Huxley continued to popularize his understanding of progress as embodied in, and driven by, the exceptional human organism. In a series of articles in the *Sunday Times*, he described the human person as a microcosm of evolutionary history in three senses: humans were the evolutionary apex of perception and with their scientific instruments they observed even the smallest thing and the object farthest away; humans were the evolutionary apex of intelligence and had come to understand what they saw in the world; and finally, humans embodied the process of evolution that had gone on for so many millennia. Taken together, "all the unrealised possibilities of life on this planet are now concentrated in the human species. Man is the condensation of the process of evolutionary advance, which began with stardust and continued through amoeba and ape; human will and human action are the agencies for realising that advance."[1]

Besides his sole contributions, Huxley also drew the community of evolutionary humanists closer together through common publication projects, in which he tried to make his colleagues realize his precise vision. That this was not always easily achieved may be exemplified for *Evolution in Progress* (Huxley, Hardy, and Ford 1954), where difficulties emerged with regard to the contribution of Huxley's new friend, the Jesuit priest and paleontologist Pierre Teilhard de Chardin. Both Teilhard and Huxley identified incorrect elements in classical humanism and dangerously reductionist approaches in science. For both men, evolutionary history was literally alive within the human body; there was a phylogenetic connection between all humans and of the entire living world. However, although Huxley could agree with Teilhard's general understanding of evolution,[2] he rejected his metaphysics, expressed in such notions as humanity's "enroulement sur elle-même."[3] Teilhard, like Huxley, believed in a "racial"-economic-social-cultural convergence of human history, but thought that there was a physical force underlying this process that drove the course of cosmic evolution. This rendered his conception deterministic. It was also finalist, because the process would end in a state akin to the Christian ultimate universe of love. Huxley and his friends were therefore not at ease with the text Teilhard submitted for their volume.[4]

The largest such collaborative project was the "History of the Scientific and Cultural Development of Mankind," which arose from Huxley's directorship of UNESCO. It was to contain the entire "memory" of humankind from prehistory to the present and emphasize the cultural achievements of the "human race," dealing with war and politics only in so far as they influenced cultural and scientific progress.[5] As vice president of the UNESCO commission for the endeavor, Huxley intended to show that this history could only be understood within the evolutionary framework. He shared with Ralph Edmund Turner, who was the chairman of the Editorial Committee and later editor of the UNESCO International Commission for the project, the belief "that 'History' is a continuation on the human level of the general process of evolution, and that fulfilment (including enjoyment as well as material survival) may be regarded as the main aim of future human development." The historical trend that had most obviously enhanced fulfillment was the increase and better organization of knowledge, "which can both be practically applied and can help us in understanding ourselves and our role, and should serve as the basis for ideals."[6]

Huxley therefore also wanted to secure the Yale historian's cooperation for his private attempt at presenting every human sphere from the perspective of evolutionary humanism in the volume *The Humanist Frame* (Huxley 1961a). *Private attempt* is not entirely correct, however, because it is clear that although Huxley alone signed as editor, he was cooperating closely with Nicholson, and the book was regarded as a statement in association with the Idea Systems Group.[7] Huxley thought

Turner should contribute a chapter on history and humanism to show that the evolutionary view extended the ethical and moral principles in time. He should show how history could help us understand the inevitable conclusion that humans had a moral duty not only to their posterity but to the whole evolutionary process out of which they had arisen and which they had to continue. Turner's chapter for *The Humanist Frame* never materialized, but Huxley pushed Nicholson, who had read history at Oxford, in similar directions.

Furthermore, Huxley criticized the contribution of the historian, philosopher, writer, and educationalist H. J. Blackham for expressing the opinion that history has no goal. Blackham clarified "that in the same sentence I am saying that humanist thinking *is* historically directed towards definite achievements, and this is dissociated from a 'goal of history,' which is generally understood as 'historicism,' some form of teleology or some overall law of historical development. It is the difference between 'our' goal and a goal of history."[8] With this misunderstanding resolved, Huxley also secured a contribution from Waddington, whom he instructed to shed light on biology as an integral part of cultural history, with idea systems influencing the development of scientific theory. However, Waddington felt it was quite beyond his horizon to explain the importance of a humanist frame for the history of biology; he focused on the opposite, the importance of the history of biology for a humanist outlook.[9] We have seen that Huxley thought of Waddington as an ally not only in the creation of fundamental biological knowledge but also in the popularization of the science-centered view on society (mostly due to Waddington's *The Scientific Attitude* of 1941). In fact, the influential developmental biologist, embryologist, geneticist, paleontologist, and philosopher thought of himself as one of the main torchbearers of scientific humanism, when he wrote to Huxley that "I always like to think that you, and I following you, kept the lught [*sic*] of scientific humanism burning a bit."[10]

Huxley's scientific or evolutionary humanism, which had humans at its center as the apex of evolution and the agents of history, also gained important followers abroad. Simpson and Dobzhansky increasingly took up the torch of evolutionary humanism for the cause of progressive human development. Despite certain differences between Huxley and Dobzhansky, such as their views on the impact of natural selection on current human evolution, the definition of fitness, or the future role of religion, they were largely on friendly terms and shared a worldview. The

Simpsons and Huxleys were friends, and although Simpson was skeptical about the notion that evolutionary progress was no longer possible in any animal line, he, too, was in agreement with Huxley's outlook in general, and the two commented on each other's writings.[11] Indeed, Huxley, Simpson, and Dobzhansky retold and rewrote the narrative of human evolution, the history of its conception and its meaning within the broader synthesis, for diverse academic and nonacademic audiences, as if to inscribe it deep into the scientific and wider historical cultures.[12] Their publications built a tight network of intertextuality. In a letter to Huxley, Simpson described the interdependence of their ideas as follows: "Much of it [a paper Huxley prepared for the American Genetical Society] says more successfully rather nearly what I tried to say in my recent book 'The Meaning of Evolution.' The parallel is not particularly coincidental, since I have of course studied your work with care and have been profoundly influenced by it."[13]

The synthesists agreed that with the evolution of humans by means of neo-Darwinian mechanisms, a new era had dawned. In humans, the cosmos, or its evolution, had become conscious of itself, and humans were therefore ethically bound to apply the growing scientific knowledge about the world to its progressive development. Here is one example of how Dobzhansky played on this Huxleyan theme: "Most remarkable of all, he [man] is now in the process of acquiring knowledge which may permit him, if he so chooses, to control his own evolution. He may yet become 'business manager for the cosmic process of evolution,' a role which Julian Huxley has ascribed to him, perhaps prematurely" (Dobzhansky 1956, 88, citing Huxley 1953, 149). The adverbial caution is rhetoric. Huxley was far from seeing his grand aim realized, and Dobzhansky, like Simpson, believed that humans were influencing the course of biological and cultural evolution—willfully or not.

To fend off accusations of anthropocentrism, Dobzhansky cited Simpson. If a fish were questioned on the order of nature, it would hardly doubt the exceptional position of humankind: "I suspect that the fish's reaction would be, instead, to marvel that there are men who question that man is the highest animal. It is not beside the point to add that the 'fish' that made such judgments would have to be a man!" (Dobzhansky 1956, 86, citing Simpson [1949] 1967, 285–286). At times the circle of mutual authorization was closed, as when Simpson quoted Dobzhansky quoting Simpson: "Seminally, I wish to quote in somewhat shortened and altered form a passage from an earlier work of mine which

Dobzhansky also chose to quote at the end of one of his books: Man has risen, not fallen. He can choose to develop his capacities as the highest animal and to try to rise still farther, or he can choose otherwise. The choice is his responsibility, and his alone. . . . Evolution has no purpose; man must supply this for himself" (Simpson 1964/1966/1967/1969, 148, citing Dobzhansky 1956, 134, which is a citation from Simpson [1949] 1967, 311). For the synthesists it was certainly true that *Homo sapiens* had descended from apes, but where humans came from (*descent*) was less important than where they had proceeded and were heading (*ascent*). Their thinking about history was focused on "the uniqueness of man" and the possibility of "his" purposeful progress. Evolution was turned into a world-unifying principle that served to reconcile "modern man" with the universe.[14]

It can thus be observed that in general, scientific and evolutionary humanism gained momentum in the postwar years, and it was also increasingly internationally organized. Blackham, who was one of the driving forces behind the postwar institutionalization of British and international humanism, observed to Huxley that "there is undoubtedly a new and interested public for these ideas."[15] Huxley presided over the first congress of the International Humanist and Ethical Union in 1952. The IHEU was to incorporate humanist organizations that had been sprouting internationally. It was recognized as an organization that had consultative status with UNESCO. In 1963, Huxley became the first president of the British Humanist Association (it had been preceded by the Humanist Council as a representation of the Ethical Union, the Rationalist Press Association, and the National Secular Society). In 1962, Huxley had been elected Humanist of the Year by the American Humanist Association; the following year the honor passed to Muller—the American geneticists with whom Huxley was acquainted since his time at Rice Institute and who had served as president of the association from 1956 to 1958.[16] Muller and Huxley stayed friends for life—if mostly at a distance—and they shared the particular evolutionary humanist outlook that Muller endearingly called "the 1001 nights of biology, philosophy, personalities, current events and literature."[17]

But while scientific and evolutionary humanism were increasingly consolidated, the changed international political situation also presented a challenge, even if Huxley was less negatively affected by the Cold War than such early scientific humanist allies as Haldane and Hogben. As a longtime member of the Communist Party and supporter of the "Lamarckian" agrarian program under Trofim Lysenko that led to the prosecution of geneticists in the Soviet Union, Haldane found himself in a difficult situation and moved to India in 1956. Hogben, too, was alienated, though not to the same degree as Haldane, since he was not a party member but had pursued his own brand of socialism (or indeed scientific humanism) (Werskey 1978, 313–314, 321–322). Huxley had been an important critic of Lysenkoism (e.g., Huxley 1949), and it was his liberal democratic ideals that were most in harmony with postwar Western liberal orthodoxy. However, with decolonialization, civil and minority rights movements, the mandate of affirmative action in the United States, the second wave of feminism, the youth movements, and so forth, the developments in many ways outran Huxley's ideal of equality of opportunity. The notion of a meritocracy, in which the development of the biology and culture of as many individuals as possible was optimized, stood squarely in the landscape of demands for the insurance of equality of performance through sociopolitical measures (Selcer 2012). In an international climate that wanted equality as a fact rather than as mere possibility, the strong emphasis on the reality of biological variability (or rather inequality), even if context-dependent, was set against what Huxley perceived as misguided cultural determinism (Sommer 2010a).

When Huxley issued yet another collection of "essays of a humanist" (1964), which was translated into German (*Ich sehe den künftigen Menschen: Natur und neuer Humanismus*, 1965b), the international press paid tribute by covering it extensively. But the book smacked of evangelism. The commentator in the *Peterborough Examiner* (Ontario), though generally impressed and in agreement, felt that Huxley "tends to preach."[18] In fact, religious readers raised a storm after Huxley had recited his belief in the replacement of existing established religions with an evolutionary one, with a reverence for life and the belief in knowledge, at the 1959 Darwin Centennial in Chicago. It may therefore be of little surprise that others still welcomed the essays of the close to eighty-year-old Huxley as "a legacy of non-conformity in a changing world" (my trans.).[19]

However, some critics aimed at the core. The commentator in the *Western Mail* was "quite baffled that a man of Sir Julian's erudition could remain satisfied for long with optimistic sugar-and-spice conceptions of perfectibility," after all, "as long ago as 1913, [Huxley] reminds us, he publicly formulated 'the concept of a critical point between the biological and the human or psychosocial phase of evolution.'"[20] German readers were also growing tired of Huxley's sermons on evolutionary humanism: "Julian Huxley is a very old man, and one of the peculiarities of old age is, in many cases, a certain rigor of the mind, a fixed clinging to certain ideas, a stubborn refusal of all facts that could tamper with an opinion once adopted" (my trans.).[21] Humanism in general seemed outdated, "as musty, dusty and fusty as the word 'rationalist' had already become by the beginning of the century." The grand general statements, the polymath posture, and "all these liberal-minded 'syntheses'" were seen as useless for "our time."[22] Huxley had become "a voice of the past," and his evolutionary humanism was exposed as "the ordinary liberal consensus," owing "little to evolutionary theory."[23] Artists and scholars resented Huxley's renewed declaration that everything was based on evolutionary biology. A German reviewer humorously exclaimed that "Oh, if only we were all Darwinists and had studied biology! All those problems we poor historians, sociologists, and philosophers struggle with would be solved: God and the world, freedom and fate, soul and body" (my trans.).[24]

Huxley was even out of tune with some of his fellow humanists. They were taken aback by his promotion of negative eugenics in the form of voluntary sterilization as a population policy and a family planning measure,[25] but his affirmation of positive eugenics seemed even worse. Huxley was hooked by Muller's sperm bank project. Flanked by scientific and humanistic counseling, Muller thought, the new technology of conserving sperm without deterioration would pave the way to "germinal choice" of intellectually creative and socially oriented donors. Huxley saw his vision of eugenic insemination by preferred donors come true: couples would choose the implantation of inseminated eggs from extraordinary individuals.[26] Indeed, he seemed fascinated to the point of upraising "so-called genetics and eugenics to the gospel of his world religion" (my trans.).[27] For many readers, Huxley's eugenic vision provoked quite a different prayer: "God save us from this future man who according to a present man by the name of Julian Huxley is yet to be bred" (my trans.).[28] As these commentators recognized (e.g., also Rayner 1966), in view of the possibilities of artificial insemination, Huxley had not only become less cautious about the ability to raise the overall fitness of the population, or about the social desirability of such a project; he now also increasingly found himself on the hereditarian side of the nature-nurture debate.[29]

Thus, notions such as the (genetic) perfectibility of humankind, the synthesis of all knowledge on the basis of evolutionary biology, and of one common human history were under attack. At the same time, the synthesists, including Huxley, Dobzhansky, Simpson, and also Mayr, found their holistic view and the special place they ascribed to humankind threatened by the rise of molecular biology. In order to set the stage for part 3, where the interpretation of phylogeny and history from the level of the gene will take center stage, I briefly retrace the development of this confrontation. In fact, the early decades of the twentieth century had already seen the introduction of new methods for analyzing the history and phylogeny of humankind. Anthropologists (e.g., Keith 1927) were aware of the work of the British bacteriologist H. Nuttall (1904) who tested the reciprocal immunological reactions of blood sera of different ape species and humans. He found that humans were very closely related to the apes, with a particular proximity to the chimpanzee and gorilla.[30]

Furthermore, the discovery of human blood groups led to the analysis of their distribution in human populations. During World War I, Ludwik and Hanka Hirszfeld used the fact that different nations were brought together under the Allied forces to carry out such research. They concluded that there were differences in the proportion of the four ABO blood types for the populations they analyzed and that the differences correlated to their geographical distribution (L. Hirschfeld and H. Hirschfeld 1919). From the beginning, the blood group studies were associated with issues of "race" from the sampling to the presentation of results. Hirszfeld described it thus: "We had to speak in a different way to each nation. It was enough to tell the English that the objectives were scientific. We permitted ourselves to kid our French friends by telling them that we could find out with whom they could sin with impunity. We told the Negroes that the blood tests would show who deserved leave; immediately, they willingly stretched out their black hands to us" (L. Hirszfeld, "The Story of One Life," unpublished trans. of *Historia Jednego Zycia* [1957], quoted in Schneider 1996, 282).

At the same time, we have seen that Huxley, Hogben, and Haldane had perceived in the new genetics of heredity a tool to deconstruct the biological basis of race, and Jenny Bangham (2013, ch. 1.2) has shown that in the aftermath of the demonstration that the blood groups were inherited through one genetic locus with three different states or alleles (A, B, and O), Hogben and Haldane became interested in blood group studies. Huxley's attention, too, was caught. The new blood group research was understood to be more scientifically robust than physical anthropology, with which it shared an interest in intrahuman classification and phylogeny. In *The Science of Life*, for example, Huxley and the Wellses situated the attempt to classify and establish the biological kinship between human groups in the older field of physical anthropology, but declared the attempts at reconstructing national and/or racial histories by means of anthropometric surveys of living humans a failure. However, predicting the ambition of later population geneticists, they suggested that it might nonetheless be possible in the future to determine the various "racial" contributions to the genetic inheritance of an individual (Wells, Huxley, and Wells 1929/1930/1931/1934, 1448–1449).

In *We Europeans* (1935), Huxley and Haddon reproduced blood group frequency maps and envisioned a future in which such maps could be made for all the important genes that distinguish human groups. They pointed out that the data on blood group frequency distributions so far did not suggest an unambiguous and coherent "racial" history. They conceded that the data might be in agreement with the notion that humans once were divided into groups sufficiently distinct to be called "races." But they emphasized that it provided evidence for a great amount of intermixing throughout human history. Human groups had several different ancestries, and "the only way to measure the 'genetic relationship' of ethnic groups would be by ascertaining the quantitative values of their coefficients of common ancestry, which would be based entirely upon the statistical methods of probability theory" (ibid., 134). Thus, while they attributed great potential to this kind of research because "differences between groups can be assessed in terms of genes" (ibid., 127), Huxley and others believed that human population genetics would disprove the concept of pure, clear-cut "human races" upheld in the sciences and beyond (see also Morant 1939).[31]

In 1931, Haldane had already estimated the potential of human population genetics for the reconstruction of history and phylogeny such "that no anthropologist who wishes to take a large view of human origins can possibly neglect it" (Haldane 1932a, 63–77 ["Prehistory in the Light of Genetics"], 77). Blood groups were genetically determined and presumably not under natural selection. In contrast to the traits that the physical anthropologist would be concerned with, which were the outcome of relatively fast adaptations to different environments, the distribution of blood groups reflected the diversification of humankind into separate groups that migrated across the globe (and intermixed). For both Haldane and Huxley, this kind of research would allow the determination of the proportional ancestral makeup of contemporary mixed populations. It was the key to human history, ancestry, and biological kinship (Gannett and Griesemer 2004, 147–155 on Haldane). However, Huxley and Haddon (1935, 264) cautioned that this assumption presupposed the existence of originally unmixed ancestral groups, which they called a purely hypothetical inference.

On the contrary, some anthropologists welcomed the new data as a confirmation of established "racial categories." This was the case for the American anthropologist Carleton Coon in his treatise on the "human races," which has so often been pronounced blatantly racist by historians of science (and contemporaries like Dobzhansky):

> In studying racial differences in living men, physical anthropologists are now relying less and less on anthropometry and more and more on research in blood groups, hemoglobins, and other biochemical features. This is all to the good because the inheritance of these newly discovered characteristics can be accurately determined. In them, racial differences have been found, differences just as great as the better known and much more conspicuous anatomical variations. Being invisible to the naked eye, they are much less controversial than the latter in an increasingly race-conscious world. To me, at least, it is encouraging to know that biochemistry divides us into the same subspecies that we have long recognized on the basis of other criteria. (Coon 1962, 663)

Human population genetics thus continued to hold different promises until the 1960s (and, as we will see, beyond): for some as a technology against race in the sense of a clearly demarcated category, and for others as a less offensive tool for "racial classification." However, if some scientists perceived the blood group studies derived from medical techniques as scientifically more robust and politically less dangerous (Silverman 2000), the claim that biochemical and molecular evidence was superior to morphological data was eventually viewed as a threat by those who understood themselves as organismic biologists. Generally speaking, Huxley, Mayr, Simpson, and Dobzhansky were beginning to feel uneasy about the (potential) breakthroughs in the wake of the identification of DNA as the molecular basis of heredity. As Betty Smocovitis (1996, ch. 5) has shown, in the mid-1950s, "the architects of the synthesis" considered their struggle to unite biology as a success. By the 1960s, they felt their goals in jeopardy because of biochemistry and molecular biology.

In fact, the year 1959 might be taken as a turning point. It was a year full of anniversaries. While this suggests consolidation, such commemorations can also signal the necessity for a clear statement of tradition and identity in view of competition or crisis. This ambiguity seems to have pertained to the Darwin Centennial of 1959 at the University of Chicago. It was to be a celebration of the achievements of the modern synthesis. Huxley was deeply involved in its organization and stayed at the university as visiting professor. The event was a "who's who" of the life sciences, including Dobzhansky, Mayr, Simpson, Muller, Waddington, Wright, Nicholson, Ford, Leakey, Tinbergen, Sherwood Washburn and many more. A selection of the distinguished panelists was asked to send in papers prior to the conference. Following a request by the chairman of the Darwin Centennial Celebration Committee, Sol Tax, Huxley commented on all of these. He had hoped that the conference would bring together biology and anthropology under an evolutionary perspective. However, to his dismay, the papers showed that the majority of social and cultural anthropologists as well as archeologists did not (yet) work within an evolutionary framework.[32]

Another centenary was celebrated at the Museum of Comparative Zoology at Harvard, where Mayr was positioned. Mayr wanted to use the celebrations to draw attention to the continuing importance of natural history museums for biology, as well as to the decisive role they had played in the past. He wrote to Huxley, who was invited to talk on the occasion, saying that "with our biological laboratories specializing to an increasing extent in the area of molecular biology, the University museum here at Harvard as well as in other places becomes more and more a center for the study of the animal as a whole [such as in systematics, biogeography, evolutionary biology, ecology, and behavior]. The existence of a vigorous and forward-looking university museum is essential for a balanced training of the zoology student."[33] And indeed, other exponents of the fields Mayr mentioned also expressed apprehension vis-à-vis the status of molecular biology. Nicholson, for example, later remembered with regard to that period that "in comparison with other, increasingly sophisticated, laboratory research, purporting to yield definitive and permanently valid results, ecology long tended to be viewed as little better than marginally scientific. The work of Tansley, Pearsall, Elton and others in Britain, and their peers in other countries, won grudging recognition, but was almost pushed aside in the excitement about molecular biology. This threatened in the late 1950s to relegate the biology of living animals and plants in nature to the status of a mere sub-science" (Nicholson 1987, 121).

This suggests that those like Huxley and Mayr, who regarded themselves as the integrators of the knowledge of life, felt to a certain degree marginalized by what they perceived as the growing power of the reductionist science of molecular biology.[34] Mayr agreed with what Huxley had "always preached in the past that a one-sided support and emphasis on the type of biology that can be carried out in the experimental laboratories cannot achieve a harmonious growth of biology as a whole."[35] At the same time, Huxley's concern for the centennial demonstrates that they resented the fact that anthropology had not yet been reformed in accordance with the modern synthesis. Indeed, it was around the subject of humans that the differences associated with the various approaches, including the molecular, became most apparent. Mayr thought that the subject of humanity would receive the greatest impetus from researchers based at or associated with a natural history museum, because of their more historical approach. And the synthesists believed that in order to understand humanity, biological anthropology needed to be integrated with cultural anthropology, and that genetic and cultural reductionism had to be balanced, because although biology was important, "man is more than a DNA's way to make more of DNA of a particular kind" (Dobzhansky 1963, 138).

Dobzhansky, just like Huxley, Mayr, and Simpson, was set against a reductionist perspective on life that considered molecular explanations to be sufficient (e.g., Delisle 2008, 227–236). However, due to his groundbreaking work on drosophila genetics, Dobzhansky did often speak of the gene as the most fundamental organismic level. Huxley distanced his thinking from Dobzhansky's in a review of *Mankind Evolving* (1962b), arguing that it was written too much from the geneticist's perspective; it was not sufficiently holistic and integrative.[36] Huxley believed Mayr's book *Animal Species and Evolution* (1963) to be "the most important study of evolution that has appeared for many years—perhaps even since the publication of *The Origin of Species* in 1859."[37] Mayr agreed with Huxley's critique of *Mankind Evolving*, but he also agreed that they were all in general accord where the larger picture was concerned. Congratulating Huxley on the introduction to *The Humanist Frame* (1961a), Mayr wrote that "I am quite intrigued how we evolutionists are inexorably forced into contact with history, philosophy and humanism. Simpson, Dobzhansky, yourself, Rensch, and several others all trend in the same direction, each one in his particular way. I think we biologists have sometimes been remiss in not stressing more emphatically our close contact with the humanities. Sir Charles Snow typically represents physics when he speaks of the two cultures separated by an un-bridgeable gap. This would be true only if there were no biology."[38]

Two things are particularly important here. Biology was seen to bridge the apparent rift (Snow 1959) between the lands of the "hard" sciences and the lands of the humanities because it was understood as a holistic approach. This meant that humans could not be reduced to biochemical or molecular characteristics or processes. Rather, in the light of evolution, they constituted the toehold of the bridge that connected nature to culture. Huxley had indeed always stressed the fact that humans were mainly cultural beings. In 1951, he had written to Bernhard Rensch, "the German synthesist" with whom he discussed evolutionary trends and progress, that because of the new and special phase of human cultural evolution, humans should be classified as completely separate from the other organisms: Realm Psychobionta, Phylum Psychozoa, Class Homines, Order Verbales.[39] This view clashed with the new molecular understanding of life, and it was challenged by the new science of *molecular anthropology*, a term introduced in 1962 for the study of primate phylogeny and human evolution through the genetic information contained in proteins and polynucleotides (Zuckerkandl 1963).

Electrophoretic comparisons of serum proteins and studies of the immunoreactivity of primates, for example, led to the proposal of a new systematics, with the chimpanzee and the gorilla as part of the family *Hominidae*, and the orangutan alone in the family *Pongidae* (Goodman 1963). Furthermore, proteins could be sequenced and their differences measured, and the hypothesis of the molecular evolutionary clock was developed (Zuckerkandl and Pauling 1965a, 148). It assumed that molecules such as proteins and genes evolve like organisms, and that two groups of organisms that at one point shared a common ancestor would also, at that point, have shared a molecular makeup. After separation, an ancestral molecule would have evolved independently in the two lines of descent while maintaining structural homologies. On the assumption of a relatively regular change in protein sequence and the calibration of the clock by means of a divergence event known from the paleontological record, a more recent date for the divergence of the human and ape lines was calculated than was generally agreed on (Zuckerkandl and Pauling 1962, particularly 198–206). The history gleaned through comparisons between homologous molecules in contemporary organisms thus consisted of information about the approximate time of existence of a molecular ancestor, its probable amino acid sequence, and the lines of descent along which given changes in sequence had occurred.

These developments went along with the notion that the method of protein sequencing was the most powerful and provided "the most precise and the least ambiguous insight into evolutionary relationships and into some of the fundamental mechanisms of evolution" (Zuckerkandl 1963, 244). The method would only be trumped by DNA sequencing that was not yet technically available. Genes, even more so than proteins, were "a 'cleaner' material for phyletic investigations than morphological characters" (ibid., 260). The understanding was established of "molecules as documents of evolutionary history" (Zuckerkandl and Pauling 1965b). When asking "the questions where in the now living systems the greatest amount of their past history has survived and how it can be extracted" (ibid., 357), the answer was in the genes. They were the *primary semantides*, in which "there is more history in the making and more history preserved than at any other level of biological integration" (ibid., 360). This conception of history and phylogeny in terms of evolving gene sequences (at a more or less constant rate) certainly stood in contrast to the synthesists' integrative and human-centered approaches. Indeed, the statistical nature of mutations was often analogized to the Poisson distribution of radioactive decay. The reactions of the synthesists against the molecular approaches therefore revolved around such fundamental issues as the meaning of *history*, *anthropology*, and *biology*. Simpson (1964/1966/1967/1969, 9–10) emphasized that biology, in contrast to the physical sciences, was inherently historical. He demanded that if molecular biology wanted to be anything but a contradiction in terms, if it wanted to make statements about the living world rather than the chemical world, it would have to deal with the contingency and the uniqueness of historical objects and events. A truly *biological* and certainly an *anthropological* molecular approach could therefore neither go along with exact-scientific epistemic virtues (such as repeatability and predictability), nor neglect the role of the organism and its environment (Sommer 2008a).[40]

Nonetheless, molecular anthropology was not only accompanied by the aura of an exact-scientific objectivity because it studied human evolutionary history on the molecular level and in the laboratory, but also because it became amenable to elaborate statistics and automatic computation.[41] In fact, the story of the molecularization of research on human phylogeny and evolutionary history has to be seen in conjunction with the processes of informatization. And this is where Luigi Luca Cavalli-Sforza comes in, since he was one of the founders and has remained one of the leading figures in a geneticized, mathematized, and computerized human population genetics throughout his career. He is the third key figure of my story. Starting out from blood group studies, he proceeded to genetic sequence comparisons for the analysis of modern human evolution and intrahuman kinship as envisioned by the early molecular anthropologists. For Cavalli-Sforza, too, the gene was *the* historical document: "Considering the great variety of information provided by living organisms, it is clear that the type of data will affect both the method of treatment and the validity of results: the higher the correlation of data and genotype, the greater the validity is likely to be. Information on nucleic acid and protein structure comes first in the scale of relevance and that on phenotype measurements last" (Cavalli-Sforza and Edwards 1967, 233). As we will see, to the prioritization of the genetic and molecular approach, Cavalli-Sforza added the notion of the objectivity of mathematics and computational analysis: "The introduction of automatic computing methods in phylogenetic analysis has the advantage of objectivity" (ibid., 253).

The synthesists were alarmed by the advances in molecular anthropology as well as by the field's self-assertion, and they felt the impact of molecular approaches on science policy. Huxley was annoyed when his suggestion to make Leakey a member of the Royal Society was rejected. He and Simpson attributed this to the fact that discoveries in paleoanthropology were not considered worth the honor. Huxley lamented: "We are suffering severely in this country from the—I was going to say arrogance, but will tone this down to over-enthusiasm—of the molecular biologists, who are crowding out many other branches of biology, notably those concerned with studies in the field, including palaeontology."[42] In 1967, the year Huxley turned eighty, Mayr, too, took stock in a letter to his friend. Once again, he referred to the curious situation the synthesists found themselves in because they insisted on the importance of biology to human affairs and at the same time refused to reduce them to it:

> It always appalls me when economists, lawyers, historians, philosophers and others attempt to plan for the future, socially and spiritually, without the slightest regard for biological considerations. You are one of the few biolo-

> gists who has virtually dedicated his life to make the public aware of the significance of biological discoveries. In a way the impact of biology has had a curious set-back as a result of the magnificent victories of molecular biology. To the outsider they suggest that physics and chemistry is the A and O of all science. We will have to make a double effort to restore the influence of organismic biology and to make better known the evolutionary trends that culminated in that unique psycho-social organism Man.[43]

As Mayr's Huxleyan rhetoric suggests, the common threat presented by molecular biology and its reductionist notion of life seems to have closed the synthesists' ranks, and Huxley publicly maintained their ideological tenets and celebrated some of their heroes, including himself (Huxley and Kettlewell 1965; Huxley 1965a, 1966, 1970, 1973). Mayr eventually set about creating an archive of the synthesis's history and planned a commemorative event. However, by that time Huxley had suffered a stroke and had an infected eye as a result of a cataract operation. Juliette Huxley answered some of the questions Mayr posed to her husband in letters, and Julian finally managed to get to the questionnaire. But it was Mayr who carried on the interpretation of the synthesis and the demonstration of its significance.[44]

Huxley's life had been marked by repeated exhaustion and depression. Up to his death on 14 February 1975, he and his collaborators had continued the ideological fight against complex dangers such as technoscientific warfare (nuclear, chemical, biological), oppressive political systems, overpopulation, the exhaustion of natural resources, poverty, and so forth. For Huxley, the molecular view of life had been part of the problem. Such a reductionist view of humankind amounted to a denial of the future. It precluded an understanding of these threats, and it also precluded the finding of solutions. In its reductionist approach and internalist limitation, science contributed to the evils of the world. Only in its synthetic capacity of enlightening all human concerns in their interdependence could it be the solution (Sommer 2010a, 66). However, we will see in part 3 that the focus on molecules as the carriers of a history within by no means signified an avoidance of synthesis, totalizing projects, and human-centered general histories, even recipes against exactly these threats. Indeed, we will find notions of human progress that are quite similar to some of Huxley's ideas. Furthermore, while Cavalli-Sforza stood for the "reductionist" approach to human phylogeny and history, he emphatically made the argument, pioneered by Huxley and his friends in the interwar years, that the knowledge gained from genetics undermines racism. At the same time, while Huxley had warned against the use of preexisting, culturally defined groups and labels in studying the biological affinities of human populations as well as against thinking of the phylogeny of these populations in terms of a tree, the development of genetic, mathematical, and computational methods for building phylogenetic trees was one of Cavalli-Sforza's central concerns and achievements.

PART III

History in Molecules

Luigi Luca Cavalli-Sforza (1922–) and the Genographic Network

Today, *the* most encompassing project on history within is arguably the Genographic Project. On the one hand, it consists of the well-publicized endeavor of reconstructing the history of entire humankind by means of population genetic analyses of indigenous peoples worldwide: "Since its launch in 2005, National Geographic's Genographic Project has used advanced DNA analysis and worked with indigenous communities to help answer fundamental questions about where humans originated and how we came to populate the Earth. Now, cutting-edge technology is enabling us to shine a powerful *new* light on our collective past" (https://genographic.nationalgeographic.com, accessed 2 Aug. 2015). On the other hand, the project combines this research into "our collective past" through comparative analysis of "indigenous DNA" with commercial genetic services that promise information on one's private genetic history. In this sense, there were in mid-2015 over 700,000 participants in the project: "The results give you an unprecedented view of your lineage. You will discover the migration paths your ancient ancestors followed thousands of years ago, and learn the details of your ancestral makeup—your branches on the human family tree" (https://genographic.nationalgeographic.com/about, accessed 2 Aug. 2015).

At the end of part 2, we encountered the molecular gaze and the molecular desire associated with the molecularization of biology that had been driven by the application of physical and chemical methods and instruments in the 1950s and 1960s.[1] As the Genographic Project renders obvious, this gaze and desire have come to shape popular historical culture. When the term *molecular anthropology* was coined in 1962, technologies for protein sequencing had only recently become available, and the genetic code had just begun to be "deciphered."[2] Today, DNA markers are used to analyze the processes of evolution, to map genes, and to reconstruct the migrations and kinship of human populations. In this last aspect, the field that is variously called *anthropological genetics*, *genetic anthropology*, or *human population genetics* is concerned with the relationship between population history and genetic variation in humans. Thus, the term *genetic history* emphasizes the fact that these molecular-genetic approaches are used in historical studies. They have become possible owing to recombinant DNA technologies in the 1970s that made feasible the isolation, cloning, and detailed analysis of specific DNA sequences. Restriction enzymes of high specificity became available, which allowed the comparison of individual human DNA at the sequence level. These enzymes cut DNA at precise locations defined by longer sequences, so that individual differences in the resulting fragments were due to mutations in these restriction sequences. Such variations were therefore called restriction fragment length polymorphisms (RFLPs). Eventually the innovation and marketing of laboratory inventory, such as polymerase chain reaction (PCR) technology for the cloning of DNA segments, rendered the comparative analyses of nucleotide differences in DNA sequences (single nucleotide polymorphisms, SNPs) less money- and time-consuming. The dream of directly extracting from the sequences of genes and genomes the human history that had long since been inscribed into the primary semantides had come true. What was needed were regions of DNA that showed high enough mutation rates to function as carriers of the very recent history of human diversification. If the authority of the best historical document had moved from bones and organisms to blood groups, to protein and finally DNA structure, the race would now take place between different kinds of DNA.

These developments from the population genetics of blood groups to the triumph of the DNA-based phylogenies and migration maps, also in the public sphere, can be reconstructed on the basis of the Italian-born geneticist Luigi Luca Cavalli-Sforza' work. Cavalli-Sforza was certainly one of the prime movers in the application of statistical and computational tools to the genetics of human diversity and history. Following him through the history of human population genetics leads from the early development of scientific tools to genetically analyze human phylogeny to the grand stories suggested by the increasing amount of data and sophisticated research techniques, and beyond to the moral behind these stories. We find a very different notion of evolution at work than that of Huxley's evolutionary humanist project. Cavalli-Sforza focused on the random spread of fortuitous mutations. As we will see in chapter 11, in his work on genetic phylogeny and evolution, history was modeled on the process of Brownian motion.

How does one get from the study of a process that resembles Brownian motion to a series of historical events behind which stand human actors—to the narratives that Cavalli-Sforza was interested in? One way in which he achieved such a translation was through synthesis of knowledge from diverse approaches to human history. This time it was to be based on the mathematical and genetic approach. Another powerful tool was the use of images that are culturally deeply entrenched: the tree and the map. Once genetic data was expressed in human population trees and migration maps, molecular randomness seemed to suggest meaningful origin stories, and the information could travel more easily. It is with these tools in hand that Cavalli-Sforza and his collaborators hoped to introduce genetic human history into other disciplines and to a wider readership. If such imagery was eventually successful to the degree of dominating the visual language as well as the rhetoric of the Genographic Project (see quotes above), Cavalli-Sforza's attempts at what some scholars perceived as unfriendly takeovers also met with resistance. The gap between the new biological approach and cultural anthropology and the humanities seemed wider than ever.

For many critics, human population genetics and its techniques of visualization could still raise the questions that Huxley had posed: Who is to be sampled? And if people are sampled as representatives of "preexisting groups," how are the groups to be defined and chosen? As we will see, in population genetic projects, people continued to be grouped into geographical and traditional racial categories such as "Caucasian" and "aboriginal Australian." Furthermore, while the graphic of the tree suggests simple family tree logic, this is rather far from the truth. Cavalli-Sforza and others produced computer-based diagrams of genetic distances or sequence differences. Thus, the people sampled were taken to be representative of regions, nations, or ethnicities. The samples were then transformed into genetic and molecular data, and the data was mathematically processed on the basis of theoretical assumptions. In the resulting images, representations of abstractions from DNA define group identities and their interrelations.[3]

Like the paleontological and organismic approach, the genetic approach is concerned with who "we" are and where "we" come from. It is therefore never neutral. What work did Cavalli-Sforza himself hope his history would do in the world? What was its intended place among other kinds of histories that also informed individual and group identities? What were its politics? Cavalli-Sforza tackled such questions as how knowledge is vertically and horizontally transmitted, how culture is maintained and transformed, and how tradition shapes our identities and capacity for development. In chapter 12 I argue that the purpose of mathematically capturing the transmission and change of culture was to possibly consciously influence it; and the knowledge from human population genetics would have to play a decisive role in progressive cultural evolution. It is through an engagement with this aspect of Cavalli-Sforza's work that we will learn the motivation behind his effort of carrying his science into society. Cavalli-Sforza regarded his work, reminiscent of Osborn's and Huxley's projects, as a contribution to a larger human endeavor of heritage care and transmission, with regard to both blood samples and cultural history. This leads to the totalizing projects in which he was involved and that will be of concern in chapter 13: the Human Genome Diversity Project and the Genographic Project. The projects were rendered meaningful within the semantics of genetic and cultural heritage preservation, and they not only encompassed the establishment of research programs, infrastructures, and working standards. They also encompassed the production of narratives and images of "the great human diasporas."

The Human Genome Diversity Project was a nearly natural outgrowth of Cavalli-Sforza's human population genetics, and the Genographic Project heavily drew on Y-chromosomal research carried out at his Stanford University lab. Indeed, the Y chromosome was yet another epistemic thing about to rock the cradle of human evolution. And in a certain sense, in this final part of the book, I tell the story of the rise to power of the Y chromosome—of the sequences from which scientists read the male part of the human story. Telling the story that ends with the Y chromosome as the star of the popular Genographic Project will involve all the issues that surfaced with the onset of molecular anthropology: the notion of objective and neutral knowledge grounded in genes and genomes as well as in the methods of their analysis; the claim that the DNA sequence, and particular DNA sequences, are the fundamental objects for the reconstruction of human history; and the seemingly paradoxical argument that a science that is interested in the genetic differences between human groups could undermine racism. These assumptions and the consequences they seemed to entail have been opposed throughout the history of genetic anthropology by proponents of other anthropological approaches and by indigenous groups who were to be sampled for the reconstruction of "our" history. Human history and diversity were negotiated within and between disciplines and within and between human communities as part of a politics of science and identity that is related to colonial history and globalization.

Therefore, what remains invisible in genetic maps and trees or the accompanying narratives is not only the development of research programs and the complicated processes involving standardized laboratory practices, technologies, local labs, collaborations between labs and scientists, and so forth, that can turn fragments of DNA into markers for ancestry. Also absent are the tight connections between the lab and the field. As Amade M'Charek (2005, chs. 3 and 4) has shown for another object that will figure prominently in my account, mitochondrial DNA, questions ranging from what needs to be sampled (DNA, blood, placenta) to what amount and from which populations are closely intertwined with global politics of ethnicity and gender. This complexity is enhanced through the novel interdependence among research, market, and processes of collective and individual identification that the Genographic Project stands for, and which I have captured with the expression *the genographic network* (Sommer 2010f).

Instrumentalizing the knowledge coming out of science, scientists and nonscientists have established companies that sell genetic ancestry tests to growing publics. Laboratories and research projects are thus connected to markets in which history and identity are commodified, also in books, films, and other media and genres (John Comaroff and Jean Comaroff 2009). For molecular biology more generally, Soraya de Chadarevian and Harmke Kamminga (1998, 2) have shown "the powerful role that molecules have played in forging links between the laboratory, the clinic, industry and wider social interests." As we will see, with somewhat different players, this is certainly true for genetic history, if we, with de Chadarevian and Kamminga, understand "DNA" to encompass not only the instrumentations and techniques and the specific molecular qualities they render readable, but also the series of references to the "molecular-technical complex" leading from the laboratories to companies, markets, and individuals and populations globally.

The genographic network takes center stage in chapter 14: on the one hand it refers to the specific interpenetrations of science, identity politics, and industry that constitute the Genographic Project. On the other hand, it stands for these interrelations more generally. *Genographic* is a contraction of *genetic geography*—the expression Cavalli-Sforza used for his scientific research— and I read it in the wider sense of the geographies of genes, of the instable relations in space and time between entities at once biological, social, and discursive. *Genographic* captures the geography of globally connected researchers, scientific technologies and spaces, sponsors, companies, populations, and real human groups and individuals who are research subjects (in the position of the authoritative subject and subjected) as well as potential customers and consumers. *Genographic* is about the geography of "the gene itself," the sequence of what we call nucleotides on a string of what we call DNA, as well as its study. It refers to the distribution of human genes on earth as much as to the representations of that distribution and their effects. The research into the geography of human genes creates trees and maps that bring labels into a chronological and spatial order. In doing so, it interferes with geographies of identification "out there." *Genographic* is thus also the graphics of the gene up to the level of exhibiting "The Great History of Human Diversity." In the final chapter, I analyze this exhibition by Cavalli-Sforza and others as the culmination of his work on the human diasporas. It constitutes yet another node in the network; it is about the performance of genetic history to the point of the stream of visitors (re)enacting it.

In all these aspects, genographic activities are a project. The totalizing project of deciphering *the* history of humankind will never come to an end, and the knowledge it produces is meant to serve the advance of that history. The project is about the preservation and freezing of history in cell lines as much as it is about a service to a progressive development of humanity—a project in self-discovery and mutual understanding. Cavalli-Sforza's genetic geography is actually meant to create a living history that interferes with identities; it is an act of politics, of establishing relations in space and time. As we will see in the course of this section, and as I discuss in chapter 15 in particular, the tree structure shaped his genetic geography beyond the phylogenetic diagrams that have a clearly identifiable dendritic iconography. Tree thinking lies at the heart of this kind of human population genetics. As Marcus Feldman, the long-term collaborator of Cavalli-Sforza, pointed out, “Luca never lost his interest in trees, in phylogenetic trees.”[4]

CHAPTER ELEVEN

Human History as Brownian Motion, or How Genetic Trees and Gene Maps Draw Things Together

In this chapter I provide one possible account of how Cavalli-Sforza and collaborators charted their way across the human genome to the Y chromosome and to the big cooperative project that the Human Genome Diversity Project represents. This will include their own accounts, which shed light on the fact that they also had to build a community of researchers. In drawing things together (Latour 1986) with computer algorithms that produced population trees out of allele frequencies and sequence data, they forged a network of kin among themselves. The importance of origin narratives and histories of "great men and discoveries" for the formation of new and the renewal of disciplinary identities, as well as for scientific and public recognition, has gained the attention of historians of science (e.g. Abir-Am 1999; Browne 2005). Such studies have engaged with the role commemorative publications, conferences, exhibitions, art, and so forth, play in this process. At the same time, through oral history projects, historians of science have become active players in these practices of commemoration. They facilitate memories of scientists about "how I got there," rich with moments of inspiration, well-renowned father figures, and claims of priority. Finally, with human population genetics, we are always also confronted with a different kind of "memory practice." Nadia Abu El-Haj (2012) has shown how Jewish population geneticists predominantly work on Jewish genetics not only because it seems legitimate but because of a desire to know their history. More generally, as human beings, population geneticists are drawn to their work by an interest in their past and they may refer to themselves as historians (Sommer 2010f, 2012a).

As we will see, besides networks of institutions and scientists, making genetic history involved accumulating material and data, as well as processes of standardization and cascades of translations and abstractions that resulted in inscriptions that could circulate (Latour 1987, chs. 2–4). It involved collecting blood samples from living people from across the globe and their transformation into a particular kind of data. Mathematical and software tools were developed that translated the data into kinships between different entities than those sampled. Group labels that accompanied the samples and data, some of which had a particular legacy in racial anthropology and colonial history, were thereby revamped as populations in genetic terms (Gannett and Griesemer 2004). This issue will occupy us throughout the following chapters, because the process characterizes human population genetics as well as its commercialization, in which the population labels are finally turned into brands (Sommer 2012a). In this chapter, I am mainly interested in how Cavalli-Sforza and collaborators developed tree-building and mapping tools to turn the enormous complexity of the genetic data into supposedly easily readable, stable, and combinable images. Seemingly paradoxically, while these visual products rendered the difficult science amenable to other disciplines and wider readerships and turned the molecular noise into meaningful signals, in building them, Cavalli-Sforza and his collaborators in fact reduced human evolution to chance effects. At the same time, through the synthesis of genetic data with knowledge from other historical fields, trees and maps were accompanied by and stood for increasingly comprehensive, detailed narratives of modern human evolution.

According to one possible account, the story of Cavalli-Sforza's engagement with the science of human population genetics began during his studies in medicine at the University of Pavia in fascist Italy. Already at this time, he had in retrospect shown an interest in population genetics, because one of his teachers, Adriano Buzzati-Traverso, was a follower of the Dobzhansky school of drosophila genetics. Cavalli-Sforza identified Dobzhansky as an early influence on his later work. He pointed to *Genetics and the Origin of Species* (Dobzhansky 1937) as the first book he read on the subject, and later became acquainted with Dobzhansky personally. He also attributed his choice of genetics as a research field to an encounter in 1942 with Nikolai Timofeev-Ressovsky, the Russian population geneticist then at the Kaiser Wilhelm Institute for Brain Research in Berlin-Buch. After the war, Cavalli-Sforza worked at the Institute for Serotherapy of Milan, where he had the opportunity to experiment on the effect of mutagens on bacteria. The International Congress of Genetics in 1948, at which he presented the results of this research on bacterial genetics, was a turning point. It was there that he met the famous geneticist and statistician R. A. Fisher, then professor of mathematical and statistical genetics at the University of Cambridge.

Fisher offered Cavalli-Sforza a job and he began to set up a bacterial genetics lab at Cambridge. However, human population genetics was also in the atmosphere. Fisher had long been interested in the population genetics of blood groups, a topic that as we have seen had been taken up by Huxley, Haldane, and Hogben in the early 1930s. After Fisher had become Galton Professor at University College London, he established the Galton Serological Laboratory at the Galton Laboratory in 1935 (Schneider 1995; Bangham 2013, chs. 1.3 and 2.1). There, he cooperated with R. R. Race who later published, with Ruth Sanger, *Blood Groups in Man* (Race and Sanger 1950). In 1943, Fisher accepted the chair of genetics at the University of Cambridge, where he and Race were joined by A. E. Mourant, who will concern us below. At the first International Congress of Human Genetics, over which Fisher presided, he predicted a great synthetic effect in human population genetics and expressed his hope for an international effort in the global study of humankind that reminds one of the call for the Human Genome Diversity Project decades later (Fisher 1956).

However, Fisher had not succeeded in creating permanent jobs for Race and Mourant, and Cavalli-Sforza returned to Milan in 1950 because Fisher had problems financing his bacterial genetics research. Nonetheless, Cavalli-Sforza had joined the circle of blood group population genetics, and a few years after his departure, the molecular structure of the gene was described as deoxyribonucleic acid at Cambridge. These may be among the reasons why Cavalli-Sforza moved from his successful *E. coli* research into human genetics (he was among those describing the F-Factor in the transmission of genetic information between bacteria). However, even prior to his stay in England, he corresponded with Haldane about the statistical expression of population genetic processes, and in particular the effect of selection on gene frequency distributions. Haldane believed that natural selection in favor of heterozygosity was the chief cause for variation in natural populations, including humans.[1] On the contrary, Cavalli-Sforza was particularly interested in the role of random genetic drift, the first statistical treatment of which he attributed to Fisher. A grant by the Rockefeller Foundation eventually allowed him to work on the effect of drift on the human genetic variation found in a population in the neighborhood of the University of Parma (i.e., genetic variation between villages that is due to chance effects calculated on the basis of village size and migrations between villages). Cavalli-Sforza and his collaborators showed that in smaller villages, the distribution of blood types was more strongly affected by genetic drift and there was more variation than in larger ones—just as the theory of the influence of hazard in gene distribution would predict (Cavalli-Sforza, Moroni, and Zei 2004; L. L. Cavalli-Sforza and F. Cavalli-Sforza 2005).

In his population genetic research on Italy, Cavalli-Sforza communicated and collaborated with Mourant, who had directed the Galton Laboratory Serum Unit in Cambridge and successively led several serological labs in London. Mourant carried out influential work on human blood groups and other polymorphisms from the mid-1950s to mid-1970s.[2] Scientists turned to him for identification of sampling populations, information on sampling techniques, and provision of sera; they asked him for experts on particular subjects; they told him about their population genetic projects and expeditions; and they sent him their samples to be analyzed in his lab. Most of all, they drew on the blood group data that Mourant and his team centralized in tables and books. Theirs became the largest collection of offprints containing such data in the world, and Mourant published data on the human blood systems in a series of books. When, at the beginning of the 1960s, Cavalli-Sforza was given a professorship of genetics at the University of Pavia, he felt that enough genetic data from populations around the world had been accumulated, not least through the work of Mourant's team, to tackle the problem of a phylogenetic tree. Establishing a cooperation that would continue up to the mid-1970s, he asked Mourant for information.[3] This makes clear that geneticists often worked with data produced by others, rather than themselves collecting blood. This also meant that the identification of populations had been done by someone else, and as we will see, the inconsistent use of population labels in the literature could lead to difficulties.

Mourant had also received the records of the American immunologist William C. Boyd, who had engaged in the compilation of blood group data (Schneider 1996). However, while Boyd saw in the blood group studies a robust means of classifying "races" (e.g., Boyd 1939,

1952, 1963; Darlu 2007, 114), Mourant, in his monumental *The Distribution of the Human Blood Groups* (1954, 1), described them as a kind of "scientific anthropology" that could replace the anthropological racialism of old. Nonetheless, his effort to centralize the blood group data cemented rather than undermined the practice of reifying group labels. Figure 21 represents a table by way of which Cavalli-Sforza asked Mourant for information that was written directly on the sheet. While Mourant, for example, could not help with regard to the genetic serum protein marker Gc (group specific component) in "Basques," he could point to the literature for data on Tf (transferrin) typing for this group, and he even provided some data. With regard to "W. Germany," to give another example, he noted down average frequencies for the Tf alleles. He also provided Cavalli-Sforza with tabulations that he had made for the new edition of *The Distribution of the Human Blood Groups* (1976). As figure 21 illustrates, the genetic data was associated with a range of population labels from regional and national to ethnic and "aboriginal." These groups were taken for granted. On their unquestioned existence, trees of human relations were built.[4]

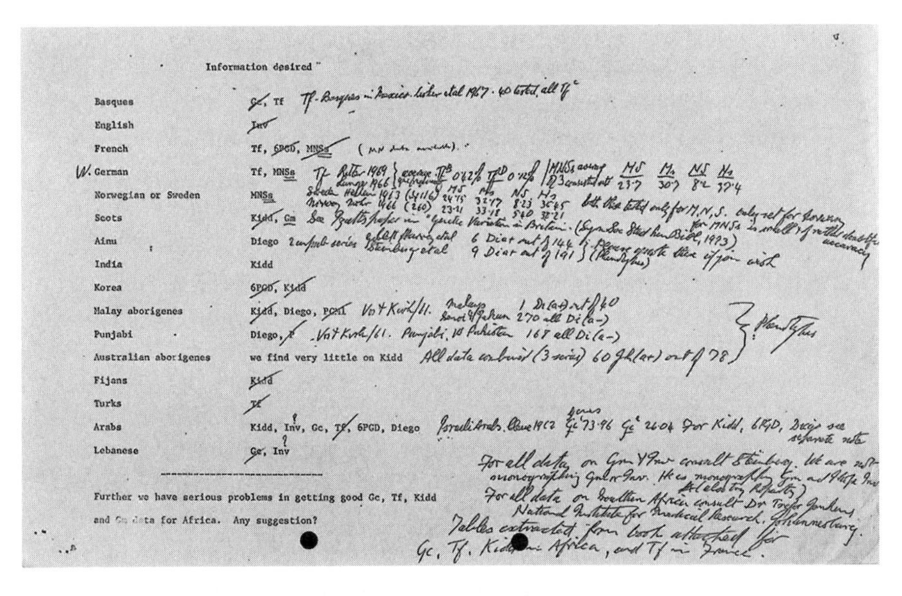

Information desired

Basques	Gc, Tf
English	Inv
French	Tf, 6PGD, MNSs
W. German	Tf, MNSs
Norwegian or Sweden	MNSs
Scots	Kidd, Gm
Ainu	Diego
India	Kidd
Korea	6PGD, Kidd
Malay aborigenes	Kidd, Diego, PGM1
Punjabi	Diego, P
Australian aborigenes	we find very little on Kidd
Fijans	Kidd
Turks	Tf
Arabs	Kidd, Inv, Gc, Tf, 6PGD, Diego
Lebanese	Gc, Inv

Further we have serious problems in getting good Gc, Tf, Kidd and Gm data for Africa. Any suggestion?

FIGURE 21. Information on blood group systems desired, Wellcome Library, London, Sforza, L. L. Cavalli, Shelfmark PP/AEM/K.121, Box 33, Reference number b17736894, correspondence 1965–1973: Mourant to Cavalli-Sforza, 1 Oct. 1973

The difficulties of sampling were considerable, in particular in the case of "remote" populations, because the blood needed to reach a refrigerator in time. This certainly is one reason why Cavalli-Sforza desperately needed data for "Africa" in the early 1970s (see figure 21). Indeed, Joanna Radin (2013) has shown that it was the prewar industrial production of refrigeration machines and the postwar knowledge of how to prevent damage to cells in freezing and defrosting that made the International Biological Program feasible (1964–1974). Reliable power sources in the labs and a system of blood transportation to "remote" Pacific locations that had been established for transfusion for wounded soldiers during the war allowed the scientists from over fifty countries involved in the program to make a concerted effort to collect blood from those they considered to be relatively pure peoples, genetically speaking. In fact, Cavalli-Sforza himself collected blood samples in the Central African Republic, "mostly from pygmies living in the forest."[5] Mourant gave him advice on the kind of container and anticoagulant to use. If he were to determine the blood groups for Cavalli-Sforza, he would need the samples within a week of their taking. Rather than by airmail, Cavalli-Sforza should send them as air freight, refrigerated in thermos flasks in waterproof containers filled with ice. A couple of years later Cavalli-Sforza was thinking about another trip to Ituri (Congo) to meet his friend, the anthropologist Colin Turnbull, and collect blood from "pygmies" whom he considered to be "purer" than those of West Africa. This time, the blood samples were to be stored in liquid nitrogen.[6]

In general, however, Cavalli-Sforza was less the field than lab biologist and statistician. To develop this line of research, he brought to Pavia another student of Fisher's. Like the effort of collecting data, this interest required collaborations, and Anthony W. F. Edwards himself reminisced thus about the connections between him, Cavalli-Sforza, and another researcher of concern in this chapter: "There is a great sort of strange connectivity which works between Walter Bodmer and John Edwards [Anthony's brother] and Luca Cavalli-Sforza and me, which involves bits of Cambridge, bits of Stanford of course, bits of Pavia, family bits inevitably, Walter Bodmer is my son's godfather, just to add to the confusion; Oxford, because John succeeded Walter in Oxford. So it's quite a sort of big academic family."[7] As part of this "academic family," Edwards made important contributions to statistical and computational human population genetics indeed.

Edwards had taught himself to work with computers at Cambridge, and the Italian Ministry of Education, too, invested in computers that appeared on the market in the early 1960s. The University of Pavia received a room-size Olivetti Elea 6001. Cavalli-Sforza and Edwards developed statistical methods for constructing evolutionary trees and Edwards wrote computer programs for them (least-squares based on an additive tree; minimum evolution/maximum parsimony [Edwards]; and maximum likelihood after Fisher). For the method of maximum likelihood that compares statistical hypotheses for their likelihood (which is proportional to the probability of the data given the hypothesis), Edwards wrote EVOTREE. Both men had their statistics from Fisher's *Statistical Methods for Research Workers* (1925), and for Edwards (2009, 10), they aimed at the "embedding of modern phylogenetic analysis in the Fisherian tradition of stochastic model building and efficient statistical estimation by maximum likelihood and the use of genetical data." However, even though EVOTREE would soon be overtaken by other programs, in the beginning it was the other two solutions—distance matrix and parsimony—that were taken up by the community. The more complicated maximum likelihood required more time.[8]

With the help of Mourant, Cavalli-Sforza and Edwards analyzed published data on twenty alleles related to the five major blood group systems from fifteen human populations (three per continent). They like to emphasize that this cooperation resulted in "the first evolutionary tree of human populations" (Cavalli-Sforza 2009a, 8; see also 1992) (figure 22). Another claim to priority is that they wrote the first computer program that allowed the representation of genetic variation in a principal components synthetic graphic analysis from a matrix of genetic differences, initiating the method's use in genetic anthropology. Principal component analyses are used to detect structure in the relationships between variables, to discover patterns hidden in data. They reduce the dimensionality of data without losing a significant amount of information. To reduce the complexity in the data and the number of factors possibly affecting it, these analyses search for those factors that affect the data most. The first principal component thus accounts for the greatest variance in the data, in our case the largest amount of the genetic variability found, and so forth. In the ideal case, the tree and principal component graphic analysis would be closely related, with the first split in the tree corresponding to the separation of populations by the first principal component. Later on, maps would be drawn on the basis of principal components, in which gene frequencies were transformed into scaled

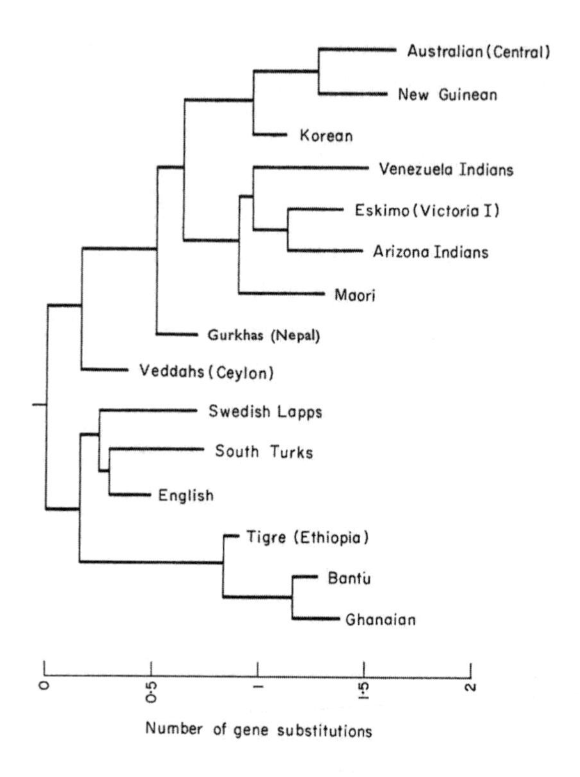

FIGURE 22. Population tree from blood group polymorphism frequencies produced by parsimony, reprinted from *Genetics Today. Proceedings of the XI. International Congress of Genetics, The Hague, The Netherlands, September 1963*, Vol. 3, edited by S. J. Geerts, Oxford: Pergamon, Cavalli-Sforza, L. L., and A. W. F. Edwards, "Analysis of Human Evolution," 923–933, 1965, p. 929, figure 5

deviations from the sample mean and plotted on the geographical regions from where the samples originated. However, Edwards had already invented another global mapping technique: that of projecting a phylogenetic tree based on human genetics onto a world map. While Cavalli-Sforza and Edwards's principal component analysis indicated a somewhat different pattern, the way they projected the as-yet-unpublished tree from figure 22 onto a map resulted in a sweep of the lines that does not suggest a travel out from Africa (figure 23). One might read the picture as placing the cradle of humankind in the area of today's Iran (Edwards and Cavalli-Sforza 1964; Manni 2010, 260, for a discussion of the controversies around principal component analysis, 246–250).

We are clearly in the prerevolution period of molecular anthropology. In the early diagram of figure 22, things were drawn together differently than in the consensus that would come to emerge. When Cavalli-Sforza and Edwards arbitrarily chose the root of the tree so as to produce two more or less equal parts, the result was a main split between "Europeans"/"Africans" and "Asiatics" (Cavalli-Sforza and Edwards 1965, 929). It was only in the early 1980s that Cavalli-Sforza collaborated in a study in which the "Caucasian" samples clustered more closely with those labeled "Oriental" and "Am. Indian" than with the "Bantu" or "Bushmen" (Johnson et al. 1983, 267). The archeological as well as mitochondrial-DNA evidence of the 1980s would suggest a common root of humankind in Africa. As we will see, the consensus on the general scheme of the human story was thus facilitated by the revolution in human population genetics through DNA sequence analysis.

The early trees and the ways in which they rendered human kinship and history differed from the later more popular ones based on mitochondrial-DNA or Y-chromosome sequencing (beyond a general increase of available data). Firstly, they were built from classical markers, or those polymorphisms (different alleles) that had been discovered by immunological and electrophoretic analysis of gene products such as proteins, and not on the differences in the nucleotide sequence of the DNA (discrete data). Secondly, these trees did not trace the genealogies of DNA sequences (the historical sequence of individual mutations), but

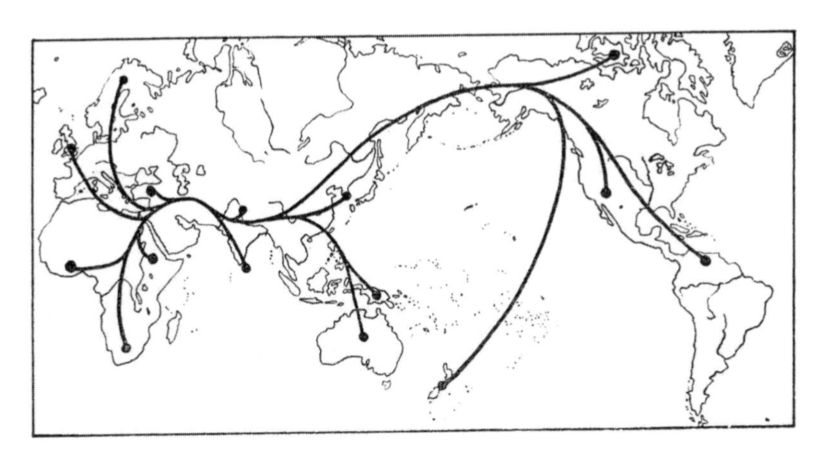

FIGURE 23. The tree of figure 22 projected on a map, A. W. F. Edwards and L. L. Cavalli-Sforza, "Reconstruction of Evolutionary Trees," in *Phenetic and Phylogenetic Classification*, ed. V. E. Heywood and J. McNeill, 67–76, on 75, figure 1 (London: Systematics Association, 1964) with kind permission of the Systematics Association

were based on the average genetic differences between entire sampling populations. In the simplest form, the genetic distance between two populations was measured by the difference in the frequency of a gene. The assumption was that the genetic difference between two populations was on average proportional to the time that had elapsed since their separation (genetic distance); but there was also a spatial dimension in that this genetic distance was taken to be proportional to distance in space.

However, this linearity might have been tempered with, for example, when a small group split from a population and became isolated through migration. This would have meant that it carried with it a sample of genes that was not representative of the mother population. Because of such possibilities, the difference between the populations under comparison had to be established as the average of the differences in frequency for a large number of genes (the law of large numbers). In the end, the trees resulting from the genetic data taken from peoples globally were meant to approximate a history of successive population splits related to the geographical distribution of actual living populations across the world.[9] Several assumptions had to be made: each population evolved independently; the cause of divergence between populations was mainly random genetic drift; and populations split at random and into two daughter populations identical to the parent. Evolution was thus conceived as "a branching random walk" with a constant probability of branching and a constant rate of walking (Cavalli-Sforza and Edwards 1967, 256; also Cavalli-Sforza 1966).[10]

It is not irrelevant that during this time, Cavalli-Sforza and Edwards also made the personal acquaintance of the Japanese population geneticist Motoo Kimura. In early 1960s, Cavalli-Sforza invited Kimura to Pavia to discuss his differential equations for gene frequency change that was modeled along the lines of Brownian motion (Edwards, interview, 29 June 2014). Kimura also played an essential role in the development of the neutral mutation idea. In 1968, he published his theory that the speed of evolution at the molecular level was determined by the rate of mutation (Kimura 1968; Dietrich 1994). This was based on the assumption that mutations mostly have neither a positive nor a negative selective effect for their carriers; they were described as selectively neutral. This allowed the assumption that alleles would spread randomly through a population (drift). The theory was in line with Cavalli-Sforza's and his colleagues' findings from the Parma valley, research they would continue and eventually summarize in *Consanguinity, Inbreeding, and Drift in Italy* (Cavalli-Sforza, Moroni, and Zei 2004).[11] The notion of silent or neutral changes in the DNA as the most direct record of evolutionary history began to take increasing ground. As Cavalli-Sforza's scientific biographers had it, "deleterious or advantageous mutations tell us more about the history of the environmental challenges that organisms have had to meet over time. Thus, neutral mutations under the control of drift are more useful in tracing the history of organisms themselves" (Stone and Lurquin 2005, 65). When Cavalli-Sforza and Edwards developed tools for building human kinships on the basis of these assumptions, the complex realities of human intermixture and of human agency more generally were lost. What remained were trees that reduced human history to "a branching Brownian-motion process" (Cavalli-Sforza and Edwards 1967, 256).

Cavalli-Sforza was well aware of the many problems associated with constructing phylogenetic trees from genetic data. At one point—reminiscent of Huxley's cautioning—he even suggested that it may only work for populations that are geographically far apart, because otherwise "instead of a 'tree' one may have to estimate a 'network'; such methods do not yet exist" (Cavalli-Sforza 1973, 96). Cavalli-Sforza also emphasized that the assumptions underlying distance analysis differ from our understanding of kinship. While kinship denotes genetic similarity by descent or inbreeding, genetic distance measures genetic difference due to nature (or chance). Furthermore, he was cautious about the linearity between kinship and geographical distance as well as between genetic difference and distance in time (for example, the possibility to assume regular average mutation rates) (Ibid., 90–91, 95; see also, for example, Mountain and Cavalli-Sforza 1997). However, Cavalli-Sforza and his colleagues developed tests for the "treeness of data," for the above assumptions of independent and constant evolutionary change (Cavalli-Sforza and Piazza 1975). Although it seemed doubtful that entirely isolated populations existed, and even though a reticulate model of modern human evolution might be developed and tested, the appropriateness of the tree iconography to render modern human phylogeny was strongly affirmed (ibid., 157–158, 163). Thus, while aware of the shortcomings of the tree iconography as representation of human population kinship and history, Cavalli-Sforza continued to employ it and thereby strongly influenced the development of human population genetics (Olson 2006, 172). There can be no doubt that the simplicity and elegance with which tree building rendered the world's messiness appealed to him. While paleoanthropology and comparative anatomy were better at capturing some of that messiness, population genetic methods suited him owing to the beauty of objectivity: "Whether such trees are based on fossil or living specimens, they may often be criticized for having a subjective element. The purpose of this paper is to show how suitable evolutionary models can be constructed and applied objectively" (Cavalli-Sforza and Edwards 1967, 233).

However, physical anthropologists had issues with the new approach to anthropology and its superiority claims. At a symposium in 1971 on "methods in anthropological genetics," problems inherent in the methods for building population trees, such as the fact that genetic distance measures could be the result of other processes than common ancestry, and the fact that they did not take into account gene flow through intermixture between populations, were severely criticized. Arguably, molecular tree building had produced nonsensical results, such as a close phylogenetic relationship between Australians and Koreans in the diagram by Cavalli-Sforza and Edwards reproduced as figure 22. It seemed that "in all of the studies on a world scale, the distance measures were applied in ignorance of real history; they are like surface collections in archaeology" (Spuhler 1973, 446). Thus, when the new tools to draw humankind together entered the territory of other disciplines, fundamental issues that we will encounter again and again were raised: What constituted history? What was in fact captured by the supposedly "branching random walk" (Cavalli-Sforza and Edwards 1967, 256)? At the same time, some physical anthropologists welcomed the new genetic/statistical/informational tools and adapted them to their purposes. Different tree-building models were experimented with, and the majority of symposium participants seemed to agree that gene frequency data was the best data for the study of human evolution; they also considered principal component maps such as produced by Cavalli-Sforza and Edwards good indicators of clusters of related groups.

The highly effective tools of gene mapping and genetic tree building did not only enter other disciplines. Throughout his scientific career, Cavalli-Sforza has also written for a larger reading public, and in these English, Italian, French, and German publications, the new visual language of human population genetics was key. Besides his groundbreaking work with Edwards, he cooperated with the Stanford mathematician and geneticist Walter Bodmer, yet another student of Fisher's and a member of Edwards's "big academic family." Together they wrote two books aimed at students and larger audiences. Bodmer had visited Cavalli-Sforza in Italy several times, and the collaboration continued during the academic year 1968–1969 when Cavalli-Sforza and his family were "trying out" Stanford University and the surrounding culture before he accepted the offer of a professorship of genetics there.[12] *The Genetics of Human Populations* (1971) was the first result of this cooperation. It appeared when Bodmer had already moved to the University of Oxford and Cavalli-Sforza took up a permanent position at Stanford, where he would remain long past his retirement age. In the book, the belief in the hierarchy of objects and approaches was given clear expression: "Biochemical variations depending on protein structure or immunological differences are much nearer to the origin of the long chain of cause and effect that starts with the gene and ends with a measurable trait. There is, thus, much less opportunity for other gene differences and environmental effects to obscure the picture" (Cavalli-Sforza and Bodmer 1971, 703–704; see also Reardon 2005, 69–71). Again, one gets the impression that instead of being about human history, history is what had to be removed from the picture—no environmental effects desired. Their second book—*Genetics, Evolution, and Man* (1976)—was a broad treatment of the genetics of heredity and evolution, population genetics, and heredity and environment; it, too, emphasized the fundamental nature and importance of the genetic knowledge and approaches. Similarly, in his single-authored *Elements of Human Genetics*, Cavalli-Sforza introduced a "wide audience" to the mysteries of genetics, a science that "uses more abstract thinking than most other biologic disciplines" (Cavalli-Sforza 1977, ix).

Around this time, the molecular desire, the dream to found human phylogenetic trees on "the origin of the long chain of cause and effect that starts with the gene" (Cavalli-Sforza and Bodmer 1971, 704) came true. DNA sequencing allowed for a "reading" of the gene in the sense of a linear sense-carrying unit. In genetic anthropology, this triumph was associated with a particular semantide: the circular cytoplasmatic mitochondrial DNA (mtDNA). Cavalli-Sforza understood the importance of mtDNA analysis for the study of human evolution early on. The fact that this extracellular DNA is transmitted by the mother alone had been revealed at Stanford, and Cavalli-Sforza's lab published an mtDNA study in 1983 (Johnson et al. 1983; L. L. Cavalli-Sforza and F. Cavalli-Sforza 2005, 217–219). They analyzed two hundred individuals from five different populations. After they digested the DNA with restriction endonuclease, the fragments were separated on horizontal agarose slab gels and visualized through autoradiography. The digestion with different enzymes provided thirty-two restriction fragment length polymorphism (RFLP) patterns, or morphs, which combined to thirty-five mtDNA types. The different mtDNA types thus constructed were shown to be related by steps of mutations and the reconstruction of this sequence was understood to provide insights into the biological history of humankind.

The researchers found that the major ethnic groups or "races" exhibited quantitative and qualitative differences in types. When represented in a parsimonious tree (i.e., assuming the minimum number of necessary mutations to account for the data), closely related mtDNA types clustered according to geographic regions (figure 24). The scientists also used the morph frequencies to produce an average linkage tree—the first such tree based on mtDNA (figure 25). When equal evolution rates were assumed for all branches, the tree separated the "Bushmen" first from all others. If they estimated the root from the three central mtDNA types in figure 24, then both "Bushman" and "Bantu" were

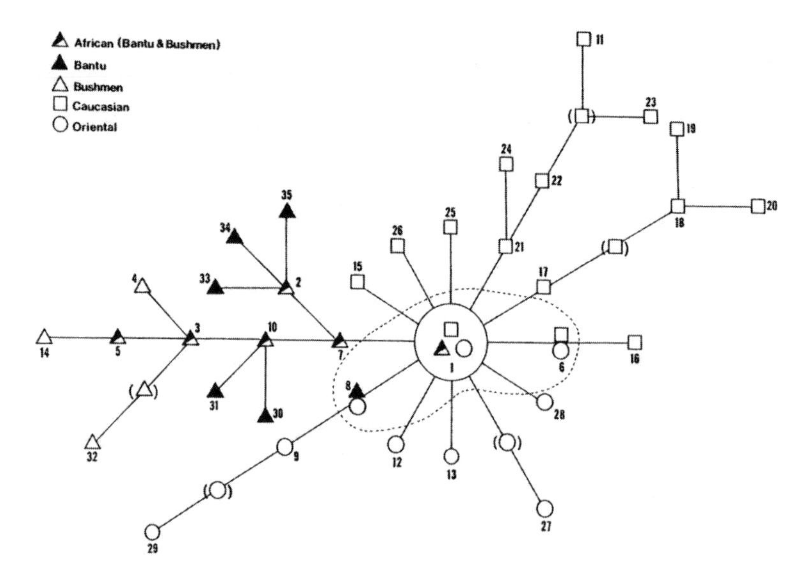

FIGURE 24. Tree of mtDNA types produced by parsimony, M. J. Johnson, D. C. Wallace, S. D. Ferris, M. C. Rattazzi, and L. L. Cavalli-Sforza, "Radiation of Human Mitochondria DNA Types Analyzed by Restriction Endonuclease Cleavage Patterns," *Journal of Molecular Evolution* 19, 1983, 255–271, on 264, figure 6, with kind permission from Springer Science and Business Media

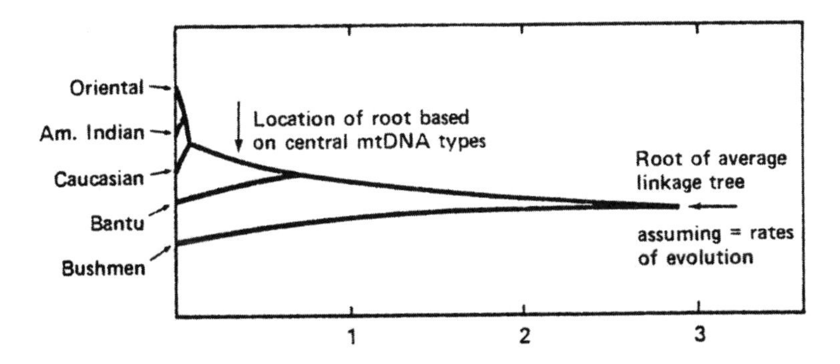

FIGURE 25. Average linkage tree based on genetic distance measures for mtDNA types, M. J. Johnson, D. C. Wallace, S. D. Ferris, M. C. Rattazzi, and L. L. Cavalli-Sforza, "Radiation of Human Mitochondria DNA Types Analyzed by Restriction Endonuclease Cleavage Patterns," *Journal of Molecular Evolution* 19, 1983, 255–271, on 267, figure 7, with kind permission from Springer Science and Business Media

on one branch and the rest on another. Thus, the authors not only concluded that there was a high correlation between mtDNA type and the ethnic origin of an individual; they also planted the root of the tree in Africa (Johnson et al. 1983; also Bonné-Tamir et al. 1986).

However, where mtDNA studies were concerned, Cavalli-Sfoza's team was outrun by Allan Wilson's lab at Berkeley. Wilson's lab used the more efficient molecular analysis. The method had been pioneered by Wes Brown, who was brought to Berkeley by Wilson to continue the development of his mtDNA technology (see, for example, Brown, George, and Wilson 1979; Brown 1980). The Wilson team also carried out a more extensive analysis. Indeed, the results coming out of the Berkeley lab caused an earthquake that reverberated far beyond the anthropological communities. Rebecca Cann, Mark Stoneking, and Wilson sequenced the mtDNA of approximately 150 people from what were called *African*, *Asian*, *aboriginal Australian*, *Caucasian*, and *New Guinean* populations (RFLP maps). The results were published in an article in *Nature* that contained a tree based on mtDNA types (figure 26). The tree Cann et al. published was a computer-built diagram of sequence differences in mtDNA molecules, based on maximum parsimony and midpoint rooting (the middle of the longest distance between two mtDNA types is taken as the root). In the process of producing this tree, the people sampled and the regions and "races" they were taken to be representatives of were reduced to DNA types. However, the tree makes clear that there are "multiple lineages per race" (Cann et al. 1987, 33). In fact,

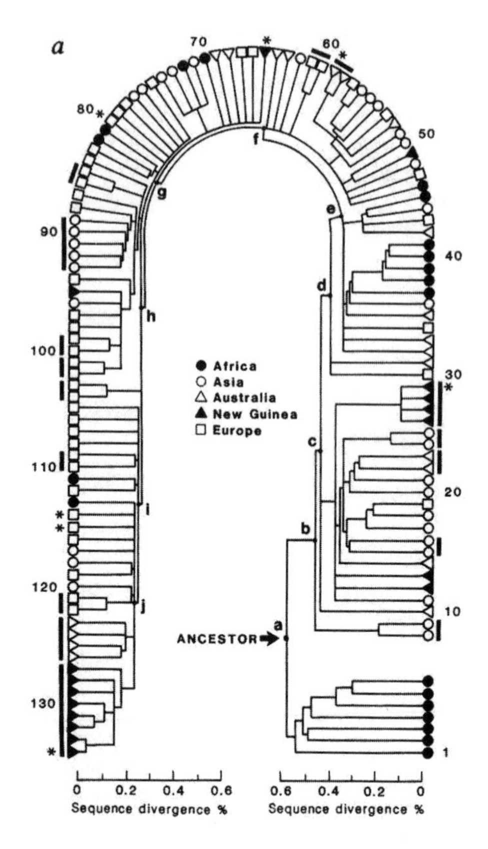

FIGURE 26. mtDNA tree, reprinted by permission from Macmillan Publishers Ltd: R. L. Cann, M. Stoneking, and A. C. Wilson, "Mitochondrial DNA and Human Evolution," *Nature* 325, 1987, 32–36, on 34, figure 2.1, with kind permission from Nature Publishing Group, http://www.nature.com/

although, or possibly because, it is more refined, the Cann et al. tree is messier than the Johnson et al. tree of 1983 shown in figure 24. Neither of these trees suggests clear-cut racial groups, however, because they visualize individual mtDNA types; it is in the population trees such as the Johnson at al. tree shown in figure 25 that the individual scatter is hidden behind population labels.

Beyond phylogeny, the results suggested a scenario of human evolution. They supported the interpretation that all human mtDNA referred back to a female who had lived in Africa as recently as some 200,000 years ago. The African Eve theory or recent African evolution model that emerged from this data was a prehistory of human origin, migration, separation, and struggle reminiscent of the biblical narrative (and innumerable other tales of human quest). Modern humans had originated in Africa (that is, modern *Homo sapiens* evolved from archaic forms in Africa), from where they began to migrate some 100,000 to 140,000 years ago to eventually spread across the globe. In the process, the modern human newcomers completely replaced archaic *Homo sapiens* (including Neanderthals) in Asia and Europe. Although the Cann et al. paper was not the first genetic anthropological study based on mtDNA, it was their story that roared through the press. However, the association of the African Eve theory with the female line and with an African origin also triggered sexist and racist reactions that purportedly took issue with the science of the research (Cann 1997). Nonetheless, for many, it became *the* story of human origin. The out-of-Africa origin model has been retold innumerable times since and to diverse audiences, so that it has become a living part of nonscientific as well as scientific historical cultures. In following this process, we will see how the story evolved.

The Cann, Stoneking, and Wilson paper (1987) certainly gave genetic anthropology a boost, and a revolution was proclaimed. There was more involved than the advent of genetic anthropology proper, equipped with both effective technologies and the right kind of DNA sequences to study the phylogeny and history of modern humans. This becomes obvious from *The Human Revolution* volume of 1989 that came out of a conference organized by the archeologist Paul Mellars of the University of Cambridge and the paleoanthropologist Chris Stringer of the British Museum of Natural History in London. The international conference brought together specialists from human evolution, archeology, and molecular genetics to discuss the revolutionary developments in molecular biology and their meaning for the study of human evolution. Further studies presented at the conference confirmed both the general conclusion that tools for DNA analysis were the key to human history and the more specific results that modern humans arose in Africa and subsequently populated Eurasia and America (Lucotte 1989; Rouhani 1989; Wainscoat et al. 1989). Thus, *human revolution* also referred to this heroic narrative of human becoming and the conquest of the globe. The human revolution encompassed both, the model of rapid emergence of modern humans in the course of evolution and the revolution brought about in the sciences of human origins by "the fast-developing field of what might be termed 'palaeogenetics'—historical reconstruction from present-day genetic data" (Mellars and Stringer 1989, 1).

In fact, Cann attributed greater authority to the analysis of mtDNA than to traditional morphological studies or even molecular studies of nuclear DNA and proteins. Generally speaking, genes of living humans were better suited for the reconstruction of phylogenies than fossils: "There is no direct evidence that any individual in the fossil record with a particular phenotype and genotype left genes in modern descendents [*sic*], yet geneticists operate with 100% certainty that genes in modern populations have a history that can be examined and will trace back in absolute time to real ancestors. This asymmetry demonstrates the inherent power of genetics to deal with evolutionary issues" (Cann 1988, 127). More particularly, mtDNA was the perfect epistemic object for analyzing modern human evolutionary history and diversity owing to three main virtues attributed to it: it was relatively simple to analyze; it evolved rapidly, because it had no efficient repair system and therefore a relatively high mutation rate; and it was inherited strictly maternally and was thus assumed to lack the complexities caused by recombination. While mtDNA consists mainly of coding segments responsible for factors in the oxidative phosphorylation system of the cell's energy-generating organelles (mitochondria) and different kinds of RNAs, there is also a so-called control region that has mostly regulatory functions. It was mainly this approximately 1,100 base-pair long segment—and in particular its two hypervariable regions (HVR I and HVR II)—that was used in historical studies.[13]

Despite this claim of priority of interpretation by geneticists, the revolution also entailed new interdisciplinary formations. The molecular analyses presented at "The Human Revolution" conference supported one of two models of modern human origins dominant in paleoanthropology and prehistoric archeology. Molecular anthropology corroborated the tenets of the out-of-Africa model: Africa had been the cradle of modern humans, since the molecules suggested a major African-Eurasian/African divide; modern humans had evolved relatively recently, as suggested by the molecular clock (regular fixation of neutral mutations); and human "racial" divergence was an even more recent phenomenon, since it was confirmed that genetic variation within human populations was greater than among populations (see, for example, Rouhani 1989). Proponents of the out-of-Africa model of modern human origins, among them the editors of *The Human Revolution* volume, therefore welcomed genetic anthropology as an ally against rival theories (see also, for example, Stringer and Andrews 1988).

To the contrary, in his contribution to the conference volume, the paleoanthropologist Milford Wolpoff (University of Michigan) pointed at the difficulties of the recent African evolution theory. He criticized Cann and her colleagues for basing their molecular reconstruction of modern human origins on the assumption of an mtDNA divergence rate that was too high. Using the same mutation rate to calculate the point of anthropoid-hominid divergence, the human and chimpanzee lineages would have split as recently as about two million years ago, when australopithecines had long since evolved. Numbers attained on the basis of a lower mutation rate seemed more likely. Wolpoff emphasized that this conjecture was supported by results from DNA hybridization and estimates from paleontology that ranged between five and eight million years for the anthropoid-hominid split.[14] As a proponent of the so-called multiregional evolution scenario, Wolpoff wanted to make the point that rather than human "racial" divergence being as recent as to coincide with the existence of *Homo sapiens*, the process had begun earlier, in the aftermath of the supposedly first migrations out of Africa by *Homo erectus* (Wolpoff 1989).

As Wolpoff's argumentation shows, the proponents of the multiregional as well as those of the out-of-Africa model of modern human origins saw their views corroborated by some of the molecular studies as well as by the fossil and archeological record. Both theories might therefore profit from the aura of objectivity surrounding the technologies, methods, and objects of genetic anthropology. At the same time, since the controversy revolved around the age of human "racial" diversity, there were mutual accusations of racism. It was also a question of which story made for "a more satisfying" (ibid., 98) kind of history, of which reading of the genetic traces would support or undermine current racial stereotypes as the basis of racial antagonism. Wolpoff insisted that the multiregional model that built on "evidence of long-lasting contact and cooperation" and suggested "a persistent shifting pattern of population contacts and shared ideas may provide an even stronger biological basis for accepting the unity of all humanity" than "a scientific rendering of the story of Cain, based on one population quickly, and completely, and most likely violently, replacing all others" (ibid.; Sommer 2008a).

These aspects of the revolution in genetic anthropology—the production of novel kinds of phylogenies based on the genealogy of DNA sequences; the synthesis of knowledge from diverse disciplines that led to an encompassing historical scenario but went along with conflict; and the question of "race"—also marked Cavalli-Sforza's subsequent work. Confronted with the other Bay Area lab's great success with mtDNA research, his team shifted focus and published insights into phylogeny based on autosomal single nucleotide polymorphisms (SNPs) and microsatellites, opening up microsatellite techniques to human population studies (Bowcock et al. 1994; Mountain and Cavalli-Sforza 1994; Underhill, interview, 21 Jan. 2013).[15] This allowed them to refine the out-of-Africa model. In fact, in collaboration with Ken and Judy Kidd of Yale University, Cavalli-Sforza's Stanford lab systematically searched for DNA polymorphisms in nuclear genes and produced a reference panel of population samples from 1984 onward. The samples from "Pygmies" of the Central African Republic and from Zaire as well as the "Caucasoid" and "Chinese" samples from the Stanford and San Francisco Bay Area were collected by Cavalli-Sforza himself (Cavalli-Sforza, Menozzi, and Piazza 1994, 89–90).

The Melanesian cell lines from the anthropologist Jonathan Friedlaender's field research were also added to the collection. Friedlaender had been part of the Harvard Solomon Islands expedition in 1966–1967 as a Harvard graduate student, photographing, fingerprinting, and measuring subjects, as well as learning how to take blood samples. In his PhD dissertation, he attempted to link the biological variation on Bougainville to small-scale migration patterns across villages that he reconstructed on the basis of parish records and interviews. This project was among other sources inspired by Cavalli-Sforza's genetic microanalysis in the Parma valley and his reliance on language variation as a correlate for genetic populations. In the course of Friedlaender's expeditions to the South Pacific, the focus on molecular and genetic data grew stronger, and the techniques moved from blood-group and serum-protein to mtDNA analysis (Radin 2009; Sommer 2010e).

In the early 1990s, 2,000 autosomal-DNA polymorphisms had been described on the basis of such samples in the form of lymphoblastoid cell lines. It was at this point that Cavalli-Sforza and his Italian colleagues presented the highly influential survey *The History and Geography of Human Genes* (Cavalli-Sforza, Menozzi, and Piazza 1994). With his coauthors Alberto Piazza and Paolo Menozzi, eventually professors of human genetics in Turin and ecology in Parma respectively, Cavalli-Sforza had carried out an analysis of data on gene frequencies in Europe beginning in 1977. The three then extended this genetic approach to geography and history to the rest of the world, the result of which was

History and Geography. The book marked a kind of crossroad in the history of genetic anthropology. On the one hand, it was still predominantly based on classical markers. On the other hand, there were the mtDNA and autosomal gene sequence analyses. Overall, while the material allowed the authors to experiment with trees, at the time of work the available genetic data was far from satisfying. Not all geographic areas were equally represented, and for Cavalli-Sforza, Menozzi, and Piazza, the genetic data presented problems beyond its relative scarcity.

The results published in *History and Geography* were largely arrived at on the basis of data published by other scientists. In these cases, the authors could not decide how to define populations before sampling; this had already been done, and not always to the authors' satisfaction. In published records, populations were sometimes insufficiently identified, solely by terms for indigenous populations (such as Australian Aborigines), nations, or even continents. Cavalli-Sforza, Menozzi, and Piazza were confronted with a database of samples identified with 1,915 different population names. Their own choice of language as the basis for population identification mainly came in when they reduced these to 491 by pooling populations according to linguistic criteria. For the analysis of the population structure of each continent, they further reduced the number. Finally, to yet further increase the average number of genes for the comparison of populations globally, they pooled again to reduce the number of populations to 42 (Cavalli-Sforza, Menozzi, and Piazza 1994, 20–22).

Cavalli-Sforza had long since maintained that the best way to define an indigenous population for genetic sampling was by language. A shared language was taken to represent a clearly identifiable social group that largely inbreeds. The work on the global classification of languages that Cavalli-Sforza's Stanford colleagues Joseph Greenberg and his former student Merritt Ruhlen were carrying out thus provided him and his collaborators with an inventory and atlas of populations worldwide (Ruhlen 1991). Furthermore, all the languages of the world were assumed to have originated from a single source and to have changed in ways similar to genetic evolution. The genetic and linguistic evolutionary trees were understood to show a high level of agreement. Therefore, the origin of a fully developed language was assumed to correspond with the expansion from Africa that, as genetics had established, had been initiated by a single "tribe." In reconstructing the expansion of human populations across the world, the authors of *History and Geography* mapped the distribution of genes on models for the fission and spread of languages such as developed by the archeologist Colin Renfrew (1987) as well as by Greenberg and Ruhlen. Beyond this use of linguistics, the analysis of genetic data was synthesized with knowledge from archeology, (paleo)anthropology, and climatic, ecological, and human history. To arrive at dates for fission and migration events, the genetic distances between populations were compared with archeological dates for first settlement of continents, and archeological knowledge provided other important insights for the timing and reconstruction of movements. Cavalli-Sforza and his colleagues assumed that mass migrations had been set in motion by an increase in the size of populations and by technological innovations. At the beginning of the journey out of Africa stood the first true language, and at the end of the Paleolithic, agriculture triggered the expansion of the Indo-European speakers into Europe (Cavalli-Sforza, Menozzi, and Piazza 1994; Manni 2010, 255; Cavalli-Sforza 1997, 7721–7724).[16]

These aspects of *History and Geography* provoked critique. The claim of one global linguistic tree met with opposition to the degree that Cavalli-Sforza made out a mafiosi tendency among Americanists to adhere to language "splitting" (L. L. Cavalli-Sforza and F. Cavalli-Sforza 2005, 280). But the identification of biological populations on the basis of languages and other cultural traits also met with criticism. Speaking for many others, the anthropologist Scott MacEachern (2000) welcomed *History and Geography* as a watershed in the analysis of human genetic variation. However, as an Africanist, he was particularly disturbed by what he took to be an ahistorical and monolithic understanding of "indigenous tribes" (who were simply taken as keys to a humanity before 1492). He emphasized that to the contrary, many of these groups had been shaped by complex colonial practices and constituted identities in flux and in encounter. Furthermore, in *History and Geography* there appeared to be a wild mix of population labels, so that these "tribes" stood side by side with North African nationalities such as Algerians and Libyans (see also Braun and Hammonds 2012).

MacEachern generally had issues with the ways in which data from different disciplines was integrated. He felt that the various sources of data were made to correspond even though "the procedures used by Cavalli-Sforza et al. to associate genetic patterning with cultural events in the past are quite unsystematic, involving attempts to fit secondhand knowledge of archaeological, ethnographic, and linguistic reconstructions into a Procrustean bed of genetic patterning" (2000, 370). In this context MacEachern further observed that "in some areas, and especially in the tracing of population movements over evolutionarily significant time spans, the [genetic] method is held to be superior to those other approaches [i.e. archeological, linguistic, historical, and physical anthropological]" (ibid., 359). This had consequences for one's understanding of *interdisciplinary research*, and MacEachern criticized the fact that "Cavalli-Sforza, Menozzi, and Piazza evade the problem of interdisciplinary analysis noted above by defining it out of existence, assuming that ethnicity, language, and genetic inheritance are today shared characteristics of well-demarcated, easily defined human populations and that these characteristics generally covaried in the past as they are held to covary in the present" (ibid., 362). Because of the scientific authority of genetics, other commentators also feared "that students of archaeology, genetics and physical anthropology will simply assume that this is the academic 'bible' on the subject and will absorb many of the speculative passages without criticism" (ibid., 377, quote from Morris 1988, 5, in comment by Alan G. Morris). This was despite the fact that "the majority of published work in human genetics which claims to elucidate historic and prehistoric processes has had no input from other specialists" (ibid., comment by Mark Pluciennik).[17]

Other discussants were sympathetic toward Cavalli-Sforza and collaborators' attempt at synthesis. In a published interview, the author Benjamin Anastas identified it as the unification of "the disparate fields of genetics, sociology, anthropology, linguistics, and archaeology into a single practice: the pursuit of a living past" (Anastas and Cavalli-Sforza 2002, 189). And indeed, Cavalli-Sforza's goal of synthesis was premised on the unity of science—and by inference of the cosmos. To illustrate this unity, he gave the example of the practice of copying manuscripts, a process that started in the Middle Ages and in the course of which small mistakes would have accumulated, allowing the philologist to trace the history and origin of a text. This analogy to genetics did not render human population genetics more humanistic in methods; rather, for Cavalli-Sforza it indicated that some humanistic research was fairly scientific (ibid., 194). But Anastas raised another issue when he described Cavalli-Sforza's history as a living history. To understand why this is the case, we have to look at Cavalli-Sforza's attempts at translating the knowledge from human population genetics into narratives potentially meaningful for larger reading audiences.

In fact, also in his popular accounts, Cavalli-Sforza not only rendered the grand sweep of modern human evolution but also continued to emphasize that genetics was fundamental to its reconstruction. *The Great Human Diasporas: The History of Diversity and Evolution* was one of the books on which he collaborated with his son, the philosopher Francesco Cavalli-Sforza. It appeared in English translation the year after *History and Geography*. They asked, "Which features reveal the history of humankind?" (L. L. Cavalli-Sforza and F. Cavalli-Sforza [1993] 1995, 115). The answer was given in the form of a narrative of the anthropological quest as a journey deeper and deeper into the human body to "the History in our Veins" (ibid., 109) and deeper still to the genes. On this journey, we move from considerations of environment, phenotype, and natural selection, to the true documents of human history and gain in scientific robustness through the methods available to analyze genetic data (ibid., 118). However, despite their insistence on the most fundamental level of organismic organization, the Cavalli-Sforzas promised to the nonspecialist reader an understanding of "our cultural and genetic heritage" (ibid., x). The synthetic history on the basis of human population genetics was a living history, merging the history written into "our" bodies with cultural developments. Similar elements characterized *Genes, Peoples, and Languages*, which first appeared in Italian and French (1996), then in 1999 in German, and eventually in English in 2000. Here, Cavalli-Sforza treated the difference between the paleoanthropological and the genetic approach to human history as *querelle des anciens et des modernes* (Cavalli-Sforza [1996] 1999, ch. 2). The reader learns that when the study of human evolution relied on fossils, it was not very far advanced; indeed, it was a hopeless endeavor before genetics. This was not only due to the scarcity of hominid fossils but also to the fact that bones were a messy source, as their evolution had been under the influence of nongenetic factors, while genes changed according to known and precise laws.

On these premises and out of the arduous and complicated processing of various kinds of data accomplished in *History and Geography*, Cavalli-Sforza popularized a human journey that would in its essentials be retold again and again: modern humans left Africa some 100,000 years ago and began to expand across the globe via the Middle East. One branch moved along the southeastern coast all the way to Australia (more than 40,000 years ago); another migrated through Central Asia (possibly some 60,000 years ago) and from there into southern Asia, Europe (ca. 35,000 years ago), as well as into the New World (between

35,000 and 15,000 years ago). Following this general outline, in *History and Geography*, the detailed multidisciplinary histories were treated for Africa, Asia, Europe, the Americas, and Australia/New Guinea/Pacific Islands (Cavalli-Sforza, Menozzi, and Piazza 1994). The integrated human history and geography that had begun to be supported by research into the variability of DNA sequences was also visualized in genetic trees (distance analysis, average linkage) and gene maps (principle component or gene frequencies), both in the accounts intended for experts and those for lay audiences (see figure 27). While the trees were meant to summarize and emphasize the historical information, the maps achieved the same for geography. Once again, both kinds of information were most effectively combined by projecting a tree on a map in the tradition of the first visualization of this kind that Edwards and Cavalli-Sforza had published in 1964 (see figure 23)—only by now the origin was clearly placed in eastern Africa (see figure 28) (Sommer 2015).

However, *History and Geography* was not the end of Cavalli-Sforza's achievement. Rather, the time of the book's publication marked the beginning of something new. The Cavalli-Sforza team's attention was finally directed toward the novel work on the Y chromosome (Casanova et al. 1985). Like mtDNA, the nonrecombining region of the Y chromosome was understood to have the advantage of allowing the reconstruction of individual genealogies, or rather the genealogy of mutations, be-

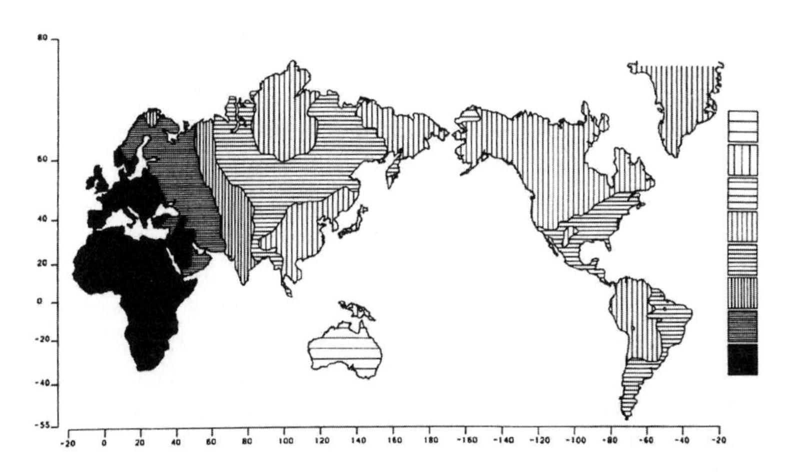

FIGURE 27. World map of the first principal component, Cavalli-Sforza, Menozzi, and Piazza 1994, 135, figure 2.11.1

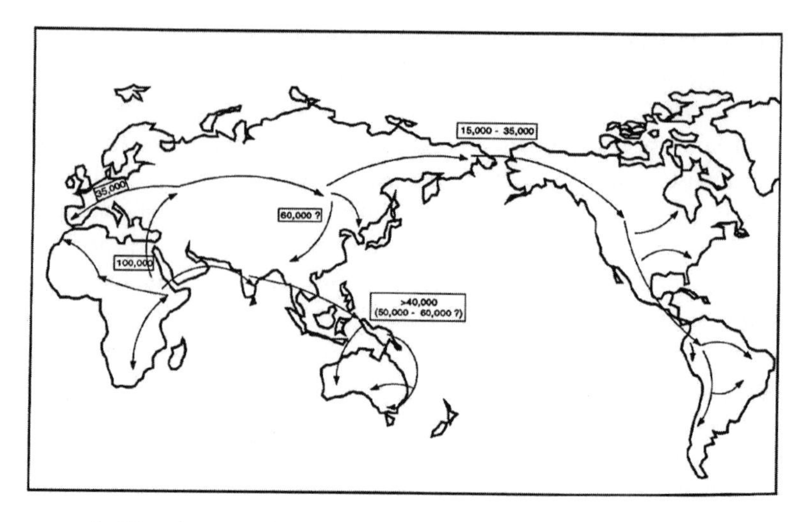

FIGURE 28. Migration map of modern human populations, Cavalli-Sforza, Menozzi, and Piazza 1994, 156, figure 2.15.1

cause it is handed down exclusively from father to son. Michael Hammer, who had worked on his PhD with Wilson at Berkeley, had also carried out postdoctoral research at MIT with David Page in a lab that focused on the Y chromosome. Hammer moved to the University of Arizona in 1991, where he studied variation on the Y chromosome as a model system to explore human evolution. Hammer's team was among those who were working on the identification of Y-chromosomal polymorphisms and they were beginning to implement a Y-chromosomal phylogeny (see, for example, Jobling 1994; Hammer 1995; Hammer and Zegura 1996).

Nonetheless, there were still few polymorphisms known and the methods to detect them were laborious. This was a chance for the Stanford team to enter an as yet largely uncharted chromosomal territory. One of the scientists at Stanford had published a paper in *Science* (Vollrath et al. 1992) on his research with Page and others at MIT in Y-chromosome mapping. Cavalli-Sforza was intrigued by the possibilities it suggested. However, the main player in this story is Peter Underhill, who came to Stanford in 1992 for a position in the core facility (service lab). Prior to that, Underhill had carried out research in marine biology, but owing to the recombinant DNA revolution decided to acquire postdoctoral training in molecular biological techniques such as cloning and

DNA sequencing at SRI International in Menlo Park, California. While he was the expert for sequencing in the core facility at Stanford, Cavalli-Sforza asked Underhill to move to his own lab to exploit the newly acquired automatic DNA sequencer. Together with the technician working the PCR machine, they began looking for Y-chromosomal SNPs. It took them over a year to get only one ("African") marker.

Then, in 1993, Peter Oefner brought a particularly efficient method of discovering DNA polymorphisms to Stanford (HPLC, a chromatography technique). Underhill joined forces with Oefner, and they identified many Y-chromosome markers by means of the new technique (SNPs that were marked with an M if discovered at Stanford). They described a number of so-called haplotypes and haplogroups, groups of related sequences that are defined by shared mutations and are interesting because of their regional specificity. Such haplogroups appear at different frequencies in different populations and are therefore used to establish phylogenies and to reconstruct population histories. In a shift of the kind I have discussed for the Cann et al. tree (figure 26), a haplogroup can also refer to the descendants of a single individual who first showed a particular marker (SNP). In common parlance, haplogroups are therefore branches of the human family tree. Because the Y chromosome is only transmitted from father to son, in this case, haplogroups are used to define worldwide patrilines. They establish relationships between male molecules that come to stand in for kinship between populations. They are also employed to retrace the early population movements. In this respect, the haplogroup research on the Y chromosome, too, eventually supported the recent-African-origin story of human evolution. In parallel to the mtDNA system, the underlying assumption is that all actual molecules carried by men all over the world today go back to one ancestral Y chromosome carried by one individual man—Y-chromosomal Adam (Peter Underhill, interview, 12 Jan. 2013).[18]

With this we are at a point in the history of genetic anthropology where there was in place a quite robust consensus on the general outline of modern human evolution, which had also entered historical cultures in narratives and images for wider readerships. In chapter 14 we will see how this relates to the Human Genome Diversity Project and the Genographic Project—how the first grew out of the research effort that went into *History and Geography* and how the second was a continuation of Cavalli-Sforza's Y-chromosomal work. However, prior to that, I have to engage more closely with some of the assumptions underlying Cavalli-Sforza's endeavor to have his science enter other disciplines and historical cultures through popularization and publicized large-scale projects. More specifically, I have to deal with his understanding of cultural evolution to learn what kind of role he envisioned human population genetics might play in the process.

CHAPTER TWELVE

Cultural Transmission and Progress

Where Cavalli-Sforza's work on cultural evolution was concerned, he again entered a long-term cooperation, this time with the mathematician and population biologist Feldman, who had worked on his PhD at Stanford (depts. of mathematics and biology) with Bodmer as one of his advisers. In the early 1970s, there were regular seminars on mathematical biology, particularly genetics. In one of the early meetings, Cavalli-Sforza, who was by then a member of the faculty, gave a talk on cultural evolution, writing equations on the blackboard. Feldman was intrigued, and the event initiated an intensive collaboration (that would be continued for more than twenty years), during which they spent evenings at Cavalli-Sforza's house trying to figure out how the subject of cultural evolution could be quantified (Feldman, interview, 22 Jan. 2013). The outcome was the impressive *Cultural Transmission and Evolution: A Quantitative Approach* of 1981 (Cavalli-Sforza and Feldman 1981).[1] Although aimed at a larger audience, the book is far from an easy read. It is full of equations that demonstrate the attempt to adapt the mathematics of population genetics developed by Fisher, Wright, Haldane, and others to the realm of culture.[2]

This approach therefore differs from that of Huxley or Conklin, both of whom extrapolated from the patterns and trends they perceived in the fossil record to evolution at the cultural stage. Huxley and Conklin also drew inferences for culture from the principles of hereditary transmission, but in a verbal way. In fact, most likely Huxley would have rejected Cavalli-Sforza and Feldman's approach as too reductionist, given that he even objected to the notion that mathematical population genetics was the fundamental approach to biological evolution. On the contrary, Cavalli-Sforza and Feldman considered language too ambiguous to get a grip on culture. Furthermore, biological mechanisms such as natural selection were more important in their consideration. Culture appeared as an adaptive complex that was not only under cultural but also natural selection. Nonetheless, we will encounter some striking similarities between Huxley's and Cavalli-Sforza's outlooks, and I will argue that their aims were similar, that both wanted to understand the workings of culture to think about how progressive intervention might be rendered more effective, and that they regarded knowledge about the evolutionary past as a fundamental contribution to cultural evolution.

While Huxley worked with mentifacts or memoids and Richard Dawkins (1976) had by then coined the term *memes* for cultural entities, the geneticists Cavalli-Sforza and Feldman introduced the notion of second-order organisms. These encompassed all cultural objects from art and technology to entertainment, and from languages to values and beliefs. They were completely dependent on the first-order organisms, that is, cultural animals such as humans. Cavalli-Sforza and Feldman then transferred the mechanisms of biological evolution to the second order, arriving at a parallel level of cultural evolution that nonetheless remained indirectly affected by the evolutionary processes acting on the first-order organisms. An important difference was that at the cultural level, change in traits of second-order organisms or the appearance of new second-order organisms could occur at random through a copying error or a chance discovery, but might also come about on purpose through the conscious response to a problem by innovation. Once in existence, the new or changed trait might or might not be diffused in space and transmitted over generations, and to varying degrees (existing in more than one state).

This could involve a process of selection. The authors thought that natural and cultural selection were largely in harmony because the structural prerequisites for culture in the human body—mainly in the brain—had evolved under natural selection. Human beings would often think or behave adaptively. However, while innovations and conscious choices might be biologically adaptive, Cavalli-Sforza and Feldman also made out cultural traits that were neutral or had a negative selective effect on the biological level of evolution. Not every trait that was culturally successful had to be a biological advantage in fitness. Furthermore, the effects of a second-order organism might seem culturally beneficial before it becomes obvious that the long-term effects are not. Maladaptive traits could only be eradicated by cultural selection once they were recognized as such. This is where I see the role of science in the scheme; it may consist on the one hand in making the processes at work in cultural evolution explicit; and on the other hand in providing knowledge about the selective value of cultural traits, for example, in proving that smoking is bad for one's health.

These observations bring us closer to the question of how the Cavalli-Sforza/Feldman theory of cultural transmission might relate to the search for the possibility of progressive human evolution that I refer to in the title of this chapter. As possibly *the* realm of conscious and goal-directed problem solving, science is crucial to a certain conception of progress present in *Cultural Transmission*: "In spite of this, there would be consensus that positive cultural evolution of human society is possible, and, at least for technology has certainly occurred" (Cavalli-Sforza and Feldman 1981, 366). Success was here measured in the Darwinian sense, that is, species were successful if they survived and multiplied. In humans, such success depended on technological innovation (in food production, transportation, military, communication), and more recently on the experimental method (in engineering, chemistry, and modern physics). From this perspective emerges a progressive history along the lines of scientific-technological development. Like Conklin and Huxley, Cavalli-Sforza believed in the growth of knowledge under the conditions of specialization with accompanying integration (or multidisciplinarity and synthesis), even if the task of spreading it demanded the work of martyrs. Citing Haldane, Cavalli-Sforza consoled his sister and brother scientists that there had never been a major scientific discovery that was not regarded at first as an insult against some divinity (L. L. Cavalli-Sforza and F. Cavalli-Sforza 2005, 342). Ultimately, however, he agreed with Huxley and Osborn that science could do as much as fiction: "Even so, when I try to imagine what science and technology can do in the remote future, I find it very difficult to foresee serious limits to our knowledge. All the imaginations of Jules Verne have come true, plus many more" (Cavalli-Sforza 2009d, 35).

Cavalli-Sforza believed that the scientific spirit was a universal human instinct that had driven humankind to invent new cultural traits to meet diverse challenges throughout their history. He certainly regarded "science proper" as a collaborative and self-correcting process directed toward the discovery of truth, as a continuous evolution toward a universal explanation of all observable phenomena—even if that goal may never be achieved (L. L. Cavalli-Sforza and F. Cavalli-Sforza 2005, 312,

320–321). We have already seen that he attempted to contribute to the development of a universal explanation where the phenomena of modern human evolution were concerned: "What is significant about Cavalli's work is that he incorporates data from multiple sources, using, for example, linguistic data to cross-check data based on genetic variation. This has enabled him to serve as a kind of grand synthesizer in the study of human prehistory" (Stone and Lurquin 2005, 20). The data from the more scientized approaches to linguistics and archeology could thus be incorporated into the picture arrived at by the "hard" methods of mathematical and molecular population genetics. For Cavalli-Sforza, the understanding of human prehistory came closer to universal explanation once it was described in terms of the "more rigorous sciences": "The major factor influencing the length of arguments among researchers is uncertainty, which tends to be greater in biology than it is in physics, and in physics greater than in mathematics. In anthropology the uncertainty reaches a peak, engendering a great deal of discussion and criticism" (L. L. Cavalli-Sforza and F. Cavalli-Sforza [1993] 1995, 261).

Cavalli-Sforza seems to have feared that cultural anthropology in particular could not keep up with the progressive kind of knowledge production in genetics: "In terms of theory, cultural anthropology is and always has been in flux. Unlike the hard science of genetics (and also less like the subdisciplines of archaeology and physical anthropology), it does not appear to make cumulative advances in the collection of data or the understanding of its subject—culture—but rather reflects intellectual trends and fashions in the broader humanities" (Stone and Lurquin 2005, 18). It might have been exactly to remedy this situation that Cavalli-Sforza developed with Feldman his own theory of cultural evolution on the basis of the hardest of the sciences—mathematics. Through quantification, the study not only of biological but also of cultural evolution should become progressive. Cavalli-Sforza attempted to open up "the possibility for cultural anthropology to be cumulative in its theory as well as its substance" (ibid., 113). This understanding of progress, as well as the claim that the unification of academic fields through mathematization might facilitate it, was largely the reason for the negative selection against Cavalli-Sforza's ideas in other knowledge cultures.

Cavalli-Sforza himself reasoned that cultural anthropologists distrusted quantification and mathematics, and he diagnosed a suspicion of thinking of culture in terms of evolution and progress caused by the guilty history of their discipline. They were still haunted by the outdated and racist notion of linear evolution and progress from primitive to civilized cultures (see for example L. L. Cavalli-Sforza and F. Cavalli-Sforza 2005, 268–269 and 277–278). However, although Cavalli-Sforza claimed not to judge what he perceived as different stages toward civilization, he nonetheless saw in the hunter-gatherers such as the "Pygmies" models for imagining life in the Upper Paleolithic. Further along these lines, South American indigenous peoples, for example, were "a little closer to certain Neolithic cultures, or even to some contemporary more primitive African farmers, and probably represent a somewhat later stage in the development of society, with greater sedentism, larger villages, and more intervillage competition" (Cavalli-Sforza 1986, 424–425). But even though "in terms of his anthropological interest in how culture evolves, he was drawn to these societies as representing the hunter-gatherer mode of subsistence, which is similar to the general human hunter-gatherer adaptation in the Paleolithic" (Stone and Lurquin 2005, 74), the Cavalli-Sforza/Feldman model was not developed from ethnographic data.

Indeed, cultural anthropologists reacted against the attempt at scientizing their field, and critics in particular resented the fact that quantification and mathematization functioned in the Cavalli-Sforza/Felman theory to arrive at a supposedly universal picture of cultural evolution. Such an attempt seemed to be driven by the same desire as human population genetics itself, for which the paleoanthropologist Ian Tattersall has observed that "it's a very internally coherent system, and we have very reductionist minds and are very receptive to this kind of illusory coherence. [Now?] the world actually turns out to be a much more messy place than the mathematicians would like it to be" (personal communication, 28 Dec. 2012). Culture is arguably even more "messy" than biology, and in a broad treatment of existing theories of cultural evolution that in various ways draw on biological evolution, including the one presented in *Cultural Transmission*, Joseph Fracchia and Richard Lewontin have offered this sweeping critique: "Cultural evolutionary theories are carefully constructed, logically consistent, and very neat. . . . But this formulaic treatment is fully inappropriate to the labyrinthine pathways, the contingent complexity, the many nuances, and general messiness of history. . . . Rather than being so flexible as to accommodate any historical sequence, they are too rigid in structure to be even plausible. They attempt to mimic, for no reason beyond the desire to appear scientific, a theory from another domain, a theory whose structure is anchored in the concrete particularities of the phenomena that gave rise to it" (Fracchia and Lewontin 1999, 77–78).[3] In the eyes of the biologist as well as the historian, the attempt to capture the processes of cultural transmission with tools developed in population genetics amounted to blatant reductionism, and it smacked of evolutionism and by inference Eurocentrism.

However, there were in fact other reasons for both Cavalli-Sforza and Feldman to turn to cultural evolution than the desire to scientize cultural anthropology (and history), and thereby to facilitate cultural progress. The attempt to understand cultural evolution was also provoked by claims that social structure and cultural potential were genetically determined. The 1960s did not only see the civil and minority rights movements, the demand for the realization of equality in all domains, and the rise of cultural relativism. With "race" riots against immigrants and anti-Semitic attacks toward the end of the 1950s, the new decade also witnessed a renewed racism in Europe and the United States (Selcer 2012, S180). By the late 1960s, it found one expression in the resurfacing idea of "race-related" differences in IQ, and Feldman and Cavalli-Sforza were at the epicenter of the controversy. At the opposite end of the spectrum from cultural anthropologists in the debate about the relation between biology and culture were those like the Berkeley psychologist Arthur Jensen who answered the question, "how much can we boost IQ and scholastic achievement?" (1969) with "negligibly." Another prominent figure in the debate was the Stanford professor William Shockley, who in Cavalli-Sforza's and Feldman's view correlated "race" and IQ without taking into account social inequality, for example, regarding access to knowledge. Feldman was horrified by Shockley's extreme stance that encompassed the endorsement of the sterilization of African American women (Feldman, interview, 22 Jan. 2013).

Cavalli-Sforza and Feldman countered such research with a model for determining the heritability of traits like IQ that incorporated the interaction of the genotype with parental cultural transmission—a model that was subsequently elaborated. This led to lower estimates of the heritability of IQ. Most importantly, as Hogben had already done, they pointed to the limited understandings of heritability evidenced in some studies. These studies missed the point that the variability of genotype expression depended on the environment. Echoing Hogben, Cavalli-Sforza and Feldman emphasized that high heritability of a trait said nothing about its genetic determinacy in another environment. From this, it also followed that the very attempt to relate the familial distribution of IQ or the measurement of its heritability to particular groups such as "races" was flawed. Their clarifications of these issues were aimed at counteracting policy suggestions based on claims like Jensen's, such as "racial" segregation in education (e.g., Cavalli-Sforza and Feldman 1973). Cavalli-Sforza did not challenge the claim that there existed a difference in IQ between American "whites" and "blacks." What he did challenge was the claim that this difference was genetic rather than environmentally induced. In other words, he doubted that the society of equal opportunity the triple-H had dreamt about had materialized. Indeed, it was still not known which environment would be most favorable for the intellectual development of the largest number of phenotypes. The interrelations between biology and culture still needed to be better understood—not in order to legitimize but to progressively change social realities (Bodmer and Cavalli-Sforza 1976, 511–515, 674–690; Cavalli-Sforza 1977, 97–98, 116).

Seen from the genetic-determinist side of the spectrum, one may regard Cavalli-Sforza's understanding of cultural evolution differently. In fact, like Huxley, he recognized a major break in the advent of language and conceptual thought—a break from a mainly biological to a mainly cultural evolution—and he understood culture to be "the heritage of knowledge accumulating over generations, or, we may also say, culture is what we learn from others and affects our behavior." Nonetheless, the Cavalli-Sforza/Feldman theory has been read as reductionist in the sense of explaining culture not only mathematically, but also genetically. However, a genetic explanation of culture is not the same as the attempt to capture processes of cultural transmission through the adaption of mathematical tools developed in population genetics. And I have already pointed out that although Cavalli-Sforza and Feldman conceptualized cultural changes in analogy to genetic mutations, there was one particularly important difference: "They [the cultural changes] are often directed toward a specific aim" (Cavalli-Sforza 2009b, 15). Cultural evolution could be goal directed, because rather than the genetic structure, the malleable neural structure was taken to be the biological substrate of culture.

Certain coevolution systems developed in sociobiology and evolutionary psychology encompass the notion that adaptive complexes that still play a role in the behaviors of contemporary humans are the result of adaptation by natural selection to Paleolithic natural and social environments. Cavalli-Sforza, on the other hand, seems to have thought that modern cultures remained the environment for which natural selection evaluates the fitness of traits. Even more so than Huxley, who had severed cultural evolution nearly completely from biological evolution, this view approaches the one held by Dobzhansky and Simpson: culture may influence genetic development. Perhaps even reminiscent of Osborn's notion of organic selection (or preadaptation), this process superficially mimics a Lamarckian inheritance of acquired characteristics (on Dobzhansky and Simpson, see Sommer 2010a). Cavalli-Sforza, for example, explained the rapid spread of genetically based lactose tolerance in adults with the increased adoption of a milk diet (Cavalli-Sforza 2009b, 16–17). Taken to its ultimate consequence, this means that a consciously planned development of culture might also bring about certain biological adaptations to it.[4]

Nonetheless, it seems that in the course of his career, Cavalli-Sforza's thoughts approached the ideas of Huxley also with regard to natural selection. Biological evolution—its mechanisms of natural selection, drift, and mutation—was increasingly expected to have nearly no impact on the future of humankind due to medicine, the dense population of the globe, and the future possibility of knocking out mutations. Future human evolution would thus have to be understood in terms of cultural change. However, according to the Cavalli-Sforza/Feldman theory (1981) also on the cultural level of development, there are cases of cultural transmission where no choice may be involved. Analogous to drift in biological evolution, some cultural traits will spread randomly, or they are handed down from parents to children without much or even any selection. Indeed, in *Cultural Transmission* we learn that transmission from parent to offspring, which is referred to as vertical transmission, may come close to imposition or unconscious adoption. This means that negative traits that are in this way vertically transmitted will be the most resistant to change. In fact, Cavalli-Sforza has made out such cultural handicaps to progress in the religious beliefs of "primitive cultures" as well as in certain political motifs (Cavalli-Sforza 2009b, 19).

On the other hand, Cavalli-Sforza and Feldman (1981) identified modi of cultural transmission that are more easily exploited to introduce and spread "good" traits. These include oblique transmission, or forms of between-generation communication that do not involve parent and child. Even more effective in this regard is the typical cultural mode of diffusion that consists in horizontal transmission—communication between people of the same generation. Indeed, horizontal transmission was regarded as so effective that it rather seemed to mimic the spread of an infectious disease than the biological hereditary process. In general, Cavalli-Sforza and Feldman maintained that the more widespread a trait, and the more people one carrier can reach, the faster the trait may travel. They stressed that in the current age, the modern technologies of communication allowed for unprecedented speed in the horizontal proliferation of cultural traits. We may add to this the impact of a charismatic scientist, and the conscious interference with cultural evolution comes into reach. It thus appears that Cavalli-Sforza acted in agreement with his own understanding of cultural transmission when he invested much time and energy in the education in and popularization of his science.

While Cavalli-Sforza and Feldman in *Cultural Transmission* thought about mechanisms of directive change in human evolution, they did not suggest toward which aims culture should be directed, and they drew no definite ethical and political conclusions from their models. However, such suggestions and conclusions can be found in other publications by Cavalli-Sforza, from which it becomes obvious that one monstrous second-order organism that he has long endeavored to slay—and that took center stage in the IQ debates—is racism. A closer look at this issue is illuminating with regard to the part he carved out for population genetics in cultural evolution. In his early book with Bodmer, written for a wider audience, he developed a narrative that would be retold with variation throughout his career. In *Genetics, Evolution, and Man* (1976), Bodmer and Cavalli-Sforza explained that there exist no pure races, that there had always been some admixture, that the distribution of most traits is clinal, and that frequency distributions change with traits. This last fact was driven home visually by maps of blood group distributions among "aboriginal populations of the world"; their confusing patterns show the messiness of the genetics of populations (ibid., 562–574). This was not to say that the genetics was meaningless, however. Rather, Bodmer and Cavalli-Sforza suggested that unlike trees based on anthropometric data that might agree with our intuitions about race because they group parallel adaptations to similar climates, genetic human population trees could be counterintuitive (ibid., 584–587). At the same time, on the basis of genetics, humanity might nonetheless be subdivided into "three large races": "Africans, Caucasians, and Easterners." Thus, Bodmer and Cavalli-Sforza maintained that "the broadest geographic groupings do correspond to a large extent with the available genetic data" (ibid., 574).

Decades later, Cavalli-Sforza opened *Genes, Peoples, and Languages* by deconstructing racial prejudice and the biological basis of a particular understanding of race under the heading of genes and history. As I took the German version to hand (Cavalli-Sforza [1996] 1999), and began to read with my European hat on, the opening chapter and its insistence on the German term *Rasse* seemed strange, at times even offensive. Cavalli-Sforza had by then spent most of his professional life in the United States and, as so often happens, transferred the seemingly omnipotent category of race to other cultural contexts without paying attention to their specificity. However, the point is that we find repeated the story of a physical anthropology of race that had been based on superficial and meaningless differences such as hair type and skin color, which in the end failed to define clear-cut races. Then came genetics that similarly failed to document clear-cut races and certainly pure ones. Instead of the races of physical anthropology, the nucleotide sequences eventually revealed the secret of history. For historical reconstructions, those molecular differences that are not affected by natural selection turned out to be more useful than those that tell how a group of humans had adapted to a particular environment. The traces of history in the gene had been left by chance. They are hidden away from the context, located in the biologically unimportant random mutations that, when measured over fairly long periods and averaged out over data from an adequate number of genes, are seen to occur reasonably regularly. While in the scientific journey from skin to DNA we lose a particular understanding of race, we gain a new understanding of history. Or, as Feldman (2010) has suggested, thinking along the lines of race should give way to an understanding of identity and diversity according to ancestry.

Lisa Gannett and James Griesemer (2004, 147) have in fact described this shift already for the blood group work in the first half of the twentieth century. They arrive at the conclusion that we are not dealing with a replacement of race by ancestry, but with a redefinition of race along genealogical rather than ecological terms.[5] Cavalli-Sforza declared the characters of physical anthropology to be meaningless for a new way of grouping people according to traits that have not been under natural selection. This new grouping practice therefore runs contrary to everyday notions of race as something visible to the eye and expressed in behavior. However, as we have seen, he nonetheless worked with certain group labels of old and confirmed categories such as "Caucasians." Furthermore, despite the accepted fact that within-population genetic variation exceeds that between populations, population trees were built on the latter's basis.[6] In this regard, Edwards (2003) has explained that in his famous paper, Lewontin (1972) only looked at loci independently. In other words, the "treeness" that Cavalli-Sforza and others found in the human genetic variation is based on "hidden" correlations between loci (Cavalli-Sforza and Piazza 1975). According to this reading, the circa 15 percent between-populations variation is of classificatory significance, because not all the characters studied are independent. With each such locus added to the comparison, the between-populations variation rises (though not in relation to the within-population variation). In fact, Lewontin himself has emphasized that although his numbers for within- and between-populations variation have often been confirmed, it does not follow that individuals do not genetically cluster according to geographical regions. With Feldman he also maintained that it is possible to assign an individual to a particular geographical region of origin on the basis of his or her genome (Feldman and Lewontin 2008).[7]

Still, Cavalli-Sforza thought that the universality of racism could be countered by acquainting people with the reality of the genetic history and present: "I do not know, but I believe a strong educational effort to eradicate racism is one of the most urgent needs. It may be impossible to totally eliminate it, but it should be possible to reduce the criminality with which it is constantly associated" (Cavalli-Sforza 2009b, 19). Just as Huxley had demanded that mentifacts be brought under conscious and scientifically informed cultural selection to achieve cultural progress, Cavalli-Sforza has worked toward the destruction of certain second-order organisms that he considered a threat to progressive development. Besides racism as well as violence and ignorance in general, he has identified poverty, population increase, and drug abuse as the main problems facing humankind—interlinked problems that partly had already troubled Huxley and his peers (Cavalli-Sforza [1996] 1999, 224–227). At the same time, the above considerations suggest that Cavalli-Sforza also tried to infuse the system with new cultural traits that he considered favorable to human progress. It is like a nucleotide substitution consciously brought about: "history" for "race." In general, one is strongly reminded of Huxley and the Idea Systems Group's notion that progress could be brought about by adapting idea systems to knowledge about the natural world. This required the cultural work of rewriting history from the evolutionary perspective and of making that history's implications for our identities and future possibilities inform historical consciousness.

Also where the ultimate aim of cultural progress through scientific enlightenment is concerned, there are parallels between Huxley's and Cavalli-Sforza's thinking. Already in *Cultural Transmission* (1981), Cavalli-Sforza and Feldman claimed that whereas natural selection favored traits that increase survival and reproduction, cultural selection was oriented toward immediate and future self-satisfaction. For Cavalli-Sforza, the goal of cultural evolution consisted in achieving as high a degree of self-fulfillment for as many people as possible. Indeed, some of Cavalli-Sforza's publications may be described as self-educators for the citizen with the very purpose of instructing him or her on how best to achieve self-fulfillment through the adoption of the "right" kinds of cultural traits. This was certainly his intention with *Chi siamo*: *La scienza della felicità* (1997), on which he once again collaborated with his son. It appeared in German as *Vom Glück auf Erden: Antworten auf die Frage nach dem guten Leben* (F. Cavalli-Sforza and L. L. Cavalli-Sforza [1997] 2000). It is a guidebook to *happiness* (English for *felicità* and *Glück*). The first part relies on the expertise of Francesco and treats philosophy, ethics, and religion in their historical and global dimensions to acquaint the reader with different ideas about the good life that humans have come up with. The second part, more Luigi Luca's realm, is a treatment of the great impact the natural sciences have on current answers to the question about the right way of living. In the third part, the perspectives from the "two cultures" are brought together to tackle fundamental problems of life in today's world for which science is identified as the source as well as the possible solution. From the outset, self-fulfillment is placed in the hands of the individual, and the happiness of the individual is presented as harmonious with the interest of the group. In agreement with Huxley's aims, what the authors most wished for was that human beings—as individuals and groups—could understand the world as completely as possible in order to enjoy its plenty to the full (ibid., 15).

Part two of *Vom Glück* begins with a tour of human evolution. Cavalli-Sforza explained that the biological evolutionary process had produced an incredible variety of organisms and enormous complexity that culminated in the human being, with whom there had arisen the possibility of directed progressive cultural evolution through the capacity to create innovations to solve specific problems and meet definite demands. He emphasized that the process of enlightenment had made a fast move forward with the step from the hunter-gatherer way of life to agricultural subsistence. The new lifestyle entailed a better knowledge of the world and it was biologically a successful strategy. Continuing the

Huxleyan tune, both father and son suggested that while at the time of writing there existed incredible possibilities to influence ourselves and our surroundings, humans had shaped not only their own bodies, minds, and social structures, but also their living and nonliving environment since the invention of agriculture. Already in their joint publication *The Great Human Diasporas*, the Cavalli-Sforzas argued that starting from agriculture and livestock farming, genetics "has made giant leaps forward and has revealed the nature of life itself. It has given us extraordinary power to modify living organisms, even if few of the potential applications have so far been developed, and mainly in the field of medicine. However, it is clear we are on the threshold of a new era" (L. L. Cavalli-Sforza and F. Cavalli-Sforza [1993] 1995, 262). Reminiscent of Huxley's evolutionary humanism and Hogben's planned ecology, Cavalli-Sforza elsewhere suggested that "with the modern triumph of communication cultural evolution is becoming more and more the directional force of human evolution, and genetic evolution may well end up be [*sic*] completely under its control. Even the evolution of animals and plants is undergoing intense acceleration as a result of cultural evolution in humans" (Cavalli-Sforza 2009b, 17).

In *Vom Glück*, Luigi Luca and Francesco Cavalli-Sforza had some thoughts to share with respect to a genetic evolution under cultural control through gene technology (F. Cavalli-Sforza and L. L. Cavalli-Sforza [1997] 2000). In their guidebook to happiness, the Cavalli-Sforzas lamented that although medicine could cure more and more diseases that were at least partly genetically induced, genetic therapy was a thing of the future. Thus, the "genetic load"—as Muller once called it—was on the rise, and with it the cost of medical treatment. Some counterbalance to this trend was made out in genetic screening during pregnancy, which, in combination with the abortion of the severely handicapped, was wholeheartedly embraced; it even appears as a responsibility (ibid., 282–286, 289–294). While they distanced themselves from classical eugenics, the Cavalli-Sforzas thus welcomed preventive abortion, gene therapy, and the genetic manipulation of animals and plants (ibid., 303–306). Nonetheless, several decades after the later writings of the triple-H, eugenics as a truly biological science was still seen to lie in the future. In fact, the Cavalli-Sforza father and son warned against the tendencies of predictive genetic medicine. Genetic diagnostics was explained as always impaired by the role of the environment, and it was a statistical issue. Moreover, the awareness of certain genetic predispositions for diseases might be harmful to people. It might be personally upsetting and lead to discrimination on the part of employers and insurance agencies.[8]

For Cavalli-Sforza, too, the present possibilities consisted in the change of cultures; and, as history could tell, social progress toward happiness had been made: slavery had been abolished, working conditions had become more humane, and at least in the West, exploitation in general had decreased. More could be learned from the deeper past. Although agriculture had set in motion an explosion in knowledge, it had actually meant reduced levels of happiness—an understanding that Cavalli-Sforza claimed was gained by anyone who had come into contact with hunter-gatherers. In fact, the experiences Cavalli-Sforza himself made when collecting blood samples for genetic analysis from "African Pygmies" had been one of the incentives to study culture in the first place.[9] Through "the pygmies" he came to realize that life in the Paleolithic must have been better: there had been adventure, enough spare time for play, enjoyment, and dance, as well as a great degree of social equality. In contrast, in *Vom Glück*, he compared the advent of agriculture to the expulsion from paradise, bringing hard work, disease, private property, exploitation, famine, war, and population density (F. Cavalli-Sforza and L. L. Cavalli-Sforza [1997] 2000, 174–175). We have already encountered this kind of nostalgia that is expressed in several of Cavalli-Sforza's publications in Osborn's and Huxley's understandings of the Paleolithic Cro-Magnons and the "more traditional African lifestyles" respectively (see, for example, L. L. Cavalli-Sforza and F. Cavalli-Sforza [1993] 1995, ch. 1; Cavalli-Sforza [2010] 2011, chs. 6 and 7). And also for Cavalli-Sforza, it was impossible to reverse history; the solutions to today's problems had to be found using today's means. At the same time, the Paleolithic Eden might serve as a model, and he argued for a new democracy—"new" in comparison to the "old democracy" of hunter-gatherers—as the only real counterbalance to relations of exploitation that had their roots in agriculture and that had been globalized in the age of European expansion.

Like Huxley and Conklin, Cavalli-Sforza thought that only in the ideal democracy could there be a true balance between individual rights and responsibilities and true equality of opportunity. Just like the triple-H, he spoke in favor of variation and wanted that variation brought to bear on human progress by having every individual fill the place in society he or she is best suited for owing to nature and nurture. In the ideal democracy, everyone would be free to develop his or her personality to the full and to strive for the greatest satisfaction possible, so long as this did not happen at the expense of the overall level of satisfaction. In 1976, Bodmer and Cavalli-Sforza had referred to this process of creating a society that favored maximization of satisfaction with one's life as "equimaximation." It seemed clear that "the need for a great variety of skills in modern society makes the existence of such variation [as the result of gene-environment interaction] an asset and not a liability if proper consideration is given to it. The aim of genetics is not just the bare study of mechanisms of biological inheritance, but more generally, the analysis of variation" (Bodmer and Cavalli-Sforza 1976, 704). At times it seems that for Cavalli-Sforza, among the "advanced nations," the United States was closest to the ideal of a meritocratically organized democracy, with checks and balances that reined in the power of politicians. American universities might serve as a model for a society in which, despite competition, collaboration is a dominant factor, and members are given opportunities to try their ideas and visions and are ultimately judged by their success. In such a democracy, transnationally institutionalized, he perceived the way to global happiness (F. Cavalli-Sforza and L. L. Cavalli-Sforza [1997] 2000, 325–331; 2005, ch. 15).

Fortunately, in striving after this goal, human nature was not only foe but also friend. In *Vom Glück*, Cavalli-Sforza explained that evolution had not only come about by struggle and competition. Besides the drive for survival and reproduction, in organisms living in groups, there was also a social instinct. The relevance of these three factors to the book's title consisted in the fact that we are biologically programmed to experience happiness in the fulfillment of nature's imperatives: life itself, reproduction, and sociality. Cavalli-Sforza allowed that there were in reality people who found or thought they could find their happiness elsewhere—in death, a life without children, or the life of a hermit—but these were discounted as reactions to abnormal situations (ibid., 245). Like Huxley, Cavalli-Sforza also believed that humans were equipped with a biological basis for altruism, and if the world was to solve the problem of overpopulation and all the consequences it might entail, such as loss of resources and natural habitat, epidemics, famines, and war, they had to enhance this natural tendency (ibid., ch. 9).[10]

In sum, for Cavalli-Sforza the goal of progressive cultural evolution should consist in making humankind happy. Progress meant a net increase in happiness—in dignity, peace, and possibilities for the realization of one's potentials. Although he considered it possible that there are genes for happiness (F. Cavalli-Sforza and L. L. Cavalli-Sforza [1997] 2000, 315), in order to make their carriers happy, these genes depended on the right cultural environment transmitted over generations. Like Huxley's purpose in life, for Cavalli-Sforza happiness was ultimately not found but made. The science of happiness would then be the endeavor to understand the conditions for happiness in order to advance them; it would study the history and diversity of humankind from the Paleolithic time; and it would have to include philosophy, religion, politics, genetics, neurobiology, and ecology. Again reminiscent of the Huxleyan fulfillment society, the Cavalli-Sforzas held that on the basis of the knowledge thus created, society should help each individual in finding his or her perfect place in it, for example, through tests that bring to light talents and affinities, and the mutual aid in finding happiness would have to be extended to nonhuman organisms on the basis of a billion years of common evolutionary history (ibid., 367–382; also L. L. Cavalli-Sforza and F. Cavalli-Sforza [1993] 1995, epilogue; Caprara et al. 2009).

Like Huxley in his total evolutionary cosmology, Cavalli-Sforza saw at work a relay, in which each generation inherited through parents, teachers, and communication technologies not only its biological constitution and its culture, but also its natural environment. And this natural heritage, too, had to be taken better care of. Cavalli-Sforza imagined that in western Europe, where people had learned from many wars, where population size decreased, where an economic crisis seemed to stay for good, and where many felt that happiness lay not in wealth alone, a new way of thinking might arise that would lead to a conscious management of the natural and cultural heritage as well as to an investment into research, art, and practical intelligence as the means of progress. Just as for Osborn and for Huxley, for Cavalli-Sforza, while hunter-gatherer communities might serve as model and inspiration, it was the "advanced" world that was ultimately seen as the critical force for further progress (F. Cavalli-Sforza and L. L. Cavalli-Sforza [1997] 2000, 331–338).

The perceived need to cultivate the cultural and natural heritage also motivated the mega-projects that are of concern in the next chapter. They involve the aim of collecting and preserving the genetic as one aspect of the cultural heritage. This endeavor—which was felt to be urgent—was driven by the belief that the cultivation of the common genetic legacy as a source of knowledge about origins, histories, and identities, as well as the distribution of this knowledge, would further mutual tolerance and ultimately progress. "Understanding history and evolution," the Cavalli-Sforzas wrote, has much to contribute to progress and happiness, because "to develop our personality harmoniously, we need to study and respect individual variation, be it cultural or biological," and because "humanity deeply needs to understand itself better and learn to exploit its cultural inheritance in far better ways" (L. L. Cavalli-Sforza and F. Cavalli-Sforza [1993] 1995, 262). However, the attempt to use the mapping technologies in the appropriation of indigenous peoples' genetic history globally proved to be more difficult than expected, and the meaning of *history* remained contested between diverse scientific and more traditional understandings.

CHAPTER THIRTEEN

The Geography of "Our Heritage"

From the Human Genome Diversity Project to the Genographic Project

The Human Genome Diversity Project (HGDP) was not a cooperation built from scratch to begin global sampling. Rather, it was the attempt to establish ongoing collecting and research on a much wider scale. The effort that *The History and Geography of Human Genes* (1994) represented, including the associated sampling, was closely related to the HGDP. In fact, as Jenny Reardon (2005, 50) has shown, a draft chapter of *History and Geography* served as the basis for formulating the call for the HGDP (Cavalli-Sforza et al. 1991). And vice versa, Cavalli-Sforza, Menozzi, and Piazza referred to the newly started project to collect representative DNA samples from all over the world in the proofs of *History and Geography.* As we have seen in chapter 11, the book was mainly based on classical markers, and with Cavalli-Sforza in his seventies, the development of the new Y-chromosomal system was only in its infancy. Thus, in the wake of the revolution through sequencing techniques especially as applied to mtDNA, *History and Geography* actually highlighted the need for the HGDP. And with the Human Genome Project under way, the researchers hoped that a small fraction of that money could be gained to boost the kind of diversity research that had marked Cavalli-Sforza's career. His interests were reflected in such HGDP aims as the investigation of the Indo-European expansion in Europe or the population migrations following critical cultural innovations like the domestication of plants and animals. The idea was to study the genetics of linguistic isolates and the relative importance of drift, selection, and admixture. This might make it possible to tackle the question of the degree to which cultural evolution parallels genetic evolution.

As spelled out in preparatory meetings of the HGDP, to this purpose, a core set of markers should be defined for which the genetic material would be tested. These markers would certainly include the mtDNA control region, human leucocyte antigen (HLA) alleles, and blood groups. Obviously, a technological infrastructure that allowed the exchange of wetware, practices, and information was required. Thus, classical markers and RFLPs would have to be adapted to PCR-based systems, a technology that the project intended to develop further in order to facilitate the faithful amplification of DNA fragments or genomes. Y-chromosomal polymorphisms were desired for complementing the mtDNA results on female lines. And microsatellite studies should accompany the analyses for those markers that had proven stable within and between labs. Traffic also needed to be regulated. For example, the work on classical markers was intended to be restricted to regional labs, because the plan was to distribute only DNA but not cell lines. Another issue was the development of data bank software. Finally, there should be a data-coordinating center that would hold the global database, in which data and the information accompanying it could be stored. The project would also aim at establishing worldwide communication channels where no Internet was yet available. Yearly international forums would bring together the executive committee with regional committees and should attract representatives of supranational organizations such as UNESCO and WHO, of funding institutions, and of the populations to be studied.[1]

The project was considered timely not only because of the new technical possibilities, but also because the genetic traces were understood to be endangered. The authors of *History and Geography* announced that "there is another reason for starting a major program in analyzing human diversity now. While our potential skills for analyzing human evolution are increasing, social changes taking place in developing countries are rapidly destroying the identities—if not the very existence—of the most important aboriginal populations. Thus, organized research efforts to save this precious information about our past have acquired a new urgency" (Cavalli-Sforza, Menozzi, and Piazza 1994, ix). "Our" past, or "our" history antedating the great migrations following 1492, was perceived to reside in the gene pools of "aboriginal populations" because they were thought of as relatively isolated (genetically unmixed). At the same time, this history that could be read from particular human groups was presented as a panhuman past: "From the point of view of genetic history, we are an endangered species, and it is essential to avoid delay before taking the necessary steps to preserve this important knowledge about ourselves" (ibid., 157). Rendering the population genetics of "aboriginal populations" in terms of the history of the entire species allowed it to be phrased within the semantic field of human heritage: "This is a critical time for organizing our efforts before we lose a unique opportunity for understanding our genetic heritage" (ibid., x).

For Cavalli-Sforza, this sense of urgency might have been increased by his approaching retirement in 1992, when he would become an emeritus professor at Stanford. The HGDP certainly represented the culmination of his working life, but he also needed research funds to maintain his lab. Finally, the urgency associated with the heritage salvation discourse resonated with wider societal concerns. The initiators of the HGDP promoted a concept of genetic heritage that analogized human gene pools to human languages and archeological and historical legacies at a time when public awareness of issues surrounding cultural heritage was at a peak, and the field of cultural heritage studies began to emerge (Smith 2007): "We must act now to preserve our common heritage" (Cavalli-Sforza et al. 1991, 490). In fact, Cavalli-Sforza and colleagues feared the genetic traces of the original human migration patterns and kinship in jeopardy owing to the same processes that are among those brought forward to explain the turn toward the past in Western publics, and in cultural studies, that began in the 1970s and culminated in the 1990s: the (perceived) acceleration of technological and social change as well as processes of globalization, mass migration, and genocide (e.g., Lowenthal 1998, ch. 1).

While the global infrastructure and scientific-technological advance made the HGDP feasible in the first place, these developments were also part of the threat to those human beings whose genetic lines were understood to connect them more or less "uncontaminated" to the past. The HGDP responded to this situation with the plan "to identify the most representative descendants of ancestral human populations worldwide and then to preserve genetic records of these populations."[2] Because the project wanted to preserve the genetic history in the bodies also of those indigenous peoples who were seen to face extinction, cell lines would have to be established, so that "the DNA would never run out" and "to make sure that most extant human populations would be represented forever" (Feldman, interview, 22 Jan. 2013). The donors would have to give blank consent in order to allow researchers to apply their increasingly sophisticated methods into the future.

The notion of heritage at play in the project rhetoric thus runs counter to the discourses in cultural studies. Recent scholarship in heritage studies warns against an understanding of world heritage as a fund of objects and events to which a historical meaning is inherent. The archeologist Laurajane Smith (2007), for example, argues that relics from our "cultural heritage" only attain meaning through the role they play in the negotiation and performance of identities. Heritage would therefore always be of an immaterial nature. On the contrary, the rhetoric of the HGDP was characterized by the understanding of human gene pools as containing authentic organic traces of a past that only needs to be read from the molecules. As the project organizers expressed it themselves, "The main value of the HGD Project lies in its enormous potential for illuminating our understanding of human history and identity."[3] However, as in any "heritageisation" process (Harvey 2007, 26), this "history and identity" would have to be negotiated—and it was likely to be controversial.

The discourse of a history and identity already fixed in DNA could draw on existing notions. The Human Genome Project (HGP) played an important role in popularizing the rhetoric of DNA as constitutive of our life as humans: "There is no more basic or more fundamental information that could be available" (Gilbert 1992, 83). The individual human being at times appeared as completely defined by the nucleotide sequence in her or his genome, so that one's identity might be carried around in the form of this code on a compact disc (ibid., 96). And the HGDP followed on the heels of the HGP. It was Bodmer—who at that time presided over the Human Genome Organisation (HUGO)—who recommended setting up a Human Genome Diversity Program under its umbrella. Up to the point of his sudden death from leukemia the same year, Wilson served as cochairman with Cavalli-Sforza, and from those scientists we have already met, not only Bodmer and his wife Julia, but also Ken Kidd, Feldman, and Piazza joined the ad hoc HGD committee (later executive committee). However, although the discourses of the HGP overlapped with those in molecular and genetic anthropology, the two have to be kept apart. The HGP aimed not to unravel the genomic differences between humans and other primates or between human groups, but explicitly to construct a universally human genome to tackle medical questions on the molecular level. In fact, chromosomes of different individuals were morphed into the universal (male) genome. According to its rhetoric, the HGP was after "the underlying human structure" to "reflect our common humanity" (ibid.).[4] Although within the HGDP confusingly similar humanist statements were made, its main focus was on intrahuman genetic difference.

At the same time, the HGDP did not completely lack the HGP's medical orientation. Cavalli-Sforza himself had been interested in genetic medicine, and he had invested some energy into such research; after all, he had been trained in medicine. In the conception of the HGDP, the hope was expressed that some diseases would be more common in and therefore more easily isolated from particular indigenous groups, and planning workshops did include medical issues such as genetic markers linked to predispositions for and resistance to certain diseases. Furthermore, the proposal to establish cell lines was also made in the opinion that medical research should not be excluded. And it was this aspect of the project that turned out to be most controversial, that was responsible for the suspicion that it constituted an attempt at neocolonial exploitation. The participants of HGDP were unable to guarantee that the effort would not generate results that could lead to commercially profitable pharmaceutical or other products. It was surmised that all they could do was promise that in the very unlikely case of the patenting of such a product, they would help to ensure that some of the profit was returned to the sampled populations.

Despite the medical interest, however, the infrastructure for the global management of our genetic heritage would mainly facilitate "purely anthropological research." The project was in fact distanced from medical approaches. At the same time, it was distanced from anthropological research into phenotypic differences between groups. As Cavalli-Sforza, Menozzi, and Piazza wrote in *History and Geography*: "None of the genes that we consider has any accepted connection with behavioral traits, the genetic determination of which is extremely difficult to study and presently based on soft evidence. The claims of a genetic basis for a general superiority of one population over another are not supported by any of our findings" (Cavalli-Sforza, Menozzi, and Piazza 1994, 19–20). Following an assumption that we have traced back as far as the 1930s, the epistemic objects of genetic history appeared not only scientifically fundamental but also politically neutral. The knowledge gained from them would lend no support to racial discrimination. This rhetoric would reach its climax with the Genographic Project. Only once there was a clear focus on the supposedly noncoding regions of DNA, as in the case of mtDNA, and even more so on nonfunctional regions, as in the case of the Y chromosome, could it gain full force. As we will see, this so-called *junk DNA* came to be presented as free from any information about the phenotype—it was finally the object that pertained exclusively to history and genealogy.

Nonetheless, also for the HGDP, Jenny Reardon (2001; 2005, ch. 5) has shown that it was such a notion of scientific objectivity, as allowing access to a socioculturally neutral kind of knowledge, that turned the well-meant initiative into a vampire project, an expression of racist and colonialist exploitation, in the eyes of the World Council of Indigenous Peoples. The belief in a politically neutral science blinded the project participants to the fact that population genetic studies also always coproduce the social and the natural order. The research project was not designed in communication with the "research subjects," so that indigenous rights organizations and some indigenous groups feared for the autonomy of First Nation Peoples. The project did not seem to be aimed at their future welfare, but only interested in the benefit to Western science and economy.[5] In harmony with Cavalli-Sforza's social goals, the HGDP participants stated that "there is a cultural imperative for us to respond to that opportunity and use the extraordinary scientific power that has been created through the development of DNA technology to generate—for the benefit of all people—information about the history and evolution of our own species."[6] However, this enlightenment notion of an objective science, the knowledge from which could lead humankind into a better future, only exacerbated concerns of neocolonialism and racism. And it is somewhat paradoxical that while the research was presented as politically neutral because of the power of technology and the fact that it was not about phenotypic differences, its results were believed to further mutual understanding and panhumanism.

Nevertheless, where its organization was concerned, the project was conceived as a centralized yet networked system. The idea was to establish regional laboratories or to expand existing ones through technology transfer to enable them to perform cell transformations and sample preservations on the material collected by local collaborators. While providing samples and cell lines to the (qualified) open access global repositories, the regional labs might retain the local lines and control over their management. As the local pendant to the central committee, there should be regional committees with their ethical subcommittees to do justice to the great differences in social, political, and economic contexts.[7] Furthermore, the project organizers also became aware that problems might arise from the clash between scientific and traditional origin accounts. They invested in solving ethical problems and forestalling the possibility of racist abuse, beginning with a workshop on ethical and human rights implications in 1993. Workshop participants felt that the issues of patenting genes and of turning the research on the indigenous samples into profit for Western companies were most urgent and difficult. Nonetheless, some anticipated that the findings of the project would harden rather than disperse people's belief in clearly defined races, and that the knowledge from the project could be appropriated to diverse political aims and ideologies. To this, Cavalli-Sforza responded with his standard account of how population genetics had revealed and would continue to reveal the biological meaninglessness of the concept of race and would thus help undermine racism—a notion the historian of science Diane Paul immediately pronounced naive.[8] Indeed, Cavalli-Sforza maintained that science was "morally clean" and only its applications might be socially problematic. Because the science of population genetics revealed the truth that racism was untenable, its social effects should be altogether beneficial.[9]

However, the concerns of indigenous groups had been exacerbated by their conceptualization in the project as stuck in a premodern past. Confronted with these people's voices, the project organizers admitted that "errors have been made" in terminology and reiterated that no judgment regarding the value of any people, culture, or language was involved.[10] Nonetheless, this stance was repeated in the document produced after the last planning workshop for the project: "Study of these populations optimizes the ability to reconstruct the ethnographic map to its state at the beginning of recorded history."[11] Again, genetic map making appears as a technology that is unconcerned by actual living people. Rather, the human world is to be re-created as it once occurred. Thereby, the globe is peopled with populations (re)constructed as they were in the past, ideally indicating their genetic relations and migrations. The paradox was that at the same time, a connection was seen to exist between the ethnographic map thus established and actual "isolated" communities, which betrayed the different conceptualization of industrialized and mixed societies and those humans who were taken to be representative of ancestral human populations.

In sum, while a public education program on the aims of the project was envisioned, the issues of how populations were to be conceptualized and sampled; how consent should be gained; how discriminations, exploitations, as well as negative self- and outside perceptions of indigenous groups should be avoided; and how gains should be shared and control given to the groups studied remained publicly debated. Whatever prognosis the experts made, many were aware that while the sampling of indigenous peoples for human population studies had gone on for decades, the HGDP brought these issues onto the world radar—or should we say map? As Ricardo Ventura Santos (2003) has shown, while in the 1960s it was still possible to suggest the analysis of "the primitives" to gain access to the human past, in the 1990s, in the international context of indigenous rights movements, the conditions for the possibility of doing human population research had changed.[12]

Despite the public affairs, Cavalli-Sforza received, among many other distinctions, the Balzan Prize of 1999 for the Science of Human Origins. He was honored as "the world's expert on human genetic diversity and what it tells us about the phylogenetic tree of human populations" (International Balzan Foundation 2009, 5). However, when revising his Balzan summary of research in 2009, he still felt hurt: "Our work has been made particularly difficult by totally unjustified attacks full of lies by a special-interest group, a Canadian NGO, which has an established network among some indigenous populations of America and Oceania. It accused the Human Genome Diversity Project of being behind totally unrelated patenting efforts of cell lines by the U.S. National Institutes of Health, of being interested in profit and consorting with pharmaceutical industries, and various other lies. The truth is that the HGDP is a nonprofit institution, and has always been against patenting DNA" (Cavalli-Sforza 2009b, 20). While Cavalli-Sforza expressed sympathy toward the demands of "a few" organized indigenous groups for a share in the profits from genetic variation, he felt that this was difficult to realize because of the high uncertainty and cost intensity of the research.[13]

However, and again in continuation of Cavalli-Sforza's earlier work, the HGDP has not only met with criticism from indigenous rights organizations. It has also been the site of struggles over authority of expertise on culture and history between the genetic and other approaches, even though "for all the criticism of the HGDP, many anthropologists and others are interested in or committed to a scientific understanding of the human species, its history and variation" (Stone and Lurquin 2005, 176). Indeed, as Reardon (2005, ch. 4) has shown, Mark Weiss, director of the

Physical Anthropology Program of the National Science Foundation (NSF), regarded the participation of the field he represented as essential for the HGDP. He considered the expertise of physical anthropologists on human diversity crucial. By the 1990s, many physical anthropologists included genetic methods in their work—genetics had become part of the anthropology curricula in the 1970s and 1980s—and some thought that they should actually lead the HGDP, because they had knowledge of both biological and cultural analysis.[14]

It was decided that the genetic sampling should be carried out by anthropologists who knew the groups, and that it should go hand in hand with the collection of linguistic, social, and health data. Nonetheless, some anthropologists felt their role was conceptualized as subservient, as collecting the data that would then be analyzed by the "real scientists." The demands for help in the sampling work strengthened the impression that cultural knowledge only served to identify the groups whose individuals were seen to carry the most relevant genes (Reardon 2005, ch. 4).[15] In spite of these concerns, the synthetic rhetoric remained part of the project's language. It continued to be hailed as an important step toward an integrative view, bringing together experts from different historical disciplines and thus bridging the gap between "the two cultures": "The primary case, therefore, for the Human Genome Diversity Project is cultural. . . . The study of genetic polymorphism in human populations creates a unique bridge between the science of human genetics and the humanities, including anthropology, archaeology, history and linguistics, and presents scientists with a unique opportunity to contribute to the world's cultural heritage."[16] That a project with the main focus on human genetic variability at the population level could be viewed as a cultural imperative and a cultural project of contributing to the world's cultural heritage is in line with Cavalli-Sforza's perspective on cultural evolution. However, also in the context of interdisciplinary cooperation, this outlook did not assuage controversies but rather rendered the different stances also on ethical issues more pronounced.

Ethical issues were discussed at the Wenner-Gren symposium on "Anthropological Perspectives on the HGDP" that was held in November 1993. It brought together geneticists and anthropologists from the four fields—physical anthropology, cultural anthropology, archeology, and linguistics. As Sydel Silverman (2002, 126–133)—then president of the Wenner-Gren Foundation—remembered, the matter was framed in terms of the anthropological legacy. Some feared the HGDP approach would confound biology, language, and culture. We have met this fear in the discussion of Huxley's and others' understanding of population studies in the interwar years. It is the fear about circularity when identifying sample donors in terms of culture and then studying their genetics as a population. It is the fear that this could result in the biologization of cultural groups whose history included discrimination and violence. In the view of many anthropologists—including genetic anthropologists—postracial and postcolonial anthropology had an ethical obligation toward holism. In a certain sense, the project also seemed to erase history rather than to reconstruct it, because it focused on groups considered uncontaminated by European expansion. The HGDP thus appeared as a continuation of nineteenth-century anthropology in the service of colonialism with respect to the way in which it classified non-Westerners as distinctive tribes stuck in the past (this argument goes back to Fried 1975). Anthropologists therefore rejected the "service role" also because of the part played by physical as well as cultural anthropology in the service of imperialism and colonialism.

Cavalli-Sforza's biographers have described the situation as follows: "Given its own particular history and consequent sense of guilt, the profession of anthropology especially fears three things: charges of racism, the wrath of indigenous groups, and attempts to deny or supersede local peoples' own understanding of themselves and their world. The HGDP has stirred up all of these fears. In short, the HGDP reminds many anthropologists of what they most dislike about their own professional history" (Stone and Lurquin 2005, 175). However, the biological anthropologist Jonathan Marks (2012) has emphasized that eugenics and populational blood group studies that attempted to define human races are legacies of genetics and population genetics. So why was Cavalli-Sforza's and others' memory not troubled by the "three things"? Indeed, Cavalli-Sforza's intensive treatment of the subject testifies to his being troubled by the history of race science and racism; and the prevalence of these topics in human population genetic discourses more generally suggests that it haunts current human population genetics just as much as current anthropology; with the difference, perhaps, that while especially cultural anthropology has reacted with a high degree of self-reflexivity, human population genetics often continues to rely on scientific objectivity.

These complex discussions, which did not strictly follow disciplinary lines, related to the particularity of American anthropology. An integrative anthropology of the four fields is a specialty of the American tradition. It had reached a turning point with the German immigrant Boas, who on the one hand was in favor of that tradition, but on the other hand criticized physical anthropology as well as the evolutionary perspective on culture. In particular, he disentangled the common conflation of "race," culture, and language (Boas 1901). With his relativist and egalitarian science and politics he thus also unleashed centrifugal forces that would distance cultural anthropology increasingly from biological anthropology. Nonetheless, in the United States, the four fields have remained more closely related (at least institutionally, educationally, and discursively) than in Europe. And the Wenner-Gren Foundation stands for promoting integrative symposia and projects. At the same time, there are important differences between a constructivist and a positivist cultural anthropology, and between processual and postprocessual approaches in archeology. Furthermore, because of the evolutionism of the past, integration with biological anthropology is increasingly questioned. Some scholars see the biocultural approaches in anthropology as a reduction of culture to biology, and they are skeptical about evolutionary regularities. They see in these methods and concepts an attack on the interpretative sociocultural approach (Segal and Yanagisako 2005). Given these developments, representatives of the Wenner-Gren Foundation and the officers of the NSF Anthropology Program were enthusiastic about the HGDP precisely because they thought it would promote four-field collaboration. However, the impact of the Wenner-Gren symposium remains questionable in view of the history of the HGDP. Even at the outset of the symposium, proponents of the project made it clear that it would go on with or without anthropology (Silverman 2002, 126–133; Reardon 2005, ch. 4; Sommer 2012b).[17]

In fact, despite the resistance also from "ill-advised anthropologists" (in Manni 2010, 252), Cavalli-Sforza continued his work in the HGDP. Besides him, Piazza, the Kidds, Friedlaender, Bodmer, Feldman, and Peter Parham and Alice A. Lin of the Stanford lab were among the members of the Human Genome Diversity Panel (HGD Panel) (Cann et al. 2002). The Diversity Project samples were stored at the Centre d'Etude du Polymorphisme Humain (CEPH) in Paris with the aim to provide noncommercial laboratories with DNA (without patenting possibility) for research on genetic variation.[18] Because, according to Cavalli-Sforza, the potential sponsors had been chased away by opponents of human population genetics, these samples have mainly been collected by his personal friends.[19] A project that was officially announced in 1991 was thus beginning to bear fruit in 2002, when DNA could finally be distributed to research labs. Some two decades after the call for the HGDP, the HGD Panel included a collection of lymphocyte cell lines from more than 1,000 anonymized individuals (there is no phenotypic information, only on the population from which the samples came), belonging to fifty-two ethnic groups from the five continents. As one outcome, a group of Stanford researchers mostly from the Department of Genetics made available on the Internet and published in *Science* the results of an analysis of the 948 unrelated individuals of the HGD Panel collection for 650,000 DNA nucleotides known to be most variable in humans (Li et al. 2008).[20] This represented the largest set of data available on the genetic variation of any species, and its full statistical analysis was estimated to occupy the researchers for some time to come. In early 2013, more than a hundred labs had accessed the HGD Panel DNA (Cavalli-Sforza 2005c, 2009c; also Manni 2010, 253; Feldman, interview, 22 Jan. 2013).

The HGD Panel has not remained the only project of its kind with which Cavalli-Sforza was associated; he was, for example, involved in one concerning the genetic heritage of Italy. Upon retirement, he again spent time in his home country. Still feeling very much connected to it, he set up an Italian Genome Project (in which he cooperated with Professor Gianna Zei of the Institute of Molecular Genetics of the National Research Council). Following the model of the HGDP, lymphocytes from the Italian Association of Voluntary Blood Donors from donors of provinces that were taken to be representative of the different Italian regions on the basis of anthroponymic methods were collected. This means that in accordance with *Consanguinity, Inbreeding, and Drift in Italy* (Cavalli-Sforza, Moroni, and Zei 2004), surnames were used to determine the ethnic origins of the individuals, once again correlating cultural and genetic inheritance and identification. BioGenomic Technology produced cell lines that would provide the material for research on the population genetics of Italy, but also for medical genomic research (Cavalli-Sforza 2009c).

As a further contribution to a particularly Italian cultural evolution, Cavalli-Sforza began to write science textbooks for Italian secondary schools with his son Francesco because he was concerned about the quality of science education, especially mathematics. He also scientifically directed the encyclopedia of Italian culture—understood as the collection of everything that might be relevant to the nongenetic part of personality and group-identity formation. It celebrates the 150th anniversary of the Italian union. Although consisting in humanistic history and archeology, the twelve large volumes of *La cultura Italiana* (Cavalli-Sforza 2009–2010) are meant to be just as scientific as the collection and analysis of cell lines: "This work tries to tell us objectively who we were and who we became, with highs and lows, and tries to answer many questions that one may ask oneself" (Cavalli-Sforza 2009c, 27). The encyclopedia was meant to ensure the cultural transmission of Italian social and intellectual history. If translated into English it might also give diaspora Italians—American Italians were particularly close to Cavalli-Sforza's heart—a sense of pride in their origin. Or, to use the concept of Benedict Anderson (1992), it might kindle long-distance nationalism.

These examples point to the fact that, while emphasizing the importance of human population genetics for a feeling of panhuman kinship, the definition of group identities was fundamental, and group-identity formation was valued. To facilitate biohistorical individual and group identifications, both the genetic and cultural heritage needed to be cultivated. Cavalli-Sforza thought that there should be maps of cultural objects comparable to the maps of gene frequency distributions and genetic distances: "This is why it is helpful to delve into the history as well as into the geography of objects, techniques, customs, values, and in general of everything related to knowledge that everyone of us acquires in the course of his life and that serves to guide him" (my trans.).[21] Regardless of scale, the collection and preservation of history was important beyond a nostalgia for the past: "And if it is important to preserve the memory and to fix it so that it does not get lost, it is not only for sentimental reasons; it is also because there is a lot to learn from history. . . . It is not impossible that some might profit from it to find new ideas that allow the modification of our social behavior in positive ways" (my trans.).[22] Comparable to Huxley's ideas of a superhuman memory and the evolution of adaptive memoids, for Cavalli-Sforza the appropriation of "our" memory, a living past, might allow the creation of positive second-order organisms that are favorable to a progressive cultural evolution.

With these observations we are set squarely in the realm of yet another project that will occupy us for the remainder of this chapter and organize the next: the Genographic Project (GP). It, too, intertwines the preservation of biological and cultural heritage and concerns the provision of history and identity for group formation. The big-budget project was initiated in 2005 and is supported by the mass media expert National

Geographic Society, the database specialist IBM, and the Waitt Family Foundation. Carried out at different universities globally, the GP aims at reconstructing the history of humankind on the basis of the analysis of the genetic variation found in so-called indigenous and isolated peoples worldwide. Even though the GP is more ambitious than the HGDP because it set out to cover many more populations and to include far more individual samples, with this organization and goal, the GP can be seen as a successor to the HGDP. The *genographic* in *Genographic Project* is a contraction of what Cavalli-Sforza has called *genetic geography*, and there are strong connections between the GP and Cavalli-Sforza's wider research context, including the HGDP. There are similarities in the conception of interdisciplinarity, and there persists the notion of an objective and apolitical science. Finally, the GP is also associated with the aim of societal progress and the nostalgia of "the noble savage as our ancestor." It is not surprising then, that the controversies between disciplines and communities continued—not least with regard to different notions of history.

There are also personal ties between the HGDP and the GP. The GP's leader, Spencer Wells, had carried out research on population genetics and evolution for his PhD at Harvard University under Lewontin. Subsequently he went to Stanford to work in Peter Parham's lab on the immunological system of HLA as a postdoctoral fellow. He visited central Asia and analyzed the population samples for HLA gene variation, while Underhill studied their Y-chromosomal genetics. Wells then joined Cavalli-Sforza's lab to finish his work (Underhill, interview, 21 Jan. 2013). He continued his research on central Asian populations at Oxford University, where he headed the Wellcome Trust Centre for Human Genetics. Following a position in an American biotechnology company, he turned freelance scientist, writer, and filmmaker. All in all, this represents an ideal combination of expertise to lead the highly visible and popular GP.

Regarding the continuities in research between the HGDP and the GP, we have to return to the Y chromosome. Besides Wells, it is the star of the project (and to a lesser extent mtDNA). Wells became the public face of the molecule when—as a prelude to the GP—he popularized the Y chromosome in *The Journey of Man: A Genetic Odyssey* (2002) (literally *man*, because of the Y chromosome). By the time the human odyssey was being retold within the GP, many researchers had come to see in the Y chromosome—or rather its nonrecombining region—*the* new historical document. In fact, the Y chromosome had been made the focus of a concerted scientific effort in the Y Chromosome Consortium (YCC). The same year that Wells cherished it in *The Journey of Man* (2002), the YCC started its website that, together with the newsletter, would keep the community updated on the YCC Repository, new polymorphisms in the nonrecombining region, and changes in nomenclature. This use of media would help to better coordinate the knowledge about the genetic variation in the Y chromosome that "by virtue of its many polymorphisms, . . . is now the most informative haplotyping system" (YCC 2002, 339). Thus, before Wells created a particular aura around the Y chromosome for the GP, a group of researchers became aficionados.

The YCC had been initiated by Hammer and Nathan Ellis at the University of Arizona, and like the GP, it took inspiration from the HGDP: "The results of typing the same set of DNAs at many Y-specific loci will be pooled into a common database (the Database), in a similar fashion to the CEPH international collaboration." The goal was not only to set up a database of genotypes at the University of Arizona. As with the HGDP, the aim of the YCC was to establish lymphoblastoid cell lines representing indigenous populations worldwide at the Laboratory of Human Genetics of the New York Blood Center (twenty to thirty cell lines from each of the major population groups, that is, "African, Caucasoid, Asian, Amerindian, Oceanic, and Australoid"), and to provide the laboratories searching for polymorphisms with DNA that the University of Arizona would extract from the cell lines expanded in New York. Priority was to be given to those polymorphisms that had arisen only once in human evolution and were thus most useful in tracing human histories. Information on the language, the location, and self-identification of the indigenous peoples sampled was to be provided, and if possible accompanied by photographs. The idea was to determine the ancestral sequence for each marker site by studying these sites in the great apes and ultimately to construct a haplogroup and haplotype phylogeny showing sequence and time of branching.[23]

The members of the YCC were the scientists who contributed to the identification of Y-chromosomal polymorphisms and their combination in a genealogical tree of haplotypes and haplogroups. Underhill served as genotyper and on the nomenclature committee. At one point, Renfrew—the archeologist Cavalli-Sforza drew on and a member of the GP advisory board—organized a workshop to resolve issues of nomenclature. The Kidds and Cavalli-Sforza provided cell lines. At a round table discussion in 1993, Judy and Ken Kidd actually brought up the possibility of joining efforts with the HGDP in the collection of cell lines. This possibility was reconsidered at the meeting of the YCC Ad Hoc Committee. Simon Whitfield had been an observer at a European Genome Diversity Project meeting and he suggested that Julie Bodmer, Piazza, Alec Jeffreys, and Svante Pääbo might assist the YCC in expanding the number of cell lines in the Repository. Moreover, the fact that the HGDP participants had discussed ethical issues made the YCC members aware that their research results might collide with the ethnic and religious beliefs of peoples sampled, so a committee was planned to consider the matter.

The first Y-chromosome workshop was held in 1994 and showed that a group of researchers felt themselves connected by the Y chromosome—the epistemic object that they mapped, and the variation and evolution of which they studied. The meeting was considered "a smashing success!" And Ellis wrote that "the next day we all went our separate way nursing our adult hangovers or proceeding to our adult airplanes or buses or both."[24] While this meeting of the Y-community was endowed with fun, as always in genetics, there were less satisfying tasks. With the growing of the Repository, Hammer and Ellis began the tedious project of producing DNA from the cell lines. In fact, before Wells would weave the threads of Y chromosomes of contemporary indigenous men into the human odyssey, Ellis spun an "Elliad" about their scientific harvesting. In a newsletter, he expressed his hope that their godlike deed of producing DNA and "the arrival of DNA from the YCC will spur some individuals to new heights of altruism, causing them to submit more cell lines to the Repository and that others will join in battle against the deadly DNA polymorphisms that lurk with uncanny invisibility on the woeful Y chromosome."[25]

The DNA samples were sent out to the YCC-associated labs that had discovered markers for which the samples could be tested. The reward for mapping the Y chromosome, for making collaboration more efficient, and for standardizing markers and nomenclature, was a "single most parsimonious phylogeny" of the Y chromosome (YCC 2002, 339). Y-chromosome research and the Y chromosome had come of age (Jobling and Tyler-Smith 2003). It seems that the visualization of the Y-phylogeny was also becoming more stabilized, because the authors settled on the tree diagram as used by Cavalli-Sforza and Underhill (or actually Cavalli-Sforza and Edwards), rather than on the alternative

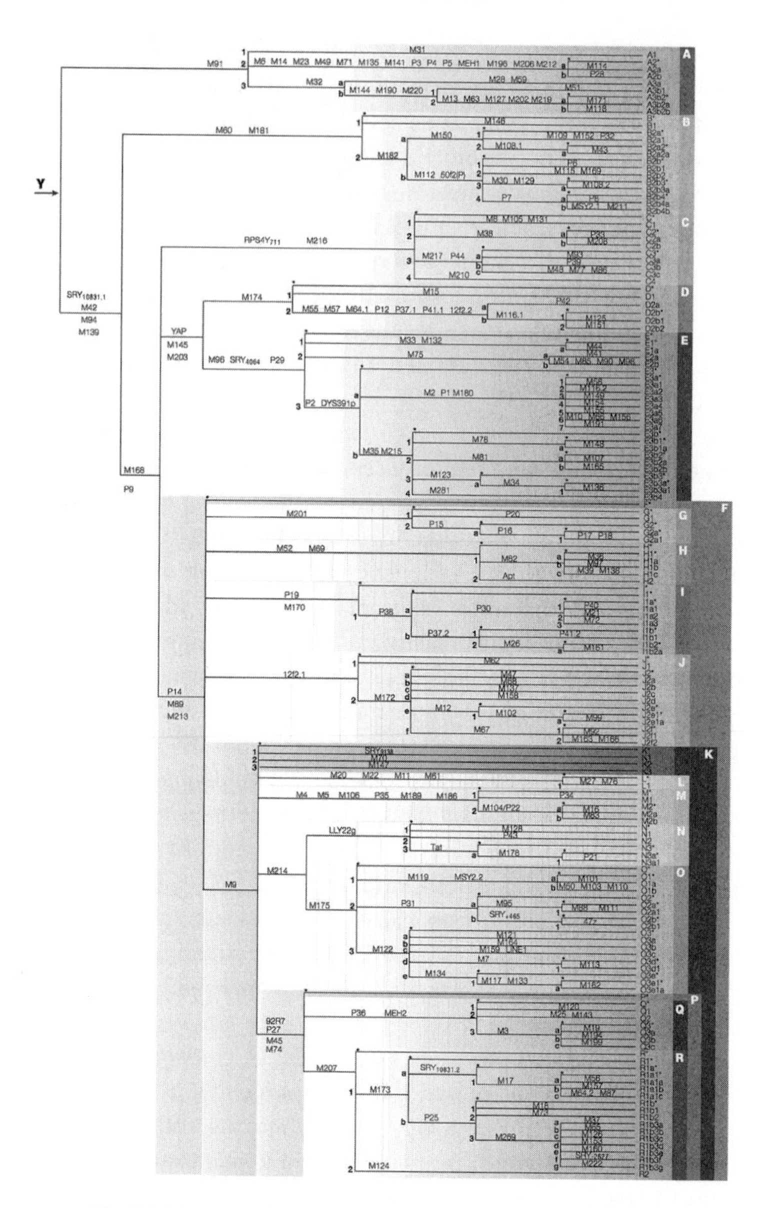

FIGURE 29. The YCC human Y-chromosomal tree showing the haplogroups, reprinted by permission from Macmillan Publishers Ltd., M. A. Jobling and C. Tyler-Smith, "The Human Y Chromosome: An Evolutionary Marker Comes of Age," *Nature Reviews Genetics* 4, 2003, 598–612, on 602, figure 3, with kind permission from Nature Publishing Group, http://www.nature.com/

structures Hammer and colleagues had published (see, for example, Karafet et al. 2002). Unfortunately, owing to its size, it was impossible to print the tree upside down, which had been the suggestion of Cavalli-Sforza to honor Darwin's emphasis on descent (as in *The Descent of Man* 1871) (see Cavalli-Sforza 1966 for such a tree; Underhill, interview 2, 7 Feb. 2013). Figure 29 shows a slightly modified YCC-tree with markers noted on the branches and haplotype and -group labels at the top of branches.

In this tree, by far the largest number of markers carries the *M* for Cavalli-Sforza's Stanford University lab. Accordingly, all the markers Wells centrally discussed in *The Journey of Man* (2002) start with an *M*. In fact, negotiations between the GP and Stanford University took place with regard to Stanford's participation in the project, but these talks failed because Stanford could not agree to the constraints the GP wanted to put on the data. So all Wells could do was, in the acknowledgments of *The Journey of Man*, to chiefly thank Underhill, "whose careful work on the population genetics of the Y-chromosome has allowed me to tell this story" (ibid., 197). In its essentials this story had already been told in *The History and Geography of Human Genes* (Cavalli-Sforza, Menozzi, and Piazza 1994), and Cavalli-Sforza and Feldman also summarized it in an article that took stock of the achievements in human population genetics in 2003 (Cavalli-Sforza and Feldman 2003). Accordingly, besides Underhill, in his popular writings on the GP, Wells paid homage to Cavalli-Sforza as a key figure in making such an endeavor possible:

> The methods for doing this, developed over the course of half a century, have been greatly influenced by Luca Cavalli-Sforza, with whom I was lucky enough to work as a postdoctoral fellow at Stanford in the 1990s. It was Luca's insight, as a geneticist with a passion for history and a talent for mathematics, which provided us with a time machine capable of resurrecting the stories of the past from people living in the present. This book could not have been written without his intellectual presence, and it is impossible not to feel humbled while taking in the view from his shoulders. (Wells 2002, xv)

Indeed, the imagery as well as the narrative of *History and Geography* occupies a core place in the history of human population genetics and its popularization, which can be illustrated on the basis of the same three publications. Wells's migration map in *The Journey of Man*

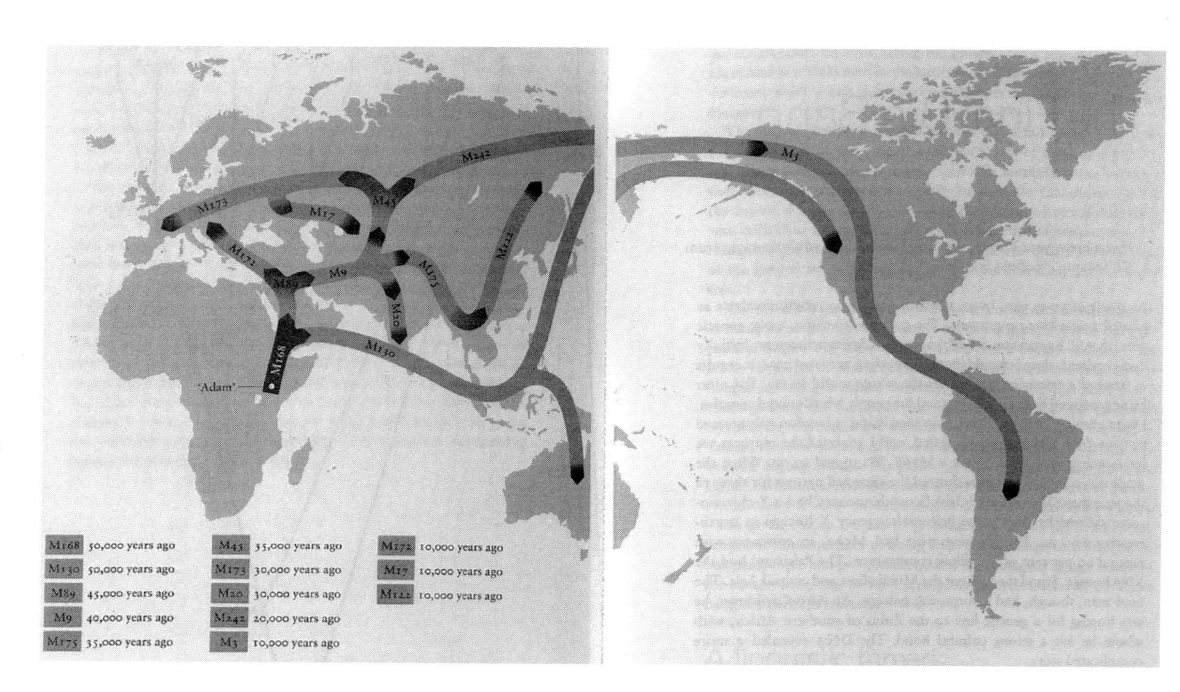

FIGURE 30. Migration map of modern human populations, Wells 2002, 182–183, figure 10

(2002) shown in figure 30 is strikingly similar to the one Cavalli-Sforza and colleagues had published in *History and Geography* (see figure 28), and that in fact is reproduced nearly unchanged in the review of genetic history by Cavalli-Sforza and Feldman of 2003 (270; see also Leslie 1999; Cavalli-Sforza 1998). Of course, the migration patterns were becoming increasingly detailed, even if, as in Wells's map, only one system was used, in this case the Y chromosome. In *The Journey of Man*, Wells therefore assured the reader that the telling of the human journey had only just begun, because "it is merely the outline of a much more detailed narrative, the whole of which will take many more years of research to decipher" (2002, 195). With the Y-chromosome research coming of age, the GP was timely. The project could draw on the necessary scientific knowledge that was in the public realm and at the same time had much larger funds than university labs to continue, popularize, and commercialize it.

Not surprisingly, Cavalli-Sforza was made president of the GP's advisory board, and one cannot help but see his signature all over the project, beginning with the conception of interdisciplinarity. Although the board was made up of an interdisciplinary group of experts,[26] the archeologist Renfrew has described his research as "archeological science," and as we have seen, he and the linguist Ruhlen at Stanford University—who was also elected to the board—were two of the main sources of inspiration in Cavalli-Sforza's parallelism between the evolution of languages, cultures, and populations.[27] Like Cavalli-Sforza, Wells emphasized that genetic history depends on other traditions in order to be meaningful in a historical sense: "The *how* and *why* questions, in addition to the *who*, *where*, and *when* that the DNA evidence provides, allow us to fill in the motives and methods of our ancestors" (Wells 2006, 62); in other words, the stories. At the same time, the hierarchy of disciplines was maintained: "And though traditional disciplines like archaeology and paleontology offer tantalizing clues, today's best prospects for breakthrough discoveries lie in genetic research" (ibid., blurb on cover).

We have seen how Cavalli-Sforza's work marked the HGDP, with which the GP thus shared many commonalities. Despite or rather because of this, the GP tried to steer clear of the issues that had troubled the HGDP. In true contrast to the HGDP, the GP is privately funded and entails the collection of DNA rather than the establishment of cell lines.[28] Distancing the GP from medical research, the endeavor is presented as "purely anthropological":

> The Genographic Project has a completely different approach [from the HGDP]. It is a *purely anthropological, nonmedical, nonpolitical, non-governmental, nonprofit international* research project involving scientists from both the developed and developing world. Our goal is to *enable indigenous communities* rather than to take from them. . . . Finally, we have said since the project launch that we see genetic discoveries as part of *the common heritage of our species*, and that *no genetic data will ever be patented* from the project. The Project is a collaborative effort between people from around the world who are interested in learning more about *our shared history.* (Wells 2006, 171–172, my emphasis)

Paradoxically, while this rhetoric is part of the effort of disclaiming the HGDP legacy, it is in fact further proof of exactly that legacy. It continues a discourse that purports a socioculturally neutral kind of knowledge. Wells stressed that the interest is exclusively in anthropology and history. However, that history is a living and shared one—a common heritage. Thus, even though processes of identification and othering have been part of the criticism leveled at the HGDP, such a living kind of history is again presented as apolitical.

In continuation of the line of reasoning we have followed through the history of molecular anthropology, part of the argument of neutrality rested on the Y chromosome. In *The Journey of Man*, which we have seen is something like a pilot episode for the GP, Wells explained why "the Y helps to place the stones, bones and languages in context better than any other part of our genetic code, and ultimately gives us the genetic answers we are looking for" (2002, xiv). This was the case because the Y chromosome not only shared all the advantages of mtDNA, such as being handed on only from one parent (in this case the father) and having nonrecombining regions. As additional advantages, the Y chromosome contained many more nucleotides but, owing to a lower mutation rate, showed a lower frequency of repeated and reverse mutations (which could tamper with the historical reconstruction). It was also characterized by a higher interpopulation variation (30 to 40 percent).[29] Within the Y chromosome, particular attention was paid to the noncoding region, a sequence that was assumed to have no function, and to which Wells popularly referred as *junk DNA*. He introduced Y-chromosomal junk DNA as the perfect epistemic object for genetic anthropology, because it was considered not to be under natural selection, stochastically accumulating changes. It was supposed to be unrelated to the phenotype and its environment (as having no function). While it thus recorded time, it was taken to have no medical potential. It did not correlate with any racial markers of old—it said nothing about physical, cultural, or psychological differences between human groups. Junk DNA therefore seemed not only neutral with regard to selection, but even more so than other genetic marker systems, politically neutral.

This is where the narrative we have encountered before, the one of the anthropological journey deeper and deeper into the human body that went along with the undermining of race, was at its strongest. Wells opened *The Journey of Man* (2002) with a chapter on the history of racial anthropology and eugenics, followed by one on the early blood group research and the arrival of true population genetics that refuted the concept of race by proving that within-populations variation is greater than that between. The heroes of the new population genetics were Dobzhansky, Lewontin, and Cavalli-Sforza. The telling of this story seems to have acquired the quality of a ritual; it cleans molecules, in this case the Y chromosome, of cultural contamination and arrives at a history that is pure nature. What appears to be a location outside culture is exactly that which renders the science and its object so culturally powerful, or to cite Sarah Franklin: "Paradoxically, then, it is because scientific knowledge is seen to be objective and universal that it can have such intimate and personal effects" (Franklin 2001, 306). Thus, in our case, the logic seems to be that only if the DNA (and the research on it) is outside culture, an objective arbiter, can it speak with authority on "race" and other kinds of personal and group identification that are so essential to our social and cultural lives.

However, in order to work for human population genetics, the narrative of the journey from cultural bias to natural (scientific) truth had to perform another "trick": "The trick is to find the polymorphisms that *do* unite us into regional groups" (Wells 2002, 48, my emphasis). The trick was to present the genomic regions analyzed as uncorrelated to physical characters and the science of their analysis as deconstructing race, only to found geographical or ethnic groups in the neutral DNA sequences. We already know about this "trick" from the time when Cavalli-Sforza and Bodmer replaced the races of physical anthropology with the major geographical groupings of "Africans, Caucasians, and Easterners" based on genetic data (Bodmer and Cavalli-Sforza 1976, 574); instead of being abandoned, these labels were thereby genetically re-created as groups that share a common ancestry. With the title of his second book that was collateral to the GP, *Deep Ancestry: Inside the Genographic Project* (2006), Wells pointed exactly to this foundation of groups in genetic ancestry. Nonetheless, *Ancestry* does at the same time point to the fact that, in contrast to racial anthropology, the aim of the GP, just like that of Cavalli-Sforza's research and the HGDP, was not primarily classification or grouping, but the reconstruction of history. Thus, in *Deep Ancestry*, Wells summarized the known Y and mtDNA history and laid out the remaining riddles the GP still had to solve—he was presenting himself as a historian: "I consider myself to be a historian. . . . By the time I applied to college, I had decided to pursue biology as a career—but focusing on its historical side, genetics" (ibid., 11–12).

In order for Wells to be a historian, the Y chromosome had to be more than a chronometer. In contrast to the arguments brought forward for its political neutrality, beyond as stochastically accumulating mutations in isolation from the environment, junk DNA had to be performed as historical document, text, and carrier of the ancestral story: "But [junk DNA] is anything but junk to those of us who use the genome as a historical document. This is our text, and it provides us with the story of our ancestors" (ibid., 15). Only with this final "trick" could Wells's popular human population genetics achieve the impossible, lose race and gain grouping by ancestry, and be both politically neutral and historically defining: "The beauty of the genetic data is that it gives us a clear, stepwise progression out of Africa into Eurasia and the Americas. The diversity we find around the world is divided into discrete, although related, units, defined by markers—the descendants of ancient mutation events. By mapping these markers on to the map of the world, we can infer details of past migrations. Following the order in which the mutations occurred, and estimating the date and any demographic details (such as population crashes or expansions), we can gain an insight into the details of the journey" (Wells 2002, 70).

The notion of junk DNA as largely independent from the environments inside and outside the body has been scientifically challenged;[30] and as already argued by Simpson, the concept of DNA as evolving independently of culture also seems to disqualify it as a carrier of history. In fact, quite contrary to the rhetoric that junk DNA is politically neutral because it is unrelated to (the history of) its carriers and their context, culture is fundamentally implicated. As we have seen, groups that are considered genetically relatively isolated have been and still are of particular interest to genetic anthropology. And this isolation is perceived as geographical *and/or* cultural—and in certain cases as purely historically and socially motivated. A tangible example exists in the ways in which Jewish marriage practices have influenced the distribution of haplotypes. Nadia Abu El-Haj (2004, 8–9) has pointed out that in such cases, junk DNA may indeed function as a cue to some historical traditions. It allows the falsification or verification of existing historical knowledge, exactly because its distribution patterns have been impacted by that history. Another interesting study uses the example of European American and African American admixture to illustrate that the history of human populations cannot be reconstructed on the basis of genetics alone, because historical and cultural factors, such as the cultural meanings of "white" and "black," racism, xenophobia, language, religion, tradition, wealth, and power, have shaped the distribution of genes: "A biohistorical approach leads to the conclusion that even when geographic and linguistic barriers fall, cultural barriers to admixture may continue to exist, and be an important factor in shaping human biological variation" (Edgar 2009, 64).

Wells of course knows this, and in *The Journey of Man* he discussed the influence of marriage customs and differential reproductive success on the distribution and divergence of Y chromosomes versus mtDNA. He also explained how the slave trade and the European expansion had led to a mixing of ancestral lines, and how these processes were accelerated in the current globalized world (Wells 2002, 174–178, 184–185). In fact, these processes constitute the sociocultural factors that in the eyes of Cavalli-Sforza had rendered the HGDP an urgent matter, and Wells ended his *The Journey of Man* with "A closing window." Echoing Cavalli-Sforza's stance, he lamented that while increased intermixture "is certainly a good thing socially, leading to the breakdown of racial stereotypes, it does mean that our genetic identities are becoming ever more closely entwined" (ibid., 193). Globalization and admixture were creating "cosmopolitan melting pots of markers" (ibid., 194). Thus, while sociocultural practices in certain cases such as Jewish marriage customs or American racism worked toward the preservation of the Y-chromosomal archive, they more often wreaked havoc. However, there remained a few nearly untouched vaults: "In a shrinking world, mixing populations are scrambling genetic signals. The key to this puzzle is acquiring genetic samples from the world's remaining indigenous and traditional peoples whose ethnic and genetic identities are isolated."[31] While Wells distanced the GP from the practice of collecting cell lines and from medical research (and by inference the possibility of exploitation), we reencounter the conceptualization of indigenous peoples as archives of historical documents that is associated with a certain nostalgia of purity, the imagining of a world lost in time when groups kept to themselves and carried clean signs of unequivocal kinship in their genomes.

These background assumptions and interests had led to controversy in the HGDP, and efforts have been made on the part of the GP to assert ethical soundness. Phil Bluehouse, a conservative spokesman of the Navajo tribe, played a key role in the project's public appearance, because he welcomed the GP despite the fact that the genetics produced a history at odds with the oral tradition of his tribe. Similarly, the encounters between Wells and indigenous groups, such as the Hadzabe of Tanzania, the Chukchi of Siberia, the Kallar of India, and Australian Aborigines, were portrayed as harmonious (see, for example, Maltby 2003; Wells 2006). In general, the transparency and ethical framework of the project were emphasized and the ethical guidelines made available online. The advisory board included an ethicist, independent local ethical committees were set up, and the public was assured that local and international ethical and legal standards were followed, as well as that all participants had given their free and informed (individual and/or group) consent to be sampled.[32]

Nevertheless, the Indigenous Peoples Council on Biocolonialism and the United Nations Permanent Forum on Indigenous Issues have voiced criticism against the GP of the kind already directed at the HGDP. The question of who can give what kind of consent had not been settled. Nor was it clear what the standards of "informed" were, and how they could be guaranteed. The fear that genetic history might undermine privileges associated with "aboriginality" persisted. And there have been mistakes. In the case of Alaska, the GP failed to gain the approval of the Alaska Area Institutional Review Board and had to return samples and suspend the work (Reardon 2009).[33] Wells showed understanding for native concerns and did link them to colonial exploitation, but from his point of view "we owe it to ourselves and to our descendants to discover what it is," the history within. Ultimately, the story of man's conquest of the earth would be most enlightening and inspiring, so that "one responsibility that we neglect at our peril is that of self-discovery. Once the document of our journey has been lost it will, like the footprints of our ancestors as they left Africa to colonize the globe, be gone for ever"

(Wells 2002, 196). When delivering a talk to an Indian audience, Wells was asked about the broader relevance of his work. Reminiscent of the call for the HGDP and Cavalli-Sforza's popular writings, Wells stressed that "we need to learn our history to understand who we are, and to speculate on where we might be going" (Wells 2010, xiv; on these issues, see Sommer 2010f).[34]

Wells addressed the question of "where we might be going" in yet another popular book, *Pandora's Seed: The Unforeseen Cost of Civilization* (2010). Again, the knowledge of the past was to show the way into a future progressive evolution. Also for Wells, the first lesson from genetic history was the indefensibleness of racism: "To the extent that we can see ourselves as connected at the genetic level, we might be able to overcome some of our prejudices" (ibid., xv). An equally important insight for Wells as well as for Cavalli-Sforza was that the changes in culture precipitated by the agricultural revolution had accelerated in the last decades. Wells asked whether we had developed the wrong kind of culture for our biological makeup: "Is there some sort of fatal mismatch between Western culture and our biology that is making us ill?" (ibid., xvi). On the one hand, Wells presented biological and cultural evolution to a certain degree in harmony. He cited evidence that human genomes had undergone significant changes through selection (for example, changes associated with lactose tolerance) within the last 10,000 years—that is, since the advent of agriculture. On the other hand, he perceived a mismatch between the biological and cultural systems. While cultural change could now occur faster than ever owing to the instant travel of ideas around the globe through the Internet, changes in a biology that had been adapted during an evolutionary time span to a hunter-gatherer way of life took place much slower.

For Wells, the resulting mismatch manifested itself in new, chronic kinds of diseases that were no longer infectious and caused by other organisms (and by inference avoidable through vaccination), but that were caused by ourselves, through the "unnatural" situation into which our own culture had put our biology: "First, the desire to eat is a basic survival instinct, so it's unnatural to try to reduce the amount of food we consume. Second, our hunter-gatherer ancestors would have found the idea of exercise for exercise's sake ludicrous" (ibid., 69). Wells believed the supposed unnaturalness of our culture also harmed our psyches. He claimed that while our bodies are damaged by a life in a culture to which they are not adapted, our minds grow ill in environments for which they have not evolved. The subjection of our personal desires to the good of the culture at large had created a Freudian *Unbehagen* ("uncanny"). Societal progress demanded the curbing of our many interests in favor of fewer and less stimulating activities. Gaining independence and control over our environments through differentiation and integration had come with the cost of enslavement of body and mind. Just like Osborn and Huxley, Wells considered overspecialization not a blessing but a curse; the technological advance had the side effect of constant overstimulation that caused stress. Finally, with a mind adapted to small group sizes, our social behavior became "decidedly unnatural" in the modern crowded spaces (ibid., 120).

These observations not only set "nature" and "culture" in opposition; they also suggest a state in the past when they were in harmony. Also as far as Wells was concerned, the hunter-gatherer way of life served as a utopia, a paradise lost through a sedentary lifestyle that brought all kinds of evil such as social stratification, conflict and war, overpopulation, disease, and the destruction of the environment with the concomitant loss of resources. The destruction of natural habitats was one of Wells's central concerns, but he felt that humanity also had the power to overcome its problems: "We are in control of our own destinies, perhaps the first generation ever to have such power, but how will we know what to do?" (ibid., 209). Looking into the crystal ball of history suggested to Wells that what happens in the future might be comparable to the Neolithic revolution. A renewed drastic change in climate might trigger technological innovations once more, most of all in alternative sources of energy and freshwater. Wells demanded transnational collaboration and—echoing Cavalli-Sforza's intergenerational relay—called for "thinking transgenerationally." Better caretaking of our natural and cultural heritage, working toward long-term goals and renewable resources, was the way into the future. Wells hoped that people were beginning to realize that they were all connected, that an action in one part of the world influenced the other parts, and that they were also united through time, inheriting the consequences of the mistakes of their ancestors and entailing a world to future generations (ibid., ch. 6).

Like Huxley and Cavalli-Sforza, Wells expressed his belief that awareness of our phylogenetic past could contribute to this new understanding, and the last chapter in *Pandora's Seed* sets off with a return to members of the Hadzabe, who according to genetics are "among the last remaining members of an ancient group that has lived in Tanzania for tens of thousands of years" (ibid., 184). I say *return*, because "to spend time living with the Hadzabe is to return to an ancient way of life." Like Huxley's travels to Africa, and Cavalli-Sforza's research among the "Pygmies," for Wells, this return to "our" origins was a healing experience: "It is a rich and varied existence, and after several days of living with them I started to feel a distinct calmness, as the worries and clutter of modern life melted away. In an odd way, it felt like returning home after a long absence" (ibid., 185). It is this urge for origins—for the possibility to return, if only temporarily or even imaginatively, to times before the agricultural revolution when our biology and culture were in harmony—that seems to constitute one motivation behind the endeavor to genetically trace roots. Wells, like Osborn and Cavalli-Sforza, drew lessons for the future from such traditional societies. Wells urged us to preserve them, so that we might profit from their myths, their ability to live in balance with nature. Saving ourselves would mean "learning from peoples that retain a link back to the way we lived for virtually our entire evolutionary history. And it might allow us to stick around for the next two million years" (ibid., 210). These living ancestors were thus supposed to reconnect "us civilized people" to our history within. And in such a reconnection lay hope for the future.[35]

In this light the GP's Legacy Fund makes perfect sense. It is the second of the three pillars of the project—if we take the first to be the population genetic research—and is aimed at preserving indigenous cultures. It is meant to raise awareness about the cultural pressures traditional peoples face. Efforts are made to ensure the survival of languages, artifacts, and techniques. The third pillar is the sale of genetic ancestry tests to the interested public. The profit thus made feeds into the research and the Legacy Fund. In fact, the GP is presented as establishing a tripartite exchange between researchers, research subjects, and customers in industrialized societies. While the indigenous populations are the donors in the research, they are the receivers in the legacy project, and while the customers have to pay for their DNA tests, the results of these tests are informed by the genetic study of indigenous communities: "Funded by a portion of the net proceeds from the sale of the Genographic Project Participation and DNA Ancestry Kits, as well as by public donations, the GLF [Genographic Legacy Fund] creates a circle of benefit between the research side of the project, the public participants, and indigenous and traditional communities, resulting in a shared, positive, and ongoing legacy for the Genographic Project."[36]

The GP's commercial pillar is of concern in the next chapter, and we will see that it is clearly part of the philosophy of the project outlined above. Indeed, Wells thinks that we all need "our Hadzabe," because the "desire to find a small community within the dizzying demographic cacophony of the modern world is nothing new. Whether inspired by fundamentalism or not, it is something we do all the time, even in the most modern places" (Wells 2010, 203). In *Pandora's Seed*, Wells thus suggested that religious fundamentalists and terrorists are offering ways of communality and myths that feel natural to us. They provide us with simple stories about who we are and where we come from. One may speculate that genetic history is brought in position against such attacks on secular humanism and liberal democracy, offering identities to groups and personal myths that are deemed better suited for the "modern age." As becomes clear in the next chapter, this "modern" way of imagining community around genetic ancestry is only possible with the aid of the Internet, web- as well as lab-based companies, data banks, and communication platforms such as chats and discussion forums. In fact, Wells compared groups such as those built via Facebook with hunting and gathering communities. Both average around 130 members, "and perhaps the deep connection between our ancient hunter-gatherer brains and the worlds we create online can show us a path through the ideological thicket, pointing a way toward a new *mythos* in the twenty-first century" (ibid., 205). In other words, through genetically testing for group membership and virtually building community around genetic markers, we might yet render "the ancestors within us" at ease in our modern culture. In this way, all of us might return to "the Hadzabe."

CHAPTER FOURTEEN

The Genographic Network

Science, Markets, and Genetic Narratives

Cavalli-Sforza has referred to his science as *genetic geography*. It is about mapping gene distributions, the genetically based population tree, and population movements onto the surface of the earth. This approach is captured in the *Genographic* of the GP. The adjective also encompasses *geographic* as in *National Geographic*. *Genetic geography* or *Genographic* links Cavalli-Sforza's and others' work that we have followed from the 1960s to the grand scheme of global projects with the old-style popular geography and mass media distribution the National Geographic Society stands for. To refer to these connections, I have introduced the term *genographic network* (Sommer 2010f). I have already shown the interconnections between the HGDP, the YCC, the GP, and other genetic geographic projects. As will be of concern in this chapter, through the GP, genetic ancestry companies—firms that sell DNA tests to find out "who you are and where you come from"—also form part of the genographic network. The GP has its own Public Participation and DNA Ancestry Kit with which it appeals to the public to have their DNA analyzed and their data added to the project's data bank. The DNA of customers who buy the GP's kit is analyzed by the genetic ancestry company Family Tree DNA. Through Family Tree DNA, the GP is also connected to the companies AfricanDNA and the Swiss-based iGENEA.

The particular network under concern here is only one instantiation of a more global phenomenon. In a world in which not only scientists but also commercial companies may use genetic data published in academic journals, share (reference) databases and computer programs, and hold similar assumptions about their work, genographic networks connect social spheres. Out of publically funded efforts might develop commercial DNA enterprises, as in the case of Rick Kittles's African Ancestry Inc. that arose out of the African Burial Ground Project, which received $5 million from the NIH.[1] University-based scientists establish DNA start-ups (Sommer 2008b), or function as scientific advisers to ancestry companies. To use an example from the context of this book, both Underhill and Feldman served as scientific advisers to 23andMe. When coming up with an ancestry product to complement their genomic medical service, Underhill assisted in the selection of Y-chromosome markers for which to test from those available in the public domain and he provided the company with information on frequency distributions. Joanna Mountain, who had worked on mtDNA in Cavalli-Sforza's lab, joined 23andMe in the early stages when it was clear that she would not get tenure at Stanford (Underhill, interview, 21 Jan. 2013; Feldman, interview, 22 Jan. 2013). Furthermore, university-based scientists and company-employed ones may cooperate in scientific projects and publications (for an example of a collaboration between members of Stanford University, Ancestry.com DNA, and other institutions, see Moreno-Estrada et al. 2013). At the same time, the practice of popularizing and commercializing genetic ancestry and history is often viewed askance by academic practitioners and criticized for its blunt overstatements of results with little regard for accuracy.[2]

Many of the medical genomic companies that sell tests for genetic predispositions for diseases offer ancestry information as a side, or the other way around, as the network under concern here illustrates. Family Tree DNA has founded the medical service DNATraits (Pálsson 2012). In such constellations, genetic history becomes tied more closely to the bigger genomic medical system. One famous example to illustrate this is the Icelandic biotechnology company deCODE Genetics (incorporated in the United States) that combined these two branches early on. As Michael Fortun (2008) discusses, the company was given the monopoly on the Health Sector Database. The database encompassed genealogical data, medical records, and genotypic information on Icelanders. The company had the exclusive rights to mine this rich information and to commercialize the results in return for paying a fee to the Icelandic government. In 2007, deCODEme was launched as a direct-to-customer service for information on genetically based health risks and genetic ancestry. The selling of a nation's gene pool to a private company led to much criticism. While Fortun especially denounces the undemocratic process that led to the agreement, the Icelandic Supreme Court has since ruled it a violation of the constitution. Furthermore, deCODEme has stopped operations.

It is this direct-to-customer medical genomics in particular that has alarmed American states and federal institutions to the point of initiating investigations. The NIH has established a database with information on the validity and usefulness of genetic tests, and a nonprofit biobank company has launched a study on the effects of genetic test results for predispositions on customers' behavior and health (Lindee 2013, 187–188). The company 23andMe has not only faced a class action; the Food and Drug Administration (FDA) has asked it to cease its medical services because their security and usefulness have not been demonstrated. It no longer offers the health-related genetic reports. Characteristic of the genographic networks, customers have been part of these developments. They criticize providers of medical DNA tests for coming up with contradictory results (Peikoff 2013). Another concern is the use companies like 23andMe will make of the enormous amount of genetic information. 23andMe carries out its own research on the database and it has received an NIH grant to further develop its web-based database and research engine—but would it ever sell genetic information on groups and individuals to insurance or pharmaceutical companies (Seife 2013)? This question has partly been answered: 23andMe has begun to sell its direct-to-customer genetic medical services in Canada and Britain, and it has come to agreements with major pharmaceutical companies. Genentech will sequence the whole genomes of about 3,000 people with Parkinson's disease in 23andMe's database to develop new treatments—and it develops its own therapeutic unit for drug development.[3] Further ethical issues are involved in medical genomics, and as we will see, commercialized individual and group identification through genetic ancestry testing, too, comes with concerns.

Obviously, for the new genetic medicine and history, computers of a very different kind than the ones Cavalli-Sforza and Edwards ran their early algorithms on are instrumental. Since then, bioinformatics has led to an understanding of life in general through the logic of computers. Bioinformatics is associated with big data, big money, and big institutions, but also with novel kinds of interdisciplinary constellations (Stevens 2013). Here, as well as for the connections between science and market, the Internet is key: "Technologically, scientists and entrepreneurs increasingly turned to the new types of online databases that could be accessed due to the proliferation and increased capacity of the Internet, linking labs, researchers, and companies. The networking of labs and databases also enabled the sharing of scientific knowledge globally. The number of databases in existence either in the academy, publically funded institutes, or private biotechnology companies grew enormously" (Chow-White and García-Sancho 2012, 144). Although the attempts at developing protein- and DNA-sequence databases reach back to the 1960s (Strasser 2010), we have seen for the HGDP that it was partly motivated by the new information technologies. Thus, on a much smaller scale, the Internet also renders possible the globalization and commercialization of genetic history. For the genographic networks, the advances in sequencing, information, and communication technologies that have dramatically increased the amount of DNA data, international collaborations, and networked databases need to be considered in conjunction.[4]

The genographic networks are grouped, one might say, around online databases and characterized by virtual paths through the Internet, but also by the movement and storage of wetware: there is traffic in data and knowledge, but also in bodily material. In previous chapters, we have, for example, seen how the processes on the one hand of networking between scientists and labs and on the other hand of the development of the Y chromosome as a genealogical system on the basis of a sample repository and database were mutually constitutive. In this chapter, we will see how with the commercialization of the Y chromosome, the network more obviously connected to markets. Again, what Steve Sturdy has observed for biomedicine therefore partly applies to genetic history: "The characterization and production of molecules is intimately dependent upon, and deeply implicated in, the formation of networks of collaboration and exchange of materials and techniques, both within biomedical research laboratories, and between laboratories and various other social contexts in which such molecules are put to use" (Sturdy 1998, 273).

The novelty in the ways in which science, economy, and also politics intersect is thus linked to tendencies of globalization driven by communication and information technologies (Thacker 2005). However, as we will see, the genographic network is a glocalized structure. Publicly available information is drawn on and cooperations transcend national boundaries, while research is situated in and products are tailored to specific cultural contexts and historical backgrounds. Thus, the circulation of techniques, objects, data, and knowledge goes along with the transformation of meanings, and in what follows, I am particularly interested in the stories that travel with the representations of molecules. In the genographic network under concern, the commercial DNA companies' websites and databases link customers who exchange diverse kinds of knowledge and build genetic communities, and this process of bonding and identification is dependent on meanings. How do DNA sequences become identity-forming historical narratives? What effect might the specificity of genetic history have, as scientific and biological as well as historical knowledge in this process? Finally, there are also the cultural bottlenecks and obstacles to circulation constituted by identity politics, technology, language, and also law.

To begin with, genetic data banks of ancestry tracing companies tend to be proprietary, and the circulation of information and substances is legally regulated. This may be illustrated with the genographic network under concern. Family Tree DNA has its headquarters in Houston. Michael Hammer's lab at the University of Arizona, which we have met in connection with the YCC, carries out the tests not only for the services offered by Family Tree DNA (where Michael Hammer is chief scientist) and for the GP, but also for AfricanDNA and for iGENEA. The money from these lab analyses finds its way into Hammer's scientific research. However, the fact that the actual samples travel through the network has created problems in the case of iGENEA. There has been an investigation by the senior public prosecutor into iGENEA's practices with regard to the privacy of data. When the substances (cheek swabs) traveled from Switzerland to the United States, were they accompanied by personal data? The Swiss company denied this. It claimed that even while in America, the samples and the genetic data were handled according to the more restrictive Swiss law (Sommer 2012c).[5]

Where the types of commercial services offered in this genographic network are concerned, all four providers work with Y-chromosomal and mtDNA tests. While some of these tests aim at situating customers within the panhuman family tree through haplogroup determination, others are tailored to specific markets. Owing to its global outlook, the haplogroup tests seem to suit the GP in particular. As we have seen, haplogroups are defined as the branches of the human family tree that has its root in our molecular first parents in Africa, and it is the goal of the GP to reconstruct the history as well as the current geographical distribution of the haplogroups. Family Tree DNA, on the other hand, is especially respected by family genealogists in the traditional sense of the word—people who work on their immediate family tree rather than their place within a global human history. This great American craze, as Nadia Abu El-Haj has discussed, is especially prominent among Jewish communities. As a result, the company-owned commercial genetic data bank—one of the largest of its kind—is particularly rich in Jewish genetic information. Furthermore, while I have shown elsewhere that genetic ancestry testing companies might be spin-offs from university research and researchers, and while they make money out of the information published in scientific journals and in public databases, El-Haj observes how Family Tree DNA, exactly because many of its customers are concerned with their close family connections, carries out its own research to detect genetic markers that are useful for the purpose (Sommer 2008b; Abu El-Haj 2012, 145–151).

Family Tree DNA offered so-called Walk through the Y projects (subsequently replaced by Next Generation Sequencing Big Y Tests). If a group of customers of the same haplogroup wants to detect new markers, the company will sequence an extensive region of their Y chromosome for money.[6] In this way, customers actually provide their own contributions to genetic history as "citizen scientists." Genetic ancestry and especially Y-chromosome aficionados have formed integrated groups for their chromosomal lines, for their haplogroups, exchanging information, mining scientific data, and communicating with scientists in the project of completing their trees online.[7] The most coordinated effort is the International Society of Genetic Genealogy, a nonprofit organization run by its members who try to keep the Y-chromosomal trees up to date. Because the members also add information that came out of personal ancestry testing, they render this otherwise inaccessible data available to scientists working at universities and other independent institutions. Thus, the circle of knowledge transfer is closed, and scientists may use this information in their research.[8] Finally, Family Tree DNA also sells its sequencing products (for the detection of new markers or the testing for known ones) to researchers and publishes a Y-tree based on its data. In sum, this is a very different picture from the early attempt of the YCC to pool information and synchronize efforts between the few labs that had established markers (Underhill, interview, 21 Jan. 2013; interview 2, 7 Feb. 2013).

The third genetic ancestry provider in this genographic network was founded by the Harvard professor for African and African American

Studies, Henry Louis Gates Jr. AfricanDNA specifically caters to African Americans. The genetic technologies are brought to bear on the reconstruction of the fragmented histories of those whose families have been affected by the slave trade and slavery. While the reconnection to an origin in Africa works with a genetic understanding of ancestry, Gates's marketing strategy is to emphasize the historical expertise of the team who enrich the information retrieved from DNA sequences with the study of historical sources: "Sometimes the tests yield multiple exact tribal matches, making it necessary for historians to interpret the most plausible result. AfricanDNA, which I co-founded with Family-TreeDNA, offers this service."[9] For the paternal Y-chromosome test it is thus stated that "our historians and scholarly advisors use this information, in combination with historical evidence concerning the slave trade, to identify the region of Africa where your lineage most likely originated and to share information with you about that region and the ethnic groups found there."[10] In fact, Gates has made cunning use of his status as an expert on African American history to popularize this particular genetic history. He has hosted and coproduced shows that staged the reconnection of well-known African Americans to their African origins. On these shows as well as in his PBS series, genetic evidence and historical sources are neatly intertwined to create meaningful narratives of origin. Congressman John Lewis could, for example, be witnessed weeping when the man who had marched with Martin Luther King Jr. learned that some of his ancestors were among the first black people to register to vote in his state (Newton 2014, 32).

While the GP's Participation Kit focuses on global ancestry, Family Tree DNA on family genealogy, and AfricanDNA on African ethnicity, iGENEA has developed packages for the European context. Suspecting a Continental European market for genetically reconstructed ancestry, the Swiss company Gentest.ch (limited liability company, 2002) set out to enrich the range of their DNA tests for paternity, kinship, and forensics with genetic history. To this purpose, the product line iGENEA was created and given its own web presence. Like African-DNA, Gentest.ch secured the cooperation of Family Tree DNA. With the agreement between the firms, Gentest.ch received access to what it identifies as the largest genetic database for DNA-ancestry tests. This database was opened to iGENEA customers to search for DNA matches, and they could join the GP's commercial branch for a reduced price. Since Gentest.ch was liquidated in 2012, iGENEA has been turned into a company. Subsequently, customers with a premium and expert test (the whole mtDNA is sequenced, respectively the highest Y resolution provided) had free participation in the GP.[11]

On the basis of Y-chromosomal and mtDNA tests, iGENEA "origins analysis" is said to provide information on the customer's haplogroup (Stone Age), "ancient tribe" (*Urvolk*, antiquity, 900 BC to AD 900), and region of origin (Middle Ages, AD 500 to 1500). The haplogroup determination potentially addresses customers from all over the world, while the genetic identification of one's "ancient tribe" and region of origin is more specific. However, Gentest.ch did not have priority with regard to catering to European historical cultures. The British university start-up Oxford Ancestors has since 2000 offered DNA tests to determine one's Paleolithic European tribe—that is to assign people to the line of a European clan father and/or clan mother (called the Seven Daughters of Eve). Even more specifically, Oxford Ancestors' Tribes of Britain package is aimed at British male customers who are interested in whether they may be of Celtic, Anglo-Saxon, or Viking origin. The iGENEA services are nonetheless new. They assign the customer's maternal and/or paternal genetic line not only to the Celts or Vikings, but also to Germanic tribes, Illyrians, Slavs, Iberians, Scythians, Arabs, Berbers, Persians, Turkic Peoples, Finno-Ugric Peoples, Baltic Peoples, Huns, Gepids, Alans, Vandals, and other "ancient tribes" whose genetic profile is said to have been isolated.

As with the services, the ways in which customers engage with their genetic identification differ widely individually and between contexts. This is actually also true for the noncommercial population genetic research, by which groups do not necessarily feel marginalized (Pálsson 2008). On the contrary, genetic knowledge about group history and identity may become part of political movements for recognition (Soodyall 2003, 205; Egorova 2009). Furthermore, not all research is carried out top-down. Some groups or group members actively seek the services of population geneticists (Brodwin 2005, 149–159; Egorova 2011). Not only the interaction between the researchers and the people sampled, but also the expectations with respect to the genetic finds and the impact of the genetic knowledge on self-conception and perception from the outside differ in each case (Abu El-Haj 2004; Parfitt and Egorova 2005, 2006). Similarly, people have various purposes for taking direct-to-customer genetic ancestry tests and they associate different hopes with the results. The Freedmen of Oklahoma ("Black Indians") have referred to genetic ancestry tests to determine their percentage of Native American descent in the context of their appeal to be recognized as tribal citizens, and a group of African Americans has used genetic ancestry test results in the attempt to litigate a slavery reparations case (Hamilton 2012). At the same time, other African Americans use genetic information to reclaim their non-African ancestry: "For some groups, PGH [personalized genetic histories] can be valued as another form of evidence that can be used to reclaim history, culture, and knowledge that is denied to individuals" (Shriver and Kittles 2008, 209).

In Alondra Nelson's (2008a, also 2008b) study on African American customers, individuals showed some flexibility in the appropriation of the genetic data. They made use of diverse sources of genealogical knowledge in their construction of biographical narratives. In general, the genetic information is playfully woven into personal memories; in other instances, a contradiction between expectation and genetic information can prove disruptive; and in yet other instances, the genetics is judged relatively unimportant when diverging from what is experienced as a much stronger existing socioculturally determined identity (see, for example, Santos et al. 2009; Sommer 2010f; Scully, King, and Brown 2013, 932–935). For instance, after a person has been genetically linked to Thomas Jefferson, the fact that a family member of that person fainted when visiting Jefferson's home may suddenly make sense. On the other hand, customers may express their unwillingness to admit to having African American or Jewish relations. The white supremacist Craig Cobb, for example, referred to the supposedly "sub-Saharan African" part of his genome as "statistical noise" (Newton 2014, 33). Consternations of this kind continue to provoke racist and anti-Semitic statements on genetic ancestry testing forums. They contradict hopes like those of Wells and Gates, who is a well-known author of books on the history of racism and sees in the popularization of genetic ancestry tests a tool to undermine notions of racial purity.

For the genographic network under concern, iGENEA can serve as an example for these diverse processes. To begin with, customers have to make sense of their test results, which may be easiest (though not necessarily less disruptive) with regard to one's more immediate kinship. We have seen that Family Tree DNA is particularly suitable for family genealogy projects, but this kind of service is offered by most genetic ancestry companies. iGENEA advertises its surname projects thus: "A surname project enables you to find people with whom you share common ancestry and can exchange information (such as the family tree). Therefore, the information content of your family history quickly increases. In contrast, you can also exclude namesakes who do not belong to your family."[12] This is on the one hand the aspect that renders the gendered nature of ancestry tracing most obvious, because the comparison of Y-chromosome sequences between men with the same or similar surnames is a family-tree-building practice that plucks the women completely off the branches. On the other hand, the genetic information seems most straightforward and the scientific results most powerful in the family research. DNA test results are here given the authority to expand but also to trim the (male part of the) family tree in order to reduce the social to the biological order (see also Nash 2007).

The example of surname projects already points to the fact that also in personalized population genetics, genetic identity building is unlikely to be done in isolation. In fact, like the geographic origin projects, the surname projects are self-organized by customers in order to get into contact with namesakes and collectively engage in family tree research. Internet contact platforms and social media have long become an integral part of everyday life, be it for conventional partner search or for the organization of subcultures or elite networks. And iGENEA draws on these sociocultural practices more generally when, like other genetic ancestry tracing companies, it offers its customers the ability to search the large Family Tree DNA data bank for so-called genetic cousins. If desired, e-mail services constantly inform the iGENEA user of new customers who show a low- or high-resolution genetic match. This is a tool to build community among genetic peers in a wider sense, and to exchange knowledge of and experiences with (genetic) history. Through these technologies, networks based on genetic kinship are established that may even lead to real-worldly contacts. And there are ways to increase the information on one's genetic identity beyond the company by which one has been tested through public access databases and websites as well, for example, in the case of mtDNA, through mtDNAlog, mtDNA Concordance, and DNAPrintlog (Smolenyak Smolenyak and Turner 2004, 183–211). Finally, as suggested on the iGENEA website, there is the possibility to really delve into the scientific expert literature.

The scientific nature of human population genetics provides for a rich cognitive aspect in genetic ancestry tracing, with its high-tech procedures, rigorous mathematical logic, and statistical tools. iGENEA emphasizes this hard-scientific nature and the certainty of results.[13] However, the genetic information these processes provide lacks the narrative aesthetics so essential to historical meaning.[14] This presents a particular challenge where deeper histories are concerned. In the case of iGENEA, test results in the form of mutation patterns are only accompanied by sparse interpretation support. Nonetheless, where the iGENEA products tailored to the European market are concerned, there is good evidence that people are able to tap existing knowledge to make sense of the genetic information. This is possible because in the (research and) services, it is well-known labels such as *Celtic* that are genetically redefined: "By ancient tribes, we refer to groups of people from the ancient past which are defined not only by their specific language, culture and history but also by their specific DNA profile. What is important is not the common linguistic ancestry or the historical-anthropological categorisation, but the individual genetic characteristics of a people."[15]

This confusing explanation represents a complex mixture of cultural identification and its genetic fixation: for example, to establish a Celtic marker, people living today in areas of Brittany, Scotland, and Ireland are sampled. These people are taken as representative of ancient inhabitants because of, among other things, their use of Gaelic languages. So, as we have seen for population genetic research more generally, language and culture do matter for genetic identification, as iGENEA elsewhere acknowledges: "Genetic research has shown that in 90% of cases, peoples who have a common linguistic origin also have more genetic commonalities than peoples who merely inhabit neighboring territories."[16] Cultural aspects and genetic markers are further confounded because the tribal labels are often highly controversial within archeology and history and because there exist many old myths about them in popular knowledge. This is of course the power of labels like *Celts* or *Vikings*. It turns them into brands (on the notion of genetic-historical brands, in particular discussed for "the genetics of the Celts," see Sommer 2008b, 2012a).

As I have shown elsewhere (Sommer 2012c), the engagement with iGENEA's test results triggers a wide range of reactions from playful identity building to indifference and aggression and inner conflict. Many customers, however, manage to make their autobiographical experiences and self-understanding emerge as a natural continuation of the deeper stories carried in their genes. These customers give testimony to their ability to draw on existing living history to make personal narratives out of their DNA results: "To my great surprise, we learned through a DNA-

Test that we are descended from the sea-faring Phoenicians. Moreover, we learned that our ancestors came from the modern-day Lebanon/Syria with their city states suchas [*sic*] Tyre and Carthage, sailing to Italy, and must have migrated to Switzerland from there. It is really fascinating to find out that genes can tell us so much. It also seems personally fascinating to me that I always felt drawn to the Mediterranean as a child. Maybe there's more passed on in our genes s [*sic*] than we think."[17] Someone else posted on the online Viking discussion forum, where customers exchange information and are provided with answers from a company "expert": "Many thanks for your rich answers and for my 'desired result.' It is somehow strange. Since I have been about 20 (now 38) I have been drawn to the North. . . . My interests were growing more strongly towards the Vikings. Bought a lot of literature and somehow a curious familiarity arose at old Scandinavian sites. Perhaps the genes store more than we know" (my trans.).[18]

Like many others of some 400,000 people using iGENEA's services toward the end of 2012 (according to the company's own information), these customers have internalized the metaphor of a historical narrative written into our DNA and have entangled the genetic identity with their autobiographical memories, their desires, longings, and tastes. The gene is here a mystical object, through which an individual can inscribe himself or herself into a fantastically present past. As the above quotes indicate, in doing so, people combine ancestry tracing by DNA with other products from a genealogy and living-history industry such as historical exhibitions, books and TV documentaries about history, historical novels, mythos films, (re)enactments, historical sites, and history parks. iGENEA customers also exchange information on such sources to provide the "ancient tribes" to which they have been genetically ascribed with stories.

Accordingly, the iGENEA genetic tests are regarded as funny Christmas and birthday presents, as yet another cyberspace hype in the times of Facebook, where the company advertises: Chat with your genetic cousins found through the company's data bank, exchange information on your genealogy and expertise of human population genetics with your virtual friends, and enhance your genetic data with living history. In fact, at one point the iGENEA Facebook site contained cartoons that make fun of genealogy and a newspaper clipping reporting that iGENEA has proven that Napoleon is of Arabic descent. Genetic identity and history here are not destiny; if anything, they are commodity and project (Rose

2007). Paul Rabinow's ([1992] 1996, 102) definition of biosociality in the context of biomedicine has to be changed only slightly to fit these biohistorically oriented groups: such groups have specialists, laboratories, narratives, traditions, and pastoral keepers to help them experience, share, intervene, and understand their *past*. This is in agreement with the self-image of iGENEA as service to a lifestyle society, in which the younger generations are no longer willing to spend time in libraries, not to mention archives. The company provides ready-made products that answer to the living-history boom—a customized exciting past spiced up by means of DNA technology. The company explains its success on the basis of a hunger for roots and history, which, although an anthropological universal, is aggravated "in a world of nearly unlimited interconnectedness of persons, of globalization, of cosmopolitans" (my trans.).[19]

However, while I have given examples of playful personal genetic history making, there are certainly also more serious incidents, exactly due to the fact that the preexisting labels, history, and myths that are geneticized might come with ideological baggage. In particular, the iGENEA statistics for Macedonian and Greek percental "tribal ancestry" have led to violent arguments on the company's discussion forums, and their tests for Jewish ancestry have even provoked an interpellation to the Swiss federal executive. In such cases, deep ancestry and current nationality or religion may be confused in problematic ways; for example, with regard to the genetic test for Germanic roots, the understanding at times surfaced in diverse media that there is something like a true "Germanness" underlying today's Germany, and that some Germans are "more German" than others because of their genetic ties to particular ancient tribes. Here is a remark of an iGENEA-produced "Germanic-descendant": "I have received my result a few days ago. I would have been pleased by any result, Celt or Viking or whatever. But to have Germanic roots and thus to feel like an original inhabitant [*Ureinwohnerin*] is also appealing" (quoted in Sommer 2012c, 128). Even though according to the rhetoric of iGENEA, genetics proves that there are no pure nations or "races," it is at times one particular Ur-folk (*Urvolk*) that becomes defining for a national identity. In one instance even the iGENEA spokesperson wrote: "Yes, a Macedonian can say that he [*sic*] is ancient Macedonian originally. . . . Only an analysis of your DNA can provide us with an absolutely certain answer" (my trans.).[20] In sum, the genetic production of groups may have a certain subversive tendency when the new labels cut across any well-established ones (think of Bryan Sykes's Paleolithic tribes called the Seven Daughters of Eve). However, it is hard to imagine how research on and testing for "Jewish-ness," "Celtic-ness," "Germanic-ness," or "ancient Macedonian-ness" should undermine anti-Semitism, racism, and nationalism.[21]

The danger of essentialism obviously not only enters when the genetic knowledge is commodified. Human population genetics, and in particular projects such as the GP, may well serve to teach the world about the common genetic and cultural legacy, but it is fundamentally about "us" and "them" groups, and it may thus be partly further motivated by, and even create the desire for, identification and differentiation along these lines (Reardon and TallBear 2012, S234). On the one hand, scientists like Cavalli-Sforza and also partly Wells have identified xenophobia—and racism, sexism, and nationalism as expressions of it—as the enemy of human happiness and progress. In *The Great Human Diasporas* (L. L. Cavalli-Sforza and F. Cavalli-Sforza [1993] 1995, 241–244), the Cavalli-Sforzas attributed some of these tendencies to an innate desire for "*Us* groups" as extensions of the self, such as the family, the firm, the football club, or the nation or "race." This "drive for identity"—as Bodmer and Cavalli-Sforza called it in 1976 (596)—was said to grow stronger in time through positive feelings toward the group and through prejudice, jealousy, as well as feelings of inferiority or superiority toward other groups. As such, xenophobia was identified as a chronic disease with its own pathogenesis that demanded education and other social policies. And once again one finds the argument that human population genetics works against xenophobia through the insight of a recent common origin and that global communication and migration contribute to the process through cultural and genetic exchange. On the other hand, Cavalli-Sforza has devoted his career to uncovering and popularizing the history of human diversity. He not only partook in the HGDP but was also associated with the GP that includes commercial DNA testing. In other words, while leading a campaign against the human tendency to form rigid "us" and "them" groups, he, like Wells, has also facilitated identification according to some such groups.[22]

Accordingly, this paradox is a structuring element of the grand genetic narratives of "the human family" à la Cavalli-Sforza and Wells, the production and circulation of which are particularly relevant in order to turn the commercial haplogroup tests into tools for identification. These narratives suggest panhuman belonging on the basis of a recent common origin and close genetic kinship. Simultaneously, they create "us" and "them" groups as the story unfolds. As we have seen, iGENEA's tests for ancient tribes genetically re-create existing categories that may go along with national and ethnic myths (Sommer 2008b; 2012a). AfricanDNA provides the customer with historical information to render the test results meaningful. However, for the haplogroup tests that the GP, Family Tree DNA, AfricanDNA, and iGENEA offer, a particular kind of additional information is required because the tests are about deep human pasts that are less generally known beyond the fact of an African origin and a general idea of subsequent dispersal. Therefore, the GP, and especially Wells, create deep human histories on the respective websites and offer books and films to bring DNA sequences to life.

Indeed, taken together, the GP and Family Tree DNA are a veritable meaning machine. Besides the web presence at National Geographic, IBM, and Family Tree DNA, the project and its insights into haplogroup history are promoted through the commercial media of the National Geographic Society. Wells's popular books and DVDs on the genetic history of humankind offer a rich fund of historical narrative. They are part of a print and film market of genetic histories the titles of which indicate a place in the history industry: *The Seven Daughters of Eve* (Sykes 2001); *Mapping Human History: Genes, Race, and Our Common Origins* (Olson 2002); *Adam's Curse: A Future Without Men* (Sykes 2003); *Out of Eden* (Oppenheimer 2003); *The Real Eve: Modern Man's Journey Out of Africa* (Oppenheimer 2004); *The Origins of the British: A Genetic Detective Story* (Oppenheimer 2006); *Before the Dawn: Recovering the Lost History of Our Ancestors* (Wade 2006); *Blood of the Isles: Exploring the Genetic Roots of Our Tribal History* (Sykes 2006); *Abraham's Children: Race, Identity, and the DNA of the Chosen People* (Entine 2007), and so forth.

Wells's own *The Journey of Man: A Genetic Odyssey* (2002) and the accompanying documentary *The Journey of Man: The Story of the Human Species* (Maltby 2003), broadcast by PBS and coproduced by the National Geographic Society, are revealing because Wells finished them just before the GP was officially launched. By 2000, according to Wells, genetics was ready to rewrite history—the history embodied in the blood of us all. Like the titles in the above list, in the retelling Wells relied on religious metaphor and origin narrative: "The human race began 60,000 years ago with a single family in an African valley" (Wells 2006, blurb on cover). As "host" of the *Journey of Man* documentary, Wells also personally reenacted the odyssey, leaving his lab and his family behind to find the answer to the riddle of our common origin. Wells emphasized that he was dealing with a living history from "*living* genomes": "Every one of us is carrying his or her personal history book around inside us—we simply need to learn how to read it" (Wells 2002, xvi). Like an Australian songline, the living genomes are said to take us back to times before collective memory. In the book as well as the film, those with a direct link back to Adam and Eve are found in southern and eastern Africa: "let's meet the family" (Maltby 2003).

The Y-DNA thread leads to the San, with their click language that indicates early separation from the other human language groups, and to the Hadzabe and Sandawe, who, too, provide evidence of this old language family. The San and other "direct link[s] back to our earliest human ancestors" (Wells 2002, 57) serve Wells for a description of genetic Adam and Eve: "So, the picture that emerges is of a dark-skinned (although perhaps not as dark as some Africans today), reasonably tall, thin person—perhaps with an epicanthic fold" (ibid., 59). Indeed, in the film, the "Bushmen" or San are presented as the original ancestors of humanity, in danger of finally being lost.[23] Thus, physical and behavioral features are mapped onto the purportedly phenotypically irrelevant junk DNA. Wells's treatment of the San is reminiscent of nineteenth-century physical anthropology, in which the "Bushmen" already figured as some of the most primitive survivors from human ancient history. Wells describes them as "someone who wouldn't look that out of place today dressed in a suit and sitting opposite you on the train" (ibid.). In this image, the anthropological stereotype of the Neanderthals, who have in more recent times often been reconstructed in contemporary Western apparel, is conjured up. Why opposite me in the train and in a suit? The image creates remoteness and familiarity at the same time: they are our distant ancestors but they are still us—*us* meaning the readers, audience, and customers who represent industrialized societies.

Reconstructed as our ancestors to provide meaningless DNA sequences with faces, in Wells's book and film, the San appear as among those who did not make it out of Africa. Wells explains that the spread out of Africa at the onset of the Upper Paleolithic coincided with a revolution in mental, social, and technical as well as artistic capacities. Harking back to Cavalli-Sforza's ideas, he follows those researchers who bring this revolution in connection with the appearance of modern language through genetic changes. According to Ruhlen (who, as we have seen, adheres to the common-origin model of language and is on the GP advisory board), the babbling of today's children is genetically programmed and a trace of the common ancestral language, the evolution of which passed the threshold of syntax at that time. This resulted in greater survival skills that might have catalyzed a population expansion that in turn—together with climatic deterioration and a reduction in game—could have been the incentive for migrations. Wells argues for a Huxleyan absolute progress, with the human brain structure evidencing an overall increase in complexity, freedom from and control over the environment, as well as a development toward adaptive flexibility rather than narrow specialization. Indeed, with regard to the earlier first steps out of the forest, there is also an element of the Osbornian "lush-jungle morons versus demanding-environment geniuses." In Wells's popularization the challenges of a life in the open could only have been met by a brainy ape—orangutans prefer to stay in the forest. This improvement of the brain enabled the development of "complex culture": "Without the early sparks of it, our hominid ancestors would never have ventured beyond the African forest margin into the savannah. And without having it in spades, we would never have survived what we encountered when we moved out of Africa into Eurasia, around 50,000 years ago" (Wells 2002, 90)—*we?*

Such renderings of the "great leap forward"—as Jared Diamond (1992) has called Mellars and Stringer's (1989) human revolution discussed in chapter 11—make the out-of-Africa model susceptible to false imaginings. It sometimes sounds as if the revolution consisted in leaving Africa, in leaving behind those who did not leap forward (Proctor 2003). For those who left Africa, the story according to Wells continues as follows: some populations migrated along the southern coast of Asia all the way to Australia. Thus, the phylogeny of Y-chromosomal markers that become associated with populations spreads out step by step on the map of the globe (see figure 30); and it continues to be accompanied by physical anthropological categorization. Australians are said to be dark-skinned, and while most southeastern Asians "today would be classified as 'Mongoloid' peoples," there are also "so-called Negritos living throughout south-east Asia who closely resemble Africans" (Wells 2002, 74). These people "have many features that link them with the Bushmen and Pygmies of Africa, including short stature, dark skin, tightly curled hair and epicanthic folds." Among them, "the Andamanese . . . represent a relic of the pre-Mongoloid population of south-east Asia—'living fossils,' if you will" (ibid., 75). The specters of racial anthropology are here again conjured up to produce representatives of a past revived for today's mixed and industrialized populations, particularly in the West.

However, "we" are not descended from these ancestors of the Australians. While they traveled along the shorelines, "the interior was left to the more active hunters, who would have had to move great distances to obtain the resources they needed to survive—animals, plants and water. They are the ones who made the leap into the unknown beyond the coast, into the wilds of interior Eurasia" (ibid., 95–96). Following prey into the Middle East, the "inheritors of the Great Leap Forward" (ibid., 99) permanently settled there about 45,000 years ago. From there, they ventured into India and conquered the central Asian steppes and beyond, accumulating new genetic markers on the way. Wells's genetic odyssey continues at some 35,000 years ago, when a significant number of Aurignacian modern humans reached Europe. They were a subset of the central Asian branch that replaced the Neanderthals. In fact—Osborn would have liked this—central Asia plays quite a crucial role in Wells's account:

> During the thousands of years they spent on the grasslands of central Asia they almost certainly underwent a period of intense cultural adaptation to this difficult environment. This period took the place of the hundreds of thousands of years of Neanderthal biological adaption—what had given them their short, stocky frames. As recent migrants from tropical Africa, Upper Palaeolithic humans initially would have been ill equipped for life in the northern hemisphere. The central Asian steppes served as their apprenticeship, in a sense—preparing them for life in the most inhospitable environments on the planet. The caves of western Europe must have seemed relatively benign after the howling winds of the frozen Kazak grasslands. (ibid., 133)

Part of the central Asian "clan" would also move into Siberia and from there into the Americas via the Bering Strait.

Thus goes the heroic strenuous journey of (some) men, which—along the lines of a branching tree—constantly creates difference between those who remained and those who moved on from a certain place, and between those in the more congenial climates and those who were hardened in harsh environments. Obviously, it is ultimately the story of "Wells's own family," who conquered Europe and brought about the end of the short and stocky Neanderthals. This story of differential becoming is supported not only by genetics, but by interdisciplinary evidence:

"The stones and bones seem to agree with the DNA" (ibid., 140), and so apparently does the knowledge about climate and from linguistics. Wells drew on Greenberg's as well as Ruhlen's theories of language evolution, and he summarized Cavalli-Sforza's comparison of the genetic with the linguistic tree (ibid., 162–163). While he discussed knowledge from many other fields, including child psychology and history of science, he in general most strongly relied on Cavalli-Sforza's work, particularly *The History and Geography of Human Genes* (Cavalli-Sforza, Menozzi, and Piazza 1994), but also *The Genetics of Human Populations* (Cavalli-Sforza and Bodmer 1971), *Neolithic Transition and the Genetics of Populations in Europe* (Ammerman and Cavalli-Sforza 1984), and *Cultural Transmission and Evolution* (Cavalli-Sforza and Feldman 1981). As further evidence of the close ties between Cavalli-Sforza's earlier genetic geography and the narratives circulating in the global genographic network, to his readers, Wells recommended the more popular *Genes, Peoples and Languages* (Cavalli-Sforza [1996] 2000).

Such ties are also evident in Wells's *Deep Ancestry: Inside the Genographic Project* (2006). Its story and the way it was visualized in the documentary *The Human Family Tree: Tracing the Human Journey through Time* (Cohen, Bacon, and Wells 2009) are particularly rich as a potential source of sense generation for customers of genetic ancestry tests. They are revealing with regard to the glocalized structure of the genographic network; that is, regarding the question of whose history is reconstructed and for whom others are turned into ancestors, or to whom the commercial haplogroup tests are addressed. Wells begins his narrative in the "melting pot" of New York City. He presumes that its uprooted and mixed inhabitants yearn for their lost history, of which he sets out in search. In the accompanying film that actually came out the year before the projected end of the GP (in 2015, it is still ongoing), we see him taking cheek swabs from people at a Queens street fair. It is their Y-chromosomal and mtDNA that is staged as leading us back in time. The DNA of Carrie—the dancer with a Slovenian background—establishes the connection to the oldest lineage that in the documentary is impersonated by dark-skinned actors playing Paleolithic humans. Carrie's marker is that of scientific Eve—and now we see a "Bushmen" woman. Wells explains that the scientists compared DNA of those who left Africa with those who never did. The first branch within Africa are the Hadzabe who when shown in the film are commented on by Wells as living like "we" did for most of our history. This pattern structures the entire documentary: the DNA of one Queens person after another brings the epos of human becoming, migration, and diversification a step closer to the present and thus to New York—the Americas are the last space to be conquered. The personal stories of the people of Queens are thereby seamlessly woven into the histories of their "ancestors" that are at times played by actors and at times represented by living indigenous groups.

At one point Wells explains the genetic results to a group of Filipinos to whom we are led by the DNA of Aeta from Queens. He tells them that they are still African-looking because they have been isolated and did not admix with the later conquerors from central Asia. They have not gone through the testing and hardening—even whitening—ground of that hot spot. Again reminiscent of Osborn's story, that of Wells culminates in the European Cro-Magnons who developed true art and whom we access through the DNA of a New York fashion designer. Both Alma and the musician Eamon of Irish descent stand in the artistic tradition of these Paleolithic people. Are we to think that the DNA actually handed down that artistic creativity over tens of thousands of years? We have already seen how physical features become linked to genetic markers in the production of living ancestors, and in the case of artistic potential, mental traits and behavior become associated with a genetic line. This is not the last instance I discuss in which the supposedly neutral DNA sequences—the silent mutations—are inscribed with physical, mental, and behavioral traits, but for now, I stay with the journey, because it is not that straightforward after all.

In fact, instead of one exodus out of Africa, the documentary *The Human Family Tree* (2009) presents a rather intricate story of back-and-forth movements between the Sudan and the Sahara, and repeated migrations out of Africa by various routes. Archeological evidence complicates the story further and sometimes stands at odds with genetics. The genetic archive is even presented as incomplete: the Y chromosome and mtDNA tell the tale of only very few ancestors, and not all genetic lineages have survived in today's genomes. Wells further informs us that many questions have not yet been answered and remain controversial, such as the peopling of the Americas. When watching him explain these complications, I began to ask myself whether, with the advances of the genetic research revealing ever more and finer branches on the tree and migrations that are not unidirectional, the story and map might not end up as a dizzying maze. Will the genetic geography eventually cover the globe with a net of human relations that seems without focus and direction—as it had been suggested by Huxley and Haddon some seventy years earlier?

In the film, Wells suggests something else. He ends in a park in New York where all the participants have gathered to hear the results from the genetic analysis. He has them stand on the lawn in flocks according to the diverse haplogroups their DNA has been ascribed to. The haplogroups are arranged in the park space according to a projection of the globe. Each group is thus placed in the geographical region its genetic markers have been traced to (we actually see the continents virtually drawn around the groups on the screen)—the participants are at the place of their supposed origin. Then Wells choreographs the groups to literally reenact backward their ancestors' journeys across the world. One group after another merges into the moving participants until finally, all unite in Africa: "one giant family reunion." The message is twofold: the genomes of Queens show that DNA sequences contain our history of diversification but also that we are increasingly mixed— "traits that formed over thousands of generations are being wiped clean in just a few." Again, the nostalgia for a former genetic order is given expression at the same time as the hope that knowledge of our genetic history as well as the reality of intermixing may further mutual tolerance and understanding. This hope is expressed by Wells and the Queens people at various points in the film: "We" may have split into diverse groups in populating the earth from a common origin in Africa, but in the end, we reunite in Queens.

That there are serious tensions in this melting pot vision of genetic intermixture becomes obvious through the African Americans who participate in the experiment. When the haplogroups retrace their journeys across the lawn, just before the final reunion, the participants are divided into two blocks: those who remained in Africa and those who did not. At this moment, the camera catches the faces of those who "stand on" Africa; we feel awkwardness: African Americans playing the first modern humans in the Paleolithic who never made it out of Africa. The African Americans represent the common ancestors of the rest of the Queens crowd—with one exception: the African American men who have been placed with the European group. A man of Greek origin finds this odd, which triggers the response from the "black officer" that "I'm questioning my own self." Another African American among the Europeans is Dave Reed, a fashion model whose naked upper body the camera has lost no opportunity to treat us with. In contradiction to the visible skin color, we are told that one of Dave's ancestors must have made that journey through the central Asian bleaching machine. Dave, with his muscular torso, carries a "white" Y chromosome. That the historical reality of slavery brought that Y chromosome into Dave's cells is smoothed over, however, when Dave says that the genetics breaks all the molds on race. Other potentially disruptive moments, such as when Mehmet, who is a Muslim, is genetically identified as Ashkenazi Jew, are downplayed by statements like "You almost feel closer somehow" or "We are all family." In the end, the messiness of history is straightened out and the story is purged of its violence when told through the lens of genetics.

Furthermore, the movement of the people across the lawn according to the genetic markers that have been identified in them suggests that "our" history sets out from indigenous peoples and ends in places like Queens. It seems that it is the populations in highly industrialized regions who carry the entire story in their DNA. However, although in the GP and through Family Tree DNA they are invited to participate in the reconstruction of "our heritage" by having their DNA analyzed, this DNA is actually considered less informative owing to its high admixture. So why is the data gained through the ancestry tests added to the larger project? Marketing strategy is clearly involved, because the project wants to present itself as participatory, as creating history for the people and by the people. At the same time, even though it is often stated that the commercially analyzed DNA is not informative because of its donors being largely part of very admixed Western populations, this is not how the GP saw it: "Following successful typing and reporting of the genotyping results, each participant may elect to donate his or her anonymous genotyping results to Genographic's research database. The magnitude of the project and its worldwide scale offer a unique opportunity to create a large, rapidly expanding, standardized database of HVS-I haplotypes and corresponding coding-region SNPs" (Behar et al. 2007, 1084). In 2007, 95 percent of the GP kits had been ordered from the United States and western Europe, but this did not make this mtDNA data bank useless. However, instead of having been the source of the historical research, it was used to refine genotyping, analysis, and reporting tools for the process of building genetic trees. In other words, while "isolated DNA" was the main epistemic object, "admixed DNA" could serve as technical object, even if the two might hardly be clearly separable in actual practice.[24]

In this way, the GP has created from global research a living history for Euro-American historical cultures: it is not only living because it is presented as alive in hunter-gatherer peoples; it is also a living history because it concerns "us," because "we" partake in its preservation and reconstruction, and because "we" carry traces of it in our own DNA. Through the purchase of a haplogroup-DNA test, each individual customer becomes part of an imagined genetic community (Simpson 2000)—complete with historical narratives and traditions (Anderson [1983] 2006). The GP's, Family Tree DNA's, AfricanDNA's, and iGENEA's sales of haplogroup tests certainly profit from the sense-imbuing narratives, films, and images that are part of the genographic network, if possibly not to the same degree. The Swiss company's strategy to meet the supposed demand of its European customers for transparency regarding the science behind its services, while mainly outsourcing to other areas of the genographic network the aesthetic-narrative dimension so central to historical meaning, has at least one disadvantage. In discussions of customers on blogs disappointment has sometimes been voiced about the poverty of the information received from iGENEA—information that is mostly of a numeric kind and carries little meaning for the customers. For that reason, one iGENEA customer advised others to take the DNA test directly with Family Tree DNA. But the same customer also pointed to a potentially serious obstacle to the circulation of knowledge: except for the iGENEA website, the information in the genographic network is in English. However, with the German market expanding not least because of iGENEA, a German translation of Wells's *The Journey of Man* has joined Cavalli-Sforza's translated books.[25]

The genographic network is thus a global structure with strong marketing foci on particular historical cultures within the United States and Europe. However, Wells has been careful to suggest that even for those not included in the "we" and "us" of his stories, who might have strong counternarratives, the DNA could become the carrier of myths: "And we have the narrative about creation and in that creation there is a divvying up of information and knowledge and we place that information-knowledge into the sacred colors . . . [ellipsis in original] and we're talking about how migration occurred. . . . We talk about personas, we talk about deities, we talk about relatives—all over the place. And when we're talking about those things we're actually remembering it because it's already imprinted within us in our DNA and our RNA" (Genographic indigenous representative Phil Bluehouse of the Diné, "Indigenous Representatives Talk about Their Migratory Histories," Genographic Project video clip, quoted in TallBear 2007, 418). Thus, the GP's indigenous representative Phil Bluehouse seems to share with some iGENEA customers the notion that we can actually remember histories that have happened before our times because we carry them in our DNA. In this notion, a complex and labor-intensive process of translation that relies on an impressive infrastructure is literally reduced to a reading of a historical document that contains stories and carries them across generations. Genetic history indeed functions as "new *mythos*" (Wells 2010, 205).

Even though that history in the gene is regionally, ethnically, and individually specific, we have witnessed the emergence of a standard grand narrative of human evolution. Genetic and other evidence on human history is generally seen to converge on the great human journey—a journey to which I again turn in the next chapter as it is staged at yet another node in the genographic network. In doing so, the tree as a structuring device in human population genetics takes center stage. I have discussed its constructive role in bringing into being global human relations in chapter 11. Especially over the course of this chapter, we have seen it structure the visualizations and narratives of the great human diasporas. Indeed, it is the tree that represents so powerfully a panhuman belonging and at the same time clearly demarcated "us" and "them" groups. The tree, which functions as a logo for iGENEA and gives the name to Family Tree DNA, is the condensation of the population genetic narrative of unity in diversity.

CHAPTER FIFTEEN

The Genographics of Unity in Diversity

As we have seen throughout this part of the book, the research into the geography of human genetic markers creates trees that are mapped in space and time; in doing so, it interferes with geographies of identification "out there." *Genographic* is thus also the graphics of the gene from autoradiography up to the level of performing the human odyssey, for example, in a film or by reenacting the movements of "our" ancestors on a lawn in Queens. In this chapter, I examine these genographics for the Italian touring exhibition "Homo sapiens: The Great History of Human Diversity." First staged in Rome from 11 November 2011, the touring exhibition moved to Venice, after which it opened its doors in Trento in October 2012—where I went to see it. On the brochure for the exhibition, there was a beautiful photograph of sand dunes marked by a trail of footprints—the human trail the reconstruction of which Cavalli-Sforza and others devoted their careers to. The exhibition that was under the aegis of the Ministry of Cultural Heritage and Activities and that carried the icon of UNESCO represented a microcosm of Cavalli-Sforza's global lifework. As Ian Tattersall, curator of anthropology at the AMNH and collaborator in the project, put it: the exhibition "fits with his [Cavalli-Sforza's] human genome project that he's been doing for such a long time, and it's in many ways a sort of—at least for the public—a kind of a culmination of the work he's been doing for many years" (personal communication, 28 Dec. 2012). With the exhibition, Cavalli-Sforza and Telmo Pievani (University of Milan Bicocca) wanted to enable people to recover their origins and relive the ways in which their ancestors populated the earth. In my discussion of the museum exhibit, I thus attempt to combine the analysis of the representation of space—the way in which it was conceptualized and imagined by the producers—with that of the actual space of representation as well as the experiences and thoughts it engendered (Lefebvre [1974] 1991).

The site of "Homo sapiens" was multilayered. There were not only guided tours through the exhibition for all age groups on offer but also educational workshops—among other topics on the issue of "race" and racism and the "tangle of civilization." A calendar informed people of international workshops and events at which they might broaden their understanding. Furthermore, there was a teacher's guide intended to allow continuing engagement with the issues raised in the exhibition either in class or in the family. Beyond possible lessons, it contained a list of further readings, including the Cavalli-Sforzas' *Chi siamo: La storia della diversità umana* (L. L. Cavalli-Sforza and F. Cavalli-Sforza 1993), and suggestions for children, even from the realm of prehistoric science fiction, such as the movies *Quest for Fire*, *The Clan of the Cave Bear*, *Ice Age*, and *Jurassic Park*. In the electronic version of the teacher's guide, there were links to the exhibition website and to those of the American Museum of Natural History and the Natural History Museum in London. The exhibition's own website—www.homosapiens.net—was a rich experience, featuring an exhibition trailer, a video clip on the human journey, as well as explanatory text on the science, the administration, and the content of the exhibit. Finally, you could order the catalogue or browse through the photo gallery.

Where the representation of space was concerned, Cavalli-Sforza's curatorship shaped the exhibition project from the scientific approach to the message and the ways of conveying it. The brochure announced "the first exhibition in the world that tells the story of mankind through a large multidisciplinary fresco: an international project involving more than 50 museums, universities and libraries from 9 different countries." Cavalli-Sforza himself was described as the expert "who for decades has probed the most hidden recesses of the depths of the history of human diversity, uniting molecules, fossils, cultures and languages in a coherent overall framework of evidence" (brochure, English version). Behind the exhibition stood different institutions from both sides of the "two-culture divide."[1] There was also an international and interdisciplinary scientific committee with experts in linguistics, archeology, paleontology, genetics, demography, but also sociology, and history and philosophy of science. Given the flavor of the project, it may not come as a surprise that Francesco Cavalli-Sforza and Wells were on this committee. In accordance with the multidisciplinary approach, the catalogue contained chapters on the human journey from the point of view of paleoanthropology, ethnography, linguistics, and pedagogy, followed by chapters on the exhibition sections "longing for Africa," "many modes for being human" ("loneliness is a recent invention" in the brochure), "genes, peoples and languages," "traces of lost worlds," "all relatives, all different: the intertwined roots of civilization," and "Italy, unity and diversity" (Cavalli-Sforza and Pievani 2011).

Through the exhibition, the curators wanted to explain both the groundbreaking interdisciplinary approach and the story of how the "kaleidoscopic mosaic of current human diversity" (brochure, English version) came about. The exhibition was intended to convey an emotional, profound, and dramatic narrative by guiding the visitors through six successive stages of the human journey. It performed the history uncovered by the genetics of human populations through objects, fossils, artifacts, tools—or rather casts and copies thereof—models of hominids and animals, reconstructions of scenes and events, documents, immersive video and photo installations, interactive displays, and of course maps. There were captions and explanatory panels with an abundance of graphics to allow a reading at different levels of attention and scientific literacy. There were quotes on the walls, and the lighting and sound were intended to generate an artistic atmosphere. The galleries contained only a few objects, which were more or less meant to speak for themselves. Like the museum and zoo exhibits of Osborn and Huxley, this exhibition was created in the spirit of Maria Montessori with the goal that the harmony of art and science, beauty and nature, expressed in the exhibits would facilitate the internalization of "our origin and history," or rather remind visitors of the history that they already carry within (Cavalli-Sforza and Pievani 2011). As we will see, this history was told through the dialectics of unity in diversity of "the human family tree." The exhibition as multilayered space conveyed this dynamic as the driving force behind "our" history of diversification along what could appear as the only possible and progressive lines. However, as the tension-rich expression of the "kaleidoscopic mosaic" in the above quote suggests, the more static forms of representing current human diversity, such as the mosaic, were at times subverted by the shifting arbitrary patterns of the kaleidoscope.

Entering the Museo delle Scienze—the actual space of representation—the visitor was told that we have been traveling for two million years, populating the world in successive movements from Africa. After meeting the first upright primates (australopithecines), one learned about the earliest diasporas of *Homo*, visualized by arrows and paleoanthropological/archeological sites on maps. Supporting my impression, Tattersall referred to the exhibits that tell about hominid evolution prior to the appearance of true *Homo sapiens* as a prologue (personal communication, 28 Dec. 2012). The exhibition conveys the message that the most important part of the story began with mitochondrial or African Eve, the mother to whom we can all be genetically traced back. In the exhibition, the hypothetical carrier of the mtDNA from which all present human mtDNA descends was rendered three-dimensional in the form of a full-body sculpture. The woman sat on a stone, her hands on her pregnant womb. Her expression was ailing, a kind of pietà. She was the centerpiece of the exhibition, binding the visitors as well as the exhibits together (figure 31). The colored band ornamenting all the panels

FIGURE 31. Mitochondrial Eve, "Homo sapiens: The Great History of Human Diversity," Trento, my photograph, 16 Oct. 2012

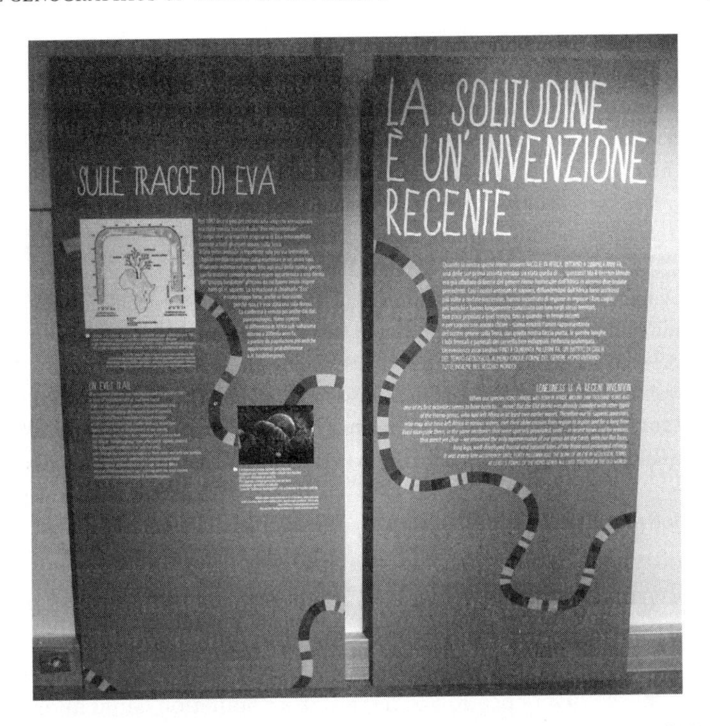

FIGURE 32. Mitochondrial Eve panels, "Homo sapiens: The Great History of Human Diversity," Trento, my photograph, 16 Oct. 2012

in the exhibition was the thread of mtDNA that leads from her to everyone on this planet; it was the umbilical cord connecting her to the child in her body and through it to all of us (figure 32).

Beyond the allusion to Mary and Jesus, the exhibit evoked trees of Jesse or of dynasties that in the European Middle Ages sometimes grew out of an ancestor's body. Although the tree usually emanated from the chest or even the genitals of the founding father, there was also a tradition of having it spring from a woman's womb (Klapisch-Zuber 2004, 94–101). In the context of the African Eve exhibit, an alternative kind of substrate for the tree was also visualized. The brown-skinned Eve was flanked by the Cann et al. genetic kinship diagram (see figure 26) that encircled a tree proper—is it an oak?—growing out of the African continent (see figure 32). As Wells put it elsewhere: "If we imagine the genetic relationships among modern mitochondrial diversity as an actual tree—say a large oak—then the root and trunk, and the branches that are closest to the ground, are all found in Africans" (Wells 2002, 70).

The trailer on the exhibition project website actually zoomed in on African Eve's womb, thus identifying it with that "fertile African soil" from which we all arose.[2] In this system of intertextuality and intermediality, our common ancestor, African Eve, appeared as the soil and the root of the tree, but the tree had branches that were neatly separated, dividing the human world into Europe, Asia, Australia, New Guinea, and Africa. Thus was set in motion the play of unity in diversity that we have found a structuring principle of human population genetic knowledge.

As the master icon in conveying unity in diversity, the tree organized the documentations as well as the exhibition space. In the catalogue, there was Darwin's sketch prefaced with "I think." Although this diagram does not suggest directionality—and is in fact not a tree (Voss 2007, 95–163)—the visual analogy between Darwin's structure of organismic evolution and the exhibition's topic of human evolution facilitated an understanding of the splits also in the case of modern humans in terms of speciation. Similarly, the maps on which the history of human migrations was represented by a treelike system of arrows mostly conveyed a sense of dispersion rather than encounter and reunion.

In the brochure, one could read: "As Darwin had already predicted, the tree of the diversification of the Earth's population could allow us to understand the structure of the tree of languages" (brochure, English version). On the exhibition panel that shows the correspondences between the genetic and linguistic trees was also the information that the Khoisan or "Bushmen" are the oldest genetic lineage, "representatives of the longest-living forms of humanity," but that they are "not frozen into 'ancestral' constitutions" (exhibition panel). In addition, a Survival International film was shown that claimed the existence of "uncontacted peoples" who have never met with other populations. Like chimpanzees that are called *naïve* because they have never met any humans, *uncontacted* peoples were seen from the perspective of the outsiders who have managed to contact everyone else. However, these "uncontacted peoples" were also said to choose to continue their better lives, and it was emphasized that these peoples are not primitive and do not live in the Stone Age. At the same time, photographs and artifacts of indigenous peoples appeared side by side with prehistoric cultural goods, and it became clear that not everyone partook in the population-multiplying machine of the agricultural revolution with which "THE GREAT VOYAGE OF DIVERSITY STARTS ONCE AGAIN" (exhibition panel). Narrating and visualizing "The Great History of Human Diversity" in trees thus allowed turning some indigenous peoples into "our" ancestors by rendering their branches as reaching back furthest in time. They represented a nostalgic past that we have already encountered at different nodes in the genographic network, when human groups were still "uncontacted" by all the evils that came in the wake of the agricultural revolution. Simultaneously, they occupied the end of branches; they were part of the mosaic of contemporary human diversity.

The presentation of earlier hominid diversity followed similar lines. The engagement with the relationship between Neanderthals and modern humans in the exhibition, too, wavered between suggesting intermixture and highlighting alterity, between close genetic kinship and a great significance of the genetic differences. However, the human Paleolithic revolution—the advent of true art, sophisticated ornamentation, complex burial rituals, musical instruments, and other indications of a symbolic mind and great intelligence—was unequivocally presented as having taken place with the Cro-Magnons and in Europe. And again with regard to deeper hominid history, the exhibition on the one hand emphasized the coexistence of different hominid species and genera. On the other hand, in the catalogue, there was an image of a serial development from what could be a chimpanzee via *Australopithecus*, *Homo erectus*, *Homo neanderthalensis*, to *Homo sapiens*—suggesting only one line of descent (Cavalli-Sforza and Pievani 2011, 7, see also 5). In particular such linear representations of hominid evolution, but also tree- or candelabra-like depictions of diversification, made "our story" appear ultimately unidirectional, progressive, and even final. In fact, Tattersall sees the exhibition as an expression of a main interest in linear, progressive development—that is in grades rather than clades (personal communication, 28 Dec. 2012).

The dialectics of unity in diversity was once again captured in the last installation in "Homo sapiens" that recapitulated the three journeys "out of Africa" on video by way of maps with archeological sites and arrows indicating migrations. Overall, the message of the exhibition was that "We find UNITY IN DIVERSITY at every level, from primary emotions to languages, from physical features to cultures." And this unity in diversity appeared as a dynamics of general progress—if not to the same degree for all of us: "FROM THOSE UNCERTAIN STEPS IN THE LAETOLI TUFF, 3.75 MILLION YEARS AGO, TO THAT OTHER HUMAN WALK THAT OPENED UP A NEW FRONTIER, ON THE MOON, WE CERTAINLY HAVE COME A LONG WAY!" (exhibition panel) Tree thinking and imagery made it all possible: diversity that is rooted in unity, differentiation that goes along with general linear advance, and the survival of old branches among more progressive and younger ones. In a Huxleyan, or evolutionary humanist way, human biological and cultural diversity was celebrated in the exhibition and its importance for progress explained through a look at prehistory. However, nearly a century after Conklin's and Huxley's warnings, the tree structure continues to convey human groups as neatly separate and their history as having taken place in isolation. It gives a tessellate rather than kaleidoscopic image of human diversity.

As biologically, culturally, socially, and politically productive practice, tree building has strong roots in European history. This history was taken up by the exhibition project with a reproduction of a traditional tree of life mosaic from the floor of an Italian cathedral in the catalogue. The tree of life was a central inspiration in the history of the iconography of the family tree. The latter has its main roots in the European Middle Ages where it gave shape to sacred genealogies and legitimated aristocratic power and the transfer of social capital. In the middle of the sixteenth century, the iconography of the family tree as a tool in social strategies began its career beyond princely and royal dynasties, culminating in the adoption by bourgeois families and branches of knowledge in the nineteenth century. Among these branches was anthropology (Klapisch-Zuber 2004; Pálsson 2007, ch. 3; see also Rheinberger and Müller-Wille 2009, 69–77). Like the family tree, the evolutionary tree continues to found social relations and cultural meanings in biological descent, ultimately removing culture by seemingly laying bare the underlying umbilical cord of genetic ancestry: "Even the most technical, machine-driven inscriptions of molecular genetics are grounded in the social complexity of the pedigree, which is nature-culture, and which represents a signal case of the employment of cultural resources to achieve erasure—of the cultural" (Lindee 2003, 43). With its erratic movements between the sacred and the secular and the scientific and the mundane, tree building is a cultural technology for producing a natural order in biology as well as culture; it creates the appearance that the cultural is in fact (aligned with) the biological.

This process was also at work in an educational activity suggested in the teacher's guide of the exhibition project that consisted in reconstructing the hominid family tree "paleoanthropologically": "Cut out the skulls and glue them on a panel, attempting to reconstruct the phylogenetics using the data provided in the table."[3] Children were given images of skulls to each represent a species and information such as the species name, its period of existence and geographical origin, as well as its height (without range) and maximal cranial capacity. The exercise did not convey the internal diversity of the species and thus suggested that the species tree is equivalent to the family tree. As in the case of placing different human populations on separate branches, or in the pruning of the family tree through genetic testing, the gluing of hominid skulls on paper that the children here were asked to practice is a technology for cleaning up a messy (natural) history. The cultural technology of tree building achieves classification and the definition of genetic relations (sociality), the allocation of geographical relations (spatiality), and the founding of these in time, and in biology (historicity).

The aspects of classification and geography were at the center of a particularly intriguing piece of the exhibition. In one room, projected on the wall, the visitors saw an ever-changing arrangement of photographs of individual human beings that together represented current human diversity. In front of that kaleidoscope of human faces, there was a screen with a selection of these images. The visitor was asked to attempt to group the thirty-six photographs into up to ten "races" using such criteria as skin color and hair texture. I very quickly ordered them into six groups, only to be informed that people doing the "race test" did not at all agree on the number of groups. Of these visitors, 0.3 percent even put the images all in one group. Quite likely owing to the way in which the test was designed, a large majority (36.6 percent) of the visitors who had done the test up to that point chose ten groups. This information seemed to undermine the notion of race because the number of groups visitors had produced appeared random, unless it was understood as determined by the task.

However, in the next step, the virtual test aimed at correcting some of my possible grouping mistakes. First, I was informed that there were two representatives of the San among the pictures—had I put them in the same box? That did not catch me out, because I had grouped them together. Secondly, there were two Burmese; again, I had correctly put these in the same box (see figure 33). It was only then that some of my mistakes began to show. In other words, despite the subversive potential of the "race test," it seemed to make a strong point that there was one correct way to group the thirty-six pictures. This impression was enforced by the penultimate projection that showed my categorizations mapped on the world: with a few aberrations, my groups formed geo-

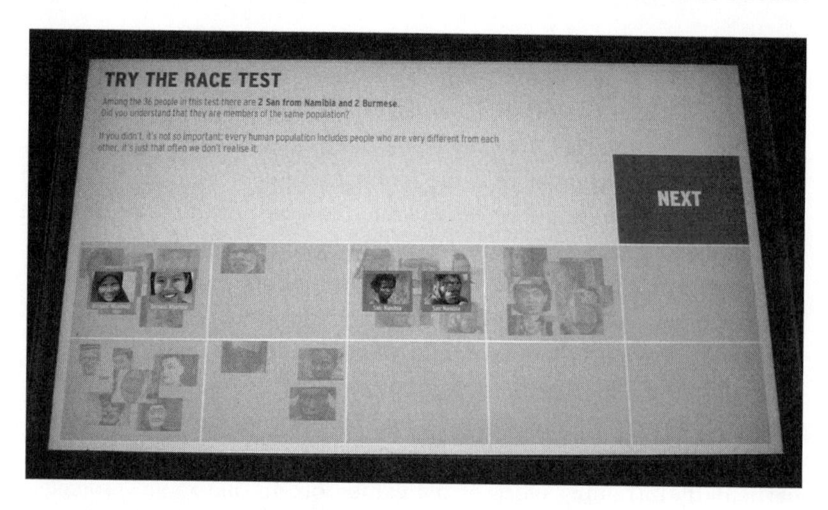

FIGURE 33. "Try the race test," "Homo sapiens: The Great History of Human Diversity," Trento, my photograph, 16 Oct. 2012

graphical regions. Even though the last screen denied that races exist, to "experiment with your personal 'racial' catalogue" may remain a dangerous thing, especially if the genie in the machine will then check "how close it comes to reality" (installation in exhibition). In the end, human diversity again appeared like a mosaic rather than a kaleidoscope. The same may hold true for the pendant of this exhibit in the teacher's guide. Students were asked to sort people into "races" on the basis of physical characteristics in the hope that they would find the task impossible—but what if the children managed to group them according to geographic origin?

Nonetheless, one installation in particular seemed to transcend the notion of a fixed tree that binds human kinds to each other in a historical root and through clearly defined relationships in the present. The installation consisted of two circles, one on the ground, and a corresponding one on the ceiling. On the ground, you could step on images of plants and animals. At that moment, the universal genetic kinship system above you—the "one great tree of life" (exhibition panel)—began to shape-shift, zoomed in on the corresponding position in the "tree," and told you what percentage of your genome you share with the chosen organism: 44.2 percent with *Drosophila melanogaster*. This might seem quite a significant amount, but as you explored the system of genomic relatedness, zooming in at different locations, the numbers constantly changed their meaning. So if you stepped on the chimpanzee, you would learn that your genomes are 99 percent identical. In this ingenious interactive installation, the game of unity in diversity received a twist through the inclusion of ever more "others": we may all be very similar, and some may indeed be more similar than others; but as in the infinite chains of signifiers of the language system, in this genetic system, the degrees of closeness and distance are in flux and the meanings of similarity and difference emerge in a dynamic network only to be already something else. It is a system of *différance* (Derrida 2001). While in the exhibition project there seemed to be a dialectics at work that reproduced the imperial narrative of progress and the simultaneity of the nonsimultaneous, at least in this installation the play of *identity* and *otherness* suggested that these are not binary opposites but relative terms, the meaning of which depends on one's changing place in a system of signifiers (see figure 34).

Finally, the "Homo sapiens" exhibition was not simply the internal vision or space of Cavalli-Sforza and his collaborators (Firstspace) or an objectified social, or physical space (Secondspace). It was a perceived,

FIGURE 34. The genetic kin system, "Homo sapiens: The Great History of Human Diversity," Trento, my photograph, 16 Oct. 2012

experienced, but also lived space, in which visitors might be subjected and which they transformed in the process (Thirdspace, after Soja 1996, ch. 2). I therefore tried to get a glimpse at this lived space beyond my own immersion. The resulting twenty-four interviews point to the fact that the exhibition was visited for various reasons. Teachers of history and science hoped to gain insights for their classes; tourists and locals became interested through advertisement; and families sought a common occupation. Some people studied the panels extensively. Others did not always have the chance to go into detail because of their children, and a number of visitors had their experience disrupted by school classes and tours, while others were part of such classes or tours.[4] A schoolteacher considered the exhibit too difficult for children and not sufficiently interactive. Indeed, prior knowledge brought to the museum ranged from hardly any to considerable, which also influenced the understanding gained from the site, as did beliefs and general background. Some visitors had learned about human population genetics through books (including Cavalli-Sforza's), magazines, museums, and films.

The genetics, this "wonderful tool to understand history," with its ability to tell us where we come from and from whom we descend, was a fascinating aspect for most visitors. A biologist couple even stated that it could reveal "almost all" about our history. For another couple, it could establish continuity between past and present, and one woman in her thirties expressed the belief that knowledge of that past would help us see our way into the future. Also for most visitors, the message included aspects of unity and diversity. Some commented on a newly gained understanding that different hominids had lived at the same time, while others emphasized that science shows how similar we all are: "We are all part of the same big family"; "we're all the same." In combination with this thought, a woman remarked on how beautiful it is that we are nonetheless so different—these differences were understood by several visitors to be meaningless. Others took home the message that there are diverse "races" but that they all have a common root. However, one man thought that the human bond was more apparent in the similarity of certain facial expressions and behaviors than in genetics, and a couple saw it best documented in "the human spirit." Another visitor found the notion of African Eve—of just one common ancestor and only in Africa—strange; someone expressed astonishment at her looks—he had not imagined her like that. Nonetheless, the social and ethical message against racism and for the idea of human connectedness was generally understood. One woman said that Eve carried an *Urzelle* ("primordial germ cell") that had been handed down all the way to our living bodies. Another woman commented on the importance of changing our view of Africa through learning about its diversity and evolutionary importance. For some, "the story of humanity," the insight that through our past and our DNA we are all connected to each other—and even to vegetables—was very personal and affected the way they thought of themselves.

The "tree of life" installation discussed above seems to have been a favorite with many. It broke open the African Eve message of a pan-human multiculturalism that to a certain extent was built on the opposition to premodern human and nonhuman species. Several visitors pointed out that genetics reveals a close relationship among humans, animals, and even plants. The more complex presentation of the tree of life as a dynamic genetic system could inspire idiosyncratic redefinitions, as when a science teacher concluded that two individual humans differ more than humans and monkeys (this was followed by the statement that humans share a great part of their DNA with bananas). Two women also pointed to the shifting system of relationships beyond the installation: Neanderthals had still been ancestors when the women had gone to school but had since ceased to be. These two women actually considered the whole exhibition too ideologically underpinned; the scientific facts could also be read differently. An anthropologist and a person with a degree in natural science found the African Eve story of human becoming too simple: often researchers try to arrive at a simple origin solution. Other visitors were more impressed by the australopithecine Lucy than her modern counterpart African Eve. Lucy's full-body reconstruction in the exhibition was entwined in the narrative of a former much more significant diversity of the dimension of species and genera.

Clearly, visitors made their own histories of human diversity. They were subjected to the exhibition structure but they also subjected the exhibits to their own interpretations. The representation of space, the space of representation, and the experienced space interacted when visitors brought the static exhibits alive by drawing on their preexisting knowledge and beliefs. Also where the "Homo sapiens" project is concerned, the accounts of origins and histories remain diverse and contested, and they may essentialize human groups but can also open up and set in motion kinship systems. They may produce mosaics or kaleidoscopes of (the history of) human diversity, or waver between the two, which is accentuated by the fact that "Homo sapiens" is a touring exhibition; its conception and realization as well as the experience of it will be different for each location. Trento could provide limited space and the museum had small rooms with rather low ceilings. The cramped quarters were no doubt part of the reason why visitors often experienced the exhibition as too crowded. Other locations were and will be more spacious.

While the traveling exhibit is expected to communicate the knowledge of our common human history around the world, it also allows for the possibility of integrating a specific regional section. It is "an Italian product exportable around the world" (brochure, English version). In Trento, the "Italy—Unity and Diversity" section was placed in the most beautiful room, with wonderful paintings and ornaments on the ceiling, communicating to the visitor that this exhibit was about a particularly cultured group of humans. A story of hybridity and circulation was performed, however: "Many of the everyday objects we consider as ours are in truth a gift given to us by the interwoven paths of human diversity" (exhibition panel). In this room, in accordance with Cavalli-Sforza's theory of cultural evolution, one gained the impression that culture supersedes biology and follows its own rules. At the same time, in the catalogue, the exhibition, and the teacher's guide, Italian children were made aware of their country's biological as well as cultural variety. Its geographical position, its shape, and the stream of peoples entering and leaving the peninsula have "given us a unity in diversity and a diversity in unity that is unparalleled anywhere else in the world" (Teacher's guide, 11). Italy is here what Queens was in Wells's story, the culmination of a long history that can be told along the tree structure but ends in a melting pot.

Postscript

He needed humans to rise to their station: conscious and godlike, nature's one shot at knowing and preserving itself. Instead, the one aware animal in creation had torched the place. —*Powers (2006) 2007, 72*

In this book we have encountered encompassing phylogenetic and evolutionary projects. Osborn wanted casts of all the known hominid types and tools, and he sent teams in search of the bones of the yet unknown first ancestor, so that vertebrate and human evolution in its entirety might be presented to museum visitors and audiences beyond. Everyone should share his way of throwing light on the present from the past, in every field from political science and politics to science fiction. Huxley wanted to preserve the worldwide "evolutionary heritage"—the animal and human diversity—and build cultural archives to ensure the continuation of knowledge production and global distribution; he aimed at the development of a world society along the lines of his evolutionary humanism. Cavalli-Sforza wanted blood samples from human populations all over the world as well as the preservation of cultural history in order to reconstruct the grand sweep of modern human origins, evolution, and migration across the globe. He hoped the knowledge from human population genetics and history more generally would give a positive impetus to human cultural evolution.

The narratives that Osborn, Huxley, and Cavalli-Sforza created out of data and information from various sources were epic. Humans had been the makers of their evolutionary destinies in the past, and they had to consciously drive evolution toward progressive aims in the present. The scientists' texts, exhibits, and images were produced and set in circulation to provide meaning and orientation, to provoke transformative experiences and ultimately to perfect the individual and the "races" or the species. At the center stood the belief in a particular kind of historical science and the unity of knowledge on its basis. All three argued against "overspecialization"—differentiation had to be accompanied by integration. Each regarded his epistemic objects as more important than other scientific and scholarly sources and viewed the approaches of paleontology, evolutionary biology, and human population genetics superior to the respective alternative scientific and scholarly methods. In the end, Osborn and Cavalli-Sforza thus shared many of the beliefs and goals that Huxley subsumed under the labels of *scientific* and *evolutionary humanism* and that were associated with a particular notion of progress through dissemination, appropriation, and implementation of the history within.

The pursuit of this humanist aim was facilitated and at the same time complicated by major societal changes. By 1920, specialization also structured American society from the assembly line to the university. The scientific method was understood as a universal tool for the optimization of all areas. Concomitantly, between 1860 and 1920, every aspect of life had become "expertized": there were special guides and magazines on the handicrafts and businesses, on cooking and housekeeping, on health, and so forth; and schools were established that gave out degrees. Even religious and spiritualist movements began to imitate scientific higher education by importing similar organizational structures and vocabulary such as *Christian Science*. With this development, learning and the expert gained unprecedented importance; expert professors not only served on government boards and commissions; they also shared their expertise on topics of public interest from sports to child rearing. But there was also resistance to the new specialists such as psychologists and anthropologists who defined people's IQ and the meaning of "race" and criminality. Most resistance came from religious circles that were not only threatened by evolutionary thought but by the entire system of education that was about to take authority over the moral education of their children out of their hands (Oleson and Voss 1979). In the United States as well as in Europe, these transformations of knowledge cultures coincided with the second peak in the communications revolution, with the introduction of rotary printing and lithographic and photographic techniques in the mass publication press, the replacement of steam power by electricity, cheap paperback books and pulp magazines, and mass circulation of daily newspapers. The circulation of news was facilitated not only by the telegraph but also by railways and ships—means of transportation that changed perceptions of the world and self (Lightman 2007, ch. 1).

Therefore, if for Osborn and Huxley the history of humankind and its meaning for the future needed to be rewritten in order to provide guidance in a world marked by transformations, they could profit from the authorization of science and the possibilities of communication, but they also found themselves in great competition over the power of definition with regard to the meaning and purpose of life. The "scientization" of society went along with a certain demystification, disenchantment, and mechanization. The differentiation of knowledge cultures and the loss of influence of traditional values unleashed centrifugal forces without as well as within the sciences. If Osborn had seen these processes mirrored in early genetics and the experimental sciences in general, and if Huxley had to imbue the meaningless evolutionary process of his grandfather with new moral authority while he and his friends fought the "reductionist" molecular approach to life, Cavalli-Sforza found himself in the context of poststructuralist turns in cultural anthropology and beyond.

In their adherence to a belief in a core set of common values and a shared history and purpose, Osborn, Huxley, and Cavalli-Sforza emphasized individual responsibility, albeit guided by a deep understanding of "racial," respectively panhuman, belonging. Tony Bennett (2004, 27–32, 120) sees in evidence at the AMNH in Osborn's time a new liberalism of the fin de siècle. Whereas in classical liberalism, self-government had been restricted to the middle and upper classes (the poor were subjected to state-controlled surveillance), proponents of the new liberalism wanted to extend self-monitoring and self-development to all ethnicities and classes, and in some cases also to women and children. For some liberals, the state had a certain responsibility in this moral and educational guidance of individual self-government. In the United States, besides the new popular schooling and progressive regulation, philanthropy was particularly important. Philanthropists in Osborn's time and in his circle of family and friends invested in elite and general education through colleges, universities, libraries, museums, and zoological parks. However, Osborn's educational goal was ultimately conservative if not reactionary. He positioned the men and women of the Stone Age against what he observed as degenerative trends; he offered images and narratives for sense making, identity formation, and perfection of self that should remind the "races and sexes" of their natural state and place.

While Huxley's efforts in education, popularization, and organization also aimed at the empowerment of the individual through the right evolutionary consciousness, his notion of the perfect society differed widely from Osborn's. In the Britain of the interwar and war years, he still saw too much laissez-faire at work and he advocated comprehensive regulation particularly through science-guided analyses, planning, and intervention. Of the three scientists, Huxley was by far the most outspoken about how the future ideal society should be brought about through central and local, public and private organization. His perspective from evolution did not suggest an ideal society structured by "race," class, and gender. Evolutionary trends and the mechanisms of genetics suggested a new democratic world order based on equality of opportunity in environments that were amenable to the greatest number of people. At the same time, even if among his friends were socialists and communists, his own outlook was still a brand of liberalism. Finally, Cavalli-Sforza shared Huxley's belief that individuals had to become conscious of the evolutionary past that they incarnated in order to feel connected to all humankind and beyond. Equality of opportunity was an aim not yet reached, and in the ideal democracy, everyone would be free to develop his or her personality to the fullest and to strive for the greatest satisfaction possible—so long as this was not at the expense of the overall level of satisfaction. Perhaps owing to his Italian background, he thought that in such a meritocratic democracy transnationally organized, checks and balances would have to keep in bounds especially state power; he at times had an American model in mind.

However, there was another model: the hunter-gatherer society. If Osborn put considerable effort into reminding people of the Cro-Magnons—that Paleolithic race with the most beautiful bodies, refined minds, and artistic skills, who had been one with their surroundings, Huxley recognized in contemporary traditional societies a particular potential for progressive development. While Osborn tended to emphasize the importance of a strenuous life to progress in past and present, Cavalli-Sforza wanted modern societies to remember the lost happiness and innocence of their ancestors who were now embodied in such populations as the "pygmies." Cavalli-Sforza and his colleagues promoted an antiracist humanism of the kind that Huxley and some of his friends had promoted from the interwar years on, while diversity of another kind remained key. And despite their proclaimed panhumanism, both Huxley's and Cavalli-Sforza's evolutionary perspectives were Eurocentric. Nonetheless, judged against his time, Huxley's critical awareness in the late 1920s and 1930s was considerable, also where colonial legacies were concerned. The politics of identity at work in Osborn's, Huxley's, and Cavalli-Sforza's science and societal endeavors were as different as the historical cultures from which they arose and at which they were aimed. At the same time, all three felt that there was a certain urgency to their task of preserving the natural and cultural "heritage" as the basis of transformative experience, which added to the "anti-modern," or at least nostalgic, sentiment that seems to have accompanied their unfailing optimism for future progress.

All three men considered the possibilities of a planned development of the human, national, and/or "racial" genetic makeup. For Osborn, the implementation of his notions of eugenic birth selection and mechanisms to force the "races" to keep to themselves were central goals, and Huxley, too, was engaged in eugenics institutions, even if he came to criticize "classical eugenics" fundamentally. Cavalli-Sforza approved of genetic screening and abortion in case of a diagnosis of severe handicap, but like Huxley during part of his career, he thought of biomedical intervention in the service of human progress as mainly a matter of the future. "Euthenics," the shaping of the social, economic, and cultural environments, seemed more promising. Finally, already in Osborn's time the preservation of natural habitats that seemed to have been formative of modern humans during evolution was a great concern. Huxley and his circle shared the understanding of "nature" as therapeutic, but came to develop a global picture of human interference with the "natural" environment and "biodiversity." Along the lines of what Huxley and H. G. Wells called *applied ecology* and Hogben referred to as planned ecology, Cavalli-Sforza held that, like it or not, since the agricultural revolution humans had also significantly shaped their natural environments. It was time that they took on their role as trustees of life to the fullest and consciously interfered with the natural as well as the cultural world in order to preserve and develop their plenty. In the 1920s and 1930s, with the growing realization of the global extent of the threat to nature, notions were therefore developed that in recent decades have been revived under the label *Anthropocene* (Crutzen and Stoermer 2000; Zalasiewicz et al. 2008)—an anthropocentric phylogenetic-ecological perspective wonderfully captured in Richard Powers's novel *The Echo Maker*, from which I have chosen the epigraph.

Today, the torch of evolutionary humanism is also carried on by sociobiology, even if the economized biological theories stand in contrast to the evolutionary humanism of a Huxley, Simpson, or Dobzhansky. As has been observed by scholars like Donna Haraway (1991, 57–67), this world of selfish and strategically acting, profit-oriented organisms and molecules, with the gene as unit of information, that sociobiologists promoted in their popular texts, was modeled on communication theory and technology as well as postwar capitalism. Nonetheless, representatives and proponents of sociobiology and evolutionary psychology—E. O. Wilson, Richard Dawkins, Steven Pinker, and Daniel Dennett—have been or are public figures of humanist organizations. Despite the obvious differences, such as a more reductionist approach and at times a more strongly determinist perspective on human sociality and behavior, many of the tenets of an evolutionary humanism have remained intact. Evolutionary biology appears as the *Leitwissenschaft*—or should we say the new religion that renders God a delusional, albeit successful meme (Dawkins 1976, ch. 11; 2006)? It is the evolutionary view that has the authority of interpretation over our identity, origin, and history. From the start, that is, when the term *sociobiology* was introduced in the 1940s to mainly describe the biological explanations of human social behavior, the approach was conceived as multidisciplinary, encompassing biology, psychology, and sociology (Scott 1950). Redefined by Wilson (1975), *sociobiology* came to mean the natural scientific foundation of sociology, even ethics. Soon enough, it seemed that sociobiology was finally the key to unite political science, law, economics, psychology, and anthropology under a new approach to the human. Also from the beginning, sociobiologists invested in (and profited from) bringing their science to life in society, where there has been the whole spectrum of reactions since, from complete appropriation to vehement rejection (Segerstråle 2000).[1]

Like the biologically based histories of my protagonists, sociobiological treatises renegotiate the relationship between "nature" and "culture" and by inference between the natural sciences and humanities. We have seen that at Osborn's and—despite C. P. Snow's (1959) verdict—even at Huxley's time of writing, the relationship between natural and humanistic approaches had not yet been so radicalized that exponents of the domains would not have united under the outlook of a new humanism. To date, scholars such as the philosopher Michael Schmidt-Salomon may call themselves evolutionary humanists and engage in its organization and popularization. For Schmidt-Salomon ([2005] 2006), the manifesto of evolutionary humanism consists in working toward a *Leitkultur* adequate for the times, founding ethics naturalistically, and integrating science, philosophy, and art. The evolutionary humanist must remind people of their ape origin, of all three great slights science has dealt to humankind, in order to make them reject belief and tradition in favor of a rebuilt humanism.

Beyond new aficionados of evolutionary humanism from the humanities, scholars have attempted to synthesize from their side: historians like Daniel Lord Smail (2008) have demanded that history be expanded to encompass evolutionary history and that historical approaches be based on the neurosciences; art historians like David Freedberg and Barbara Stafford have attempted an explanation of aesthetic phenomena on the basis of the knowledge about mirror neurons (Freedberg and Gallese 2007; Stafford 2007); and literary scholars have applied sociobiological theory to the analysis of the behavior of fictional characters (see Eibl 2010). Such attempts at alignment with and borrowing from the natural sciences have to be seen in the light of the increasingly relentless struggle for resources (reminiscent of the problems of the historical sciences, especially the museum sciences, felt by Osborn, the synthesists, and even Cavalli-Sforza)—a successful example is the digital humanities, which have adapted the approach of big data mining. However, the historian of science Roger Cooter (2012) has warned that under the current knowledge regime—which he diagnoses as characterized by the rule of the life and information sciences, while the humanities are considered redundant where they do not prove ancillary or applicable—such borrowings can only be to the disadvantage of the cultural approaches and their disciplinary autonomy.

It is also in the context of these issues that we need to recall the processes of naturalization, denaturalization, and renaturalization that we have found at work in the course of this book. Attempts at analogizing cultural to biological characteristics and processes, or to base the first on the latter, are never definitive or consensual. This historicity and multiplicity of determinations of the relations between "nature" and "culture" point to what I might call *boutique humanities*, when syntheses are built through integration of that knowledge from other fields that is compatible with the scientific framework, a pattern that we have observed in the work of all three protagonists of this book. Correspondingly we might speak of *boutique biology* when from the store of biological knowledge are chosen those tidbits that seem to support one's scholarly and/or political point of view. Some scholars and social scientists, for example, welcome the developments toward epigenetic approaches because they seem to suggest more holistic understandings of life, because they seem to lessen the grip of genes and increase the power of environments with regard to who we are (Meloni 2014). However, in drawing on epigenetics in a discussion of social aspects one accepts the relevance of biological explanations to an understanding of the sociocultural in principle, that is, regardless of whether the natural-scientific pendulum might again swing toward greater genetic determinism.

For about two decades, many scholars have been noticing an alignment of "nature" and "culture." For molecular and especially genetic and genomic approaches to life there has been observed an implosion of the realms (Haraway 1997, 102), the enterprising up of life itself (Strathern 1992, 39), and a remaking of nature after culture and as biosociality (Rabinow [1992] 1996). The cultural anthropologist Gísli Pálsson (2008) has thus declared the theoretical dichotomy of nature and culture obsolete in genomic medicine. And in genomic anthropology, he sees "a hybrid field that increasingly involves anthropologists in the fusion of the 'social' and the 'biological' in the wake of genomic studies" (ibid., 546). Genetic anthropology, however, is not interested in modifying its objects to shape them after culture, of interfering with life for new cultural uses; the field is interested in DNA in its most "original" forms as traces of a past state of human biological relations and of migration histories that are perceived as endangered by the processes of globalization. At the same time, genetic anthropology may share with other natural-scientific approaches the attitude of a culture of no culture (Traweek 1988, 162). Especially in the process of demarcation from other anthropological fields and in public and popular discourse, not only the production of genetic history but "history itself" may be conceptually freed from culture and become "pure nature" in the silent mutations of "junk DNA." Although in biotechnological practices, in genomic medicine, and in systems biology, "nature" and "culture" can be seen as increasingly isomorphic, it seems relevant to analytically retain the distinction in order to ask how the relation between "culture" and "nature" is performed at different sites and at different times. Along these lines, Sarah Franklin (2003, 68) has emphasized: "We argue that the category of the natural remains central to the production of difference, not only as a shifting classificatory category, but through *processes* of naturalization, de-naturalization, and re-naturalization" ("We" refers to Franklin, Lury, and Stacey 2000, 9–10).

The importance of looking at such processes of naturalization, denaturalization, and renaturalization has recently become evident in the controversies around the science journalist and writer Nicholas Wade's *A Troublesome Inheritance: Genes, Race, and Human History* (2014). In the book, Wade argues that modern human evolutionary history has led to genetically based differences in mental traits and behavior between human groups who have been largely isolated and adapted to different environmental challenges. Rather than the phenotypically meaningless, purely historical, population genetic differences that Cavalli-Sforza and colleagues emphatically were after, in Wade's book we encounter claims for genetically determined racial disparity in IQ, political institutions, and economic and general societal success. In fact, related arguments have never completely vanished. Following the controversies of the 1970s discussed in part 3, there was the controversy provoked by *The Bell Curve* (Herrnstein and Murray 1994). The psychologist Richard Herrnstein and the political scientist Charles Murray claimed that differences in social status between classes and between ethnic groups could no longer be best explained by differences in socioeconomic privilege. They instead believed that the American society of their time was largely socially stratified according to differences in intelligence that were, to a high degree, due to inheritance.

Reminiscent of the wide spectrum of transformations that we have witnessed Osborn's, Huxley's, and Cavalli-Sforza's knowledge undergo in the course of circulation, it seems that whatever data comes out of the lab, it can be interpreted as increasing as well as decreasing the importance of biologically based "racial" or group differences (which in itself does not necessarily render a stance more or less racist). Although the out-of-Africa model goes along with a human "racial" history of less than 200,000 years, it can nonetheless be combined with the belief in essential differences among what is understood as the human races. This is also exemplified by the book *Race: The Reality of Human Differences* by Vincent Sarich and Frank Miele (2004, particularly chs. 7–9). The authors argued that the relatively small genetic differences among the human "races" rendered morphological, behavioral, and mental differences more rather than less significant, because they were the effect not of chance but of natural selection. They reaffirmed the notion of a brain size–intelligence link that in their view established a hierarchy in mean intelligence from "African" to "White" to "Asian." Believing African Americans to have a lower mean IQ than "white" Americans, they denied the usefulness of any group-based affirmative action, while they held that "race" could be used as a factor in selecting the best-qualified person from a pool of job applicants. In fact, they thought that the global market was already structured in important ways by meritocratic processes, because "the best and brightest move to the top—and to the United States" (ibid., 240).

In contrast to Huxley, Hogben, and Haldane, as well as Cavalli-Sforza, Sarich and Miele mostly wanted to make the argument that a genetic meritocracy had largely been realized and that it was and would be strongly racially stratified. This was not the first time that Sarich scandalized not only his academic peers but the public as well. At the beginning of the 1990s, a war broke out at Berkeley, in which students, professors, and the media were involved, around what Sarich's critics deemed an unscientific, racist, sexist, and homophobic introductory course in anthropology (Selvin 1991). Ideas such as those promoted in the *Bell Curve* as well as Sarich's strong emphasis on the role of genes in even complex sociocultural processes were renewed incentives for Feldman and Cavalli-Sforza to stress the importance of cultural transmission. They kept up the fight against the "extremely hereditarian explanation of inequality in educational achievement" (Feldman, Otto, and Christiansen 2000, 62; also, for example, Feldman 1993). And critics of Wade continue to draw on Cavalli-Sforza, especially on his "disproving" of the notion of genetic races (Allen 2014).

Nonetheless, new technologies in genomics might change the conditions for possible performances of the relationship between histories within and without. I have already alluded to epigenetic research and thinking. In fact, Cavalli-Sforza thought that one advantage the smaller HGDP would have over the ambitious and more precise GP is in this realm: "However, the HGDP has an important advantage because in the GP there are no cell lines, whereas in the HGDP we can study those cell lines and many aspects of cell physiology. This century will be the century of epigenetics, and the HGDP will be pivotal once again" (in Manni 2010, 253). Most likely Cavalli-Sforza was thinking of medical research: what will come of the challenge that the rise of epigenetics presents to genome-wide medical association studies that do not take into account the role of regulatory mechanisms (Feldman, interview 2, 7 Feb. 2013)?

But will an understanding of the genome as only one factor in a multilayered interactive organismic system also impact how the embodiment of evolutionary history and biological kinship are perceived? Will our history and identity sometime be read in methylation patterns and other epigenetic factors instead of, or besides, from the sequence of the genome?

The above considerations also hint at a change that has already impacted the complex of technologies, required skills, and epistemic objects in population genetics, and this is the trend toward the analysis of whole genomes: population *genomics* is the keyword. Computers have become the stores and miners of huge amounts of data, and information technology and bioinformatics specialists are the experts of the novel approaches. These changes are recognizable within the GP, where the work of population genomicists has become more dominant and a whole-genome analysis kit GP02 has been developed to commodify the most recent advances. Innovations in sequencing, computer storage and power, as well as mathematical modeling have affected the last epistemic and political object, as well as popular star, introduced in this book—the Y chromosome. With enhanced sequencing technologies and medical whole-genome association studies, Y-chromosome sequences are churned out. There certainly is no longer a shortage of Y-chromosome markers. On the contrary, Y-chromosome nomenclature needs to be revised again because there are so many markers that the names are getting cumbersome, and more markers are published daily in such efforts as the 1000 Genomes Project. The most parsimonious Y-chromosome tree is thus being exploded and fine-chiseled with branches, especially at the tips, where family genealogists advance the knowledge. This of course also points to the fact that "old" epistemic objects do not simply disappear. The Y chromosome remains an "elegantly simple system" because it is this perfect tool in the service of the male desire to expand genetic genealogy and clean the family tree from intruders (Underhill, interview 2, 7 Feb.; also interview, 21 Jan. 2013).[2]

Although the genetic technology of (family/population) tree building is unlikely to lose its charm soon, the question suggests itself whether, with technologies that sequence entire genomes for less than $1,000 in the pipeline, genomic studies will eventually render the information on migration history and ethnic ancestry so detailed that it becomes meaningless for processes of identification. If we can already use "an exclusive, custom-built genotyping chip [to] test nearly 150,000 DNA markers that have been specifically selected to provide unprecedented ancestry-related information" in order to "learn what percentage of [a customer's] genome is affiliated with specific regions of the world,"[3] will the results of the future reveal bits of DNA from many different regions and ethnicities? Will the dominant phylogenetic diagram change? Will we finally give up the image of the tree and arrive at a thinking of phylogeny in terms of Conklin's network or Huxley's river system?

In 2010, Cavalli-Sforza reaffirmed the conditions on which such trees are built, that there is a linear relationship between geographic and genetic distance of all pairs of indigenous populations (the correlation coefficient being about 0.9), and that there is a regular decrease in genetic variation within populations with their geographic distance from Africa due to the serial founder effects involved in the migrations from there. According to this model, drift and migration have determined the great majority of modern human genetic variation, with natural selection responsible for only about 10 to 20 percent.[4] Generally speaking, population tree building and principal component analysis are still prominent. Also widely used today is the "Structure" method, developed by a team including Jonathan Pritchard, who had received his PhD at Stanford under Feldman (Pritchard, Stephens, and Donnelly 2000). Structure software allows the separation of N individuals belonging to n geographically and/or ethnically distinct populations into a number of groups k that have the smallest within-population variation and the highest between-population variance. One commonly starts with dividing N into 2 k. The method assorts the individuals in just two groups. The same operation can then be repeated with 3 k and so on. The optimal number of k that is estimated in this process is dependent on N and n, which is also true for principal component analysis (Manni 2010, 249–250, 261–262).[5]

Programs such as Structure and Frappe (Stanford School of Medicine, Tang et al. 2005) enable the estimation of individual ancestry proportions and the grouping of individual genetic samples without prior determination of ethnicity. And they visualize admixture. It seems, therefore, that Huxley's vision of human population genetics has come true after all, because as we have seen, already in *The Science of Life*, he and the Wellses suggested that it might be possible in the future to determine the various "racial" contributions to the genetic inheritance of an individual (Wells, Huxley, and Wells 1929/1930/1931/1934, 1448–1449). In *We Europeans*, Huxley and Haddon had declared that

> we can thus no longer think of common ancestry, a single original stock, as the essential badge of a "race." What residuum of truth there is in this idea is purely quantitative. Two Englishmen, for instance, are almost certain to have more ancestors in common than an Englishman and a negro. For the sharply-defined qualitative notion of common ancestry we must substitute the statistical idea of the probable number of common ancestors which two members of a group may be expected to share in going back a certain period of time. Being quantitative and statistical, this concept cannot provide a sharp definition of race, nor do justice to the results of recombination. If, however, concrete values for the probability could be obtained for various groups (which would be a matter of great practical difficulty), it would provide a "coefficient of common ancestry" which could serve as the only possible measure of their biological relationship. (Huxley and Haddon 1935, 106)

Indeed, human population geneticists are increasingly interested in reconstructing the admixture events in human history. Drawing on the HGD Panel data, and on the premise "that human populations have interacted throughout history," a paper in *Science* has presented results of genetic admixture that can be correlated with such historical events as "the Mongol empire, Arab slave trade, Bantu expansion, first millennium CE migrations in Eastern Europe, and European colonialism" (Hellenthal et al. 2014, 747). But there remains a caveat, and again Huxley and Haddon already identified a difficulty with this conception. It still suggests a state in the past when there were isolated human groups, the traces of which might now be recovered in the genomes of individuals and the gene pools of populations—only once pure groups can be admixed. Such a notion of "primary subspecies" they called hypothetical, "a matter of inference only" (Huxley and Haddon 1935, 264). Nonetheless, visualizations like figure 35a at least do not suggest independent development of human groups. It is the dendritic phylogenetic diagram that conveys isolated human populations, which becomes drastically evident in figure 35b, which represents the data of figure 35a in the shape of a tree. If the information is visualized in a tree, in which the populations function as entities rather than (groups of) admixed individuals, we are prone to read a diasporic structure in which populations have differentiated but not converged (Sommer 2015).

Of course, not only molecular anthropology, but the human origins sciences in general, have become informatized and technologized with a plethora of tools such as radiocarbon dating, Uranium-series,

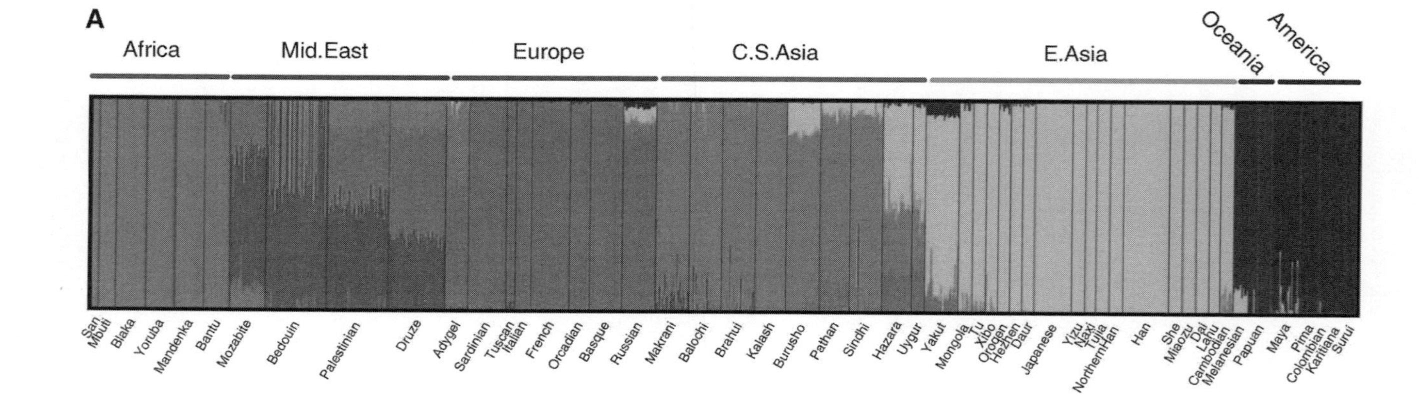
A
Africa
Mid.East
Europe
C.S.Asia
E.Asia
Oceania
America
San
Mbuti
Biaka
Yoruba
Mandenka
Bantu
Mozabite
Bedouin
Palestinian
Druze
Adygel
Sardinian
Tuscan
Italian
French
Orcadian
Basque
Russian
Makrani
Balochi
Brahui
Kalash
Burusho
Pathan
Sindhi
Hazara
Uygur
Yakut
Mongola
Tu
Xibo
Oroqen
Hezhen
Daur
Japanese
Yizu
Naxi
Tujia
NorthernHan
Han
She
Miaozu
Dai
Lahu
Cambodian
Melanesian
Papuan
Maya
Pima
Colombian
Karitiana
Surui

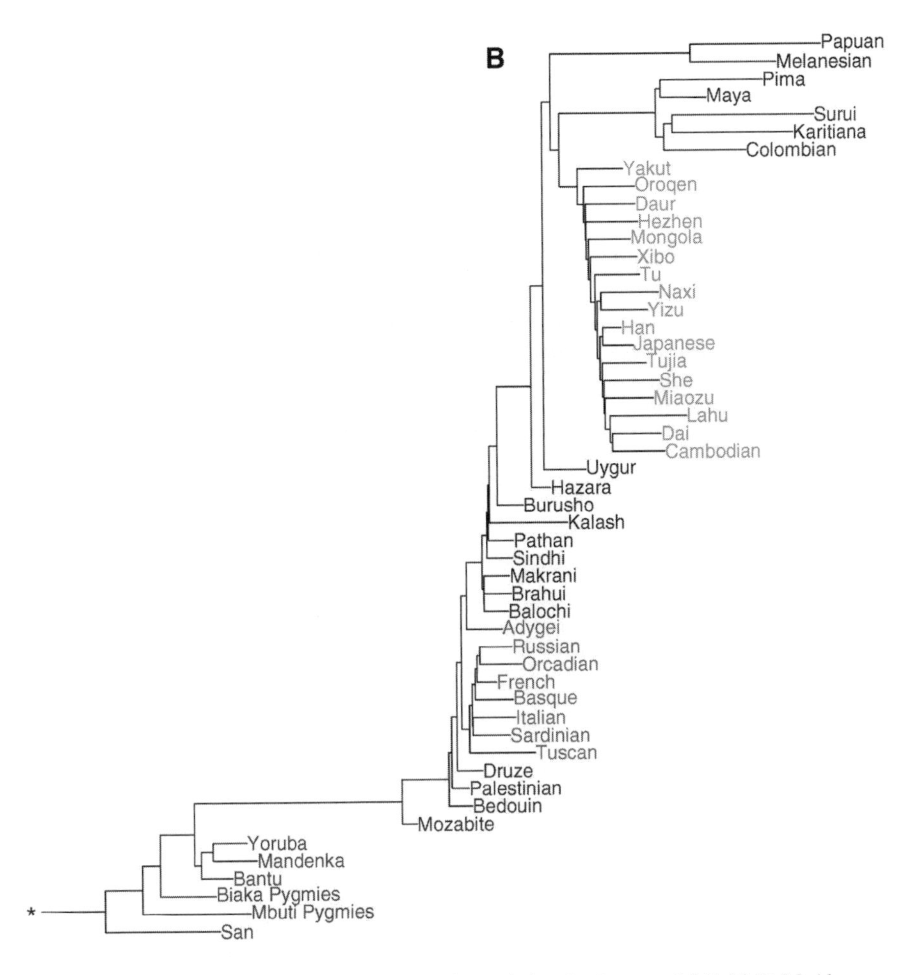

FIGURES 35A AND B. "Individual ancestry and population dendrogram," J. Z. Li, D. M. Absher, H. Tang, A. M. Southwick, A. M. Casto, S. Ramachandran, H. M. Cann, G. S. Barsh, M. W. Feldman, L. L. Cavalli-Sforza, and R. M. Myers, "Worldwide Human Relationships Inferred from Genome-Wide Patterns of Variation," *Science* 319 (5866), 2008, 1100–1104, on 1101, figure 1 A and 1 B, ©2005 WILEY-LISS, INC

thermoluminescence, isotope analysis, and other techniques applied to stone, ivory, stalagmite, sediment, pollen, plant, and bone, including the extraction and analysis of ancient DNA. And as is true for genomic history, despite this process of "scientization," there is and will be no shortage of popular stories in the future, in this case about novel "oldest fossil ancestors." Knowledge about such ancestors will continue to come alive in particular historical cultures, such as in the case of *Homo antecessor* (Atapuerca, northern Spain, 1994), which has become "the first Spaniard" in the media and in accounts of historians as well as scientists—providing an origin to Spain way beyond the legacy of the Franco regime (Hochadel 2011). Also the entertainment genres, such as prehistoric science fiction, are unlikely to lose in attraction, especially considering the novel digital film technologies—think of *10,000 BC* (directed by Roland Emmerich, 2008), which made it to number one at the box office the opening weekend and was the first film of the year to make more than $200 million worldwide. However, now as in Osborn's time, fiction may not only draw on but also be judged with reference to science, and this movie was generally regarded as archeologically wrong and full of anachronisms.

Bones, organisms, and molecules remain the objects of increasingly technological and complex sciences, and they are still translated in cascades of inscriptions up to the level of (oral and written) narratives, (moving) pictures, and spatial exhibitions about "our" histories. These stories and images circulate through knowledge cultures, through the diverse disciplines and social spheres, and they are read, watched, and experienced, in the process of which their meanings as well as the creative recipients are transformed. The very concept of knowledge cultures suggests diversity, exchange, and also controversy and rivalry. In this sense it might have a normative aspect: it is diametrically opposed to the notion of a historical monoculture of any kind.

Notes

Introduction

1. Please note that throughout this book, I use the term *race* and its derivative forms in the way they were employed by the scientists under concern. Neither their application to fossil and to recent human groups, nor the value judgments associated with them represent my own views. The same applies to such denominations as *Caucasian* or *Negroid*. Similarly, I have at times chosen to employ *man,* which was the predominant designation used for humanity by the earlier scientists under consideration, to convey their male-centered perspectives on history. I, for example, adopt the original usage in expressions like *Neanderthal Man*.

2. For an overview, see the classical article Cooter and Pumfrey 1994. Exemplary studies are Shapin 1980; Desmond 1982; Shapin and Schaffer 1985; Pumfrey, Rossi, and Slawinski 1991; Golinski 1992; Stewart 1992. The nineteenth century has gained particular attention: see, for example, Kuritz 1981; Secord 2001 (2003); Schwarz 2003; Lightman 2007. The 1980s also saw the Public Understanding of Science Movement that continued to work with a diffusionist model of popularization (Lewenstein 1992, 1995; Wynne 1995).

3. On this ambivalent relationship between scientists and the press, see also Nelkin (1987) 1995.

4. See in particular Cloître and Shinn 1985; Whitley 1985.

5. For a summary, see Wynne 1995; also Schwarz 2003, 231.

6. *Wissensgeschichte* captures the trend to expand analyses of "the popularization of science" in geography, sociocultural spheres, and time, and thus toward an interest in the circulation of objectified and mediatized scientific knowledge broadly conceived. It is associated with the more encompassing concept of knowledge cultures that includes the role of the social sciences and humanities as well as nonacademic traditions (see, for example, Kretschmann 2003; van Dülmen and Rauschenbach 2004).

7. This crisis is manifest in texts such as Nietzsche (1871) 1954; Troeltsch 1922; Benjamin 1974, 251–263 ("Über den Begriff der Geschichte," first published 1940).

8. Peter Burke speaks of a cultural turn (Burke 1997, 183; see also Burke [1991] 2001; Iggers 2007). The cultural turn may be seen as differentiated into, or as complemented by, further "turns," such as the material, iconic, practical, and spatial turns (for a discussion see Bachmann-Medick 2006)—these also form aspects of analysis in my own study.

9. The journal *History and Memory* was founded in 1989 (Indiana University Press), and, in 2008, *Memory Studies* was launched (Sage). The Routledge Studies in Memory and Narrative Series was initiated in 1998, and the series Media and Cultural Memory/Medien und kulturelle Erinnerung in 2004 (de Gruyter), to name just a few of the markers of the consolidation of the heterogeneous field. There are also research programs, lexica, readers, and anthologies (e.g., Erll, Nünning, and Young 2008; Harth 1991; Fleckner 1995; Pethes and Ruchatz 2001). On periodization see also Lenger (2005, 530–535), who defines cultural historical epochs of remembrance (following the long nineteenth century) as: (1) The period of transition during and between world wars; (2) the period following World War II marked by retrospective involvement; and (3) the time of self-reflection about the conditions and meanings of the memory boom toward the end of the twentieth century.

10. "Et s'il importe de garder le souvenir et de le fixer pour qu'il ne se perde pas, ce n'est pas seulement pour des raisons sentimentales; c'est aussi parce qu'il y a beaucoup à apprendre de l'histoire. . . . Il n'est pas impossible que certains en tirent profit pour trouver des idées nouvelles qui pourraient permettre de modifier positivement notre comportement social."

11. For discussions of these issues with respect to "the new biologies," see also, for example, Reardon 2008b; Franklin 2001.

12. Among the classics on the history of biological anthropology in conjunction with issues of race, nation, as well as imperialism and colonialism are Gould (1981) 1996; Stepan 1982; Stocking 1988; Marks 1995; Kuklick 2007.

Part I

1. This took place in the historically favorable moment of university reform and the new museum that changed the museum landscape and the role of the museum in society. Exhibitions and collections were sorted out according to discipline. There were the processes of curatorial professionalization, the establishment of career paths within museums, and interinstitutional organization, as well as the goals of preservation, research, and education. Pivotal for the new museums were also the international exhibitions. Chicago's Field Columbian

Museum of Natural History grew out of the material exhibited at the World's Columbian Exhibition of 1893. Like the exhibitions that were celebrations of the "scientization of society," museums became instruments of civic and moral education (Bennett 2004, ch. 3; on the "new museum," see Kohlstedt 2005b). The AMNH had already been founded in 1869 (building began in 1874 and the first gallery opened in 1877) in the hope that it would contribute to New York's development as a cultural center. But the collections were badly exhibited and the labels insufficient; there was no efficient management or an explicit educational program. This changed in the 1880s under the presidency of the banker Morris K. Jesup (from 1881 onward) and reached its apex under his successor Osborn (Osborn 1927a, 244, 258; Kohlstedt 1987; Rainger 1991, ch. 3; Kohlstedt and Brinkman 2004).

2. New-York Historical Society, Henry Fairfield Osborn and Family Papers, MS 474, Box 22, S. J. Woolf, "Osborn Surveys Fifty Years of Science: The Head of the Museum of Natural History Recalls the New World Man Has Created for Himself in That Time," *New York Times Magazine*, 31.5.1931, 11; see also ibid., Rosefeld, "At 70, Evolution's Champion Is Serene: Henry Fairfield Osborn, the Friend and Pupil of Huxley Surveys the Struggle to Advance Modern Science," *New York Times Magazine*, 7.8.1927, 7 (courtesy of the New-York Historical Society).

3. Ibid., *Fairfield Reminiscences of 1867, An Address by Henry Fairfield Osborn, 13 June 1935, for the Connecticut Tercentenary*, priv. print 1935 (courtesy of the New-York Historical Society).

4. However, Tony Bennett emphasizes that the AMNH was a "hybrid institution," founded by the Central Park Commission and tied into the network of education departments of the city and state of New York. Although it was privately controlled, it was public in access and driven by nonprofit motives and the use of state and city resources. Thus, "American museum practices in the postbellum period were shaped by the conjunctions of evolutionary thought and new liberalism as a part of a network of new governmental means for acting on the social in much the same way as their British counterparts" (Bennett 2004, 120). There were obvious particularities in context: In the United States, museums for a short time were as influential if not more so than the just nascent university system. Also, the population was very different from Europe's, with a large African American community, emancipated after the Civil War and migrating north, with great numbers of immigrants from Europe, and finally with a population of Native Americans defeated and dispersed from the 1860s.

5. These changes have been classically described by John Higham as the 1890s "watershed of American history," constituted by political, religious, and scientific changes that began to take shape in the 1870s and 1880s (Higham [1972] 1973, 73–102). Roderick Nash (1966; [1967] 1973) and Jackson Lears (1981) have articulated many of these interwoven aspects of American turn-of-the-century culture (see also Fabian 2008).

6. New-York Historical Society, Henry Fairfield Osborn and Family Papers, MS 474, Box 22, John Kimberly Mumford, "Henry Fairfield Osborn Directs Vast Exploration Activity as Head of Museum of Natural History. Reconstructing Knowledge of World's Past a Serious Responsibility to Distinguished Scientist," Who's Who in New York, No. 39, *New York Herald Tribune* (courtesy of the New-York Historical Society); on Osborn's involvement in the preservation movement, see also Barrow 2009, chs. 4–5.

7. See also Colbert 1968; Rainger 1991, ch. 1; Vetter 2004, 2008; Brinkman 2010.

8. Thomson has observed that dinosaurs became truly popular animals only in the 1980s. The first seed of the dinosaur mass hysteria was planted in 1897 with a lavishly illustrated article in the *Century Magazine* (Ballou 1897). On paleontological reconstruction, see also Davidson 2008; on the long-term history of the dinosaur cult, see, for example, Mitchell 1998; Sanz 2002.

9. New-York Historical Society, Henry Fairfield Osborn and Family Papers, MS 474, Box 22, Rose Feld, "At 70, Evolution's Champion Is Serene: Henry Fairfield Osborn, the Friend and Pupil of Huxley Surveys the Struggle to Advance Modern Science," *New York Times Magazine*, 7.8.1927, 7 (courtesy of the New-York Historical Society).

10. Ibid., Walter Tittle, "A Champion of Evolution," Personalities, *World's Work*, May 1928, 80–85, on 82 (courtesy of the New-York Historical Society).

11. Ibid., 83.

12. Ibid., 84.

13. It was the time of the new education movement that demanded the replacement of repetition and surveillance as disciplinary techniques with the cultivation of an independent, questioning, and self-activating approach to learning and moral development. There was a stress on teaching from objects rather than texts motivated by Johann Heinrich Pestalozzi. In the 1890s, progressive educators, intent on building a literate and numerate citizenry, were reformulating teaching methods, and the nature study movement brought natural history into the public school systems at a time when the natural sciences had gained prominence in higher education and in society. Museums were placed on the front line of education. Beyond the emphasis on adult learning, the new museum idea drew attention to the instruction of children, and the need to provide closer links with the new state-provided schooling that was developed during that time in Britain and the United States. Natural history and ethnology were considered particularly fit to the purpose (Bennett 2004, ch. 1; Kohlstedt 2005a, 2010). Museums in general were building up programs of public lectures and connections with popular schooling and universities, and by the 1890s, American institutions had gained leadership in this process of developing the educational functions. As Sally Kohlstedt has shown, a sure sign of this was that, instead of looking to Europe, American curators and directors began touring the museums of their own

country for models of best practice, and the AMNH was beginning to set the standard (Kohlstedt 1987; Kohlstedt and Brinkman 2004; Bennett 2004, ch. 5; for an early history of the museum, see Hellman 1968).

Chapter 1

1. McCosh and Osborn did not follow Galton in his notion that visual remembrance is similar to composite photography, in that the brain blends different memories of the same or similar subjects together (Galton [1883] 1907, 229–233). Rather, they took recollection to be always of a singular visual impression, which may, however, no longer be associated with any information about its origin.

2. Osborn read Théodule-Armand Ribot, who according to Laura Otis (1994, 17) held a theory of memory as transcending generations. He was also acquainted with the writing of James Sully, whom Frank Sulloway (1979, 257–275, 321–327) has shown to be among the psychiatrists who applied John Hughlings Jackson's developmental model of the mind to mental diseases before the turn of the century. Jackson, Sully, Havelock Ellis, and others "anticipated" Freud's dream theory, that is, the notion of dreams as offering access to primeval mental states of humankind. However, Osborn understood Sully and Ribot as denying the existence of a racial memory (on the concept of inherited memory between biology, literature, and psychology, see also Sommer 2005b).

3. American Museum of Natural History Library, the Papers of Henry Fairfield Osborn (1857–1935) MSS.O835 (hereafter AMNH, Osborn Papers), Box 20, Folder 12: William Milligan Sloane, Osborn to Sloane, 21.11.1921.

4. AMNH, Osborn Papers, Box 17, Folder 19: Sir Edward Bagnall Poulton, Poulton to Osborn, 16.9.1888.

5. New-York Historical Society, Henry Fairfield Osborn and Family Papers, MS 474, Box 22, Benjamin C. Gruenberg, "Henry Fairfield Osborn: America's Foremost Palaeontologist," *Scientific American* 105 (4), 22.7.1911, 79 (courtesy of the New-York Historical Society). On the trends of neo-Lamarckism, Mendelism, and neo-Darwinism in general, see Bowler 1988, 1983, especially ch. 8; 1989, especially chs. 6–7.

6. *Homo heidelbergensis* was represented by a fossil jaw from Mauer in Germany (near Heidelberg), which showed affinities to the Neanderthal. So did the remains from Northern Rhodesia (today Zambia).

7. Osborn (1900, 1910) had located centers of evolution and dispersal of fauna in the Holarctic (i.e., North America, Asia, Europe) and the Antarctic (a southern continent now mostly under water that once linked South America, South Africa, Australia, and New Zealand). The stronger focus on Asia was motivated by the work of his former student and colleague at the AMNH, William Diller Matthew, that suggested Asia as an evolutionary center of paramount impact

(see particularly Matthew [1911] 1915). Osborn was not outside the range of contemporary theorizing. Similar phylogenies and biogeographic scenarios of hominid evolution were developed by some of Osborn's colleagues internationally (see Sommer 2007a, part 2).

8. See also Rainger 1991, 124–132 and 163–169 on Osborn and evolution, and 145–149 on human evolution.

Chapter 2

1. Osborn to Scribner, Charles Scribner's Sons, New York, 29.6.1915; where no box and folder information is given, the references in the remainder of this chapter refer to AMNH, Osborn Papers, Box 99, Folders 6–10: reader reactions to *MOSA*, three eds.: 1915, 1916, 1918—scrapbook.

2. AMNH, Osborn Papers, Box 13, Folders 22–23: J. Howard McGregor, Osborn to McGregor, 27.10.1915.

3. Ibid., Box 3, Folder 23: Henri Breuil, Osborn to Breuil, 5.8.1914.

4. Osborn to Jules J. Jusserand, Ambassade de la République Française aux États-Unis, 1.12.1915.

5. Osborn to Breuil, 19.7.1916.

6. Breuil to Osborn, 24.4.1916, from Madrid; AMNH, Osborn Papers, Box 3, Folder 23, Osborn to Breuil, 6.1.1925, and Osborn-Breuil correspondence in general; Obermaier, National Museum of the Natural Sciences, Geological Laboratory, Madrid, to Osborn, 22.2.1916; see also Obermaier to Osborn, 16.4.1916.

7. Boule to Osborn, 16.2.1916; Osborn to Boule, 14.1.1916; Hrdlicka to Osborn, 17.2.1916.

8. Scribner to Osborn, 18.12.1915; Scribner to Osborn, 14.1.1916; Scribner to Osborn, 14.12.1917.

9. "Men of the Old Stone Age," *L.A. Times*, 16.1.1916; "Men of the Old Stone Age," *Union* (Springfield, MA), 30.4.1916.

10. Osborn to Roosevelt, 1.12.1915.

11. Roosevelt to Osborn, 24.11.1915.

12. See, for example, "Men of the Old Stone Age," *New York American* (NYC), 28.12.1915; *L.A. Times,* 16.1.1916.

13. "'Men of the Old Stone Age,'" *Union* (Springfield, MA), 30.4.1916; see also "Men of the Old Stone Age," *Journal* (Minneapolis), 17.1.1916; "A Little Bit of Science," *Examiner* (L.A.), 8.1.1916.

14. "Race That Lived in Europe Twenty-Five Thousand Years Ago Had High Artistic Powers," *Enquirer* (Cincinnati, OH), 3.6.1916; see also "Human Race Half a Million Years Old," *New York Times,* 5.12.1912.

15. "When Man Began to Walk on Two Legs Instead of Four," *Sun* (Baltimore), 9.1.1916.

16. "A Little Bit of Science," *Examiner* (L.A.), 8.1.1916.

17. "The Ape, the Ape-Man, the Adonis: Mysterious Appearance of Beauty in the Prehistoric European Countenance," *Current Opinion* (NYC), Apr. 1916, 265.

18. Madison Grant, Review of *MOSA*, *Geographical Review,* Aug. 1916, 167–168.

19. Regal 2002, ch. 5, on eugenics directly 106–112. To his colleagues abroad, Osborn presented the interest of the Galton Society as foremost one in evolution. He made Boule one of the first corresponding members because of his influential Neanderthal work: "This Society is a small but select research association, and we are just beginning to get together a very talented group of men, particularly to discuss the problems of human evolution" (AMNH, Osborn Papers, Box 3, Folder 19: Marcellin Boule, Osborn to Boule, 30.12.1919).

20. "A Vanished People," *School* (NYC), 6.6.1918, 398; see also, for example, "May Be Missing Link," *Journal* (Kansas City), 18.6.1916; "Men Our Mental Equals Lived 20,000 Years Ago," *Tribune* (NYC), 12.12.1915.

21. "Facts about Old Stone Age: Some with Modern Application," *Evening Sun* (NYC), 4.12.1915.

22. Charles Schuchert, "Men of the Old Stone Age," *American Journal of Science, Series* 4, 41 (242), 1916, 217–219, on 219?.

23. Cockerell, "The Long Childhood of the Race," *Dial* (Chicago), 2.3.1916, 205–206; see also, for example, C. Allen Gore, "Taking the Census 2,000,000 Years Ago," *Boston Evening Transcript,* 27.11.1915.

24. "Primitive Man," *Newark Evening News*, 24.10.1916; "Environment and Life of Men of the Old Stone Age," *North American* (Philadelphia), 4.12.1915.

25. AMNH, Osborn Papers, Box 3, Folder 23, Osborn to Breuil, 7.12.1920.

26. Ibid., Box 9, Folders 10–12: William King Gregory, Osborn to Nicholas Murray Butler, President of Columbia University, 18.11.1920.

27. John C. Merriam to Osborn, 12.11.1915; Gortner to Osborn, 2.3.1916; Naohide Yatsu, Tokyo, to Osborn, 27.7.1916; Harold Fuller de Wolf to Osborn, 16.3.1917?; Osborn to Butler, 23.11.1917. At the same time, Gregory was attaining an eminent international standing in evolution and paleontology. His Columbia courses were attracting students from all over the United States, which meant that professional instruction was continued. It seems that when Gregory was promoted to an assistant professorship of vertebrate paleontology at Columbia, Osborn granted a cutback on his own salary so the budget of Columbia would allow for Gregory's rise (AMNH, Osborn Papers, Box 9, Folders 10–12, E. B. Wilson, Columbia University, Dept. of Zoology, to Osborn, 5.5.1916). Osborn acknowledged the great assistance he received from Gregory in the production of the monograph on Titanotheres (Osborn 1929c) and other scientific work, and recognized "that my diversion to other studies has rendered my knowledge of the evolution of the vertebrates entirely out of date, while your assumption

of the Columbia University lectures on this subject formerly given by myself has rendered you entirely fitted to take up and finish this work" (AMNH, Osborn Papers, Box 9, Folders 10–12, Osborn to Gregory, 6.1.1916). Osborn simply wished Gregory would not restrict his view to morphology and adaptation, but include environment, climate, and geography, because without this information the successive adaptations through which the vertebrates passed could not be properly understood. In fact, Osborn instructed that in the case of his death, Gregory should finish the Titanothere monograph and place his name on the title page, while Matthew should continue the work on the Equidae, the Proboscidea, and the Moropus (ibid., Osborn to Gregory, 6.2.1920). Gregory signed his letters to Osborn with *Fidus Achates* (meaning Faithful Achates—the companion of Aeneas, a true friend) (ibid., Gregory to Osborn, 25.8.1925; also Gregory 1937; Osborn 1936, 1942).

28. Giddings to Osborn, 30.10.1916; Ford to Osborn, 22.9.1915; Ford to Osborn, 6.8.1915; Ford to Osborn, 28.11.1915; Ford, Review of *MOSA*, *American Political Science Review,* May 1916, 402–403; John H. Wigmore, Chairman of the Committee on Graduate Studies in France, to Osborn, 7.8.1916.

29. Coriat, "The Men of the Old Stone Age," *Journal of Abnormal Psychology* 11 (4), Oct.–Nov. 1916, 280–286, on 286; Coriat to Osborn, 31.1.1916.

30. J. A. Hagemann, M.D., Pittsburgh, "Man's Senses Decadent," reprint from *Lancet-Clinic* (Cincinnati), 20.5.1916.

31. Walsh to Osborn, 5.2.1916; Anita Maris Boggs, Bureau of Commercial Economics, Dept. of Public Instruction, Washington, to Osborn, 3.3.1916; Osborn to Boggs, 8.3.1916; Boggs to Osborn, 16.3.1916.

32. Ralcy Husted Bell, "The Philosophy of Painting," 228, enclosed in Bell to Osborn, 5.1.1916; Bell in fact devoted an entire study to the subject. See Bell to Osborn, 8.3.1916.

33. Enclosed in Theodore D. A. Cockerell to Osborn, 10.11.1916; see also Osborn to Cockerell, 27.1.1917.

34. "Stag Artists Elevated by Blushing Stage Idyl," *Sun* (NYC), 11.3.1916.

35. Theodore F. MacManus, "When Homo Sapiens Sells His Soul for a Monkey Pedigree," reproduced in George Barry O'Toole, *The Passing of Darwinism: Being A Critique in Lecture-Form of the Darwinian Synthesis*, the Democrat Print, Bowling Green, OH, 22.9.1916, 47–48, on 48.

36. See, for example, Theodore Justice to Osborn, 11.4.1916.

37. Osborn to Justice, 8.5.1916.

38. G. Frederick Wright, American geologist and Darwinian, later theistic evolutionist, Review of *MOSA, Bibliotheca Sacra*, Apr. 1916, 324–330, on 330.

39. "Men of the Stone Age," *Congregationalist* (Boston), 16.3.1916.

40. International Press-Cutting Bureau, extract from "Prehistoric Man," by the Bishop of Willochra, *Church Standard* (Sydney, Australia), 8.8.1924.

41. *America* (weekly Roman Catholic magazine, NYC), 18.5.1916.

42. George Barry O'Toole, *The Passing of Darwinism: Being A Critique in Lecture-Form of The Darwinian Synthesis*, the Democrat Print, Bowling Green, OH, 22.9.1916; H. R. Mosnat, "Going Back of the Record," *Truth Seeker* (NYC), 19.8.1916.

43. Charlotte Raton, "Niagara Falls," July 1911, enclosed in Raton to Osborn, 8.12.1915.

44. *Evening Post* (NYC), 4.11.1915.

45. "Lessons from the Past," *Herald?* (Boston), 17.12.1915.

46. "Facts about Our Oldest Great Grandfather," *Gazette* (Little Neck, AR), 19.12.1918; see also "Our Oldest Great Grandfather," *World* (magazine, NYC), 19.12.1915.

47. "Facts about Our Oldest Great Grandfather," *Gazette* (Little Neck, AR), 19.12.1918.

48. "Good Hunting," *Morning Post* (London), 21.7.1916.

49. Rheta Childe Dorr, "As a Woman Sees It," *Evening Mail* (NYC), 7.12.1915.

50. Thomas Barbour to Osborn, 11.6.1918; Hawkins to Osborn, 19.7.?.

51. Frederick B. Van Vorst, college pal of Osborn, class 1815, to Osborn, 26.5.1916.

52. Marco F. Viquez, San José, Costa Rica, to Osborn, 10.7.1917; Viquez to Osborn, 1.1.1918.

53. "New Books at the Library," *Post* (Houston), 16.1.1916.

54. Alexander Fraser, editor of *Brooklyn Daily Eagle*, to Osborn, 21.3.1916; see also Fraser's article "Revival of Interest in Stone Age Study—Indicated by Somewhat Unusual Demand for Book at Brooklyn Public Library," *Brooklyn Daily Eagle*, 26.3.1916, 71.

55. Osborn to Fraser, 24.4.1916; F. R. Kaldenberg, F. W. Kaldenberg's Sons, Manufacturers of Ivory and Pearl Work, New York, to Mrs. Osborn, 29.5.1917.

56. James Beck, Shearman and Sterling Law Offices, 12.5.1915.

57. Bashford Dean to Osborn, 22.12.1915.

58. William Henry Hayes to Osborn, 19.1.1917; Osborn to Hayes, 27.1.1917.

59. J. Usang Ly, Graduate House, Philadelphia, to Osborn, 6.1.1917.

60. Grace Nicoll to Osborn, 21.10.1917.

61. Elbert Wakeman to Osborn, 22.1.1918, with enclosure: Paul Ayres Rockwell, "The People of the Free Forest: Until the War Came the Buschkanters of Houthulst Wood Lived Like Ancients," clipping from unknown newspaper; Horace V. Winchell, Mining Geologist, Minneapolis, to Osborn, 24.4.1916.

62. John C. Merriam, Review of *MOSA*, *American Anthropologist* 18 (3), 1916, 426–429, on 428.

63. *American Historical Review* 21 (3), 1916, 561–563, on 562.

64. George Douglas, "Men of the Old Stone Age," *San Francisco Chronicle*, 20.2.1916.

65. "Men of the Old Stone Age," *Chronicle* (Poughkeepsie, NY), Mar. 1916.

66. Emma Peirce, "Men of the Old Stone Age," *Guide to Nature* (Connecticut), June 1916, ix–x, on x.

67. *Book Buyer* (NYC), Nov. 1916.

68. George Douglas, "Men of the Old Stone Age," *San Francisco Chronicle*, 20.2.1916.

69. James J. Walsh, "The Evolution of Man," *Catholic World* (NYC), May 1916, 207–218, on 210 and 214; see also ibid., June 1916, 315–322.

70. Osborn (1915) 1916, 84–86 on eoliths, 130–144, 512.

71. On the long history of Haeckel's embryological images and the controversies surrounding them, see Hopwood 2014.

72. George Barry O'Toole, *The Passing of Darwinism: Being a Critique in Lecture-Form of the Darwinian Synthesis*, The Democrat Print, Bowling Green, OH, 22.9.1916, quotes from foreword (n.p.) and 38.

73. A. S. W., "Palaeolithic Man," *Nature* 98 (2447), 1916, 45–46.

74. "Ape Ancestor of Man Not Yet Found," *Sunday*, 5.12.1915.

Chapter 3

1. On the problem of presenting contemporary ethnicities and their cultural products within the American Museum of Natural History—while "Western high culture" was and is presented across Central Park in the Metropolitan Museum of Art, see Bal 1992.

2. Lukas Rieppel (2012) has detailed the complex process of mounting dinosaurs and the controversies about dinosaur anatomy as well as the question of the scientific quality of mounts that might involve bones from several individuals if not species, the complementing of lacking parts, and the use of metal structures, plaster, and other materials.

3. There were other artists. Lindsey Morris Sterling was an important artist for the illustration of Osborn's books on the Titanotheres and Proboscidea (AMNH, Osborn Papers, Box 21, Folder 1). Others included Erwin Christman, John Hermann, Helen Brown, a Miss Walker, and Margaret Flinsch, who made models and zoo life studies for the reconstruction of fossil Proboscidea. Among the museum painters were James Fraser, William S. Taylor, Carl and Elizabeth Rungius, John P. Benson, and a large staff of artists in the Department of Preparation under James L. Clark.

4. AMNH, Osborn Papers, Box 23, Folder 31: Arthur Smith Woodward, Osborn to Smith Woodward, 12.1.1915.

5. Ibid., Folder 19, Osborn to Boule, 12.1.1915.

6. "tend à ramener *l'Homo sapiens* var. *germanicus* au niveau des *Homo ferus* primitifs," ibid., Boule to Osborn, 1.2.1915.

7. Ibid., Box 3, Folder 23, Osborn to Breuil, 16.1.1920.

8. Ibid., Box 23, Folder 31, Osborn to Smith Woodward, 13.10.1920.

9. Ibid., Box 13, Folders 22–23, Osborn to McGregor, 26.5.1921.

10. Ibid., Osborn to McGregor, 11.5.1921; Charles Fraipont, Université de Liège, Laboratoire et Collections de Paléontologie, Direction, to Osborn, 20.5.1921; Keith to Osborn, 10.6.1921.

11. Ibid., McGregor, from Paris, to Osborn, 16.6.1921.

12. Ibid., Box 16, Folder 11: Nels C. Nelson, Nelson to Osborn, 16.11.1921.

13. "Mr. Burkitt wrote to Reid Moir that Breuil officially acknowledged the Red Crag finds at the Liege congress, i.e., the Foxhall flints" (ibid., Osborn to Nelson, 28.11.1921).

14. Ibid., Box 15, Folder 15, Moir to Osborn, 21.4.1922.

15. Ibid., Folder 16, Osborn to Moir, 1.2.1927. Moir did not agree that his discoveries contradicted a common ancestry for apes and humans (ibid., Moir to Osborn, 5.2.1927).

16. Ibid., Folder 15, Moir to Osborn, 21.4.1922. Reid Moir did not share this interpretation and also disliked the use of the term *eoliths*, which he would have liked to reserve for the previous discoveries in Britain by Benjamin Harris (ibid., Osborn to Moir, 8.6.1922).

17. Ibid., Box 23, Folder 31, Osborn to Smith Woodward, 6.10.1921. The reconstruction of these events in general is based on Osborn's correspondences with J. Reid Moir (ibid., Box 15, Folders 15–17), Nels Nelson (ibid., Box 16, Folder 11), Grafton Elliot Smith (ibid., Box 20, Folders 15–16), and Arthur Smith Woodward (ibid., Box 23, Folder 31). See also Sommer 2010b.

18. Ibid., Box 23, Folder 31, Osborn to Smith Woodward, 5.5.1922; Osborn to Smith Woodward, 5.12.1921.

19. Ibid., Osborn to Smith Woodward, 13.12.1923.

20. This seemed all the more apparent because *Pithecanthropus*'s age had been revised from late Plio- to early Pleistocene, which made this most primitive form younger than *Eoanthropus*, to which Osborn now gave a Pliocene age.

21. Ibid., Box 20, Folders 15–16: Grafton Elliot Smith, Osborn to Smith, 22.4.1922.

22. Ibid., Box 9, Folders 10–12, Gregory to Osborn, 22.8.1922.

23. Ibid., Osborn to Gregory, 24.8.1922; Gregory to Osborn, 31.8.1922; Gregory to Osborn, 10.6.1928; for the story of scientific analysis and interpretation, see Gregory 1927b. On the *Hesperopithecus* story, especially in the context of the Scopes Trial and religious controversy, see Gould 1989; Regal 2002, 146–151; Clark 2008, 120–124.

24. AMNH, Osborn Papers, Box 23, Folder 31, Osborn to Smith Woodward, 5.12.1921.

25. Thinking about places of origin was of course never neutral. While on the basis of the Bible the Middle East had long been the place of origin of human-

kind, the German romantics, with their love for the exotic, shifted the view to India. This idea spread fast to other European countries and to America. The neo-Romantics and Volkish proponents, however, could not associate their nationalism with peoples of dark complexion. They created the Aryan myth, with the Aryans alone having descended from Central Asia, a place shrouded in ignorance (see Regal 2002, ch. 4 on the Central Asia Hypothesis, and ch. 6 on the Central Asiatic Expedition).

26. AMNH, Osborn Papers, Box 3, Folder 26: Barnum Brown, Osborn to Brown, 28.12.1920. Furthermore, expeditions were associated with economic motives. Brown had repeatedly been engaged in prospecting for oil and minerals. In the late 1910s, he was on a government project on oil and gas pools in the Appalachians (ibid., Brown, from Treasury Department, Pittsburgh, to Osborn, 2.1.1919), followed by a leave from the museum to travel for the Anglo-American Oil Company (later Esso) to Africa. This is not surprising when taking into account the enormous amount of time and money the expeditions cost. The one to Abyssinia (Ethiopia) consisted of a caravan of forty-five mules and horses and thirty uniformed and drilled men; it yielded several hundred pounds of fossil invertebrates (marine Jurassic). Brown also collected live specimens, and he estimated that the collection that went to the AMNH would have cost thousands of dollars (ibid., Brown, from Abyssinia, to Osborn, 10.11.1920; Brown, from Hargeisa, British Somaliland, to Osborn, 10.12.1920; Osborn to Brown, Care Thomas Cook and Son, Calcutta, India, 31.3.1922; Brown, from Kala Kahar, Punjab, to Osborn, 19.6.1922; on Brown see also Dingus and Norell 2010, with chs. 9 and 10 on Abyssinia and India).

27. AMNH, Osborn Papers, Box 1, Folder 17, Osborn to Andrews, 28.10.1924; see also ibid., Andrews to Osborn, Peking, 6.8.1924; ibid., Box 16, Folder 11, Nelson, Grand Hotel Des Wagons-Lits, Peking, to Osborn, 3.11.1925.

28. Ibid., Box 9, Folders 10–12, Osborn to McGregor, 28.12.1925.

29. Ibid., McGregor, London, to Osborn, 30.10.1925.

30. See also Gould 1995, in particular 42–60.

31. AMNH, Osborn Papers, Box 99, Folder 6, International Press-Cutting Bureau, extract from "The Ancestors of Man," *Nature* 114 (2865), 27.9.1924.

32. Bowler 1986, ch. 5; Gould 1989; Rainger 1991, 231–232; Regal 2002, particularly 154–173; Clark 2008, particularly ch. 6, and here especially 115–116.

33. AMNH, Osborn Papers, Box 11, Folders 4–7: Julian Huxley, Osborn to Huxley, 11.6.1924.

34. This passage was dictated in response to a request for a clear statement of Osborn's present position; ibid., Folder 23, Osborn's Literary Secretary to Johnston, 3.1.1928.

35. In 1925, for example, Osborn sent McGregor to Europe to inspect and photograph the new Neanderthal remains from Galilee and Ehringsdorf and to secure casts (AMNH, Osborn Papers, Box 9, Folders 10–12, Osborn to Mc-

Gregor, 28.12.1925). Earlier that year, McGregor had revisited England, Germany, Holland, France, and Spain to collect casts and make stereophotographs (ibid., Box 13, Folders 22–23, McGregor to Osborn, 4.9.1925). Besides studying the material and obtaining casts of fossils and artifacts, Osborn wanted him to make blue-pencil notes directly into *MOSA*, with the promise that his support and that of Nelson and Gregory would be acknowledged in the new edition (ibid., Osborn to McGregor, 17.9.1925).

36. AMNH, Osborn Papers, Box 9, Folders 10–12, Osborn to McGregor, 28.12.1925. In his report on the archeology, Nelson, who had been sent into the field for the season of 1925, conjectured that primitive man would have ventured into the Gobi desert only for brief intervals because of its lack of resources, and that he would have brought his tools for lack of raw material. Nonetheless, Nelson described six cultural horizons in the Gobi desert and its border regions, five of which were of commonly recognized prehistoric date: Eolithic (these flints Nelson ascribed to natural causes), Upper Paleolithic (paleoliths of Mousterian and Aurignacian type), Mesolithic (strikingly distinctive if close to the Azilian of western Europe; they named it the Shabarakh culture), Neolithic, Metallic (possibly an expansion of the mound culture of Bronze and Iron Age found by Russians on the other side of the Altais), and Mongol (ibid., Box 16, Folder 11, Nelson, Grand Hotel Des Wagons-Lits, Peking, to Osborn, 3.11.1925, with enclosure: Charles P. Berkey and N. C. Nelson, "Geology and Prehistoric Archaeology of the Gobi Desert," and N. C. Nelson, "Archeology" [copy to Matthew], submitted for publication in American Museum *Novitates* 1926, eventually published in No. 222, 1–16).

37. AMNH, Osborn Papers, Box 1, Folder 17, Andrews to Osborn, 30.4.1926; Andrews, "General Statement of Operations of Central Asiatic Expedition for the Year 1926"; ibid., Folder 18, Andrews to Osborn, Peking, 7.7.1927.

38. Ibid., Osborn to Andrews, 1.7.1927; see also Gregory 1927a; Rainger 1991, 228–239.

39. AMNH, Osborn Papers, Box 1, Folder 18, Osborn to Andrews, 28.12.1927.

40. Ibid., Folder 19, Andrews to Osborn, 31.8.1928.

41. Ibid., Osborn to Andrews, 15.2.1928; Andrews to Tung-li Yuan, Peking, 1.9.1928. The expeditions to Central Asia, too, were enmeshed with economic concerns. Andrews received offers for cooperation from an oil company and a mining syndicate, which he, however, declined. Nonetheless, the project was supported by John D. Rockefeller, J. P. Morgan Jr., and Cleveland H. Dodge (Gallenkamp 2001, 93–94). Thus, in the minds of many, the Central Asiatic Expeditions were associated with commercial interest. Indeed, with power gained in Latin America, inroads made into Asia, the annexation of the Philippines, and the open-door policy with China, Mongolia kindled imperialistic objectives (Rainger 1991, 100–104).

42. AMNH, Osborn Papers, Box 1, Folder 19, Andrews to Osborn, 21.5.1929; Osborn to Andrews, 17.6.1929; Osborn to Kellogg, 19.10.1928; Osborn to Andrews and Granger, 22.10.1928; Osborn to Andrews, 14.5.1929; ibid., Folder 20, Osborn to Andrews, 19.8.1929; quote from ibid., Folder 21, Osborn to Andrews, 21.2.1930.

43. Ibid., Box 1, Folder 21, Andrews to Osborn, 24.5.1930.

44. Ibid., Osborn to Andrews, 3.2.1930; see also Osborn 1930a; Gregory 1930.

45. See particularly *Illustrated London News*, 2.2.1929; Osborn 1929a.

46. The figure of Roy Chapman Andrews and his Mongolian "adventures" have continued to attract attention from popular writers (for children's books see Hartzog 1999; Marrin 2002).

47. AMNH, Osborn Papers, Box 23, Folder 31, Osborn to Arthur Smith Woodward, 13.12.1923.

48. Ibid., Box 1, Folder 20, Osborn to Andrews, 13.7.1929.

49. Ibid., Osborn to Andrews, 27.6.1928.

50. On the development of New York City in this time, see Hammack 1982; Ward and Zunz (1992) 1997; Reitano 2006.

51. Osborn 1927b, 240, quoted in New-York Historical Society, Henry Fairfield Osborn and Family Papers, MS 474, Box 22, Florence Milligan, "Henry Fairfield Osborn, Man of Parnassus," *Bios* 7 (1), 1936, 4–24, on 8 (courtesy of the New-York Historical Society); see also Keyes 1936.

52. See also Osborn 1912; Rainger 1991, 119–120.

53. See also New York Public Library, Rare Books and Manuscripts Division, Charles R. Knight Papers, Box 1, Folder 2, Osborn to Knight, 9.8.1920.

54. See, for example, the discussion of the panels for the Hall of the Age of Mammals and on the model of the Hall of the Age of Reptiles (American Museum of Natural History Library, Central Archives: Knight, Charles Robert, 1874–1953 [hereafter AMNH, Knight Papers], Box 245, 1065, Osborn to Knight, 16.7.1923; and ibid., Box 728, 1262, George H. Sherwood [executive secretary] to James L. Clark, 12.11.1925).

55. See, for example, ibid., Box 245, 1065, Osborn to director Frederic A. Lucas, 19.6.1923; Osborn to William Diller Matthew, 24.10.1924.

56. Knight would eventually describe his method for the case of living animals in Knight 1947.

57. AMNH, Knight Papers, Box 728, 1262, Knight to Osborn, 15.5.1920; note from Osborn to Sherwood, 18.6.1925.

58. See, for example, ibid., Box 90, 249, Osborn to Lucas, 27.6.1918; ibid., Box 728, 1262, Osborn to Sherwood, 3.7.1925.

59. Knight would be allowed to reproduce the images only under the condition that the museum had already published them, and that they accompanied articles he himself authored (see, for example, the contract for the Pampean mural, Hall of the Age of Man [ibid., Box 90, 249, Sherwood to Knight, 25.7.1918];

for the three human murals, Hall of the Age of Man [ibid., Lucas to Knight, 26.6.1919]; for the Blanco waterhole panel, Hall of the Age of Mammals [ibid., Box 245, 1065, 20.6.1923, Pliocene Panel A, signed by Knight 28.11.1923]; and for Pliocene Panel C [ibid., Box 728, 1262, signed by Knight 19.1.1925]). Already the draft of the contract for the Rancho La Brea Panel A, "The Tar Pool," for the Hall of the Age of Man, contained the clause that the copyright for the sketch would rest with the museum (ibid., Sherwood to Knight, 28.1.1921). The final and signed contract carried the date of 10 February 1921 (ibid., Box 90, 249, Sherwood to Knight).

60. Ibid., Box 90, 249, Knight to Osborn, 5.2.1915; Osborn to Knight, 7.2.1916.

61. Ibid., Osborn to Knight, 9.2.1915.

62. Ibid., Osborn to Morgan, 28.9.1916; see also ibid., Osborn to Knight, 28.9.1916.

63. Ibid., Sherwood to Osborn, 22.1.1920.

64. Ibid., Osborn to Daggett, 19.1.1920; Osborn to Daggett 22.1.1920; Osborn to Daggett 23.1.1920; Osborn to Merriam, 19.1.1920; Osborn to George Hale, 24.1.1920; Osborn to J. A. B. Scherer, president, Throop College of Technology, 26.1.1920; Osborn to L. E. Wyman, ornithologist at the LA museum, 22.4.1920; Rainger 1991, 85.

65. AMNH, Knight Papers, Box 90, 249, Osborn to Lucas, 28.12.1920; Annie Knight to Osborn, n.d.

66. Ibid., Knight to Osborn, 21.6.1921.

67. Ibid., Osborn to Knight, 24.6.1921.

68. Ibid., Osborn to Clyde Fisher about photographing and copyrighting the panels, 21.10.1921; Sherwood to Knight, 4.1.1922; Mrs. Knight to Osborn, 24.3.1922.

69. Ibid., Knight to Sherwood, 16.1.1922; 19.1.1922; 26.1.1922; Annie Knight to Sherwood, 22.4.1922; contract for Rancho La Brea panels B, C, D, 27.4.1922; Osborn to Sherwood, 5.5.1922; Knight to Osborn, 30.10.1922; Sherwood to Osborn, 29.11.1922; Osborn to Knight, 4.6.1923. In fact, Knight had already taken up work on the final painting for the hall, the "Patagonian" panel (ibid., Memorandum Knight to Osborn, 6.6.1922).

70. Ibid., Osborn to Knight, 24.6.1921.

71. Museum Archives, Natural History Museum of Los Angeles County, Director's Correspondence Collection, Bryan to Knight, 1.8.1925.

72. Bryan hoped to eventually commission Knight to paint smaller murals of individual animals for the La Brea hall, but this did not come about. In the 1920s, Knight painted animals from the Rancho La Brea site for the publications of Chester Stock, who was at the Department of Geology of the California Institute of Technology (Pasadena) and worked on the La Brea fauna (he published a book on Pleistocene life in California in 1930). Also during this cooperation, Knight would at times insist on his scientific expertise. Although Daggett had

brought up the idea of models, paintings, murals, and sculptures of the larger La Brea animals with Knight as early as the 1910s, when the museum created a Pleistocene Zoo, the sculptures of La Brea animals for the museum park were executed by J. L. Roop and Herman Beck. In the 1930s, after his work at the Field Museum in Chicago (see below), Knight once more took up restoration paintings for Stock. Knight was beginning to feel his age and hoped to be able to secure another important commission, such as mural paintings, on the Pacific Coast. But even though Stock kept trying to realize this, and in 1940 thought he might be able to commission Knight for life-size sculptures of the representative La Brea animals at the Los Angeles museum, this project ultimately came to nothing (Museum Archives, Natural History Museum of Los Angeles County, Director's Correspondence Collection, correspondence between Daggett and Knight; ibid., correspondence between William Alanson Bryan and Knight; Page Museum Archives, Chester Stock Papers, correspondence between Stock and Knight; McNassor 2011, 53–55).

73. AMNH, Knight Papers, Box 90, 249, Knight to Osborn, 14.5.1923; Knight to Sherwood, 21.5.1923; quote from ibid., Box 126, 456, "Conference between Curators Osborn and Matthew," 23.1.1924.

74. For a development of that theme and a discussion of Knight's long-term cooperation with Osborn and of Knight's own publications, see Sommer 2010c; on the racial aspects of the Hall of the Age of Man, see Porter 1983; Rainger 1991, 169–177; Regal 2002, 151–154; on its controversial nature within the context of the evolution and religion debates, see Clark 2008, ch. 6. There is a coffee-table book with many reproductions of Knight's wonderful paintings (Milner 2012).

75. AMNH, Knight Papers, Box 728, 1262, Osborn to Knight, 6.7.1925; Knight to Osborn, 25.8.1925; Knight to Osborn 12.1.1927; Knight to Osborn, 23.4.1928; Knight to Osborn, 1.4.1929; Knight to Sherwood, 29.4.1929.

76. Ibid., Knight to Osborn 17.5.1928; Sherwood to Knight, including memo from Osborn, 27.11.1929; Osborn to Mrs. Anne Knight, 22.9.1932; Annie Knight to Osborn 4.11.1932; quote from Osborn to Lucas, 3.5.1916.

77. AMNH, Osborn Papers, Box 7, Folder 38: Henry Field, Field to Osborn, 2.8.1922; Field, from on board the ship Cunard *R.M.S. Berengaria*, to Osborn, 22.7.1927; Field, Field Museum of Natural History, Chicago, to Osborn, 4.1.1928; Field to Osborn, 5.9.1929; Field to Osborn, 2.9.1932; Field, president of Field Museum, to Osborn, 20.7.1933.

Chapter 4

1. New-York Historical Society, Henry Fairfield Osborn and Family Papers, MS 474, Box 22, news clipping, John Kimberly Mumford, "Henry Fairfield Osborn Directs Vast Exploration Activity as Head of Museum of Natural History:

Reconstructing Knowledge of World's Past a Serious Responsibility to Distinguished Scientist," Who Is Who in New York, No. 39, *New York Herald Tribune* (courtesy of the New-York Historical Society).

2. On the development of New York City in this time, see Ward and Zunz (1992) 1997; Reitano 2006; on the new woman, see, for example, Hunter 2008.

3. This continued to be a mutual process. While Osborn, for example, received expert advice from Obermaier, he initiated the translation of Obermaier's *Der Mensch der Vorzeit* and had an article Obermaier wrote for *Natural History* translated (AMNH, Osborn Papers, Box 16, Folder 19: correspondence with Hugo Obermaier).

4. AMNH, Osborn Papers, Box 3, Folder 26, Brown to Osborn, 8.8.1927.

5. Ibid., Box 20, Folder 11: Upton Sinclair, Sinclair to Osborn, 1.2.1926.

6. The resulting statements found their way into *On the Meaning of Life* (Durant [1932] 2005).

7. AMNH, Osborn Papers, Box 7, Folder 15: Will Durant, Durant to Osborn, 12.6.1931.

8. Ibid., Osborn to Durant, 15.6.1931; see also Osborn 1926c; AMNH, Osborn Papers, Box 17, Folder 19, Osborn to Poulton, 26.2.1926; and Bergson (1907) 1911.

9. AMNH, Osborn Papers, Box 16, Folder 11, Nelson to Osborn, 12.11.1924.

10. See also ibid., Nelson to Osborn, 19.1.1925.

11. Ibid., Box 21, Folder 23: correspondence with Frederick Tilney. This is not to say that Tilney followed Osborn's dawn man theory, which he did not. It further testifies to Osborn's tendency to smooth out controversy and divergence of opinion that he wrote the preface to the book without so much as a hint to this difference, instead congratulating Tilney for "the linking of Man in all his parts and functions with the long lines of his ancestry" (foreword to Tilney 1928, xvi).

12. AMNH, Osborn Papers, Box 7, Folder 15, Osborn to Durant, 3.11.1926.

13. Osborn first considered the title *The Rise of Man* (AMNH, Osborn Papers, Box 16, Folder 11, note by Osborn [written by secretary] to Nels Nelson, 28.1.1925).

14. New-York Historical Society, Henry Fairfield Osborn and Family Papers, MS 474, Box 22, "Scientist Claims Man Has Existed 1,250,000 Years—Dr. Henry F. Osborn Delights Large Audience with Lecture—MAN'S ORIGIN AND GROWTH IS SUBJECT—Address is Illustrated with Lantern Slides of Strange Spots of World," *Telegraph*, 11.9.? (1927 or later), 2 (courtesy of the New-York Historical Society).

15. "A Giant and His Work," *Dixie Magazine*, news clipping without date; where no box and folder information is given, the references in the remainder of this chapter refer to AMNH, Osborn Papers, Box 96, Folders 5–7: post publication letters and reviews "Man Rises to Parnassus."

16. Vernon Kellogg (National Research Council), "Evolution of Humanness," *New York Sun*, 11.2.1928; Earnest A. Hooton, "The Rise to Humanity," *Christian Science Monthly*, 15.2.1928.

17. AMNH, Osborn Papers, Box 83, Folder 2, Osborn and Charles H. Nager of the Columbia University School of Journalism, "Who Is Responsible for Education?," unpublished manuscript of a report on the study of New York City newspapers 1764–1928.

18. Reginald Smith to Osborn, 18.1.1928.

19. "How Long Has Mankind Lived on Surface of This Old Earth?," *San Antonio Express*, Texas, 8.1.1928, 6D.

20. "Concurrent Evolution," *Portland Evening News*, 31.7.1928. On such theories, see also Bowler 1986, 112–130; Sommer 2007a, part 2; 2010b.

21. In a series of attacks on the ape theory before the American Philosophical Society, Osborn put the final solution of the controversy on the fossil hunter and explorer and more precisely on Andrews. The commentator who introduced the address reasoned that the ape theorists had the great British anthropologist Sir Arthur Keith on their side "and probably all students of human evolution in America" (Osborn 1928b, 151).

22. "Progress in the Light of Prehistory," London *Times Literary Supplement*, 5.7.1928.

23. Advertisement by Princeton University Press in *Princeton Alumni Weekly*, 3.2.1928, 483; see also Henry E. Christman, "Osborn Tells Story of Man's Gradual Ascent to Parnassus," *Albany* (NY), 8.1.1928; Christman, "Man's Evolution," *Buffalo* (NY), 7.1.1928.

24. Richard Arman Gregory, MacMillan (editor of *Nature*), to Osborn, 5.4.1928; *Courier* (Buffalo, NY), 28.1.1928.

25. "Progress in the Light of Prehistory," London *Times Literary Supplement*, 5.7.1928.

26. Osborn 1927b, 186, quoted in "The First Reader: Going or Coming?," *New York World*, 2.12.1927.

27. "The First Reader: Going or Coming?," *New York World*, 2.12.1927.

28. "Concurrent Evolution," *Portland Evening News*, 31.7.1928 (example of polygenist interpretation); "Race Deteriorates, Osborn Says, as Civilization Invites Luxury," *New York World*, 4.12.1927; quote from "Modern Man Descendant of 'Dawn Man,' Not the Ape, Says Osborn in New Book," *Brooklyn Daily Eagle*, 4.12.1927.

29. "Race Harmed by Civilization, Osborn Writes," *New York Herald Tribune*, 12.4.1927, quoting Osborn 1927b, 185.

30. Review of *Man Rises to Parnassus, the Quarterly Review of Biology* 3 (4), 1928, 570–571, on 571; "Race Harmed by Civilization, Osborn Writes," *New York Herald Tribune*, 12.4.1927.

31. Attwood? from Auckland, New Zealand, to Osborn, 20.12.1928.

32. *Lyttelton Times* (Christchurch, New Zealand), 14.4.1928.

33. "The First Reader: Going or Coming?," *New York World*, 2.12.1927.

34. "Progress in the Light of Prehistory," *London Times Literary Supplement*, 5.7.1928.

35. Grant to Osborn, 23.12.1927.

36. Grant, Review of *Man Rises to Parnassus*, *Princeton Alumni Weekly* 28 (14), 20.1.1928, 408; see also Grant to Miss Milligan (Secretary), 20.1.1928.

37. "Progress in the Light of Prehistory," London *Times Literary Supplement*, 5.7.1928.

38. New-York Historical Society, Henry Fairfield Osborn and Family Papers, MS 474, Box 22, Frederick Boyd Stevenson, "Says the 'Melting Pot' Is a Failure," *Brooklyn Daily Eagle*, 9.10.1921 (courtesy of the New-York Historical Society); see also Osborn 1921b, 1921c.

39. Review of *Man Rises to Parnassus*, *the Quarterly Review of Biology* 3 (4), 1928, 570–571, on 571.

40. New-York Historical Society, Henry Fairfield Osborn and Family Papers, MS 474, Box 22, "Eugenists Hail Their Progress as Indicating Era of Supermen: Scientists, in Congress Here, Advocate Birth Selection, Not Control, as Proper," *New York Herald Tribune*, 23.8.1932 (courtesy of the New-York Historical Society); see also Osborn 1932; Regal 2002, ch. 5.

41. AMNH, Osborn Papers, Box 11, Folders 4–7, Osborn to Huxley, 29.5.1928.

42. Ibid., Box 1, Folder 20, Osborn to Andrews, 6.11.1929.

43. *Kansas City Star*, 24.3.1928.

44. ? Nelson, Philadelphia, to Osborn, 5.11.1927.

45. *Kansas City Star*, 24.3.1928, quote from Osborn 1927b, viii.

46. A member of Saint Bartholomew's Church, NYC, to Osborn, 12.12.1927.

47. William Walker Rockwell, librarian, Union Theological Seminary, NYC, to Osborn, 27.2.1928.

48. Henry van Dyke, Princeton, to Osborn, 15.1.1928.

49. W.S. ?, from Camden, South Carolina, to Osborn, 17.1.1928.

50. F. P. LeB., Review of *Man Rises to Parnassus*, *America—a Catholic Review of the Week*, 15.9.1928.

51. Henry E. Christman, "Osborn Tells Story of Man's Gradual Ascent to Parnassus," *Albany* (NY), 8.1.1928; see also Christman, "Man's Evolution," *Buffalo* (NY), 7.1.1928; *Courier* (Buffalo, NY), 28.1.1928; Osborn 1923.

52. Maynard Shipley, "Osborn Bows to Fundamentalism," *Argonaut* (San Francisco), 28.1.1928.

53. AMNH, Osborn Papers, Box 5, Folder 32: Basil Crump, Crump to Osborn, 8.1.1930, with enclosure: S. Morgan Powell, "'Evolutioned [*sic*] as Outlined in the Archaic Eastern Records,' by Basil Crump of the Middle Temple. Luzac and Co. London," *Montreal Daily Star*, 25.10.1930 (transcribed by Crump); quote from ibid., Crump to Osborn, 31.10.1930.

54. C. H. Sherill? to Osborn, 17.3.1928.

55. "Man's Rise to Parnassus," *Seattle Times*, 14.2.1928.

56. Herbert, from Hotel Bellevue Palace and Bernerhof, Berne, Switzerland, to Osborn, 3.7.1928.

57. "Modern Man Descendant of 'Dawn Man,' Not the Ape, Says Osborn in New Book," *Brooklyn Daily Eagle*, 4.12.1927.

58. *Lyttelton Times* (Christchurch, New Zealand), 14.4.1928.

59. Translation of a review from the *Journal of the Zoological Society of Japan* 40 (476), 4.8.1928, 268–269.

60. ? from Auckland, New Zealand, to Osborn, 20.12.1928; Fulda to Osborn, n.d.; Mrs. Irene Favorite to Osborn, 9.10.1931; quotes from Dorothy Harris, Princeton, to Osborn, 22.11.1932.

61. Henrietta Ricketts?, Princeton, to Mrs. Osborn, 25.2.1928.

62. R. E. S., "Step Up," *Granta*, 17.2.1928, 286.

63. A. E. Wiggam, NY, to Osborn, 22.3.1928.

64. New-York Historical Society, Henry Fairfield Osborn and Family Papers, MS 474, Box 22, H. Gordon Garbedian, "New Acts in the Long Drama of Man's Origin," *New York Times*, 11.5.1930, Section 10, Science Radio, 1 (courtesy of the New-York Historical Society).

65. Henry E. Christman, "Osborn Tells Story of Man's Gradual Ascent to Parnassus," *Albany* (NY), 8.1.1928; see also Christman, "Man's Evolution," *Buffalo* (NY), 7.1.1928.

66. New-York Historical Society, Henry Fairfield Osborn and Family Papers, MS 474, Box 22, Rose Feld, "At 70, Evolution's Champion Is Serene: Henry Fairfield Osborn, the Friend and Pupil of Huxley Surveys the Struggle to Advance Modern Science," *New York Times Magazine*, 7.8.1927, 7 (courtesy of the New-York Historical Society).

67. Charles Johnston, "Professor Osborn Rounds Out His Splendid Trilogy," *New York Times Book Review*, 8.1.1928, 13.

Chapter 5

1. AMNH, Osborn Papers, Box 11, Folder 13: Ernest Ingersoll, Ingersoll to Osborn, 16.7.1924; Ingersoll to Osborn, 3.12.1925; Ingersoll to Osborn, 21.6.1927; Ingersoll to Osborn, 26.11.1927; Ingersoll to Osborn, 19.12.1927; Osborn to Ingersoll through Miss A. L. Seeling, 27.12.1927.

2. Ibid., Box 12, Folder 21: Louise Lamprey, Lamprey, Limerick, ME, to Osborn, 24.9.1927.

3. Ibid., Lamprey to Osborn, 10.12.1926.

4. Ibid.

5. Ibid., James R. McDonald, Secretary, Little, Brown and Company, Boston, to Osborn, 10.9.1927.

6. Ibid., Osborn to Lamprey, 25.10.1927.

7. Ibid., Lamprey to Osborn, 5.9.1928, quoting a letter from C. W. Lively.

8. Ibid., Lamprey to Osborn, 23.10.1934.

9. Ibid., Box 8, Folder 30: Emma Wheat Gillmore, Leane Zugsmith, Liveright publishing house, to Osborn, 31.8.1932; Gillmore, NY, to Osborn, 7.9.1932.

10. Ibid., Osborn to Gillmore, 24.9.1932.

11. On the story of the *Baluchitherium*'s discovery and reconstruction between different national teams, see Manias 2015.

12. AMNH, Osborn Papers, Box 7, Folder 13: Arthur Conan Doyle, Note of introduction from Sir Arthur Everett Shipley, Christ's College Lodge, Cambridge, for Edmund Preston to Osborn, 6.3.1922.

13. Ibid., Osborn to Doyle, 12.4.1922.

14. Ibid., Doyle to Osborn, 14.4.1922?; Preston to Osborn, 16.4.1922; quote from Preston to Osborn, 18.4.1922.

15. Ibid., Osborn to Doyle, 20.4.1922; Preston to Osborn (in Doyle's name), n.d. (Thursday).

16. Ibid., Box 12, Folder 25: George Langford, Langford to Osborn, 15.12.1918; Osborn to Langford, 13.2.1919.

17. Ibid., Langford, American McKenna Process Company, Joliet, IL, to Osborn, 6.1.1920.

18. Ibid.

19. Ibid., Osborn to Langford, 6.1.1920; quote from ibid., Matthew to Osborn, 15.1.1920.

20. Ibid., Langford to Osborn, 28.1.1920; see also Clark 2008, 204.

21. AMNH, Osborn Papers, Box 12, Folder 25, Osborn to Langford, 7.2.1920.

22. Ibid., Langford to Miss Mabel Rice Percy, Secretary to Osborn, 14.2.1920; quote from ibid., Gregory to Langford, 17.2.1920.

23. Ibid., Langford to Gregory, 19.2.1920; ? to Langford, 21.2.1920.

24. Ibid., Langford to Osborn, 15.3.1920.

25. Ibid., Osborn to Langford, 8.7.1920.

26. Ibid., Langford to Osborn, 18.7.1920.

27. Ibid., J. T. J., "Pre-Adamite Man," *St. Paul Pioneer Press*, 26.9.1920.

28. Ibid., Osborn to Langford, 19.10.1921.

29. Ibid., Langford to Osborn, 21.2.1922.

30. Ibid., Osborn to Langford, 14.6.1922.

31. Ibid., Langford to Osborn, 13.4.1922; Osborn to Langford, 14.6.1922.

32. Ibid., Box 8, Folder 8: Victor Forbin, Forbin, *L'Illustration*, Paris, to Osborn, 11.11.1924.

33. Ibid., Osborn to Benjamin W. Mitchell, translator, 7.2.1925; Mitchell, Philadelphia, to Osborn, 10.1.1925.

34. Ibid., Osborn to Mitchell, 8.5.1925; Forbin to Osborn, 11.11.1924.

35. Ibid., Osborn to Forbin, 10.2.1926.

36. Ibid., Box 12, Folder 25, J. T. J., "Pre-Adamite Man," *St. Paul Pioneer Press*, 26.9.1920.

37. See *The Wizard of Venus* (Burroughs 1964, published posthumously, opening chapter 1, available online as e-book scanned and proofed by Binwiped on October 24, 2002 [v1.0], http://www.vb-tech.co.za/ebook.php?title=Burroughs,%20Edgar%20R%20-%20Venus%2005%20-%20The%20Wizard%20of%20Venus&start=61, accessed 2 Aug. 2015).

38. The prehistoric romance trilogy *The Land That Time Forgot* (August 1918), *The People That Time Forgot* (October 1918), and *Out of Time's Abyss* (December 1918) that Burroughs published in the *Blue Book Magazine* (1918a, 1918b, 1918c) is available and searchable online at www.literature.org.

39. Harry Hamilton Laughlin claimed that while eugenics as a social practice was as old as humankind, eugenics as a science was the invention of the movement he supported (G. Allen 1986, 259). In this sense, Burroughs portrayed his prehistoric races that still live in harmony with (their) nature as equipped with a "healthy instinct of eugenics." In Burroughs' prehistoric fiction, "the healthy races" do not practice miscegenation, and the transformed white heroes end up coupled to the fittest females. From his analysis of Burroughs's fiction, Michael Orth (1986) reaches a similar conclusion. Although Burroughs did affirm eugenics "as a science," the Wieroos in *Out of Time's Abyss* (1918c) who reproduce according to eugenic principles serve rather as a warning than as an example. Nonetheless, despite ambiguities, it seems that Burroughs moved closer to the eugenicist thinking in the course of his career. In 1928, he commented on the famous Hickman murder trial for the *Los Angeles Examiner.* In the columns, he was upset about the plea for insanity and thought about the extermination of "naturally born killers." When writing "I See a New Race," Burroughs argued for intelligence tests for politicians and voters, and for eugenic measures to improve general intelligence (on Burroughs's ideas on eugenics, see Taliaferro 1999, 225–231, 264–267).

40. Stanford University Libraries, Department of Special Collections, M 367, Edgar Rice Burroughs to Eugene T. Sawyer, 10.3.1917.

41. New-York Historical Society, Henry Fairfield Osborn and Family Papers, MS 474, Box 22, Fritz Deverman, "A Visitor's Impression of the American Museum," *Natural History*, Sept.–Oct. 1930, 559–560 (courtesy of the New-York Historical Society).

42. AMNH, Osborn Papers, Box 9, Folders 10–12, Gregory to Osborn, 24.8.1932; quote from ibid., Gregory to Osborn, 8.12.1934.

43. Ibid., Gregory, New Orleans, to Osborn, 1.1.1932.

44. Ibid., Osborn to Gregory, 5.3.1934.

45. Ibid., Gregory to Osborn, 26.6.1930; Gregory 1930.

46. New-York Historical Society, Henry Fairfield Osborn and Family Papers, MS 474, Box 22, clipping from "Down the Saw-Dust Trail," Editorial Comment, *New Masses* (NYC), 19.11.1935, n.p. (courtesy of the New-York Historical Society).

47. AMNH, Osborn Papers, Box 16, Folder 1: correspondence with Thomas Hunt Morgan.

48. E.g., ibid., Osborn to Morgan, 10.7.1928; Osborn to Morgan, 19.11.1928; Morgan to Osborn, 6.12.1928.

49. In fact, Osborn had put his weight into finding a way for Simpson to publish his work on the Peabody Museum's American Mesozoic mammal collection. When Osborn then recruited him away from Yale, the president of the university reacted irritably: "I am sure we shall be glad to assist in the publication of Dr. Simpson's monograph, if we can arrange to do so. It is possible this would have been simpler to accomplish, if you had not stolen him from our staff. To outsiders, there seems to be no good reason why the American Museum of Natural History should absorb all the most promising men, and, while we are deeply appreciative of the great work the American Museum has been doing, our enthusiasm would be less equivocal if you did not raid our treasure house quite so vigorously" (AMNH, Osborn Papers, Box 20, Folder 9: George Gaylord Simpson, James R. Angell to Osborn, 9.1.1928). Already assistant curator of vertebrate paleontology, Simpson was well aware that the publication of the book by Yale University Press had been brought about by Osborn's interventions: "I know how largely this is due to you, and I wish once more to express my appreciation of your constant and helpful interest in my work, or rather in the advance of knowledge towards which I am attempting to work" (ibid., Simpson to Osborn, 12.11.1928).

50. Ibid., Box 11, Folder 8: Leonard Huxley; ibid., Folders 4–7, Osborn to Huxley, 23.10.1912; Osborn to Huxley, 16.11.1914.

51. Ibid., Osborn to Huxley, 27.4.1917.

52. See also ibid., Huxley to Osborn, 21.1.1925.

53. Ibid., Osborn to Huxley, 5.1.1930, and their correspondence in general.

54. Ibid., Box 42, Folder 10, Osborn to Page Cooper, Doubleday, Doran and Company, 9.1.1931.

Part II

1. "Opening of the Zoological Park," *Fourth Annual Report of the New York Zoological Society*, 1900, 77, quoted in Rainger 1991, 118.

2. See also, for example, Huxley 1950, 21; 1964, 82–84.

Chapter 6

1. Steindór J. Erlingsson (2009) has drawn attention to the critique Huxley's scientific writings received during the 1920s. Kenneth Waters (1992) provides a more positive evaluation, and speaks of the achievement of an embryological synthesis.

2. AMNH, Osborn Papers, Box 11, Folders 4–7: Huxley to Osborn, 24.11.1919.

3. Ibid., Huxley to Osborn, 25.1.1922.

4. Julian Sorell Huxley—Papers, 1899–1980, MS 50, Woodson Research Center, Fondren Library, Rice University (hereafter JSH Papers), Series V: Manuscripts, Typescripts, Notes, Box 65: 1940–1945, Folder 4: Typescripts 1942, Huxley, "British Biology To-day," typescript of report to the British Council, Nov. 1942, 1.3; correspondence between Huxley and Ford, JSH Papers, Series III: General Correspondence, see Index to Selected Correspondents in Guide to JSH Papers.

5. See, for example, Huxley 1930, a book that was based on very popular radio talks.

6. Most famously in Huxley 1914; on Huxley's ethology, see, for example, Burkhardt 2005, 102–126.

7. On the holistic and integrative philosophy within the modern synthesis in general, see Smocovitis 1996; Delisle 2009.

8. "Als Evolutionäre oder Moderne Synthese soll der historische Versuch der 1930er und 1940er Jahre verstanden werden, eine materialistische, gradualistische und selektionistische Evolutionstheorie zu entwickeln, die möglichst umfassend die Evolution der Organismen—sowohl die Transformation der Arten als auch ihre Aufspaltung, sowohl die Mikro- als auch die Makroevolution—erklärt und dabei die Ergebnisse möglichst vieler Teilbereiche der Biologie integriert. Dies gelang vor allem für Genetik, mathematische und ökologische Populationsgenetik, Systematik und Paläontologie" (Junker and Engels 1999, 11–12).

9. See, for example, Mayr and Provine 1980; Bowler 1988, 125–130; (1983) 1989, 307–318; Smocovitis 1996; Weber 1998.

10. John Peter Beurton (1999) ascribes this achievement already to the mathematical population geneticists (see also Weber 1998, ch. 1).

11. For different opinions, see, for example, Shapere 1980; Olby 1992.

12. See, for example, Dobzhansky 1955, chs. 6 and 14; 1956, 55–85 ("Who Is the Fittest?"); 1980; Dobzhansky and Boesiger 1983, ch. 3; Huxley 1936b, 87; 1942a, ch. 1.3; Mayr 1962, 1980a, 1980b, 1988, ch. 9; 1999; Simpson 1947, 1960, 1974.

13. See Ruse 1988, 107–108; 1994; also Durant 1989, 361–362; 1992. Scholars may acknowledge that the synthesists explained evolutionary progress, including the origin of hominids, by means of naturalistic principles, but criticize that progress was then valued positively on the basis of anthropocentric criteria

(Provine 1988). Scholars also make out nonmaterialist and teleological elements in some of the synthesists' writings on progress (Swetlitz 1993, chs. 2–4 on Huxley and chs. 5–7 on Simpson; 1995; Smocovitis 1996, 128–131 [see in particular footnote 98] and 144–153). For a more nuanced discussion of these interpretations and an extended bibliography on the subject, see Sommer 2010a.

14. See also, for example, Huxley 1936b, 96–98; 1940a, 14; 1942a, ch. 10; (1964) 1992, ch. 2 ("Higher and Lower," first published 1962).

15. See also Huxley 1934a, 11; 1936b, 98; 1940a.

16. JSH Papers, Series VI: Publications by Julian Huxley, Box 97: 1920–1935, Folder 3: 1922–1923, Huxley, "Heredity and Evolution," *World's Work*, Dec. 1922, 15–22, on 21. Huxley identified two evolutionary transitions: the first from the cosmic/inorganic to the biological/organic, and the second from the biological/organic to the human/psychosocial phase of evolution. On Huxley's three-phases model of evolution and the exceptional position of humans therein, see also Green (1981, 244–269); Olby (1992), whose interpretation of Huxley's understanding of progress is close to my own; and Smocovitis (1996, 144–145), who explains the importance of human uniqueness in Huxley's system in order to have both, directed progress and an understanding of evolution as mechanistically as possible. Greene (1981, 167, 171, 176) describes it as Huxley's and Simpson's central paradox that the blind mechanisms of evolution resulted in an organism that can be regarded as progressive and that has purpose. The epistemic break between organismic and cultural evolution allowed a naturalistic understanding of evolution as well as the possibility of progress and direction on the human level (ibid., 172; see also Greene 1990; Sommer 2010a).

17. See also, for example, Huxley 1950, 9; 1955, 8–10.

18. For another, early expression of these thoughts, see JSH Papers, Series VI, Box 97, Folder 1: 1920, Huxley, "Progress Resurrected," *Athenaeum*, 30.7.1920, 150.

19. See also Huxley (1964) 1992, ch. 1 ("The Emergence of Darwinism").

20. For Singer's own interpretation of Sarton's new humanism and the history of science, see D. Singer and C. Singer 1957.

21. See, for example, Huxley (1964) 1992, 76–77.

22. On Huxley's scientific and evolutionary humanism, see in particular Huxley 1941a, ch. 13 ("Scientific Humanism"); (1964) 1992, chs. 4 ("The Humanist Frame," first published in Huxley 1961a) and 5 ("Education and Humanism"); see also Divall 1992; Durant 1992; Smocovitis, forthcoming.

23. JSH Papers, Series IX: Organizational Materials, Box 113: H–"Idea Systems Group," Folders 2–7: Idea Systems Group 1950–1956, n.d.

24. Ibid., Folder 4, Idea Systems Group, "Modern Systems of Ideas and Their Adaptation to a Changing Society," 1956, 1–13, on 5.

25. Ibid., Folder 6, "Part II: A 'New Humanist Institute,'" typescript of grant application to Rockefeller Foundation, 16–19, on 16; see also Huxley 1934a. The

whole approach of developing a global idea system by analyzing the constituent parts of actual ones for integrating key concepts seems structuralist, and Claude Lévi-Strauss was one of the sympathizers of the project. The Idea Systems Group was associated with the Political and Economic Planning group (PEP), for which Max Nicholson served as chairman and Huxley as a member. On the Idea Systems Group and the efforts to finance it see also the correspondence between Huxley and Nicholson, who was also in the Idea Systems Group, and Huxley and Jacob (Bruno) Bronowski in the 1950s—in 1963, Bronowski became one of the first group of staff at the Salk Institute for Biological Studies (JSH Papers, Series III, see Index to Selected Correspondents in Guide to JSH Papers). Other members of the Idea Systems Group were A. J. Ayer (philosophy of mind), Dennis William Brogan (political science), Stephen Spender (writer), Francis Williams (author and journalist), Barbara Wootton (social studies), and John Richman (psychoanalysis) (see list of members and consultants in the application to the Wenner-Gren Foundation for a symposium, ibid., Box 23, Folder 6, Huxley to Paul Fejos, 14.11.1955).

26. See, for example, Huxley 1942a, ch. 10.5; also Green 1981, 274–295.

27. See in particular the work of Maurice Halbwachs (1925, 1950), who developed the concepts of a socially framed individual and collective memory as well as the notion of an objectified history that might influence personal ways of remembering.

Chapter 7

1. For this and subsequent efforts at institutionalization, see Smocovitis 1994.

2. JSH Papers, Series XI: Clippings, Box 136: 1909–1939, Folder 5, "News and Views: Sir Peter Chalmers Mitchell, C.B.E., F.R.S.," *Nature* 134 (3382), 1934, 280; ibid., Series IX, Box 120: University–Z, Folder 7, Zoological Society of London, *Reports of the Council and Auditors of the Zoological Society of London for the Year 1941,* prepared for annual General Meeting on 29.4.1942, Richard Clay, Bungay, 1942.

3. Ibid., excerpt from D. M. S. Watson to Huxley, 6.3.1940, in Zoological Society of London, "Letters or extracts from letters to the Secretary: (a) concerning his work in U.S.A.; (b) arising out of the paragraph in the 'Daily Telegraph' of February 28th erroneously stating that he was contemplating resignation because of pressure of outside activities; (c) constituting part of the evidence which he had secured for possible use before the Special Committee," 1942, 2–4, on 3. There were also the society's *Transactions* and the *Zoological Record*—a yearly listing of the zoological literature—as well as the project of a complete list of generic names of animals (Nomenclator Zoologicus).

4. Ibid., Series V, Box 65, Folder 4, Huxley, "British Biology To-day," typescript of report to the British Council, Nov. 1942, 1.3.

5. See also, for example, ibid., Series VI, Box 97, Folder 16: 1935, Huxley, "Science and Citizenship" (review of Daniel Hall et al., *The Frustration of Science*), *Nature* 135 (3411), 1935, 414–415; ibid., Box 98: 1936–1943, Folder 1: 1936, Huxley, "Science," *Times Review of the Year,* 1.1.1936.

6. Ibid., Series XI, Box 136, Folder 5, Beatrice E. Kidd (Secretary of the British Union for the Abolition of Vivisection, London), "Anti-Vivisectors and the Zoo," letter to the editor, *Spectator*, 7.9.1934; also J. Cain 2010, 363.

7. JSH Papers, Series VI, Box 97, Folder 7: 1927, Huxley, "Animal Experiments: What We Owe to Vivisection," *Evening Standard*, 2.5.1927; ibid., Series III, Box 9, Folder 4, Haldane to Huxley, 1.5.1927; regarding the license, see ibid., Boxes 12 and 13, Francis Hemming to Lord Onslow, 1.10.1937, and Huxley to Maureen ? on behalf of Lord Nuffield, 4.7.1938.

8. Ibid., Series IX, Box 120, Folder 7, Zoological Society of London, *Reports of the Council and Auditors of the Zoological Society of London for the Year 1941*, prepared for annual General Meeting on 29.4.1942, Richard Clay, Bungay, 1942.

9. Ibid., Series VI, Box 98, Folder 5: Reprints 1938/1939, Huxley, "Display of the Mute Swan," reprinted from *British Birds* 40 (5), 1947, 130–134; ibid., Huxley, "Nests and Broods in the Successive Seasons at Whipsnade," reprinted from *British Birds* 32 (2), 1938, 40–41. Huxley thought it likely that the hibernation state constituted an adaptation with local and individual variety (ibid., Huxley, C. S. Webb, and A. T. Best, "Temporary Poikilothermy in Birds," reprinted from *Nature* 143 [3625], 1939, 683–684). There were also opportunities to study the genetics of hybridization when crossings were acquired by the zoo (ibid., Folder 7: 1941, Huxley, "Genetic Interaction in a Hybrid Pheasant," reprinted from the *Proceedings of the Zoological Society of London* A 111, 1941, 41–43).

10. Ibid., Folder 6: 1940, Huxley, "Library of Bird Songs," *Morning Post*, 7.10.1936; see also ibid., Huxley, "What Animals Talk About," *Travel* 74 (3), Jan. 1940, 4–10, 42, on 5–7.

11. BBC, 3.5.2001, "Oral History: John Burton" (Transcription), Wild Film History: 100 Years of Wildlife Filmmaking, http://www.wildfilmhistory.org/oh/6/John+F.+Burton.html, accessed 2 Aug. 2015. On the early use of sound recordings in British and American ornithology, see Bruyninckx 2013.

12. JSH Papers, Series VI, Box 98, Folder 6, Huxley, "What Animals Talk About," *Travel* 74 (3), Jan. 1940, 4–10, 42, on 42.

13. See also Wells, Huxley, and Wells 1929/1930/1931/1934, 3.9.

14. Zuckerman/Huxley letters, 1931–1967, MS 56, Woodson Research Center, Fondren Library, Rice University. In the 1920s, the zoo had built a new enclosure for a community of Hamadryas baboons. This was anything but a

natural setting, however, and the males exhibited excessive aggression that resulted in deaths of females. Zuckerman speculated about the importance of the reproductive physiology and sexual behavior for primate and mammalian social structures. In the 1930s, he experimented on monkeys' and other mammals' sexual cycles by treating them with hormones. By 1940, Zuckerman was asked to test a new synthetic estrogen that might be given to ovariectomized women from B.D.H., Boots, and Glaxo. Also during the war, Zuckerman did research for the military that involved very cruel treatment of animals, including monkeys (Burt 2009, 164–168).

15. JSH Papers, Series XI, Box 137: 1940–1959, Folder 1, "Brains Trust No. 2," *Illustrated*, 1.11.1941, 13–16?, on 14. Huxley commented on the light this observation threw on the origins of graphic art in *Nature* (Huxley 1942b).

16. JSH Papers, Series XI, Box 137, Folder 3, John Rydon, "Beside His Picassos in Julian Huxley's Hall . . . a Painting by a Chimpanzee! Children and Apes Share a Novel Art Exhibition," *Daily Express*, 27.9.1958, 4–5; ibid., Series VI, Box 100: 1950–1959, Folder 8: 1957, Huxley, "Aping the Artist," *New York Times Magazine,* 6.10.1957.

17. Ibid., Series V, Box 64: 1935–1939, Folder 5: 1939, "The Zoo at Night," typescript, 14.6.1939. The evening openings in summer months (Wednesdays and Thursdays until 11:30 p.m.) attracted 65,000 additional visitors (Zoological Society of London library and archives [hereafter ZSL archives], Eirwen Owen, "Zoo Book," typescript, 3.10.1984, 30–84 on 1935 to 1942 [part of "The Second Phase 1935 to 1955"]).

18. Correspondence between Huxley and Ernest William MacBride in JSH Papers, Series III, see Index to Selected Correspondents in Guide to JSH Papers.

19. JSH Papers, Series XI, Box 136, Folder 5, "News and Views: Sir Peter Chalmers Mitchell, C.B.E., F.R.S.," *Nature* 134 (3382), 1934, 280; ibid., Series IX, Box 120, Folder 7, Zoological Society of London, *Reports of the Council and Auditors of the Zoological Society of London for the Year 1941,* prepared for annual General Meeting on 29.4.1942, Richard Clay, Bungay, 1942.

20. Ibid., Series VI, Box 98, Folder 2: 1937, Huxley, "The Lesson of the Dodo," *Listener*, 13.10.1937, 767–769, on 767.

21. Ibid., Box 97, Folder 15: 1935, Huxley, "Zoos in Their Educational Aspect," the University of London Animal Welfare Society, Report of 23. Conference of Educational Associations held at UCL, 2.1.1935, 115–122, on 115.

22. Ibid., Box 98, Folder 3: 1938, Huxley, "Rare Animals & the Disappearance of Wild Life," Christmas Lectures: The One Hundred and Twelfth Course of Six Lectures Adapted to a Juvenile Auditory, The Royal Institution, London, 28.12.1937 to 8.1.1938, abstracts of lectures.

23. Ibid., Series XI, Box 136, Folder 5, "Zoo Nights," *Cavalcade*, 9.5.1936, 36. See also ibid., Box 140: 1933–1959, Folder 1: 1934–1939, "So This Is 'Undignified'!!! Scenes in Pets' Corner, Now Closed by the 'Dignified' Authorities,"

Tatler 1894, 13.10.1937, which expressed the paper's indignation at the opposition to the pets' corner demonstrated by some of the fellows of the ZSL. However, there was also some public opposition to the very popular children's zoo and its precedent, the pets' corner: "He [Huxley] has announced fearlessly that 'over 70,000 persons have had their photographs taken with a chimpanzee in London Zoo Pets' Corner, and no complaints have ever been made concerning the animal sitters.' But he admits that: 'We allow children to play with lion and bear cubs, but not with adult lions and bears'" (ibid., Box 136, Folder 5, "Cruelty to Chimpanzees," *Seattle Daily Telegraph,* 6.8.1937).

24. Ibid., Series VI, Box 98, Folder 2, Huxley, "They Live Near Man," *Zoo,* July 1937, 25–26, on 26; this was the last in a series of three articles on animals' relationships to their environment, the other two being "They Live in Forests" (May 1937) and "They Live on Plains" (June 1937).

25. Ibid., Box 97, Folder 16, Huxley, "To All Girls and Boys from Julian Huxley," *Teachers World and Schoolmistress*, 18.9.1935, 879.

26. Ibid., Series VII: Travel Materials, Box 102: 1913–1944, Folder 7: Canada–USA 1939, St. Kilda 1939, USA 1939–1940, "Points out Moral in Comparisons of Ants to Man," *Evening Standard?*, 31.1.1935? At another occasion, Huxley would emphasize that war is very rare in evolution, only known in man and ants, and that even within humans, it could not have been widely practiced in prehistoric times. There was therefore no such thing as a war instinct or any biological foundation to modern warfare (ibid., Series VI, Box 98, Folder 8: 1942, "A Biologist Finds: War Is Rare Phenomenon in Nature," *Town and Country,* Summer 1942, 36–37, 64, 66).

27. Ibid., Series XI, Box 136, Folder 5, "If Children Were Fed Like Zoo Apes," *Illustrated London News?*, 20.6.1936; also ibid., " 'Nutrition' Facts as a Film," *Morning Post,* 7.10.1936; ibid., "Business Concern Becomes Nutrition Campaigner," *Food Industries Weekly,* 20.11.1936, 12–13, on 13: "Huxley made powerful statement [*sic*] when he said 'If we at the Zoo gave the animals the sort of food on which one-eighth of the population of Great Britain is living we would be considered a disgrace to the community.'"

28. Ibid., Series VI, Box 98, Folder 1, "The Nutrition Film," leaflet on the documentary spoken by Huxley, 1936.

29. Ibid., Series IX, Box 120, Folder 6, typescript of Huxley to Onslow, 4.3.1940, to be circulated to the committee. The Elephant House project was one of Huxley's crucial tests as director. The society had set up a fund in 1934, but this was not working, and the treasurer eventually took a bank loan. A layout committee was established for the general layout of the gardens, and Tecton was commissioned (as they had been for any major buildings since 1933). The company submitted several schemes and plans for an Elephant and Rhino House. A two-story building was rejected because of the costs. Huxley succeeded in interesting the maharaja of Bhavnagar, who gave £5,000, and promised £5,000

more, on the condition that it would be opened in the coronation year of 1937 to mark British-Indian cooperation, and that it would be called the Edward VIII Coronation Elephant House. The society's patron, George VI, approved of the scheme, but the council did not like Huxley's chosen architect, Lubetkin, and the treasurer thought a British architect should be hired. (The same argument would be used against Lubetkin in the case of the Rodent House.) Huxley lobbied, and after bitter arguments, Lubetkin was allowed to continue. However, in 1937, the treasurer and council voted for a second proposal by another architect for a lower price; with ongoing arguments, Tecton's proposal was approved in October pending some alterations. A year later, there were still no final drawings approved by the council. Thus, the coronation dinner and entertainment for George VI were held with Indian princes, prime ministers, ambassadors, and African authorities, but without the new Elephant House in place (now to be King George VI Coronation Elephant House); the gardens were floodlit and there was dancing in a marquee on the lawn; films were shown and a program about zoo animals on television—a technology that was then in its infancy. Building of the Elephant House only began in 1939, and the outbreak of war halted both the Elephant and Rodent House projects (ZSL archives, Elephant House letters, 1937–1947, GB 0814 SBCA; ibid., Eirwen Owen, "Zoo Book," typescript, 3.10.1984, 30–84; JSH Papers, Series III, Boxes 12–13). There had been an earlier elephant tragedy for Huxley, when the offer of London Films to give a Tusker elephant to the zoo (it was intended for Whipsnade) that should be used in promoting the film *Elephant Boy* (1937) came to nothing; at one point, Huxley had even hoped that the money from the premiere might be charitably given toward the Elephant House for Regent's Park (ZSL archives, Tusker elephant correspondence, 1936–1937, GB 0814 QCBA).

30. When Huxley opened the Liverpool school exhibition, he was honored as a man with "his own reputation as a sociologist and a scientist and as one of the most adventurous patrons of modern architecture in England in his position as secretary of the London Zoo" (JSH Papers, Series VI, Box 98, Folder 2, "Exhibition, Modern Architecture and Thomas Harris," *Journal of the Royal Institute of British Architecture, 3rd Series* 44 [13], 8.5.1937).

31. Ibid., Folder 8, "Brave New Zoo," *Illustrated*, 4.7.1942; ibid., Series IX, Box 120, Folder 7, Zoological Society of London, *Reports of the Council and Auditors of the Zoological Society of London for the Year 1941,* prepared for annual General Meeting on 29.4.1942, Richard Clay, Bungay, 1942.

32. In fact, during the winter of 1940, the opportunity of small attendance at Whipsnade was taken for killing three elderly elephants and three lions in order to avoid the costs of tending for them (ZSL archives, Minutes of Council, 1826–1998, GB 0814 FAA, Minutes of the Council, 16.10.1940).

33. Both parks suffered from air raids in 1940 and 1941 (eleven incidents, eight London, three Whipsnade; fifty-five high-explosive bombs, two hundred

incendiaries, and two oil bombs). Despite considerable material damage, there were no serious casualties to staff or visitors and few among animals (a giraffe and some antelopes died of "fright"). The headquarters of the Home Guard and Civil Defence Services at Whipsnade were in the society's office (JSH Papers, Series IX, Box 120, Folder 7, Zoological Society of London, *Reports of the Council and Auditors of the Zoological Society of London for the Year 1941*, prepared for annual General Meeting on 29.4.1942, Richard Clay, Bungay, 1942).

34. JSH Papers, Series VI, Box 98, Folder 6, Huxley, "Air Raids and the Zoo," *Spectator*, 1.11.1940, 436–437, on 437.

35. Ibid., Huxley, "Animal Behaviour in Air Raids," *News Chronicle*, 4.12.1940.

36. Ibid., Huxley, "Air Raids and the Zoo," *Spectator*, 1.11.1940, 436–437, on 437; see also ibid., Series VII, Box 102, Folder 7, "Dr. Julian Huxley Arrives With Wisecrack About Shaw," *Washington Post*, 2.12.1939.

37. Ibid., Series XI, Box 137, Folder 1, Theodora Benson and Betty Askwith, "Londoners Face Up to the *Blitzkrieg*," *Sketch* 10, 11.12.1940, 338–339. Wöbse and Roscher (2010) deem the reconstruction of the animal-human relationships in war times as particularly insightful for an understanding of animal agency.

38. JSH Papers, Series V, Box 64, Folder 5, Huxley, "The Zoo at Night," typescript, 14.6.1939, 1–4, on 1.

39. Ibid., Series XI, Box 136, Folder 5, "Zoo Nights," *Cavalcade*, 9.5.1936, 36.

40. Ibid., Series III, Box 12, Folder 5, Onslow to Huxley, 22.3.1937.

41. Ibid., Box 13, Folder 3, J. Levis to Huxley, 9.4.1938; also ZSL archives, Eirwen Owen, "Zoo Book," typescript, 3.10.1984, 30–84. For the difficulties between Huxley and the council and Huxley's zoo years in general, see the respective Minutes of Council and Minutes of the General Meetings (ibid., Minutes of Council, 1826–1998, GB 0814 FAA; ibid., Minutes of the General Meetings, 1826–1999, GB 0814 EAA) as well as Huxley's correspondence during the zoo years, especially with Richard William Alan (Earl of) Onslow (ibid., Series III, Boxes 11–16).

42. Quotes from ibid., Series VI, Box 98, Folder 1, Huxley, "Our Plans for 1936 . . . ," *Evening News*, 1.1.1936.

43. ZSL archives, reports on visits to foreign zoos, 1934–1940, GB 0814 YBAA.

44. The company had the monopoly of filming at the zoo. *The Evening Standard* wrote: "The inhabitants of the Zoo are becoming film stars. Strand Film Productions, Ltd., has been registered as a private company to make films there. One of the directors is Professor J. S. Huxley" (JSH Papers, Series XI, Box 136, Folder 5, "Film Company to Make Pictures at the Zoo," *Evening Standard*, 17.1.1938; see also ZSL archives, Eirwen Owen, "Zoo Book," typescript, 3.10.1984, 30–84).

45. JSH Papers, Series XI, Box 136, Folder 5, "Cinema for the Zoo," *Morning*

Post, 24.8.1935. In 1935, Huxley signed a contract with London Film Productions for his supervision and assistance in natural history short talking motion pictures (ibid., Series XIII: Memorabilia, Box 149: 1930–1939, Folder 1).

46. Quotes from JSH Papers, Series VI, Box 97, Folder 16, Huxley, "Making and Using Nature Films," *Listener* 13, 10.4.1935, 595–597, 629, on 597.

47. Ibid., Series IX, Box 120, Folder 6, Mary Ogilvie, F.Z.S., wife of F. W. Ogilvie, director-general B.B.C., to Huxley, 15.3.1940.

48. The details of the process that culminated in Huxley's resignation can be traced from documents at the ZSL archives (Huxley Papers [uncatalogued box, two folders containing minutes of meetings, memoranda, and correspondence relating to the "zoo affair"]; Minutes of Council, 1826–1998, GB 0814 FAA; Minutes of the General Meetings, 1826–1999, GB 0814 EAA), as well as from JSH Papers, Series IX, Box 120, Folders 6–7.

49. See news coverage in JSH Papers, Series XI, Box 137. For Huxley's zoo time in general see the respective Minutes of Council and Minutes of the General Meetings as well as the society's annual reports (ZSL archives, Minutes of Council, 1826–1998, GB 0814 FAA; ibid., Minutes of the General Meetings, 1826–1999, GB 0814 EAA; ibid., Report Zool. Soc. 1934–1937 and Report Zool. Soc. 1938–1942) and JSH Papers, Series III, Boxes 11–16. On the zoo's long-term history, see Barrington-Johnson 2005, 111–127 on the time of Huxley's directorship; also Clark 1968, 254–268.

50. E.g., JSH Papers, Series XI, Box 137, Folder 2, Arthur Keith, "Darwinism To-day," *Sunday Times*, 11.10.1942.

51. Ibid., W. B. Turrill, "Evolution," *Nature* 150 (3817), 1942, 747–749.

52. Ibid., Paul G. Espinasse, "The Modern Biologist," *New Statesman and Nation*, 24.10.1942.

53. Ibid., P. C. Koller, "Huxley, J. [S.] 1942. Evolution. The Modern Synthesis," *Animal Breeding Abstracts*, Jan. 1943.

Chapter 8

1. JSH Papers, Series VI, Box 99: 1944–1949, Folder 2: 1945, "Julian Huxley on T. H. Huxley: A New Judgment," reproduction of a BBC broadcast, Watts and Co., London, 1945, for the *Rationalist Press*, 5–32, on 32.

2. Ibid., Box 98, Folder 8, Huxley, "Message from Another Age: An Imaginary Interview between Julian Huxley and Thomas Henry Huxley," *Listener*, 15.10.1942, 501–502, on 501.

3. Ibid.

4. There was among other expressions a series of BBC talks on humanism, in which Huxley gave a compact and compelling account of his scientific version (Huxley, Murray, and Oldham 1944).

5. JSH Papers, Series III, Box 5, Conklin to Huxley, 20.4.1917.

6. On Conklin's and Huxley's understandings of biological progress, see also Swetlitz 1995. On Conklin's popular evolutionism, see also Atkinson 1985.

7. For Conklin's anti-eugenics writings from the 1910s onward, especially his emphasis of the role of environment and education in the shaping of the human individual, see Cooke 2002. As in the case of Huxley, Haldane, and Hogben, Conklin's ideas on heredity and eugenics were complex and changed over time. An analysis of some of his correspondence by Miriam G. Reumann and Anne Fausto-Sterling (2001) shows how inconsistent and even contradictory his statements—pro and contra eugenic measures—could be, and that even in the late years of World War II and after, the continuation of the "better human types" was a concern to him. Nonetheless, Conklin had joined American geneticists in their public critique of eugenics that gained force in the 1920s (Cravens 1978, 158–190; Kevles [1985] 1995, 122).

8. E.g., Haldane 1932a, 211–224 ("My Philosophy of Life," especially 223–224), 225–230 ("What I Think About," especially 229); Hogben 1939, 23.

9. Huxley actually referred to Hogben as "H" (e.g., Wellcome Library, London: Sir Julian Huxley, Shelfmark SA/EUG/C.185, Box AMS/MF/111, Reference number b1623280x, Huxley to C. P. Blacker, 4.11.1930).

10. In his autobiography, Hogben not only acknowledged Huxley's enormous influence on the biology of his time (Hogben actually took up comparative physiology of the ductless glands because of Huxley), but also gave testimony to many meetings. Hogben and Haldane went to Cambridge, and Huxley and Haldane were both at Eton and continued their friendship in post–World War I Oxford as fellows at New College. They stayed friends until Haldane's tragic death from cancer (see also lifelong correspondence between Huxley and Haldane and Huxley and Hogben, JSH Papers, Series III, see Index to Selected Correspondents in Guide to JSH Papers).

11. Also JSH Papers, Series IX, Box 117: S–Unesco 1961, Folder 2: Sierra Club 1959; Hogben 1966.

12. For a scholarly treatment of the phenomenon, see Smith 2003; see also Werskey 1971a. This "social relations of science movement" became institutionalized in the Division for Social and National Relations of Science within the BAAS (McGucken 1984).

13. See also JSH Papers, Series VI, Box 98, Folder 4: 1939, Lancelot Hogben, "I Believe," *Minibook Magazine* (NY), Dec. 1939, 17–18, on 17. Following Hogben's confession of creed, there are Huxley's (ibid., 18–19), and Harold J. Laski's (ibid., 19–20). A few years earlier, Huxley wrote of Hogben:

> Professor Hogben's ideal is a scientific humanism; he is one of the rare few who can claim to talk with authority on such a subject. In the specialized field of biological research, he has made distinguished new contributions in such different subjects as the

> biology of twins, the mechanism of colour change, the methods of studying inheritance in man, the physiology of reproduction, the behaviour of chromosomes, the biology of population growth, and the metamorphosis of tadpoles. In addition, he has expended much time and thought in remodelling biological curricula and teaching methods to bring them up-to-date, and has made extensive studies in the history of science and its social implications." (ibid., Folder 1, Huxley, Chairman's Introductory Address" [Professor Hogben's Moncure Conway Lecture], v–x, on v)

For Huxley, Hogben made clear that "biology is as important as the sciences of lifeless matter; and biotechnology will in the long run be more important than mechanical and chemical engineering" (ibid., vii). "The deduction of course is clear that we should strive for an educational system which is tendencious in so far as it encourages the practical realization of scientific humanism; on this I will not further anticipate our lecturer's intriguing conclusions" (ibid., viii). "Humanism to be worthy of the name must be not only intellectual, literary, artistic, not only altruistic and emotionally well-meaning—it must be scientific as well" (ibid., ix). "But, much more than this, we are reminded of the essential sanity induced by a biological basis for social and political thinking" (ibid., x). In the obituary for Haldane in *Encounter,* Huxley called him "truly humanist" (ibid., Box 101: 1960–1974, Folder 6: 1965, *Encounter,* Oct. 1965, 59–61, on 60).

14. On the importance of the contact with Marxist historians (at the Second International Congress of History of Science in London in 1931), see McGucken 1984, ch. 4; Fox 2006. McGucken identifies Huxley (1934b) as an exponent of this novel understanding of science as a social activity that takes place in a particular historical setting and shows that this approach stunned an unaccustomed public (McGucken 1984, 179–180).

15. Also Huxley and Andrade 1935, 296–348 (Huxley, "The History of the Science" and "Science and General Ideas"); Haldane 1932a, 43–62 ("Is History a Fraud?"), 119–141 ("The Place of Science in Western Civilization"), 199–210 ("Science and Invention").

16. On Haldane, Hogben, and Huxley as popular writers within the larger context of "the popularization of science" in early twentieth-century Britain, see Bowler 2009, especially 31–32, 221–230; on Huxley in particular, see also Kevles 1992a; LeMahieu 1992; Patten 1992.

17. There was also a variation on this theme, with the idea that genetic as well as social changes might be of short-term benefit but long-term disadvantage: "Further, as Darlington has pointed out in his recent book, 'The Evolution of Genetic Systems,' certain evolutionary changes may be of immediate advantage, but of eventual disadvantage in robbing the stock of evolutionary plasticity and adaptability. Here again there are doubtless parallels from ethics. The

short-term efficiency of ruthless State dictatorship as opposed to the inevitable long-term triumph of more humanistic systems is a case in point" (JSH Papers, Series VI, Box 98, Folder 7, Huxley, "The Relations between Science and Ethics," reprinted from *Nature* 148 [3749], 1941, 279–280, quote from p. 4 of reprint).

18. See Mazumdar 1992, ch. 4, who refers the attitudes described below for Haldane to a slightly later period in his life.

19. Wellcome Library, London: Sir Julian Huxley, Shelfmark SA/EUG/C.185, Box AMS/MF/111, Reference number b1623280x, C. P. Blacker to Huxley, 24.10.1930.

20. A problem Huxley identified early on in genetics and its application to humankind was the isolation from the branches of experimental embryology and physiology (JSH Papers, Series VI, Box 97, Folder 3, Huxley, review of R. Ruggles Gates, *Heredity and Eugenics*, 1923, *The Scientific Worker,* n.d., 13; Huxley had other issues with the book but nonetheless generally recommended it. It was over the publication of *We Europeans* [Huxley and Haddon 1935] that the two clashed irrevocably).

21. Wellcome Library, London: Sir Julian Huxley, Shelfmark SA/EUG/C.185, Box AMS/MF/111, Reference number b1623280x, Huxley to C. P. Blacker, 4.11.1930.

22. JSH Papers, Series VI, Box 97, Folder 1, Huxley, "Eugenics and Eugenicists," *Athenaeum*, 31.12.1920, 895; see also, for example, Huxley 1926c, where he claimed that, throughout democratic civilization, the environment (in the broadest sense) was relatively uniform and individual variation thus more strongly influenced by heredity than nurture. In the same lecture, he not only stated that Charles Davenport had shown that morons bred morons and criminals, criminals, but he also suggested that nations should strive for racial uniformity, to prevent racial problems from coming up.

23. JSH Papers, Box 98, Folder 2, Huxley, "Foreword," in Herbert Brewer, *Eugenics and Politics*, The Eugenics Society, Mar. 1937.

24. Ibid. Huxley's change in outlook can actually be gleaned from his role within the Eugenics Society. He was a member and life fellow of the Eugenics Society and functioned as its vice president (1937–1944) and president (1959–1962). By 1933, Huxley warned that the society needed to distance its aims from Eugen Fischer's eugenics in Germany and promoted an understanding of eugenics as focusing on the influence of social structure—an approach that at that time was considered "provocative." By 1935, he argued within the society that practical eugenics depended on equal opportunities for all classes and types (Wellcome Library, London: Sir Julian Huxley, Shelfmark SA/EUG/C.185, Box AMS/MF/111, Reference number b1623280x, quote from C. P. Blacker ["Pip," General Secretary of the Eugenics Society] to Huxley, 23.9.1933). In a pamphlet written in 1937, Huxley still argued for the voluntary sterilization of mental defectives, but he stated that otherwise eugenic measures were not urgent for the race;

what was urgent was a new organization of society and better knowledge (ibid., Shelfmark PP/CPB/A.2/1, Box 1, Reference number b18229116). The argument that heredity cannot be read from the phenotype and that measures therefore had to be social goes back to Wilhelm Johannsen and also T. H. Morgan (Paul 1995, 115–117). On Huxley's eugenics, see Hubback 1989; Allen 1992; Barkan 1992a; Sommer 2010a. Diane B. Paul (1992) warns that Huxley's reform politics were not a priori progressive, just as the classical eugenics movement cannot be reduced to political conservatism. Huxley's utopia of a genetically diverse society, in which the division of labor is genotypical, is not necessarily more socially just than the ideal of a limited number of optimized genotypes (see also Waters 1992; Wellcome Library, London: Sir Julian Huxley, Shelfmark SA/EUG/C.186, Box AMS/MF/111, Reference number b16232811).

25. In "Man's Destiny" (in 1932a), Haldane urged that humans needed to take their future evolution into their hands and that further progress was solely a matter of biology, that is, eugenics (interestingly enough, this is the only essay from the collection that has not been included in the American version [Haldane 1933]). But on the eve of the war, Haldane made a very cautious statement with regard to the contemporary and future possibilities of human genetics as eugenics due to the complexity of the hereditary processes and the nature-nurture relation (Haldane 1938; about Haldane and eugenics, see also von Lünen 2009).

26. As one reaction to the lecture, the *Evening Standard* accused Huxley of socialism, which Huxley denied, but of course, this was just another chance to distribute his ideas: "In the first place I am not an adherent of any political party, Socialist or otherwise. Secondly, my Galton Lecture was not politically biased in any way; it was devoted to a discussion of the need for relating eugenic proposals to the environment in which they would have to be operated, especially that provided by the current social and economic organization in vogue. This important theme is usually neglected by eugenists, who for the most part assume that the existing type of social environment is the best or the only possible one" (JSH Papers, Series VI, Box 98, Folder 2, Huxley, "Professor Julian Huxley and Dr. Inge," letter to the editor, *Evening Standard*, 26.7.1937).

27. JSH Papers, Series XI, Box 136, Folder 5, "Germany a Huge Laboratory: 'Vast Eugenic Experiment,'" *Morning Post*, 18.2.1936. On this, as well as for a more systematic treatment of Huxley's, Haldane's, and Hogben's views on eugenics, see Barkan 1992b, 178–189, 229–260, 296–310. The classical study is Kevles (1985) 1995; on the triple-H, see in particular 122–134.

28. On Hogben's actions and science in relation to eugenics and racism, see also Sarkar 1996.

29. However, while deconstructing the genetic basis of grouping humanity according to class—for example, correlating with differential intelligence—he maintained the existence of a few "general human races," such as "negroes" and

"whites." At the same time, also with regard to the notion of race, he emphasized overlap and internal variation and warned against correlation with ability.

30. JSH Papers, Series VI, Box 97, Folder 16, Huxley, "The Concept of Race in the Light of Modern Genetics," *Harper's Monthly Magazine*, May 1935, 689–698, on 691.

31. Ibid., Box 98, Folder 8, Huxley, "The Nazi Racial Theory," *John O'London's Weekly*, 16.1.1942. An example for the self-undermining tendency of the "racial measurement crave" is Morant (1939), for which Haldane wrote the preface.

32. JSH Papers, Series VI, Box 97, Folder 16, Huxley, "Scientific Pitfalls of Racialism," *Yale Review* 24 (4), June 1935, 667–682, on 669.

33. Ibid., 681–682.

34. Ibid., Series VI, Box 98, Folder 3, Huxley, " 'Aryan' Racial Myth Exploded as Merely 'Mystic Nonsense,'" *Examiner*, Oct. 1938. See also ibid., Box 97, Folder 16, A. C. Haddon and Huxley, "Racial Myths and Ethnic Fallacies," reprinted from *Discovery,* Sept. 1935, 1–6; ibid., Box 98, Folder 1, "Royal Institution of Great Britain, Weekly Evening Meeting," 27.3.1936, George Gaylord Simpson and Huxley, "The Race Problem" (Abstract), 1–3, where—in anticipation of the UNESCO Statement on Race—they suggested that some international inquiry should be made which would result in an impartial scientific pronouncement on the subject, under the auspices of the League of Nations, by representatives of the leading universities of the civilized world, or by some body of international scope such as the Rockefeller Foundation.

35. Ibid., Series XI, Box 136, Folder 5, Raymond Moley, "Science and Tolerance," review of Huxley and Haddon 1935, clipping from unknown source, p. 13.

36. Ibid., Box 137, Folder 1, "Huxley, Julian," *Current Biography*, n.d., 22–24, on 24.

37. Ruggles Gates for one aggressively criticized it and Huxley in person. Like others, he tried to rob the book of Haddon's authority by claiming that Haddon did not support the conclusions Huxley drew in it. However, Haddon made statements both in support and with the effect of distancing himself from Huxley's antirace propaganda (JSH Papers, Series III, Boxes 12–13, Gates to Huxley, 22.2.1937; Gates to Huxley, 12.3.1937; Gates to Huxley, 24.3.1937; Gates to Huxley, 14.4.1937).

38. JSH Papers, Series VI, Box 98, Folder 4, Huxley, "Life Can Be Worth Living," *Literary Guide*, Feb. 1939, 35–37, on 36–37; Huxley 1941a, 291–300 ("Life Can Be Worth Living," first published 1938), 299–300. Note that these political issues and anthropological discussions correlated with the reform of systematics within the synthetic biology as expressed in *The New Systematics,* which Huxley edited (Huxley 1940b).

39. JSH Papers, Series VI, Box 98, Folder 1, Huxley, "Foreword," in "History on a Racial Basis by Johann von Leers" (*Geschichte auf rassischer Grundlage*, 1934, Reclams Universal Bibliothek Nr. 7249), *Friends of Europe Publications* 42, 1936, London: Friends of Europe, 3–6.

40. Ibid., Series XIII, Box 149, Folder 3, "Still More Peers; or Tissue Culture Ltd."

41. See, for example, Huxley 1942a, ch. 10.5; also Green 1981, 274–295. On the literary and popular science writings of Julian and Aldous Huxley, J. B.S. and Charlotte Haldane, and Naomi Mitchison with regard to reproductive technology, see also Squier 1994.

42. The statement also reiterated that what was known at the time was that genetic traits did not cluster according to social groups and that human beings were the outcome of complex interactions between heredity and environment. Muller would continue to use this argumentation into the postwar decades (Comfort 2012, 130–162).

43. See, for example, JSH Papers, Series VI, Box 97, Folder 3, Huxley, "The Bases of Modern Biology: The Living Machine," *New Leader*, 9.11.1923, 11–12; and Huxley 1926c.

44. See for example also JSH Papers, Series VII, Box 102, Folder 7, Harold B. Hinton, "Huxley Sees Us All Still Undeveloped," *The New York Times*, 3.12.1939: "While the speaker warned strongly against overdoing analogy in the comparatively young field of social science, he explored the animal kingdom extensively to search out examples of the development of individuality, as opposed to mere group participation, in the advance of biological evolution. He reached the conclusion that this fact ought to serve a useful purpose to social planners, 'since it immediately exposes the fallacy of all social theories, like those of Fascism and National Socialism, which exalt the State above the individual.' "

45. Also ibid., Series VI, Box 98, Folder 9: 1942, Huxley, "The War: Two Jobs, Not One," *Fortnightly*, Oct. 1942, 221–228. On Huxley's views on reconstruction and global planning see also ibid., Folder 6: 1940, Huxley, "Science, War and Reconstruction," reprinted from *Science* 91 (2355), 16.2.1940, 151–158, in which his scheme of a development toward world citizenship draws on the lessons from biology and the evolutionary past; ibid., Huxley, "The Sociology of Planning" (review of Karl Mannheim, *Man and Society in an Age of Reconstruction*), *Nature* 146 (3688), 6.7.1940, 2–4; ibid., "Men of Science and the War" and Huxley, "Science in War," *Nature* 146 (3691), 27.7.1940, 107 and 112; ibid., Huxley, "Surplus Stores: Foodstuffs against Famine. A Case for Planning," *Times,* 13.8.1940; ibid., Huxley, "America Looks at the War. I.—Isolationism," *Times,* 15.4.1940; ibid., Huxley, "The Science of Society," *Virginia Quarterly Review* 16 (3), 1940, 348–365; ibid., Folder 7, Huxley, "Revolution by Evolution," *Listener,* 10.4.1941, 519–520; ibid., Huxley, "Armaments and Security," *New Republic,* 2.6.1941, 750–752; ibid., Huxley, "Economic Man and Social Man," reprinted from *Fortnightly*

Review (London), July 1941. For a particularly comprehensive expression, see ibid., Folder 8, Huxley, "Reconstruction and Peace," *New Republic,* 10.6.1942, first published 1941 by Kegan Paul, Trench, Trubner, copyright 1942, Editorial Publications.

46. Ibid., Folder 7, Huxley, "The Growth of a Group-Mind in Britain Under the Influence of War," *Hibbert Journal* 39 (4), 1941, 337–350, on 337–338.

47. Ibid., Series XI, Box 137, Folder 2, Leon Whipple, "Science Plus People," *Survey Graphic* (NYC), Apr. 1941.

48. Ibid., Charles Sherrington, "The Proper Study: Julian Huxley Looks at Man," *Reader's News* 6 (2), July 1943, 2–3, on 2, reprinted from *Cambridge Review.*

49. Ibid., A. A. B., "'The Uniqueness of Man': Essays by Julian Huxley," *Plan Bulletin*, Apr. 1941, 2–3, on 3.

50. Ibid., Christopher Stull, "Essays of a Great British Scientist," *San Francisco Chronicle*, 25.5.1941.

51. Ibid., Wesley Fuller, "Scientific Methods Applied to Study of Man, Society," *Boston Herald*, 28.5.1941.

52. Ibid., Christopher Stull, "Essays of a Great British Scientist," *San Francisco Chronicle*, 25.5.1941.

53. See, for example, ibid., Charlton Ogburn, Jr., "Man Stands Alone by Julian Huxley," *Book of the Month Club News* (NYC), May 1941; ibid., Robert M. Green, M.D., "Looks Toward Future World," *Transcript* (Boston, MA), 4.5.1941.

54. Ibid., D. W. Brogan, "'Eppur si Muove,'" Books of the Day, *Spectator*, 1.8.1941.

55. Ibid., Ellsworth Huntington, "Huxley and Evolution," *Saturday Review*, 12.4.1941, 5. The book nonetheless received considerable favorable attention, see, for example, ibid., "Emergent Democracy," *Times Literary Supplement,* 2.8.1941.

56. See also, for example, ibid., Series VI, Box 98, Folder 7, Huxley, "How Two Years of War Have Changed Us," *Sunday Dispatch*, 31.8.1941.

57. Ibid., Huxley, "Freedom Will Be Organised," *Democracy Marches On* 2, 5.8.1941.

58. Ibid., Folder 9: 1943, Huxley, " 'It Is Up to All of Us': Julian Huxley Sums Up the Series of Talks 'Reshaping Man's Heritage,'" *Listener* 29 (743), 8.4.1943, 407–409, on 409.

Chapter 9

1. JSH Papers, Series VIII: Conference Materials, Box 107: 1953–1970, Folder 21: International Congress of the World Wildlife Fund 1970, *All Life on Earth: Second International Congress of the World Wildlife Fund*, London, 17/18.11.1970 (WWF publication), 21.

2. Ibid., draft speech for reception of WWF medal, 17.11.1970.

3. Ibid., *All Life on Earth: Second International Congress of the World Wildlife Fund*, London, 17/18.11.1970 (WWF publication), 94.

4. Huxley had, for example, warned about the consequences of overpopulation as early as the 1920s. In 1965, he had even taken the issue to the *Playboy* reader (JSH Papers, Series VI, Box 101, Folder 6, Huxley, "The Age of Overbreed," *Playboy,* Jan. 1965, 103, 104, 106, 177, 179–181), on which grounds his friend Simpson confessed that "I find the idea of an article by you on population juxtaposed with female nudes irresistably [*sic*] charming" (ibid., Series III, Box 38, Simpson to Huxley, 19.1.1965). By the time of the Second International Congress of the WWF, the population problem was—together with pollution—identified as *the* threat to a responsible trusteeship of the human heritage (ibid., Series VIII, Box 107, Folder 21, *World Wildlife News: Second International Congress Special Report*).

5. Elton defined ecological communities and niches on the basis of food cycles and webs as well as other interactions, and he was interested in species distribution along environmental gradients, ecological succession, adaptation, and the role of numbers, such as optimum population size and fluctuations in numbers of animals (see also the new introduction by Mathew A. Leibold and J. Timothy Wootton in Elton [1927] 2011, xix–lvi).

6. On the distinction between "total" and "partial" protection of nature from human interference see Kupper 2012, 147–151. The Swiss National Park was the model for the total sanctuary, even if, as Kupper shows, the park was not left completely alone. Along with an efficient managerial staff, scientists conducting research, and a limited number of tourists, new animal species were introduced into the area.

7. On the history and significance of these Anglo-Maasai agreements, see Hughes 2006; on the history of the Serengeti National Park in Tanzania, see Bender 2007, ch. 6; on the case of the removal of Makuleke in the context of the establishment of the Kruger National Park in South Africa, see Carruthers 1995—both the Maasai and Makuleke have fought for their rights to land; for the example of the Swiss National Park, see Kupper 2012; for the more recent case of the Masoala National Park in Madagascar, see Keller 2015.

8. JSH Papers, Series VI, Box 98, Folder 3, Huxley, "Rare Animals and the Disappearance of Wild Life," Christmas Lectures: The One Hundred and Twelfth Course of Six Lectures Adapted to a Juvenile Auditory, the Royal Institution, London, 28.12.1937 to 8.1.1938, abstracts of lectures, 8.

9. While Huxley feared for the loss of local traditions through Western civilization and hoped for salvation in common evolutionary humanist ideals, missionaries could have similar perceptions, only that Christianity was seen as the compensation for the partial loss of original cultural elements. While the new generation of professional anthropologists openly criticized the impact of West-

ern civilization and embraced native customs, such "enlightened," liberal missionaries of the early twentieth century saw this as a dangerous kind of idealization of the "primitive life" that went along with the aim of returning to the past. Huxley seems to have shared aspects of both views, thus valuing the cultural achievements and richness of the Africans he encountered but at the same time regarding with suspicion attempts at simply preserving the present or even restoring the past instead of striving for optimal conditions for progressive development (see, for example, Morier-Genoud 2011).

10. JSH Papers, Series VI, Box 97, Folder 10: 1930, Huxley, "Should We Educate the Native?," *Pall Mall*, Oct. 1930.

11. A couple of years later, Huxley hoped that this would be a self-regulating process, with the slump in the price of tropical produce and the synthetic production of many substances having the effect of leaving "primitive tropical regions" to themselves. If no longer worth exploiting, native life might again be free to develop along its own lines. Huxley was dispirited by "the story of the white man's past greed and cruelty." In a review, he recommended a book on the topic to everyone in the hope that "similar horrors [were] not perpetrated elsewhere under the cloak of economics and the pretext of a civilising mission" (ibid., Box 98, Folder 2, Huxley, Review of T. H. Harrisson, *Savage Civilisation*, *The Book Society News*, Jan. 1937, 9).

12. In the aftermath of his early travels in Africa, Huxley became acquainted with the Tennessee Valley Authority and recognized in it a model case for his planning ideal (see, for example, ibid., Folder 8, Huxley, "Tennessee Revisited: The Technique of Democratic Planning," reprinted from *the Times*, 25.6.1942; for its history and Huxley's appraisal see, for example, ibid., Folder 10: 1943, Huxley, "The Origin of the TVA," *Rucksack Magazine of the Youth Hostel Organisations of Great Britain and Ireland* 11 [3], 1943, 3–5). He mentioned it in the context of the development of colonies between international and local governance as an example for balancing the expertise and control from a distance with the interests and authority of local officials and democratic decision making (ibid., Folder 4, Huxley, "Federalism and Colonies: A Possible Interim Solution," letter to the editor, *Times*, 29.11.1939, 6).

13. See also ibid., Box 97, Folder 10, Huxley, "Biology and the Biological Approach to Native Education in East Africa," *African (East)* 1134, Apr. 1930.

14. On the notion of the zoo as microcosm of nature in need of management, see also Deese 2010, 283, 289.

15. JSH Papers, Series IX, Box 114: Ins–Nature Conservancy, Folder 3, comments by Huxley on IUCN, "Wildlife Conservation in Ghana," 1–2, on 1.

16. See, for example, ibid., Series VI, Box 98, Folder 6: 1940, Huxley, "The Future of Colonies," *Tribune*, 19.1.1940; ibid., Huxley, "The Future of Colonies," *Fortnightly*, Aug. 1940, 120–130; ibid., Folder 10, Huxley, "The Social Experiment—White and Black Must Work Together," *London Calling* 210, 1943,

18 and 20. Huxley listed three options for colonial development: handing over some tropical territories to those nations that had none (like Germany), pooling all colonies immediately under an international administration, and finding a balance between increasing local rule and development and international guidance with the ultimate aim of self-governance; he advocated the last (e.g., ibid., Box 99, Folder 1: 1944, Huxley, "Colonies and Freedom," *New Republic,* 24.1.1944, 106–109). Huxley's three solutions to the problems of the colonies were not the only ones in existence, however. There was also the view that Africans should be given full rights as free citizens immediately, that they should be given civil rights, equal wages as whites, the right to buy land and free access to the market with equal prices for their produce, as well as equal taxation. In fact, a respondent to Huxley's *Tribune* article retorted that "in sober fact, the people of the countries of British South and East Africa have far fewer rights than Hitler allows to Jews in Vienna and Czechs in Prague" (ibid., Box 98, Folder 6, Norman Leys, "But What of the People?"; see also ibid., Folder 10, Huxley, "The Sky's the Limit for the Colonies," *Daily Herald,* 22.10.1943).

17. Huxley described the goals governing the nonparty organization Next Five Years Group, in which he worked with Wells, as those of economic reconstruction, social justice, and international peace. The group was against laissez-faire politics. Instead, the state should take an increasing proportion of the profits of successful businesses and redistribute them in various social services. "We cannot betray the heritage we hold in trust for posterity. If law in international affairs can as yet only grow out of the sanction of force, then we must see to it that the body in which law resides, the League of Nations, has behind it adequate force" (ibid., Folder 1, Huxley, "All Party Planning: Its Necessity and Possibility," *Service in Life and Work* 5 [18], 6d, 1936, 16–23, on 22–23). On details of the foundation of UNESCO, see Krill 1968.

18. JSH Papers, Series VI, Box 99, Folder 2: 1945, Huxley, "Science and the United Nations," in "A United Nations Educational and Cultural Organisation," reprinted from *Nature* 156 (3967), 1945, 553–561.

19. Huxley had formally taken over duties in March and was looking for staff for the eleven sections, all of which needed a senior counsellor and a counsellor. He secured Needham as senior counsellor of natural sciences (JSH Papers, Series IX, Box 117, Folder 5: UNESCO 1946, 1948, 1949, 1951–1961, "Report of the Executive Secretary to the Executive Committee—19/3/46").

20. On the International Committee on Intellectual Cooperation and the Intellectual Institute for Intellectual Cooperation as precursors of the UNESCO heritage conservation efforts, see Wöbse 2012, 278–287 on Huxley's role in them.

21. On his fiftieth birthday, he had informed the papers that he believed in the facts about evolution and in human trusteeship over their progressive application (JSH Papers, Series VI, Box 98, Folder 2, Huxley, "'What I Believe,'" *News*

Chronicle, 22.6.1937, 8; see also UNESCO 2006, part 1 ["Setting the Scene, 1945–1965"]).

22. JSH Papers, Series IX, Box 117, Folder 5, "Report on the Programme of the United Nations Educational, Scientific and Cultural Organisation," Preparatory Commission of the UNESCO 1946 Conference, 89–120; "General Conference, First Session, Held at UNESCO House, Paris, from 20 November to 10 December 1946," Paris: United Nations Educational Scientific and Cultural Organization 1947, 269–77. In fact, J. P. Singh (2011, 2) states that UNESCO still promotes Enlightenment values, and that "at its core, UNESCO reflects a scientific humanism" in the sense of Huxley's "UNESCO: Its Purpose and Its Philosophy." For more on UNESCO under Huxley's directorship, see also Sewell 1975, ch. 3; Armytage 1989.

23. Huxley's election was controversial for several reasons, one such was that he was subject to accusations of radical politics and communist affiliations on the part of the United States, an accusation that was reiterated throughout Huxley's short tenure of two instead of four years (see, for example, a draft of a reply to a publication in a U.S. magazine by one of Huxley's friends, JSH Papers, Series IX, Box 117, Folder 5, "U.N.E.S.C.O. Conference: Last Thoughts" by J. B. Priestley [from the Home Service], 16.12.1946; and correspondence between Zuckerman and Huxley 1946–8, ibid., Series III, Box 56, Folder 2; in general on the subject, see "Souvenirs de l'UNESCO par Sir Julian Huxley," in *Julian Huxley in memoriam*, UNESCO 1975, 37–39; also J. Toye and R. Toye 2010).

24. JSH Papers, Series VI, Box 99, Folder 4: 1948, "This Is Our Power . . . ," *UNESCO Publications* 273, speeches of Huxley (5–10) and Bodet (11–19) in Beirut. Unlike in the farewell speech, Huxley was explicit about his evolutionary humanism in the introduction to the annual report that was reproduced in UNESCO's *Courier*. He suggested the establishment of a single unifying idea or goal, such as the advance of world civilization, to further world peace and development, knowledge and creative activity along the lines of "evolutionary progress." There should be a new interpretation of life based on a humanistic perspective on scientific knowledge. Out of the crowd of existing ideas and interpretations of the world, there should be distilled the "view of human destiny as the crown and continuation of the cosmic process of evolution." This would include the development of nondiscriminatory principles to achieve "equality of opportunity" (Huxley 1948, 6). Since UNESCO was intergovernmental and thus limited by the sovereignty of the nation-states, an important part of its work should be delegated to particularly endowed individuals—thinkers, artists, writers, scientists.

25. JSH Papers, Series IX, Box 114, Folder 3, "International Union for Conservation of Nature and Natural Resources, Tenth Anniversary: 1948–1959," pamphlet, Brussels 1958; ibid., IUCN, brochure "How IUCN Started," Morges 1962; ibid., Folder 5, Harold J. Coolidge, "The Birth of a Union," reprinted from

National Parks Magazine, Apr./June 1949; ibid., Box 117, Folder 5, Report of the Director General on the Activities of the Organization in 1948, Paris 1948.

26. This was also the appreciation of Melanie Staerk when she wrote the twentieth-anniversary retrospective for UNESCO. Returning to Huxley's ideas as proposed in "UNESCO: Its Purpose and Philosophy" (1946), she found that even though the organization had settled on a more pragmatic program, many of the projects that had been realized were very much in harmony with Huxley's grander scheme (JSH Papers, Series XI, Box 137, Melanie Staerk, "Anniversary Reading: Sir Julian Huxley's 'Unesco,'" *Swiss Review of World Affairs*, Sept. 1966, 20–22).

27. Correspondence with René Maheu, JSH Papers, Series III, see Index to Selected Correspondents in Guide to JSH Papers; Dronamraju 1993, 189–190. For Huxley's role in the environmental movement through UNESCO and IUCN, see also Deese 2010, 2011.

28. In 1958, Huxley was also involved in the foundation of the Council for Nature and served as its vice president. The council should function as an umbrella body for the nature movement, all natural history and conservation bodies in the UK, especially focusing on common policy aims, provision of membership and funding, and medial and educational representation (JSH Papers, Series IX, Box 112: Charles Darwin Foundation for the Galapagos Islands 1964–G, Folder 3).

29. Ibid., Box 117, Folder 3, UK National Commission for UNESCO.

30. Before Huxley was commissioned by UNESCO, he actually mobilized protest against plans to reduce the size of the Serengeti National Park that had been brought to his attention by Louis Leakey, who afterward helped him come up with a realistic itinerary for his journey for UNESCO and showed him Olduvai (JSH Papers, Series III, correspondence between Huxley and Leakey from 17.1.1956 [Box 24, Folder 1] to 2.8.1960 [Box 30, Folder 2], see Index to Selected Correspondents in Guide to JSH Papers).

31. JSH Papers, Series IX, Box 113, Folders 2–7; ibid., Folder 6: Idea Systems Group, "Notes on the Idea of Ecology as Applied to Man" and "Note on Ecology"; for conservation and ecology in a humanist frame, see also Nicholson 1961.

32. Correspondence with Elton in JSH Papers, Series III, see Index to Selected Correspondents in Guide to JSH Papers.

33. JSH Papers, Series VI, Box 101, Folder 3: 1962, Huxley, "Eastern Africa: The Ecological Base," reprinted from *Endeavour* 21 (82), Apr. 1962, 98–107.

34. Ibid., Folder 2: 1961, "The Humanist Revolution," in *Third Series of Aggrey-Fraser-Guggisberg Memorial Lectures, Nov. 1961, Delivered by Sir Julian Huxley*, published by the University of Ghana, 36–47.

35. Ibid., Series IX, Box 120, Folder 3: World Wildlife Fund, 1961, A. Mark II: "We Must Save the World's Wild Life: An International Declaration," in 3rd Meeting, "Saving the World's Wild Life," Note of meeting held at 19 Belgrave Square, 16.5.1961, 1–3.

36. Ibid., "World Wildlife Funds (Save the World's Wildlife)," in "Saving the World's Wild Life," Note of meeting held 25.4.1961, 1–9, on 7.

37. "nichts anderes als eine Zusammenfassung von Julian Huxleys postkolonialer Naturschutzstrategie" (Schwarzenbach 2011, 46).

38. JSH Papers, Series IX, Box 112, Folder 3, Council for Nature, "Intelligence Unit: Monthly Press Bulletin, No. 20, October 1961," 1–3, on 1.

39. Ibid., Series VI, Box 101, Folder 2, "Huxley in Africa," reprint of series of articles published by *the Observer* and distributed by the Fauna Preservation Society.

40. On the importance of Huxley's reports on the African trip, see also ibid., Series III, Box 30, Nicholson to Huxley, 27.7.1960.

41. See Nicholson 1987 for a condensed account of the events, with 163–165 on Huxley exclusively.

42. Huxley would not tire of reiterating the value of the African mammals as the surviving exemplar of the Pleistocene climax of biological evolution and to use their popular appeal to advocate for conservation and biology (see, for example, JSH Papers, Series VI, Box 101, Folder 6, Huxley, "Serengeti: A Living Laboratory," *New Scientist* 27 [458], 1965, 504–508).

43. JSH Papers, Series III, see Index to Selected Correspondents in Guide to JSH Papers. Obviously, Huxley was less at ease with attempts to turn humans into naked apes (Morris 1967) than with discovering the advanced stage of ape "culture." At the same time, he must have been pleased when his friend Leakey wrote him about his Kenyan discoveries that confirmed the separation of the hominid and anthropoid lines in the Lower Miocene, supporting a notion Arthur Keith—and Osborn—had been propagating (JSH Papers, Series III, Box 41, Folder 3, Leakey to Huxley, 17.11.1966; on Keith see Sommer 2007a, ch. 12).

44. In December 1962, the general assembly of UNESCO adopted a resolution that dealt with the impoverishment of the aesthetic, cultural, and even vital world heritage (natural as well as man-made) and that stressed the cultural and scientific importance of wildlife. In 1963, the emperor and government of Ethiopia approached UNESCO to mobilize a mission to their country to come up with guidelines for nature conservation, a mission Huxley headed. The mission developed suggestions for nature conservation for scientific, tourist, and economic use in the form of national parks, reserves, areas where hunting and cropping would be allowed, as well as zoological and botanical gardens. And again, the global interest of scientifically managed natural resources for mass education was stressed. In a resolution of December 1963, the general assembly of UNESCO called on all member states to economically develop and conserve natural resources and to educate ("The Conservation of Nature and Natural Resources in Ethiopia," report by Huxley, A. Gille, Th. Monod, L. Swift, E. B. Worthington, United Nations Educational, Scientific and Cultural Organization, Paris, 27.12.1963, UNESCO/NS/NR/47).

45. JSH Papers, Series III, Box 37, Folder 3, Goodall, Kigoma, Tanganyika, to Huxley, 2.9.1964; ibid., Box 44, Folder 1, Goodall, London, to Huxley, 16.9.1971; see also ibid., Box 38, Folder 4, Huxley to Goodall, 11.3.1965.

46. Just as this book reaches the phase of production two years after manuscript submission, R. S. Deese's intellectual history of Aldous and Julian Huxley appeared. Deese in particular emphasizes Julian Huxley's "religious" or "spiritual biology" and situates it in the context of the long history of interrelations between human understanding of nature and religious thought (2015, 105–107). He relates Julian Huxley's attempt to create a religious cosmology without recourse to supernatural entities to the monism of Ernst Haeckel (ibid., 6, 94). The growing interest in Huxley as conservationist is not surprising in view of the contributions the Huxley brothers made to "the progressive and pastoral strains of environmental thought and activism" (ibid., 146–154) and the ecological perspective captured by the term *Anthropocene* (ibid., 2).

47. JSH Papers, Series III, Box 38, Folder 1, Maheu to Huxley, 4.1.1965.

48. In fact, in his address to the UN Conference on Human Environment, Maheu reminded his audience that the demand for an interdisciplinary and international approach to the conservation of natural and human diversity with the aim of integrated solutions that suited local contexts could be traced back to the first director-general of UNESCO. He stressed that already Huxley had promoted the ecological approach in which nature and culture were two sides of the human coin (UNESCO, "Address by René Maheu, Director-General, at the UN Conference on Human Environment," Stockholm, 7 June 1972, appended to JSH Papers, Series III, Box 44, Folder 5, Maheu to Huxley, 19.6.1972). In the course of a celebration in commemoration of Huxley (*Julian Huxley in memoriam*, UNESCO 1975), it was also stated that he had directed UNESCO toward lifelong education, conservation, alphabetization, and the application of science that have come to characterize its work. He was portrayed as a pioneer of international and intellectual cooperation and distributor of ideas, although not as a lover or talent of administration. His engagement for UNESCO's Scientific and Cultural History of Mankind (see below) was to make people see the general human achievement and to bring them closer together (also *Courier,* Mar. 1976, in which Huxley's philosophy for UNESCO was partly reprinted).

49. For a more positive evaluation, see Blue 2001.

50. Huxley was part of the group commenting on the first draft of the UNESCO Statement, and as the reactions among his friends to the first and revised statements show, there was no consensus. Mayr, and Muller along with other geneticists, had issues. Muller disagreed with the declaration that genetics did not suggest or denied differences in mental capacity between groups. Huxley was pleased with the Statement in so far as he understood it to carry the message that genetic inequality was not to mean inequality of opportunity or inequality before the law (see JSH Papers, Series VIII, Box 108: Darwin Centennial Cele-

bration 1959, Folder 7: JSH comments on papers, Muller to Huxley, 8.4.1951; ibid., Series III, Box 20, Folder 4, Mayr to Huxley, 2.9.1952; Huxley to Mayr, 12.9.1952. The literature on the history and impact of the UNESCO Statements on Race is considerable and it draws particular attention to the differences between the first and second draft with the effect of diluting the attempt of the first version to identify race as a social construct [e.g., Gayon 2003]; for a particularly thorough investigation, see Brattain 2007).

51. Huxley 1963, 712, adapted from his talk at the Ciba Foundation Symposium "The Future of Man," held in London (proceedings were being published); see also JSH Papers, Series VIII, Box 107, Folder 10: "The Future of Man" symposium 1959, William L. Laurence, Science Editor, *New York Times*, to Huxley, 7.4.1959.

Chapter 10

1. JSH Papers, Series VI, Box 99, Folder 5: 1949, Huxley, "A Number of Things," *Sunday Times*, 29.5.1949.

2. See also Huxley's introduction to Teilhard de Chardin (1955) 1959.

3. JSH Papers, Series III, Box 18, Folder 4, Huxley to Teilhard de Chardin, 16.11.1949.

4. Ibid., Box 20, Folder 2, Teilhard de Chardin to Huxley, 1.4.1952; ibid., Folder 4, Teilhard de Chardin to Huxley, 8.9.1952; ibid., Box 23, Folder 1, Teilhard de Chardin to Huxley, 31.1.1955.

5. It resulted in UNESCO's *History of the Scientific and Cultural Development of Mankind* and the journal *World History* that accompanied it (see JSH Papers, Series IX, Boxes 118 and 119; also communication with Ralph Edmund Turner in ibid., Series III, see Index to Selected Correspondents in Guide to JSH Papers). During that time, Huxley also received the UNESCO Kalinga Prize of 1953 for his intensive effort in the communication of science and its social consequences (ibid., Series IX, Box 117, Folder 6). The fact that he was not only nominated by the Institut de France but also by the Royal Society of London is slightly ironic, considering that his popular work was one of the grievances the society had voiced in deferring his election to a fellowship.

6. JSH Papers, Series III, Box 28, Folder 2, Huxley to Ralph Edmund Turner, Chairman of the Editorial Committee of the UNESCO International Commission for a Scientific and Cultural History of Mankind, 6.4.1959.

7. Correspondence between Huxley and Nicholson between 1.5.1959 and 26.2.1960 in JSH Papers, Series III, see Index to Selected Correspondents in Guide to JSH Papers.

8. JSH Papers, Series III, Box 30, Folder 4, Blackham to Huxley, 29.10.1960.

9. Ibid., Box 28, Folder 2, Huxley to Turner, 6.4.1959; Turner to Huxley,

10.4.1959; ibid., Box 29, Folder 1, Huxley to Waddington, 1.1.1960; Waddington to Huxley, 18.1.1960; Huxley to Waddington with suggestions for revision, n.d. [Jan. 1960]; ibid., Folder 2, Waddington to Huxley, 8.2.1960; Huxley to Waddington, 10.2.1960.

10. Ibid., Box 44, Folder 2, Waddington to Huxley, 15.7.1963.

11. See ibid., Box 34, Folder 3, Dobzhansky to Huxley, 14.3.1963; Huxley to Dobzhansky, 9.4.1963; for the Huxley-Simpson correspondence see Index to Selected Correspondents in Guide to JSH Papers.

12. See, for example, Dobzhansky and Montagu 1947; Dobzhansky 1955, 1956, 1962a, 1962b, 1973; Dobzhansky and Boesiger 1983; Simpson 1941; 1960, 973–974; 1964; 1964/1966/1967/1969; 1966; 1977.

13. JSH Papers, Series III, Box 19, Simpson to Huxley, 2.8.1950.

14. On these evolutionary humanist philosophies, see Smocovitis 1996, ch. 5; Sommer 2010a. On the personal relationship and the convergence and divergence of Huxley's and Simpson's thoughts on progress in evolution, see Swetlitz 1993, ch. 8, with the inclusion of Dobzhansky and Mayr see ch. 9; also Greene 1981, 167, 171, 176; Beatty 1994; Krimbas 1994; Paul 1994; Swetlitz 1995, 199 and 202–215.

15. JSH Papers, Series III, Box 35, Folder 5, Blackham, Director of the British Humanist Association, to Huxley, 18.10.1963.

16. Ibid., Box 19, Folder 12, Blackham, General Secretary of the Ethical Union, to Huxley, 9.10.1951; Blackham to Huxley, 8.1.1952ff.; ibid., Box 34, Folder 2, Blackham, now Director of the British Humanist Association, to Huxley, 14.2.1963ff.—the exchange of letters gives a good impression of the humanist movement and the internal differences about what exactly the new humanism meant; also ibid., Series VIII, Box 106: 1934–1952, Folder 15: First International Conference on Humanism and Ethical Culture, 1952; ibid., Series IX, Box 111: A–Charles Darwin Foundation for the Galapagos Islands 1963, Folder 3, American Humanist Association; ibid., Box 114, Folder 2, International Humanist and Ethical Union (IHEU), Proceedings of the 3rd Congress, 2.–7.8.1962, Oslo University—Huxley took part as vice president for the UK and first director-general of UNESCO. There were sections called "Towards the Mature Personality," "Towards Freedom in an Organized World," and "Working Parties."

17. Ibid., Series III, Box 24, Folder 5, Muller to Huxley, 5.8.1957; for their correspondence in general see Index to Selected Correspondents in Guide to JSH Papers; see also Muller 1961, 399–414.

18. Ibid., Series XI, Box 138: 1960–1975, Folder 3: reviews *Essays of a Humanist*, 1964, Ralph Hancox, "Mankind: Agent of Evolution?," *Peterborough Examiner*, 6.6.1964.

19. "Ein Vermächtnis des Nonkonformismus in einer sich wandelnden Welt" (ibid., "Suche nach dem zukünftigen Menschen," *Neue Ruhr Zeitung*, 23.5.1966).

20. Ibid., Norbert Nye, "Sugar, Spice and All Things Nice," *Western Mail*, 21.3.1964, quoting Huxley 1964, 5.

21. "Julian Huxley ist ein sehr alter Herr, und zu den Eigentümlichkeiten des Alterns gehört in vielen Fällen eine gewisse geistige Erstarrung, ein unbewegliches Festhalten an bestimmten Vorstellungen, ein stures Ignorieren aller Fakten, die der einmal gefaßten Idee abträglich sein könnten" (ibid., Huberta von Bronsart, "Huxleys evolutiver Humanismus," *Stuttgarter Nachrichten*, 11.12.1965).

22. Ibid., Philip Toynbee, "How to Give a Word a Bad Name," *Observer Weekend Review*, 22.3.1964.

23. Ibid., Anthony Quinton, "Onward from Darwin," *Sunday Telegraph*, 22.3.1964.

24. "Darwinist müßte man sein, Biologie sollte man studiert haben! Dann wären alle jene Probleme gelöst, mit denen wir arme Historiker, Soziologen und Philosophen uns herumschlagen: Gott und Welt, Freiheit und Schicksal, Seele und Leib . . ." (ibid., Helmut Günther, "Huxley, Julian: Ich sehe den künftigen Menschen," *Welt und Wort* 2, 1966).

25. Ibid., Philip Greer, "The Religion of Julian Huxley," *Humanist*, Mar. 1964.

26. Communication between Huxley and Muller in JSH Papers, Series III, see Index to Selected Correspondents in Guide to JSH Papers; Muller 1961.

27. "Die sog. Genetik und Eugenetik [*sic*] zum Evangelium seiner Weltreligion erhebt" (JSH Papers, Series XI, Box 138, Folder 3, Ursula Anders, "Julian Huxley: 'Ich sehe den künftigen Menschen,'" *Blätter für Anthroposophie*, July/Aug. 1966, 277–279, on 278).

28. "Gott bewahre uns vor dem zukünftigen Menschen, der nach der Vorstellung eines gegenwärtigen Menschen namens Julian Huxley erst noch gezüchtet werden muß" (ibid., Friedrich Deich, "Wie sieht der Mensch der Zukunft aus?," *Die Welt der Literatur* 3 [3], 3.2.1966, 1–2, on 2).

29. As early as 1937, Huxley had been asked by Arthur E. Mason to donate sperm to father his child (JSH Papers, Series III, Box 12, Folder 5, Mason to Huxley, 3.1.1937; on Huxley's enthusiasm for the possibility of eugenic artificial insemination see also his correspondence with the General Secretary of the Eugenics Society, Colin Bertram, Wellcome Library, London: Sir Julian Huxley, Shelfmark SA/EUG/C.186, Box AMS/MF/111, Reference number b16232811). The biochemist Alfred E. Mirsky of Rockefeller University was also among Huxley's critics. When Huxley defended himself against Mirsky's critique in *Scientific American*, Mirsky replied (Jan. 1965, 6–9), followed by an attack on Huxley by Dobzhansky (Mar. 1965, 8–9), possibly motivated by a somewhat critical review by Huxley of Dobzhansky's *Mankind Evolving* (1962b; see below). Mayr, who took Huxley's side, reacted with a personal note of regret to *Scientific American* (JSH Papers, Series III, Box 37, Folder 4, Mayr to Huxley, 10.11.1964;

also ibid., Huxley to Mayr, 5.10.1964; Mayr to Huxley, 9.10.1964). To be fair, by that time, Muller's project found rather wide acclaim and was among others not only greeted by Mayr but also by Haldane (Kevles [1985] 1995, 261–264). Then, the *Eugenics Review* published a letter by Muller (1965), who felt involved because, after all, "germinal choice," as he called it, had from 1925 onward been his hobbyhorse. He called on his friendship with both critics and on the Geneticists' Manifesto of 1939 (Gruenberg 1939). It had been signed among others by Mirsky, Dobzhansky, as well as Muller, and, as we have seen, had incorporated many of the views broached by the triple-H in the 1930s. It could vouch for Huxley's and Muller's deeply rooted social outlooks also with regard to eugenics. Muller had been represented by Dobzhansky and the American geneticist L. C. Dunn as favoring a eugenic establishment of a single ideal genotype. He denied this in his letters to Huxley, sending him passages from his publications in support. But Muller did find Dobzhansky's argument for the benefit of genetic variability and heterozygosis too general and linked it to speculations about the possible beneficial effects of radiation (JSH Papers, Series III, Box 33, Folder 3, Muller to Huxley, 6.7.1962; Muller to Huxley, 13.7.1962; Huxley to Muller, 16.7.1962; for a strong statement for continuity in Huxley's eugenic ideals and propaganda, see Weindling 2012).

30. However, as Joel Hagen (2010b) has put it, Nuttall was more concerned with the method of the precipitin test from forensic medicine than with primate systematics. Although a small community of researchers interested in systematic serology did develop, it did not position itself against traditional taxonomy. Rather, its members were part of museum work to the point that a central figure, Alan Boyden, referred to his workplace at Rutgers University as a serological museum instead of as a laboratory.

31. See also JSH Papers, Series VI, Box 97, Folder 16, Huxley, "The Concept of Race in the Light of Modern Genetics," *Harper's Monthly Magazine*, May 1935, 689–698.

32. Ibid., Series VIII, Box 108, Folder 7: JSH comments on papers; for a thorough analysis of the mega-event as commemorative practice, see Smocovitis 1999, and for the attempts to integrate anthropology, see Smocovitis 2012.

33. JSH Papers, Series III, Box 28, Folder 2, Mayr to Huxley, 3.4.1959.

34. Where the study of human diversity was concerned, Huxley still held great stakes in the power of human population genetics and as president of the Eugenics Society he tried to push such research (Wellcome Library, London: Sir Julian Huxley, Shelfmark SA/EUG/C.186, Box AMS/MF/111, Reference number b16232811, Colin Bertram, General Secretary, to Huxley, 9.7.1959).

35. Ibid., Series III, Box 28, Folder 5, Mayr to Huxley, 23.10.1959.

36. Ibid., Series VI, Box 101, Folder 3, Huxley, review of Dobzhansky 1962b, *Perspectives in Biology and Medicine*, Autumn 1962, 144–148. Dobzhansky was aware of Huxley's critique and also understood Huxley's comments in the Gal-

ton Lecture (Huxley 1964, 251–280 ["Eugenics in Evolutionary Perspective"]) to be directed at his own understanding of the persisting role of natural selection in current human populations and the definition of fitness simply in terms of numbers of offspring (see Dobzhansky and Allen 1956).

37. JSH Papers, Series VI, Box 101, Folder 4: 1963, Huxley, "Units of Evolution," reprinted from *Nature* 199 (4896), 1963, 838–840, on 838.

38. Ibid., Series III, Box 34, Folder 2, Mayr to Huxley, 19.2.1963; quote from ibid., Box 32, Folder 5, Mayr to Huxley, 27.11.1961; see also Smocovitis 1996, ch. 5.

39. JSH Papers, Series III, Box 19, Folder 2, Huxley to Rensch, 31.5.1951.

40. For an elaborated treatment of the topic, see Simpson 1964, ch. 7 ("The Historical Factor in Science"). On Simpson's emphasis on paleontology as a historical science in his self-assertion vis-à-vis molecular biology, see also Aronson 2002. Mayr attempted a similar differentiation with his distinction between functional sciences that are concerned with proximate causes, such as functional anatomy and molecular biology, and sciences concerned with ultimate causes, or questions of how something came about historically, such as evolutionary biology (see, for example, Mayr 1961).

41. On the controversies between molecular and organismic approaches to human phylogeny and evolution, see also Dietrich 1998. Edna Suárez-Díaz and Victor Anaya-Muñoz (2008) have discussed what they call methodological anxiety for the history of molecular evolution, or a striving for ever greater objectivity, in the search for ever more elaborate computer algorithms to adequately reflect the mechanisms of evolution in the reconstruction of phylogenies. Objectivity in molecular evolution in general is thus linked to processes of quantification (statistics) and automation.

42. JSH Papers, Series III, Box 36, Huxley to Simpson, 17.3.1964; Simpson to Huxley, 19.3.1964.

43. Ibid., Box 41, Folder 7, Mayr to Huxley, 3.10.1967.

44. Ibid., Box 46, Folder 6, correspondence between Mayr and Juliette Huxley, 1975; Mayr and Provine 1980.

Part III

1. According to Edward Yoxen (1982), molecular biology arose from a complex of factors: a new conception of life reduced to the molecular level, which was reductionist and informational, having its beginnings in genetics (theoretical population genetics); a boom in the experimental physical sciences after World War I and even more so during and after World War II; a reorganization of science funding during this period, with the Rockefeller Foundation (and later the National Science Foundation) funding new areas of life sciences research that fit

the criteria of the hard sciences; and the new post–World War II media such as television that allowed the molecular approach to the secrets of life to be broadcast. Hans-Jörg Rheinberger (1995, 2) emphasizes the importance of new technologies of visualization and the use of new model organisms (see also Kevles and Geison 1995; for detailed histories of molecular biology see Olby [1974] 1994; Judson [1979] 1996; Kay 1993, 2000; de Chadarevian 2002).

2. The English biochemist Frederick Sanger succeeded in protein sequencing in the late 1940s, and in DNA sequencing in the 1960s, developing improved techniques in the 1970s. The American biochemist Marshall Nirenberg and the German biochemist Heinrich Matthaei "deciphered" the first "codon" in 1961, and within five years the basics of the genetic "code" were understood (Judson 1992, 52–54, 59–60).

3. Although what is referred to as *tree* in human population genetics and beyond comprises very different imagery that can diverge from the icon of the tree to a considerable degree, in the following I will use the term to refer to all "tree-like" structures without using quotation marks.

4. Interview, 22 Jan. 2013. The interviews I carried out for part 3 with long-term collaborators of Cavalli-Sforza were conducted according to the semistructured interview method (for a discussion of the advantages of this method that corresponds with my own experiences, see Barriball and While 1994). With some informants I carried out follow-up interviews, so that the overall conversation could amount to several hours, with a maximum of eight and a half hours.

Chapter 11

1. Wellcome Library, London: Cavalli, Luigi, Shelfmark HALDANE/5/2/1/119, Reference number b19936278, Cavalli-Sforza, Instituto Sieroterapico, to Haldane, University College London, 28.10.1946; Haldane to Cavalli-Sforza, 31.10.1949.

2. Galton Laboratory Serum Unit in Cambridge (1945; part of the wartime Emergency Blood Transfusion Service), Blood Group Reference Laboratory at the Lister Institute in London (1946; associated with the peacetime National Blood Transfusion service), Nuffield Blood Group Centre at Royal Anthropological Institute in London (1951), then Blood Group Reference Laboratory and Serological Population Genetics Laboratory at Saint Bartholomew's Hospital in London (1965). On Mourant's building of an international network through standardization of antisera, collection of antibody types, blood-grouping expertise, and the establishment of contacts, see Bangham 2013, ch. 4.1.

3. Wellcome Library, London: Sforza, L. L. Cavalli, Shelfmark PP/AEM/K.120, Box 33, Reference number b17736328, correspondence 1953–1963;

also ibid., Shelfmark PP/AEM/K.122: Box 33, Reference number b17740186, 1973–1974.

4. For the collecting and labeling practices within a network of colonial administration, metropolitan labs, blood banks, anthropological studies, and institutions such as transfusion depots, colonial hospitals, public health centers, and university departments, see Bangham 2013, ch. 5.

5. Wellcome Library, London: Sforza, L. L. Cavalli, Shelfmark PP/AEM/K.121, Box 33, Reference number b17736894, correspondence 1965–1973, Medical Research Council, Blood Group Reference Laboratory, London, Cavalli-Sforza to Mourant, 6.10.1965.

6. Ibid., Cavalli-Sforza to Mourant, 9.1.1971.

7. Interview by Peter Harper, 12 Oct. 2004, Genetics and Medicine Historical Network, http://www.genmedhist.info/interviews/Edwards%20A, accessed 2 Aug. 2015.

8. On likelihood see also, for example, Edwards 1969, 1974. Cavalli-Sforza and Edwards were of course not the only scientists working on such problems. Besides networks and friendships, from the beginning, there were also rivalries involved (Feldman, interview, 22 Jan. 2013). Joseph Felsenstein has especially brought Joseph H. Camin and Robert R. Sokal into the picture, who he thinks developed what they called the parsimony method independently from Cavalli-Sforza and Edwards, even though the latter were first in presenting their minimum evolution principle (or rather method) to the academic public. Felsenstein regards the parsimony method as described by Camin and Sokal as the more influential, whereas the first molecular sequence parsimony method is ascribed to Margaret Dayhoff and R. V. Eck. The distance matrix method had been published by Sokal and Charles Duncan Michener before Cavalli-Sforza and Edwards, but was most effectively publicized by Walter Fitch and Emanuel Margoliash. Finally, Felsenstein emphasizes that the way Cavalli-Sforza and Edwards introduced maximum likelihood, it was unworkable (Felsenstein 2004, 123–133). As Cavalli-Sforza also recorded, it was Felsenstein himself who solved the difficulties (Cavalli-Sforza and Piazza 1975, 127). Finally, Sokal's and Peter H. A. Sneath's *Principles of Numerical Taxonomy* (1963) was contemporary with Cavalli-Sforza and Edwards's phylogenetic tree and also on phylogeny. (For another historical account of the development of the different methods, see Edwards 1996.)

9. There has been criticism of the notion that the larger the number of proteins, genes, or nucleotides analyzed, the lesser the noise from natural selection, irregular clocks, etc. (Schwartz 2006). On the assumptions associated with phylogenetic tree building (and the differences between cladistic parsimony and phenetics), see Sober 1988.

10. The expectation was that trees built with the minimum evolution method

would approach the results of the maximum likelihood method. However, for the Brownian motion model (see below), this approximation was not very good (Edwards 1996, 85).

11. In the Italian studies, records of the Roman Catholic archives (bishoprics) were used. Most informative were the dispensations for consanguineous marriages and associated genealogies to calculate inbreeding, as well as parish books of deaths, marriages, and baptisms. Further use of the data collected in the consanguinity studies was made by analyzing surnames as genetic markers to evaluate drift and migration. Full names were available from the records but also from telephone books etc. Genetic data on blood groups was collected from nearly all villages of the Parma valley and correlated with the genealogical studies. The expected genetic variation was compared to the observed variation, and the relative role of drift versus other evolutionary factors evaluated. (For example: Did the necessity of dispensation for consanguineous marriages lower the number of such marriages? What is the connection between drift and inbreeding, which both increase with smaller population size?)

12. Bodmer wrote his PhD dissertation with Fisher at Cambridge and began at Stanford in 1961, working on Human Leucocyte Antigen (HLA) data as well as crossing-over and DNA transformation in bacteria. In 1970, he returned to England for the chair of genetics at Oxford (Cavalli-Sforza 1981; interview with Walter Bodmer by Peter Harper, 6 Dec. 2007, Genetics and Medicine Historical Network, http://www.genmedhist.info/interviews/Bodmer, accessed 2 Aug. 2015).

13. Increasingly, however, phylogenetic analyses are based on whole-mtDNA sequencing. But while technologies have allowed the refining of analyses, some research has also come to challenge the three main virtues (Pakendorf and Stoneking 2005).

14. In general, although results from molecular anthropology on primate phylogeny and divergence times were much better received in the anthropological community than twenty-five years earlier, the controversies around the molecular clocks had far from subsided (see, for example, Gibbons 1995; Lewin 1985, 1988a, 1988b).

15. Autosomal SNPs are single nucleotide polymorphisms (variations) in genetic sequences of the cellular DNA that occur at appreciable and different frequencies in populations. The universal existence of satellite DNA in the sense of highly repetitive sequences, each some four hundred base pairs long, that are repeated thousands if not millions of times in the chromosomes of eukaryotic cells was described by Roy Britten and David Kohne (1968). The analysis of the pattern of microsatellite repetitions on DNA-segments (short tandem repeats, STRs) provides a unique profile of a person and determines the haplotype (see below). (On the stabilization of satellite DNA as a phenomenon in the course of DNA hybridization research, see Suárez 2001.)

16. For an earlier presentation of the approach and some of the results see, for example, Piazza, Menozzi, and Cavalli-Sforza 1981; Cavalli-Sforza et al. 1988.

17. In fact, the discussion part following MacEachern's paper seems a fair sample of the general critical discussion (MacEachern 2000, 371–380). See also Bateman et al. 1990; Terrell and Stewart 1996; Berlan and Lewontin 1998; Clark 1998; Juengst 1998; Sims-Williams 1998; Zvelebil 1998. It should be noted that Cavalli-Sforza and collaborators have tried to tackle the issue that the way in which the populations that are to be sampled are defined has an effect on the research results (e.g., Cavalli-Sforza and Feldman 1990).

18. For the description of the technique, see Underhill et al. 1997; for the Y-Adam research, see Underhill et al. 2000; in general, see also, for example, Underhill et al. 2001; Underhill 2003; Cavalli-Sforza and Feldman 2003; Underhill and Kivislid 2007; Chiaroni, Underhill, and Cavalli-Sforza 2009; see also Stone and Lurquin 2005, 139–151.

Chapter 12

1. Several joint papers appeared before and after *Cultural Transmission*: e.g., Cavalli-Sforza and Feldman 1973, 1983a, 1983b, 1984; Feldman, Cavalli-Sforza, and Peck 1985.

2. Feldman and Cavalli-Sforza were influenced by the classical population genetic models from the first half of the twentieth century in their attempt to mathematically capture the processes of cultural transmission, even if their models were not about genetics (as Feldman explained in the interview, 22 Jan. 2013).

3. Despite the much more appreciative reception in the natural sciences, including biology and population genetics, as well as in sociology and economics, there were also reviewers from these fields who were skeptical about the general relevance of statistical models from population genetics to cultural evolution (e.g., Cloninger 1981; Charlesworth 1982; Maynard Smith and Warren 1982). Nonetheless, there has been some influential uptake also by anthropologists (see in particular Boyd and Richerson 1985), and in the twenty-first century, increased transmission took place into archeology, ethnography, and psychology (Feldman, interview, 22 Jan. 2013).

4. It is noteworthy that E. O. Wilson's *Sociobiology* (1975) and *On Human Nature* (1979) appeared in the 1970s as well, while his *Genes, Mind and Language: The Co-evolutionary Process,* coauthored with Charles J. Lumsden (Lumsden and Wilson [1981] 2005), came out the same year as *Cultural Transmission and Evolution.* In fact, Feldman reported that Wilson hurried up the publication process in order to achieve this (interview, 22 Jan. 2013). The storm that broke out around what was taken to be the bad biology and ugly politics of Wilson's sociobiology, a storm that was essentially stirred by his colleague

Lewontin at Harvard, who again was a friend and collaborator of Feldman's, therefore has to be seen as another piece in the background relief of *Cultural Transmission,* even if it had no direct influence on Feldman's work. In fact, Feldman's research interests are often purely in the realm of culture. Rather than focus on the influence of genetics on culture, with regard to altruism, for example, his team asks, "What if it is not a genetic but a culturally transmitted trait?" (Feldman, interview 2, 7 Feb. 2013; on the sociobiology controversy, see Segerstråle 2000).

5. The critical literature on "race" and genetics/genomics is abundant, especially where medical research is concerned; for a recent treatment, see Bliss 2012.

6. See also the more specialized paper "An Apportionment of Human DNA Diversity" by Guido Barbujani et al. (1997) that Cavalli-Sforza coauthored and that supported earlier results that individual variation far exceeds genetic difference between populations, as well as the clinal nature of variation (using RFLP and microsatellites). Cavalli-Sforza's practice in these matters has been inconsistent. While he would argue for giving up the term *race* altogether, he continued to use it himself (e.g., "When Did the Races of Humanity Separate?," L. L. Cavalli-Sforza and F. Cavalli-Sforza [1993] 1995, 121). The same holds true for the ways in which he referred to populations, sometimes using old anthropological denominators such as *the Lapps* and at other times denouncing such uses.

7. The discussions about the significance of genetic clustering results, especially with regard to the biological existence of "races," are ongoing (see, for example, Kopec 2014; Spencer 2014; Winther 2014; and Kaplan and Winther 2014, who present an insightful discussion of different understandings of genetic variation: diversity, differentiation, and heterozygosity).

8. Already in *Elements of Human Genetics* (1977), Cavalli-Sforza had condemned eugenics, stressing the impossibility of identifying the good traits and those that were genetically determined. Further in line with the arguments of the triple-H, we encounter the fear that any kind of eugenics would mean a reduction in variability. Complex societies depended on great individual diversity.

9. The volume *African Pygmies* edited by Cavalli-Sforza in 1986 presents studies of the demography, physical and cultural anthropology, genetics, evolution, and epidemiology of the "Pygmies." Where his interest in culture is concerned, apart from the cooperation with Feldman, Cavalli-Sforza worked on the spread of agriculture in the Neolithic on the basis of Fisher's mathematics. In cooperation with the archeologist Albert Ammerman, whom he met in 1970, he reached the conclusion that agricultural techniques were partly distributed into Europe by means of migrating people—not knowledge alone (demic diffusion model; eventually published in book format as Ammerman and Cavalli-Sforza 1984). In their analysis, genetics was not yet included. This Cavalli-Sforza later undertook in cooperation with Menozzi and Piazza for *History and Geography* (1994).

10. For a more expert discussion of the evolution of altruism, see Eshel and Cavalli-Sforza 1982; Feldman, Cavalli-Sforza, and Peck 1985.

Chapter 13

1. "The Human Genome Diversity (HGD) Project: Summary Document, Incorporating the HGD Project Outline and Development, Proposed Guidelines, and Report of the International Planning Workshop, Held in Porto Conte, Sardinia (Italy), 9–12th September 1993," http://www.osti.gov/scitech/servlets/purl/505331, accessed 2 Aug. 2015.

2. "Human Genome Diversity Workshop 1, Stanford, California, July 16–18, 1992," 1, http://www.osti.gov/scitech/servlets/purl/505330, accessed 2 Aug. 2015.

3. "The Human Genome Diversity (HGD) Project: Summary Document, Incorporating the HGD Project Outline and Development, Proposed Guidelines, and Report of the International Planning Workshop, Held in Porto Conte, Sardinia (Italy), 9–12th September 1993," 1, http://www.osti.gov/scitech/servlets/purl/505331, accessed 2 Aug. 2015.

4. The first complete human sequence was intended to be a composite being (having both X and Y) from DNA of people from the United States, European countries, and Japan; it would be a multinational and multiracial second Adam (see Kevles 1992b, 18–36; on the DNA mystique see Nelkin and Lindee [1995] 2004).

5. For reviews of Reardon 2005 by a project participant, see Greely 2005; Cavalli-Sforza 2005a; on the HGDP see also Kevles and Hood 1992; Strobel 1993; Haraway 1996, 352–355; 1997, 244–254; Knoppers, Hirtle, and Lormeau 1996; Marks 2001, 368–377.

6. "The Human Genome Diversity (HGD) Project: Summary Document, Incorporating the HGD Project Outline and Development, Proposed Guidelines, and Report of the International Planning Workshop, Held in Porto Conte, Sardinia (Italy), 9–12th September 1993," 7, http://www.osti.gov/scitech/servlets/purl/505331, accessed 2 Aug. 2015.

7. Weiss, Cavalli-Sforza, Feldman, and King, "Report of the Second Human Genome Diversity Workshop, Pennsylvania State University, October 29–31, 1992," written by organizing committee for the second workshop, http://www.osti.gov/scitech/servlets/purl/505329, accessed 2 Aug. 2015.

8. "Human Genome Diversity Project: Summary of Planning Workshop 3(B): Ethical and Human-Rights Implications," USA National Institute of Health, Bethesda, 16–18 Feb. 1993, principally authored by Henry T. Greely, 18, 22–23, http://www.osti.gov/scitech/servlets/purl/505328, accessed 2 Aug. 2015.

9. "The Human Genome Diversity Project," address delivered to a spe-

cial meeting of UNESCO, Paris: 12 Sept. 1994, 10, http://www.osti.gov/scitech/servlets/purl/505327, accessed 3 Aug. 2015.

10. Weiss, Cavalli-Sforza, Feldman, and King, "Report of the Second Human Genome Diversity Workshop, Pennsylvania State University, October 29–31, 1992," written by organizing committee for the second workshop, 5, http://www.osti.gov/scitech/servlets/purl/505329, accessed 2 Aug. 2015.

11. "The Human Genome Diversity (HGD) Project: Summary Document, Incorporating the HGD Project Outline and Development, Proposed Guidelines, and Report of the International Planning Workshop, Held in Porto Conte, Sardinia (Italy), 9–12th September 1993," 4, http://www.osti.gov/scitech/servlets/purl/505331, accessed 2 Aug. 2015.

12. "Human Genome Diversity Project: Summary of Planning Workshop 3(B): Ethical and Human-Rights Implications," USA National Institute of Health, Bethesda, 16–18 Feb. 1993, principally authored by Henry T. Greely, http://www.osti.gov/scitech/servlets/purl/505328, accessed 2 Aug. 2015. To make matters worse, the project to draw blood from "cultural isolates" and ultimately also from "more mixed populations" coincided with the public debates about HIV.

13. See also the interview by Meredith F. Small (2006, 50). Cavalli-Sforza voiced the hypothesis that indigenous rights organizations stirred the controversy to get more money from sponsors. He called opponents of the HGDP and the Genographic Project (see below) haters of biology and humanity.

14. Weiss met with the HGDP organizers, and they—among them Cavalli-Sforza, the Kidds, Piazza, Bodmer, and Feldman—applied to the NSF, the National Institutes of Health (NIH), and the Department of Energy for the funding of the three planning workshops and a conference.

15. At the first workshop about sampling methods, there were only human population geneticists ("The Human Genome Diversity Workshop 1, Stanford, California, July 16–18, 1992," http://www.osti.gov/scitech/servlets/purl/505330, accessed 2 Aug. 2015; Feldman, interview, 22 Jan. 2013). At the second workshop about sampling strategies, both biological and cultural anthropologists as well as archeologists were present (Weiss, Cavalli-Sforza, Feldman, and King, "Report of the Second Human Genome Diversity Workshop, Pennsylvania State University, October 29–31, 1992," written by organizing committee for the second workshop, http://www.osti.gov/scitech/servlets/purl/505329, accessed 2 Aug. 2015).

16. "The Human Genome Diversity (HGD) Project: Summary Document, Incorporating the HGD Project Outline and Development, Proposed Guidelines, and Report of the International Planning Workshop, Held in Porto Conte, Sardinia (Italy), 9–12th September 1993," 7, http://www.osti.gov/scitech/servlets/purl/505331, accessed 2 Aug. 2015.

17. In the aftermath of the HGDP symposium, genetic anthropology and history have not been central foci at the Wenner-Gren events. However, during

the Wenner-Gren symposium "New Directions in Kinship Study: A Core Concept Revisited," held by Sarah Franklin and Susan McKinnon in 1998 (out of which the volume *Relative Values: Reconfiguring Kinship Studies* [Franklin and McKinnon 2001] resulted), the idea for a symposium called "Anthropology in the Age of Genetics: Practice, Discourse, Critique" took shape for the following year. The aim of integration was reflected in the choice of organizers: biological anthropologist Alan H. Goodman and cultural anthropologist Deborah Heath. The symposium was judged to be a success, even if on a very small scale, in that it developed the impression that exponents of different approaches to anthropology could work with each other. As prerequisites to the cooperation across the (sub)disciplines, the bringing in of complexity, the adding of humans (or *anthropos*) to the genetics, and a true transcending of boundaries were identified (Silverman 2002, 219–220). The outcome of this meeting was the volume on *Genetic Nature/Culture: Anthropology and Science Beyond the Two-Culture Divide,* which certainly features more prominently on the reading list of social or cultural anthropologists than genetic anthropologists (Goodman, Heath, and Lindee 2003; Sommer 2012b).

18. http://www.cephb.fr/en/hgdp_panel.php, accessed 2 Aug. 2015.

19. Indeed, while the regional HGDP committee for Europe that was headed by Piazza was integrated into HUGO and was supported by the European Union starting in 1992, the North American regional committee only received some funding by 1997 (M'Charek 2005, 7; see also the introduction in general for issues raised here, and the entire book for ethnographic studies in laboratories associated with the HGDP). The HGDP started with representatives from the United States, Great Britain, Italy, Germany, India, Kenya, Japan, and China, and with workshop grants from the United States and Europe, but it was planned to become a truly global endeavor, with funding from the United Nations and the ministries of wealthy or industrialized nations worldwide (Weiss, Cavalli-Sforza, Feldman, and King, "Report of the Second Human Genome Diversity Workshop, Pennsylvania State University, October 29–31, 1992," written by organizing committee of the second workshop, 4–5, http://www.osti.gov/scitech/servlets/purl/505329, accessed 2 Aug. 2015). One informant even suggested that the move to transfer the database to CEPH after the NIH had declined to fund the project reflected badly on Cavalli-Sforza's standing with some NIH representatives. If they did not want the research done not least owing to political and ethical issues in particular related to North American tribes, then Cavalli-Sforza had tried to circumvent this veto. After that, he had more problems getting grants for his research projects. There was some money coming in from the Sorenson Molecular Genealogy Foundation in Utah, especially for Underhill's work on Y-chromosomal SNPs, but the lab was eventually closed down, "forcing" Cavalli-Sforza into retirement. In retrospect, the refusal to fund the HGDP while so much money was going into the HGP has an ironic twist, consid-

ering that genetic and genomic diversity research—particularly in biomedicine—has become *the* thing, not least with the NIH's participation in the International HapMap Project.

20. The HGDP samples had been used for important population genetic research prior to that (Rosenberg et al. 2002, which was a Structure analysis of autosomal microsatellites—see postscript).

21. "C'est pourquoi il est utile d'approfondir tant l'histoire que la géographie des objets, techniques, coutumes, valeurs, et en général, de tout ce qui fait partie des connaissances que chacun de nous acquiert au cours de sa vie et qui servent à le guider" (Cavalli-Sforza 2005b, 236).

22. "Et s'il importe de garder le souvenir et de le fixer pour qu'il ne se perde pas, ce n'est pas seulement pour des raisons sentimentales; c'est aussi parce qu'il y a beaucoup à apprendre de l'histoire. . . . Il n'est pas impossible que certains en tirent profit pour trouver des idées nouvelles qui pourraient permettre de modifier positivement notre comportement social" (ibid., 244).

23. *YCC Newsletter* 1 (1), Jan. 1994, n.p., http://web.archive.org/web/20100326211959/http://ycc.biosci.arizona.edu/newsletters/YCC_Newsletter_1.pdf, accessed 19 Jan. 2013.

24. *YCC Newsletter* 1 (2), July 1994, n.p., http://web.archive.org/web/20091116165606/http://ycc.biosci.arizona.edu/newsletters/YCC_Newsletter_2.pdf, accessed 19 Jan. 2013.

25. *YCC Newsletter* 1 (3), n.d., n.p., http://web.archive.org/web/20091116165703/http://ycc.biosci.arizona.edu/newsletters/YCC_Newsletter_3.pdf, accessed 18 Jan. 2013.

26. The advisory board provides advice and oversight on matters such as funding priorities, ethical issues, and legal compliance.

27. Similarly, even though the ethics statement of Stanford professors that Cavalli-Sforza has signed encourages the funding of studies in human genetic variation that include experts from the natural and social sciences as well as the humanities, the example then given of the importance of interdisciplinarity in genomic medicine concerns the collaboration with epidemiologists and social scientists only (Lee et al. 2008).

28. "Spencer Wells: A Family Tree for Humanity," TEDGlobal 2007, http://www.ted.com/talks/spencer_wells_is_building_a_family_tree_for_all_humanity.html, accessed 3 Aug. 2015.

29. According to Wells, a graduate student of Cavalli-Sforza explained this by the fact that men stayed at home more than women, who tended to enter new families.

30. Initially, 95 percent of the genome was thought to be without function, mostly consisting of introns and intergenic DNA. Today, some scientists suspect that parts of so-called junk DNA are involved in gene regulation, in cell-genome interaction, and possibly have as yet unknown functions.

31. https://web.archive.org/web/20120511131701/https://genographic.nationalgeographic.com/genographic/lan/en/about.html, accessed 3 Aug. 2015.

32. Genographic Project, "Ethical Framework," https://genographic.nationalgeographic.com/wp-content/uploads/2012/07/Geno2.0_Ethical-Framework.pdf, accessed 3 Aug. 2015.

33. On ethical and political issues associated with the GP and other projects, see also Harry and Kanehe 2006 (who are of the Indigenous Peoples Council on Biocolonialism); Pennisi 2005; Nash 2007; TallBear 2007; and Reardon 2008a on the HapMap Project.

34. On the continuities between the HGDP and the GP and the GP in general, see also Nash 2007, 2012; while this book went into production, a monograph on genetic geography that contains a new version of Nash 2012 appeared: Nash 2015.

35. However, with the desire for the original is associated the danger also identified by Huxley to create anthropological museums: "Today, if they still follow a traditional way of life at all, it is as part of a conscious effort, effectively making themselves into living museum pieces as a cultural statement or for tourist dollars" (Wells 2010, 188). At the same time, in the reference section for chapter 2, Wells states that "a thorough—if slightly dated—review of many aspects of hunter-gatherer societies is presented in *Man the Hunter* (Aldine, New York, 1968), the proceedings of a conference held at the University of Chicago in April 1966" (ibid., 215; Lee and DeVore 1968). Indeed, the volume has been widely criticized not only for its androcentrism but also for the use of existing human populations as models for early hominid life styles.

36. "Giving Back: The Legacy Fund," https://genographic.nationalgeographic.com/legacy-fund, accessed 3 Aug. 2015.

Chapter 14

1. "Howard University Molecular Biologist Fails to Deliver on His African Ancestry Tracking System," *Journal of Blacks in Higher Education* 29, Autumn 2000, 25–28. For a detailed history of the African Burial Ground Project and its commercial outgrowth see Nelson 2011. The African Burial Ground case also points to the many ways in which the genetic historical technologies have become involved in reconciliation projects (Nelson 2012). Another example for public-private connections is Mark Shriver's work. He received public funding for his research and developed admixture marker panels as well as an algorithm that he owns with DNA Print Genomics for determining the admixture proportion of an individual (Bliss 2008).

2. There have, for example, been attacks against the British company BritainsDNA, particularly against the marketing strategies of its managing direc-

tor Alistair Moffat (see "BritainsDNA Saga," Molecular and Cultural Evolution Lab, University College London, https://www.ucl.ac.uk/mace-lab/genetic-ancestry, accessed 3 Aug. 2015).

3. Stephanie M. Lee, "23andMe and Genentech Target Parkinson's in Deal Up to $60 Million," The Technology Chronicles, *San Francisco Chronicle*, 6 Jan. 2015, http://blog.sfgate.com/techchron/2015/01/06/23andme-and-genentech-target-parkinsons-in-deal-up-to-60-million, accessed 3 Aug. 2015. It is impossible to keep up with developments in this area in a yearlong publication process (after two years of review process): In the meantime, the FDA has given permission for a particular kind of genetic carrier test (Andrew Pollack, "F.D.A. Reverses Course on 23andMe DNA Test in Move to Ease Restrictions," *New York Times*, 19 Feb. 2015, http://mobile.nytimes.com/2015/02/20/business/fda-eases-access-to-dna-tests-of-rare-disorders.html?ref=topics&_r=1&referrer=, accessed 11 Aug. 2015).

4. The first centralized DNA sequence database was developed at the European Molecular Biology Laboratory in the 1980s (on the history of protein and DNA databases see Hagen 2010a; Strasser 2011).

5. Other issues are involved here. For example, according to Swiss legislation regarding genetic analyses of humans, samples have to be taken by physicians and the identity of the person sampled has to be verified. iGENEA customers take their samples themselves in the privacy of their homes. But iGENEA has mainly been accused of not rendering the genetic samples and data anonymous and handing both on to the American company without consent of the customer, where it enters the Family Tree data bank, again without the consent of the customer. This would constitute a breach of the protection of the privacy of data (Hostettler 2009a, 2009b, 2010).

6. The markers found in that way carry an *L* in the Y-chromosomal trees because the first such search for private markers was carried out on a Mr. Littlefield's DNA.

7. See, for example, the Haplogroup G Project at https://sites.google.com/site/haplogroupgproject/, accessed 3 Aug. 2015.

8. Underhill gave the example of using such "citizen scientist" information in his work on the Ötzi genome (Underhill, interview, 21 Jan. 2013).

9. http://www.africandna.com/history.aspx, accessed 2 Aug. 2015.

10. http://www.africandna.com/tests.aspx, accessed 2 Aug. 2015.

11. The ensuing discussion of iGENEA is informed by a semistructured interview and e-mail communication with the former managing director of Gentest.ch, by participation in the data bank–based genetic kinship service, as well as by analyses of websites, blogs, and discussion forums over several years, following the changes and witnessing the policing going on.

12. http://www.igenea.com/en/surname-projects, accessed 3 Aug. 2015.

13. There are many issues involved from the scientific point of view. Among

these are that with mtDNA or Y-chromosome tests, only one ancestor per generation is traced; the quality of the results depends on the representativeness of the database with which samples are compared; a haplotype could also be inherited from a population in which it is less common (the tests are probabilistic); and some of the regions containing autosomal ancestry-informative markers may have undergone strong selection (Bolnick et al. 2007).

14. For a discussion of historical cultures along the lines of a political, cognitive, and aesthetic dimension, see Rüsen 1994a, 211–258.

15. http://www.igenea.com/en/ancient-tribes, accessed 3 Aug. 2015.

16. http://www.igenea.com/en/index.php?c=42, accessed 24 Jan. 2013.

17. http://www.igenea.com/en/customer-reviews, accessed 3 Aug. 2015.

18. "Vielen Dank für ihre reichlichen Antworten und für mein 'Wunschergebnis.' Ist irgendwie komisch. Seit ich ungefähr 20 (Nun38) bin zog es mich in den Norden. . . . Mein Interesse wuchs immer mehr Richtung Wikinger .Hab mir massig Literatur angeschaft und irgendwie stellte sich eine komische Vertrautheit an alten skandinavischen Stätten ein. Vielleicht speichern die Gene mehr als wir wissen" (https://www.igenea.com/de/forum/d/wikinger-dna-isolation/147, accessed 3 Aug. 2015).

19. "In einer Welt der beinahe unlimitierten Vernetzung von Personen, der Globalisierung, der Kosmopoliten" (Apter 2008).

20. "Ja, ein Mazedonier kann schon sagen, dass er ursprünglich ein Makedonier ist. . . . Erst die Analyse Ihrer DNA kann uns eine absolut sichere Antwort liefern" (posted on 15 Feb. 2008 on the online forum "Ex-Jugoslawien/Mazedonien/Serbien/Kroatien/Albanien/Montenegro/Bosnien," subsequently included at https://www.igenea.com/de/forum/d/ex-jugoslawien-mazedonien-serbien-kroatien-albanien-montenegro-bosnien/25, accessed 3 Aug. 2015).

21. On the narrativization and mediatization of genetic history, the associated old and new myths and brands, as well as the politics of identity concerning the science and its commercialization, see also Sommer 2007a, 269–271; 2009; 2012a; 2012c; for the case of "Celtic-ness," see Sommer 2008b.

22. However, in 2008, Cavalli-Sforza was one of the signatories of an open letter to *Genome Biology* titled "The Ethics of Characterizing Difference: Guiding Principles on Using Racial Categories in Human Genetics" (Lee et al. 2008). They cautioned that in the genetic determination of an individual's biogeographical ancestry only part of his or her ancestry is analyzed (for example in Y-chromosomal analysis)—an information that they would like to see communicated to genetic ancestry customers—and that the cultural ancestry is just as important as the genetic. They also urged researchers to explain how individual samples are attributed to ethnic or racial categories, why such categories are used, and whether they are research variables.

23. On the HGDP and for an analysis of *The Journey of Man* documentary, particularly the ways in which the differences between "indigenous myths" and

"scientific evidence" are played out, see also Wald 2006, 318–333. On the genetic reconstruction of the category of the Khoisan as "Edenic origin of humankind" more generally, see Erasmus 2013. Most interestingly, Zimitri Erasmus mentions in passing that on the basis of "their genetic ancestry" that has confirmed "aboriginality," the Khoisan have claimed First Nation status. Population genetic research and its commercialization has therefore often very complex if not paradoxical effects between subjection to research and genetic definition on the one hand, and the opening up of a subject position that comes with novel power and possibilities on the other (Sommer 2010f).

24. Hans-Jörg Rheinberger (1997) distinguishes epistemic from technical things or objects. The technical objects are the experimental conditions that embed the epistemic objects. They simultaneously embody local and temporal research contexts and participate in the shaping of epistemic things. Conversely, once stabilized sufficiently, an epistemic object might take on the role of a technical condition.

25. See, for example, http://www.kracke.org/2008/02/and-here-are-the-results, accessed 3 Aug. 2015. Benedict Anderson ([1983] 2006, 154) emphasizes the importance of a shared language for the formation of imagined communities: "Through that language, encountered at mother's knee and parted with only at the grave, pasts are restored, fellowships are imagined, and futures dreamed." On the GP see also Nash 2012; 2015, ch. 2 (which just appeared as this book went into production), where the tension between the liberal antiracism as well as an emphasis on multiculturalism on the one hand and the (re-)creation of difference and the performance of postcolonial narratives on the other, too, is carved out.

Chapter 15

1. These are Azienda Speciale PalaExpo and Codice; Idee per la Cultura; Istituto Geografico DeAgostini; the Institute for Human Evolution, University of the Witwatersrand; the Veneto Institute of Science, Literature and Arts; ISITA—Italian Institute of Anthropology; the Department of Science and Technology of South Africa; the Giuseppe Sergi Museum of Anthropology (Rome); and the Tridentine Museum of Natural Sciences (Trento).

2. www.homosapiens.net, accessed 2 Aug. 2015.

3. Teacher's guide, 6, http://www.homosapiens.net/didattica/il-dossier-pedagogico/lang/en/, accessed 2 Aug. 2015.

4. The semistructured interviews were conducted during two days with departing visitors who were willing to communicate (the reasons for choosing this method are detailed by Barriball and While 1994).

Postscript

1. The criticism from within science as well as the humanities is manifold. On Richard Dawkins and Edward O. Wilson see, for example, Ruse 1999, 122–134 and 172–193 respectively; in general, see Lewontin, Rose, and Kamin 1984; Kitcher 1985; Weber 2005, 182–210.

2. See Scully, King, and Brown 2013 on Y-chromosomal testing for Viking origins.

3. http://shop.nationalgeographic.com/ngs/browse/productDetail.jsp?productId=2001246&gsk&code=MR20936, accessed 2 Aug. 2015.

4. For a summary of the synthesis of knowledge from different fields based on these assumptions to arrive at a picture of "the great human expansion," see Brenna, Cavalli-Sforza, and Feldman 2012.

5. For a critical discussion of the program, see Bolnick 2008; on different Bayesian algorithm models that do and do not take into account admixture and geographic information in the determination of population structure, see François and Durand 2010.

References

Abir-Am, Pnina G. 1999. "The First American and French Commemorations in Molecular Biology: From Collective Memory to Comparative History." *Osiris, 2nd Series* 14: 324–372.

Abu El-Haj, Nadia. 2004. " 'A Tool to Recover Past Histories': Genealogy and Identity after the Genome." *IAS, School of Social Science, Occasional Paper* 19. Accessed 2 July 2015. http://www.sss.ias.edu/files/papers/paper19.pdf.

——. 2012. *The Science of Genealogy: The Search for Jewish Origins and the Politics of Epistemology*. Chicago: University of Chicago Press.

Adams, Mark B. 1994. "Theodosius Dobzhansky in Russia and America." In *The Evolution of Theodosius Dobzhansky: Essays on His Life and Thought in Russia and America*, edited by Mark B. Adams, 3–28. Princeton, NJ: Princeton University Press.

Allen, Arthur. 2014. "Charging into the Minefield of Genes and Racial Difference: Nicholas Wade's 'A Troublesome Inheritance.' " *New York Times*, 15 May. Accessed 2 July 2015. http://www.nytimes.com/2014/05/16/books/nicholas-wades-a-troublesome-inheritance.html.

Allen, Garland E. 1978. *Thomas Hunt Morgan: The Man and His Science*. Princeton, NJ: Princeton University Press.

——. 1980. "The Evolutionary Synthesis: Morgan and Natural Selection Revisited." In *The Evolutionary Synthesis: Perspectives on the Unification of Biology*, edited by Ernst Mayr and William B. Provine, 356–382. Cambridge, MA: Harvard University Press.

——. 1986. "The Eugenics Record Office at Cold Spring Harbor, 1910–1940: An Essay in Institutional History." *Osiris, 2nd Series* 2: 225–264.

——. 1992. "Julian Huxley and the Eugenical View of Human Evolution." In *Julian Huxley: Biologist and Statesman of Science. Proceedings of a Conference Held at Rice University 25–27 September 1987*, edited by C. Kenneth Waters and Albert Van Helden, 193–222. Houston: Rice University Press.

Ammerman, Albert, and Luigi Luca Cavalli-Sforza. 1984. *The Neolithic Tran-*

sition and the Genetics of Populations in Europe. Princeton, NJ: Princeton University Press.

AMNH (American Museum of Natural History). 1927. *The American Museum School Service: Fifty-Eighth Annual Report of the Trustees for the Year 1926*. New York: AMNH.

Anastas, Benjamin, and Luigi Luca Cavalli-Sforza. 2002. "Genes, Peoples, and Languages: An Interview with Luca Cavalli-Sforza." *Grand Street* 70 ("Against Nature"): 188–194.

Anderson, Benedict. (1983) 2006. *Imagined Communities: Reflections on the Origin and Spread of Nationalism*. London: Verso.

——. 1992. *Long-Distance Nationalism: World Capitalism and the Rise of Identity Politics*. Amsterdam: University of Amsterdam Centre for Asian Studies.

Andrews, Roy Chapman. 1921. *Across the Mongolian Plains: A Naturalist's Account of China's "Great Northwest."* New York: Appleton.

——. 1926. *On the Trail of Ancient Man: A Narrative of the Field Work of the Central Asiatic Expeditions*. With an introduction and a chapter by Henry Fairfield Osborn; with 58 photographs by J. B. Shackelford. New York: Putnam.

——. 1929. *Ends of the Earth*. New York: Putnam.

——. 1932. *The New Conquest of Central Asia: A Narrative of the Explorations of the Central Asiatic Expeditions in Mongolia and China, 1921–1930*. New York: AMNH. http://archive.org/details/newconquestofcen00andr.

Andrews, Roy Chapman, and Yvette Borup Andrews. 1918. *Camps and Trails in China: A Narrative of Exploration, Adventure and Sport in Little-Known China*. New York: Appleton.

Anker, Peder. 2001. *Imperial Ecology: Environmental Order in the British Empire, 1895–1945*. Cambridge, MA: Harvard University Press.

Apter, Joëlle. 2008. "Dank DNA-Test genetische Cousins finden," interview by Dania Zafran, *tachles* 8 (9), 29 Feb. Accessed 2 July 2015. https://www.igenea.com/docs/tachles.htm.

Armytage, W. H. G. *1989*. "The First Director-General of UNESCO." In *Evolutionary Studies: A Centenary Celebration of the Life of Julian Huxley. Proceedings of the Twenty-Fourth Annual Symposium of the Eugenics Society, London, 1987*, edited by Milo Keynes and G. Ainsworth Harrison, 186–193. Basingstoke: Macmillan.

Aronson, Jay D. 2002. " 'Molecules and Monkeys': George Gaylord Simpson and the Challenge of Molecular Evolution." *History and Philosophy of the Life Sciences* 24 (3–4): 441–465.

Assmann, Aleida. 1999. *Erinnerungsräume: Funktionen und Wandlungen des kulturellen Gedächtnisses*. Munich: Beck.

——. 2002. "Gedächtnis als Leitbegriff der Kulturwissenschaften." In *Kultur-*

wissenschaften. Forschung—Praxis—Positionen, edited by Lutz Musner and Gotthart Wunberg, 27–45. Vienna: WUV.

———. 2006. *Der lange Schatten der Vergangenheit: Erinnerungskultur und Geschichtspolitik*. Munich: Beck.

Assmann, Jan. 1992. *Das kulturelle Gedächtnis: Schrift, Erinnerung und politische Identität in frühen Hochkulturen*. Munich: Beck.

Atkinson, James W. 1985. "E. G. Conklin on Evolution: The Popular Writings of an Embryologist." *Journal of the History of Biology* 18 (1): 31–35.

Bachmann-Medick, Doris. 2006. *Cultural Turns: Neuorientierungen in den Kulturwissenschaften*. Reinbek bei Hamburg: Rowohlt.

Baker, John Randal. 1976. "Julian Sorell Huxley." *Bibliographical Memoirs of Fellows of the Royal Society* 22: 206–238.

Bal, Mieke. 1992. "Telling, Showing, Showing Off." *Critical Inquiry* 18: 556–594.

Ballou, William H. 1897. "Strange Creatures of the Past: Gigantic Saurians of the Reptilian Age." *Century Magazine* 55 (1): 15–23.

Bangham, Jenny. 2013. "Blood Groups and the Rise of Human Genetics in Mid-Twentieth Century Britain." PhD diss., University of Cambridge.

Barbujani, Guido, Arianna Magagni, Eric Minch, and Luigi Luca Cavalli-Sforza. 1997. "An Apportionment of Human DNA Diversity." *Proceedings of the National Academy of Sciences* 94: 4516–4519.

Barkan, Elazar. 1992a. "The Dynamics of Huxley's Views on Race and Eugenics." In *Julian Huxley: Biologist and Statesman of Science. Proceedings of a Conference Held at Rice University 25–27 September 1987*, edited by C. Kenneth Waters and Albert Van Helden, 230–237. Houston: Rice University Press.

———. 1992b. *The Retreat of Scientific Racism: Changing Concepts of Race in Britain and the United States between the World Wars*. Cambridge: Cambridge University Press.

Barriball, K. Louise, and Alison While. 1994. "Collecting Data Using a Semi-structured Interview: A Discussion Paper." *Journal of Advanced Nursing* 19: 328–335.

Barrington-Johnson, John. 2005. *The Zoo: The Story of London Zoo*. London: Robert Hale.

Barrow, Mark V. Jr. 2009. *Nature's Ghosts: Confronting Extinction from the Age of Jefferson to the Age of Ecology*. Chicago: University of Chicago Press.

Bartley, Mary M. 1995. "Courtship and Continued Progress: Julian Huxley's Studies on Bird Behavior." *Journal of the History of Biology* 28 (1): 91–108.

Bateman, Richard, Ives Goddard, Richard O'Grady, V. A. Funk, Rich Mooi, W. John Kress, and Peter Cannell. 1990. "Speaking of Forked Tongues: The Feasibility of Reconciling Human Phylogeny and the History of Language." *Current Anthropology* 31: 1–24.

Beatty, John. 1994. "Dobzhansky and the Biology of Democracy: The Moral and Political Significance of Genetic Variation." In *The Evolution of Theodosius Dobzhansky. Essays on His Life and Thought in Russia and America*, edited by Mark B. Adams, 195–218. Princeton, NJ: Princeton University Press.

Becker, Carl. 1932. "Everyman His Own Historian." *American Historical Review* 37 (2): 221–236.

Bederman, Gail. 1995. *Manliness and Civilization: A Cultural History of Gender and Race in the United States, 1880–1917*. Chicago: University of Chicago Press.

Behar, Doron M., Saharon Rosset, Jason Blue-Smith, Oleg Balanovsky, Shay Tzur, David Comas, R. John Mitchell, Lluis Quintana-Murci, Chris Tyler-Smith, and R. Spencer Wells. 2007. "The Genographic Project Public Participation Mitochondrial DNA Database." *PLoS Genetics* 3 (6): 1083–1095.

Bender Shetler, Jan. 2007. *Imagining Serengeti: A History of Landscape Memory in Tanzania from Earliest Times to the Present*. Athens: Ohio University Press.

Benjamin, Walter. 1974. *Illuminationen: Ausgewählte Schriften*. Vol. 1. Frankfurt am Main: Suhrkamp.

Bennett, Tony. 2004. *Pasts beyond Memory: Evolution, Museums, Colonialism*. London: Routledge.

Bergson, Henri. (1907) 1911. *Creative Evolution*. New York: Holt.

Berlan, Jean-Pierre, and Richard C. Lewontin. 1998. "La menace du complexe génético-industriel." *Le Monde Diplomatique* 12: 1, 22–23.

Beurton, John Peter. 1999. "Was ist die Synthetische Theorie?" In *Die Entstehung der Synthetischen Theorie: Beiträge zur Geschichte der Evolutionsbiologie in Deutschland 1930–1950*, edited by Thomas Junker and Eve-Marie Engels, 79–105. Berlin: VWB.

Blavatsky, Helena P. 1888. *The Secret Doctrine: The Synthesis of Science, Religion, and Philosophy*. 2 vols. London: Theosophical Publishing Company.

Bliss, Catherine. 2008. "Mapping Race through Admixture." *International Journal of Technology, Knowledge and Society* 4 (4): 79–83.

——. 2012. *Race Decoded: The Genomic Fight for Social Justice*. Stanford, CA: Stanford University Press.

Blue, Gregory. 2001. "Scientific Humanism at the Founding of UNESCO." In *Comparative Criticism*. Vol. 23, *Humanist Traditions in the Twentieth Century*, edited by E. S. Shaffer, 173–200. Cambridge: Cambridge University Press.

Boas, Franz. 1901. "The Mind of Primitive Man." *Science, New Series* 13 (321): 281–289.

Bock, Walter J. 2005. "Ernst Mayr at 100: A Life Inside and Outside Ornithology." In *Ernst Mayr at 100: Ornithologist and Naturalist*, edited by W. J. Bock and M. R. Lein, 2–16. Washington DC: American Ornithologists' Union.

Bodmer, Walter Fred, and Luigi Luca Cavalli-Sforza. 1976. *Genetics, Evolution, and Man*. San Francisco: W. H. Freeman.

Bolnick, Deborah A. 2008. "Individual Ancestry Inference and the Reification of Race as a Biological Phenomenon." In *Revisiting Race in a Genomic Age*, edited by Barbara A. Koenig, Sandra Soo-Jin Lee, and Sarah S. Richardson, 70–85. New Brunswick, NJ: Rutgers University Press.

Bolnick, Deborah A., Duana Fullwiley, Troy Duster, Richard S. Cooper, Joan H. Fujimura, Jonathan Kahn, Jay S. Kaufman, Jonathan Marks, Ann Morning, Alondra Nelson, Pilar Ossorio, Jenny Reardon, Susan M. Reverby, and Kimberly TallBear. 2007. "The Science and Business of Genetic Ancestry Testing." *Science* 318 (5849): 399–400.

Bonné-Tamir, Batsheva, M. J. Johnson, A. Natali, D. C. Wallace, and Luigi Luca Cavalli-Sforza. 1986. "Human Mitochondrial DNA Types in Two Israeli Populations—A Comparative Study at the DNA Level." *American Journal of Human Genetics* 38: 341–351.

Boule, Marcellin. 1908. "L'homme fossile de La Chapelle-aux-Saints (Corrèze)." *L'Anthropologie* 19: 519–525.

Bowcock, Anne M., Andres Ruiz Linares, James Tomfohrde, Eric Minch, Judith R. Kidd, and Luigi Luca Cavalli-Sforza. 1994. "High Resolution of Human Evolutionary Trees with Polymorphic Microsatellites." *Nature* 368: 455–457.

Bowker, Geoffrey C. 2005. *Memory Practices in the Sciences*. Cambridge, MA: MIT Press.

Bowler, Peter J. 1983. *The Eclipse of Darwinism: Anti-Darwinian Evolution Theories in the Decades Around 1900*. Baltimore: Johns Hopkins University Press.

———. (1983) 1989. *Evolution: The History of an Idea*. Berkeley: University of California Press.

———. 1986. *Theories of Human Evolution: A Century of Debate, 1844–1944*. Baltimore: Johns Hopkins University Press.

———. 1988. *The Non-Darwinian Revolution: Reinterpreting a Historical Myth*. Baltimore: Johns Hopkins University Press.

———. 1989. *The Mendelian Revolution: The Emergence of Hereditary Concepts in Modern Science and Society*. London: Athlone.

———. 2007. *Monkey Trials and Gorilla Sermons: Evolution and Christianity from Darwin to Intelligent Design*. Cambridge, MA: Harvard University Press.

———. 2009. *Science for All: The Popularization of Science in Early Twentieth-Century Britain*. Chicago: University of Chicago Press.

Boyd, Robert, and Peter J. Richerson. 1985. *Culture and the Evolutionary Process*. Chicago: University of Chicago Press.

Boyd, William C. 1939. "Blood Groups." *Tabulae Biologicae* 17: 113–240.

———. 1952. *Genetics and the Races of Man*. New York: Little, Brown.

——. 1963. "Genetics and the Human Races." *Science, New Series* 140 (3571): 1057–1064.

Brattain, Michelle. 2007. "Race, Racism, and Antiracism: UNESCO and the Politics of Presenting Science to the Postwar Public." *American Historical Review* 112 (5): 1386–1413.

Braun, Lundy, and Evelynn Hammonds. 2012. "The Dilemma of Classification: The Past in the Present." In *Genetics and the Unsettled Past: The Collision of DNA, Race, and History*, edited by Keith Wailoo, Alondra Nelson, and Catherine Lee, 67–80. New Brunswick, NJ: Rutgers University Press.

Brinkman, Paul D. 2010. *The Second Jurassic Dinosaur Rush: Museums and Paleontology at the Turn to the Twentieth Century*. Chicago: University of Chicago Press.

Britten, Ron Joy, and David Kohne. 1968. "Repeated Sequences in DNA." *Science* 161 (3841): 529–540.

Brodwin, Paul. 2005. " 'Bioethics in Action' and Human Population Genetics Research." *Culture, Medicine and Psychiatry* 29: 145–178.

Brown, Wesley M. 1980. "Polymorphism in Mitochondrial DNA of Humans as Revealed by Restriction Endonuclease Analysis." *Proceedings of the National Academy of Sciences* 77 (6): 3605–3609.

Brown, Wesley M., Matthew George Jr., and Allan Charles Wilson. 1979. "Rapid Evolution of Animal Mitochondrial DNA." *Proceedings of the National Academy of Sciences* 76 (4): 1967–1971.

Browne, Janet. 2005. "Commemorating Darwin." *British Journal for the History of Science* 38 (3): 251–274.

Bruyninckx, Joeri. 2013. "Sound Science: Recording and Listening in the Biology of Bird Song, 1880–1980." PhD diss., Maastricht University.

Burke, Peter. (1991) 2001. "Overture: The New History: Its Past and Its Future." In *New Perspectives on Historical Writing*, edited by Peter Burke, 1–24. University Park: Pennsylvania State University Press.

——. 1997. *Varieties of Cultural History*. Cambridge, UK: Polity.

Burkhardt, Richard W. 1980. "Lamarckism in Britain and the United States." In *The Evolutionary Synthesis: Perspectives on the Unification of Biology*, edited by Ernst Mayr and William B. Provine, 343–352. Cambridge, MA: Harvard University Press.

——. 2005. *Patterns of Behavior: Konrad Lorenz, Niko Tinbergen, and the Founding of Ethology*. Chicago: University of Chicago Press.

Burroughs, Edgar Rice. Oct. 1912. "Tarzan of the Apes: A Romance of the Jungle." *All-Story Magazine*: 241–372.

——. Sept. 1918a. "The Land That Time Forgot." *Blue Book* 27 (4).

——. Oct. 1918b. "The People That Time Forgot." *Blue Book* 27 (6).

——. Nov. 1918c. "Out of Time's Abyss." *Blue Book* 28 (2).

——. 1930. *Tarzan at the Earth's Core*. New York: Grosset and Dunlap.

Burt, Jonathan. 2009. "Invisible Histories: Primate Bodies and the Rise of Posthumanism in the Twentieth Century." In *Animal Encounters*, edited by Tom Tyler and Manuela Rossini, 159–170. Leiden: Brill.

Butler, Samuel. 1880. *Unconscious Memory*. London: David Bogue.

Cain, Joe. 2010. "Julian Huxley, General Biology and the London Zoo, 1935–42." *Notes and Records of the Royal Society of London* 64: 359–378.

Cain, Victoria E. M. 2010. " 'The Direct Medium of the Vision': Visual Education, Virtual Witnessing and the Prehistoric Past at the American Museum of Natural History, 1890–1923." *Journal of Visual Culture* 9: 284–303.

Cann, Howard M., Claudia de Toma, Lucien Cazes, Marie-Fernande Legrand, Valerie Morel, Laurence Piouffre, Julia Bodmer, Walter F. Bodmer, Batsheva Bonné-Tamir, Anne Cambon-Thomsen, Zhu Chen, Jiayou Chu, Carlo Carcassi, Licinio Contu, Ruofu Du, Laurent Excoffier, G. B. Ferrara, Jonathan S. Friedlaender, Helena Groot, David Gurwitz, Trefor Jenkins, Rene J. Herrera, Xiaoyi Huang, Judith Kidd, Kenneth K. Kidd, Andre Langaney, Alice A. Lin, S. Qasim Mehdi, Peter Parham, Alberto Piazza, Maria Pia Pistillo, Yaping Qian, Qunfang Shu, Jiujin Xu, S. Zhu, James L. Weber, Henry T. Greely, Marcus W. Feldman, Gilles Thomas, Jean Dausset, and Luigi Luca Cavalli-Sforza. 2002. "A Human Genome Diversity Cell Line Panel." *Science, New Series* 296 (5566): 261–262.

Cann, Rebecca L. 1988. "DNA and Human Origins." *Annual Review of Anthropology* 17: 127–143.

———. 1997. "Mothers, Labels, and Misogyny." In *Women in Human Evolution*, edited by Lori D. Hager, 75–90. London: Routledge.

Cann, Rebecca L., Mark Stoneking, and Allan Charles Wilson. 1987. "Mitochondrial DNA and Human Evolution." *Nature* 325: 31–36.

Cantor, Geoffrey. 1997. "Charles Singer and the Early Years of the British Society for the History of Science." *British Journal for the History of Science* 30 (1): 5–23.

Caprara, Gian Vittorio, Corrado Fagnani, Guido Alessandri, Patrizia Steca, Antonella Gigantesco, Luigi Luca Cavalli-Sforza, and Maria Antonietta Stazi. 2009. "Human Optimal Functioning: The Genetics of Positive Orientation Towards Self, Life, and the Future." *Behavior Genetics* 39 (3): 277–284.

Carruthers, Jane. 1995. *The Kruger National Park: A Social and Political History*. Pietermaritzburg: UKZN Press.

Casanova, Myriam, Pascale Leroy, Chafika Boucekkine, Jean Weissenbach, Colin Bishop, Marc Fellous, Michele Purrello, Gianmario Fiori, and Marcello Siniscalco. 1985. "A Human Y-linked DNA Polymorphism and Its Potential for Estimating Genetic and Evolutionary Distance." *Science* 230 (4732): 1403–1406.

Cassirer, Ernst. (1942) 2007. "Zur Logik der Kulturwissenschaften: Fünf Studien 1942." In *Aufsätze und kleine Schriften (1941–1946)*, by Ernst Cassirer, edited by Claus Rosenkranz, 357–486. Hamburg: Felix Meiner.

Cavalli-Sforza, Francesco, and Luigi Luca Cavalli-Sforza. (1997) 2000. *Vom Glück auf Erden: Antworten auf die Frage nach dem guten Leben.* Translated by Renate Heimbucher. Reinbek bei Hamburg: Rowohlt.

Cavalli-Sforza, Luigi Luca. 1966. "Population Structure and Human Evolution." *Proceedings of the Royal Society of London, Series B: Biological Sciences* 164 (995): 362–379.

——. 1973. "Some Current Problems of Human Population Genetics." *American Journal of Human Genetics* 25: 82–104.

——. 1977. *Elements of Human Genetics.* Menlo Park, CA: W. A. Benjamin.

——. 1981. "The William Allan Memorial Award." *American Journal of Human Genetics* 33: 659–663.

——, ed. 1986. *African Pygmies.* Orlando, FL: Academic Press.

——. 1992. "Forty Years Ago in Genetics: The Unorthodox Mating Behavior of Bacteria." *Genetics* 132 (3): 635–637.

——. (1996) 1999. *Gene, Völker und Sprachen: Die biologischen Grundlagen unserer Zivilisation.* Translated by Günter Memmert. Munich: Hanser.

——. (1996) 2000. *Genes, Peoples and Languages.* Harmondsworth: Allen Lane.

——. 1997. "Genes, Peoples, and Languages." *Proceedings of the National Academy of Sciences* 94 (15): 7719–7724.

——. 1998. "The DNA Revolution in Population Genetics." *Trends in Genetics* 14: 60–65.

——. 2005a. "Studying Diversity." *European Molecular Biology Organization Reports* 6 (8): 713.

——. 2005b. *Évolution biologique, évolution culturelle.* Translated by Jacqueline Henry. Paris: Odile Jacob.

——. 2005c. "The Human Genome Diversity Project: Past, Present and Future." *Nature Reviews Genetics* 6: 333–340.

——. 2009a. "Acceptance Speech (Berne—16 November, 1999)." In *Luigi Luca Cavalli-Sforza: 1999 Balzan Prize for Science of Human Origins; Excerpts from the Publications; Balzan Prizes 1999 (Enlarged and Revised Edition, 2009); Meeting the Challenges of the Future; A Discussion between "The Two Cultures" (Leo S. Olschki Editore, 2003)*, edited by International Balzan Foundation, 6–10. Milan: Fondazione Internazionale Balzan. Accessed 2 July 2015. http://www.balzan.org/upload/EstrattoCavalli_ING.pdf.

——. 2009b. "A Comprehensive Outline of My Research (1999—revised July 2009)." In *Luigi Luca Cavalli-Sforza: 1999 Balzan Prize for Science of Human Origins; Excerpts from the Publications; Balzan Prizes 1999 (Enlarged and Revised Edition, 2009); Meeting the Challenges of the Future; A Discussion between "The Two Cultures" (Leo S. Olschki Editore, 2003)*, edited by International Balzan Foundation, 10–24. Milan: Fondazione Internazionale Balzan. Accessed 2 July 2015. http://www.balzan.org/upload/EstrattoCavalli_ING.pdf.

——. 2009c. "The Developments of My Research after the Balzan Prize (1999–2009)." In *Luigi Luca Cavalli-Sforza: 1999 Balzan Prize for Science of Human Origins; Excerpts from the Publications; Balzan Prizes 1999 (Enlarged and Revised Edition, 2009); Meeting the Challenges of the Future; A Discussion between "The Two Cultures" (Leo S. Olschki Editore, 2003)*, edited by International Balzan Foundation, 25–28. Milan: Fondazione Internazionale Balzan. Accessed 2 July 2015. http://www.balzan.org/upload/EstrattoCavalli_ING.pdf.

——. 2009d. "Are There Limits to Knowledge? (Balzan Symposium 2002: Meeting the Challenges of the Future: A Discussion between 'The Two Cultures')." In *Luigi Luca Cavalli-Sforza; 1999 Balzan Prize for Science of Human Origins; Excerpts from the Publications; Balzan Prizes 1999 (Enlarged and Revised Edition, 2009); Meeting the Challenges of the Future; A Discussion between "The Two Cultures" (Leo S. Olschki Editore, 2003)*, edited by International Balzan Foundation, 29–35. Milan: Fondazione Internazionale Balzan. Accessed 2 July 2015. http://www.balzan.org/upload/EstrattoCavalli_ING.pdf.

——, ed. 2009–2010. *La cultura Italiana*. Turin: UTET.

——. (2010) 2011. *L'aventure de l'espèce humaine: De la génétique des populations à l'évolution culturelle*. Translated by Pierre Savy. Paris: Odile Jacob.

Cavalli-Sforza, Luigi Luca, and Walter Fred Bodmer. 1971. *The Genetics of Human Populations*. San Francisco: W. H. Freeman.

Cavalli-Sforza, Luigi Luca, and Francesco Cavalli-Sforza. 1993. *Chi siamo: La storia della diversità umana*. Milano: Mondadori.

——. (1993) 1995. *The Great Human Diasporas: The History of Diversity and Evolution*. Translated by Sarah Thorne. Reading, MA: Addison-Wesley.

——. 2005. *La Génétique des populations: Histoire d'une découverte*. Translated by Marilène Raiola. Paris: Odile Jacob.

Cavalli-Sforza, Luigi Luca, and Anthony W. F. Edwards. 1965. "Analysis of Human Evolution." In *Genetics Today: Proceedings of the XI. International Congress of Genetics, the Hague, the Netherlands, September 1963*. Vol. 3, edited by S. J. Geerts, 923–933. Oxford: Pergamon.

——. 1967. "Phylogenetic Analysis: Models and Estimation Procedures." *American Journal of Human Genetics* 19 (3): 233–257.

Cavalli-Sforza, Luigi Luca, and Marcus W. Feldman. 1973. "Cultural versus Biological Inheritance: Phenotypic Transmission from Parents to Children (A Theory of the Effect of Parental Phenotypes on Children's Phenotypes)." *American Journal of Human Genetics* 25 (6): 618–637.

——. 1981. *Cultural Transmission and Evolution: A Quantitative Approach*. Princeton, NJ: Princeton University Press.

——. 1983a. "Paradox of the Evolution of Communication and of Social Interactivity." *Proceedings of the National Academy of Sciences* 80: 2017–2021.

——. 1983b. "Cultural versus Genetic Adaptation." *Proceedings of the National Academy of Sciences* 80: 4993–4996.

——. 1984. "Cultural and Biological Evolutionary Processes: Gene-Culture Disequilibrium." *Proceedings of the National Academy of Sciences* 81: 1604–1607.

——. 1990. "Spatial Subdivision of Populations and Estimates of Genetic Variance." *Theoretical Population Biology* 37 (1): 3–25.

——. 2003. "The Application of Molecular Genetic Approaches to the Study of Human Evolution." *Nature Genetics* 33: 266–275.

Cavalli-Sforza, Luigi Luca, Paolo Menozzi, and Alberto Piazza. 1994. *The History and Geography of Human Genes*. Princeton, NJ: Princeton University Press.

Cavalli-Sforza, Luigi Luca, Antonio Moroni, and Gianna Zei. 2004. *Consanguinity, Inbreeding, and Drift in Italy*. Princeton, NJ: Princeton University Press.

Cavalli-Sforza, Luigi Luca, and Alberto Piazza. 1975. "Analysis of Evolution: Evolutionary Rates, Independence and Treeness." *Theoretical Population Biology* 8: 127–165.

Cavalli-Sforza, Luigi Luca, Alberto Piazza, Paolo Menozzi, and Joanna Mountain. 1988. "Reconstruction of Human Evolution: Bringing Together Genetic, Archaeological, and Linguistic Data." *Proceedings of the National Academy of Sciences* 85: 6002–6006.

Cavalli-Sforza, Luigi Luca, and Telmo Pievani. 2011. *Homo Sapiens: La grande storia della diversità umana*. Turin: Codice edizioni and Rome: Azienda Speciale PalaExpo.

Cavalli-Sforza, Luigi Luca, Allan Charles Wilson, Charles R. Cantor, Robert M. Cook-Deegan, and Mary-Claire King. 1991. "Call for a World-Wide Survey of Human Genetic Diversity: A Vanishing Opportunity for the Human Genome Project." *Genomics* 11: 490–491.

Cell, John W. 1989. "Lord Hailey and the Making of the African Survey." *African Affairs* 88 (353): 481–505.

de Chadarevian, Soraya. 2002. *Designs for Life: Molecular Biology after World War II*. Cambridge: Cambridge University Press.

de Chadarevian, Soraya, and Harmke Kamminga, ed. 1998. *Molecularizing Biology and Medicine: New Practices and Alliances, 1910s–1970s*. Amsterdam: OPA.

Charlesworth, Brian. 1982. "Cultural and Biological Evolution," review of *Cultural Transmission and Evolution: A Quantitative Approach*, by Luigi Luca Cavalli-Sforza and Marcus W. Feldman. *Quarterly Review of Biology* 57 (3): 300–302.

Chiaroni, Jacques, Peter A. Underhill, and Luigi Luca Cavalli-Sforza. 2009.

"Y Chromosome Diversity, Human Expansion, Drift, and Cultural Evolution." *Proceedings of the National Academy of Sciences* 106 (48): 20174–20179.

Chow-White, Peter A., and Miguel García-Sancho. 2012. "Bidirectional Shaping and Spaces of Convergence: Interactions between Biology and Computing from the First DNA Sequencers to Global Genome Databases." *Science, Technology and Human Values* 37 (1): 124–164.

Clark, Constance Areson. 2001. "Evolution for John Doe: Pictures, the Public, and the Scopes Trial Debate." *Journal of American History* 87 (4): 1275–1303.

——. 2008. *God—or Gorilla: Images of Evolution in the Jazz Age*. Baltimore: Johns Hopkins University Press.

Clark, Geoffrey Anderson. 1998. "Multivariate Pattern Searches, the Logic of Inference, and European Prehistory: A Comment on Cavalli-Sforza." *Journal of Anthropological Research* 54: 406–411.

Clark, Ronald W. 1968. *The Huxleys*. London: Heinemann.

Cloître, Michel, and Terry Shinn. 1985. "Expository Practice: Social, Cognitive and Epistemological Linkage." In *Expository Science: Forms and Functions of Popularisation*, edited by Terry Shinn and Richard Whitley, 31–60. Dordrecht: D. Reidel.

Cloninger, C. Robert. 1981. "The Dynamics of Social Learning," review of *Cultural Transmission and Evolution: A Quantitative Approach*, by Luigi Luca Cavalli-Sforza and Marcus W. Feldman. *Science, New Series* 213 (4510): 858–859.

Cohen, Chad, Kevin Bacon, and Spencer Wells. 2009. *The Human Family Tree: Tracing the Human Journey through Time*. Washington DC: National Geographic. Film.

Colbert, Edwin H. 1968. *Men and Dinosaurs: The Search in Field and Laboratory*. London: Evans Brothers.

Cole, Charles C. 1994. "Public History: What Difference Has It Made?" *Public Historian* 16 (4): 9–35.

Collyer, R. H. 1867. "The Fossil Human Jaw from Suffolk." *Anthropological Review* 5 (17): 221–229.

Comaroff, John L., and Jean Comaroff. 2009. *Ethnicity, Inc*. Chicago: University of Chicago Press.

Comfort, Nathaniel. 2012. *The Science of Human Perfection: How Genes Became the Heart of American Medicine*. New Haven, CT: Yale University Press.

Conklin, Edwin Grant. 1921. *The Direction of Human Evolution*. New York: Charles Scribner's Sons.

Cooke, Kathy J. 2002. "Duty or Dream? Edwin G. Conklin's Critique of Eugenics and Support for American Individualism." *Journal of the History of Biology* 35: 365–384.

Coon, Carleton Stevens. 1962. *The Origin of Races*. New York: Knopf.

Cooter, Roger. 2012. "Preisgabe der Demokratie: Wie die Geschichts- und Geisteswissenschaften von den Naturwissenschaften absorbiert werden." In *Wissenschaft und Demokratie*, edited by Michael Hagner, 88–111. Berlin: Suhrkamp.

Cooter, Roger, and Stephen Pumfrey. 1994. "Separate Spheres and Public Places: Reflections on the History of Science Popularization and Science in Popular Culture." *History of Science* 32: 237–267.

Cope, Edward Drinker. 1883. *The Vertebrata of the Tertiary Formations of the West.* Washington, DC: Government Printing Office.

——. 1893. "The Genealogy of Man." *American Naturalist* 27: 316–335.

——. 1896. *The Primary Factors of Organic Evolution.* Chicago: Open Court.

Crane, Stephen. 1893. *Maggie: A Girl of the Streets.* New York: Appleton.

Cravens, Hamilton. 1978. *The Triumph of Evolution: American Scientists and the Heredity-Environment Controversy 1900–1941.* Philadelphia: University of Pennsylvania Press.

Crew, Francis Albert Eley, Cyril Dean Darlington, John Burdon Sanderson Haldane, C. Harland, Lancelot T. Hogben, Julian Sorell Huxley, Hermann Joseph Muller, Joseph Needham, George P. Child, Peo Charles Koller, P. R. David, Walter Landauer, Gunnar Dahlberg, H. H. Plough, Theodosius Dobzhansky, Bronson Price, Rollins Adams Emerson, J. Schultz, C. Gordon, Arthur G. Steinberg, J. Hammond, Conrad Hal Waddington, and Charles Leonard Huskins. 1939. "Social Biology and Population Improvement." *Nature* 144 (3646): 521–522.

Crutzen, Paul J., and Eugene F. Stoermer. 2000. "The 'Anthropocene.'" *Global Change Newsletter* 41: 17–18.

Darlu, Pierre. 2007. "Gènes et Origines: Quelques réflexions à propos du 'Genographic Project.' " *Bulletins et Mémoires de la Société d'Anthropologie de Paris* 19: 113–124.

Darwin, Charles. 1859. *On the Origin of Species by Means of Natural Selection, or The Preservation of Favoured Races in the Struggle for Life.* London: John Murray.

——. 1871. *The Descent of Man, and Selection in Relation to Sex.* London: John Murray.

Daston, Lorraine, and Peter Galison. 2007. *Objectivity.* New York: Zone Books.

Daum, Andreas W. 1998. *Wissenschaftspopularisierung im 19. Jahrhundert: Bürgerliche Kultur, naturwissenschaftliche Bildung und die deutsche Öffentlichkeit, 1848–1914.* Munich: Oldenbourg.

——. 2002. "Science, Politics, and Religion. Humboldtian Thinking and the Transformation of Civil Society in Germany, 1830–1870." *Osiris, 2nd Series* 17: 107–140.

Davidson, Jane P. 2008. *A History of Paleontology Illustration*. Bloomington: Indiana University Press.

Davis, Janet M. 2008. "Cultural Watersheds in *Fin de Siècle* America." In *A Companion to American Cultural History*, edited by Karen Halttunen, 166–180. Malden, MA: Blackwell.

Dawkins, Richard. 1976. *The Selfish Gene*. New York: Oxford University Press.

——. 2006. *The God Delusion*. London: Bantam Books.

Deese, R. S. 2010. "The New Ecology of Power: Julian and Aldous Huxley in the Cold War Era." In *Environmental Histories of the Cold War*, edited by J. R. McNeill and Corinna R. Unger, 279–300. Cambridge: Cambridge University Press.

——. 2011. "Twilight Utopias: Julian and Aldous Huxley in the Twentieth Century." *Journal for the Study of Religion, Nature and Culture* 5 (2): 210–240.

——. 2015. *We Are Amphibians: Julian and Aldous Huxley on the Future of Our Species*. Oakland: University of California Press.

Delisle, Richard G. 2008. "Expanding the Framework of the Holism/Reductionism Debate in Neo-Darwinism: The Case of Theodosius Dobzhansky and Bernhard Rensch." *History and Philosphy of the Life Sciences* 30: 225–246.

——. 2009. *Les philosophies du néo-darwinisme: Conceptions divergentes sur l'homme et sur le sens de l'évolution*. Paris: Presses universitaires de France.

Derrida, Jacques. 2001. *Die différance: Ausgewählte Texte*. Stuttgart: Philipp Reclam jun.

Desmond, Adrian. 1982. *Archetypes and Ancestors: Palaeontology in Victorian London, 1850–1875*. London: Blond and Briggs.

Diamond, Jared. 1992. *The Third Chimpanzee: The Evolution and Future of the Human Animal*. New York: HarperCollins.

Dietrich, Michael R. 1994. "The Origins of the Neutral Theory of Molecular Evolution." *Journal of the History of Biology* 27 (1): 21–59.

——. 1998. "Paradox and Persuasion: Negotiating the Place of Molecular Evolution within Evolutionary Biology." *Journal of the History of Biology* 31 (1): 85–111.

Dingus, Lowell, and Mark A. Norell. 2010. *Barnum Brown: The Man Who Discovered* Tyrannosaurus rex. Berkeley: University of California Press.

Divall, Colin. 1992. "From a Victorian to a Modern: Julian Huxley and the English Intellectual Climate." In *Julian Huxley: Biologist and Statesman of Science. Proceedings of a Conference Held at Rice University 25–27 September 1987*, edited by C. Kenneth Waters and Albert Van Helden, 31–44. Houston: Rice University Press.

Dobzhansky, Theodosius. 1937. *Genetics and the Origin of Species*. New York: Columbia University Press.

——. 1955. *Evolution, Genetics and Man*. New York: Wiley and Sons.

——. 1956. *The Biological Basis of Human Freedom*. New York: Columbia University Press.

——. 1962a. "Genetics and Equality." *Science, New Series* 137 (3542): 112–115.

——. 1962b. *Mankind Evolving*. New Haven, CT: Yale University Press.

——. 1963. "Anthropology and the Natural Sciences: The Problem of Human Evolution." *Current Anthropology* 4 (2): 138, 146–148.

——. 1964. "Biology, Molecular and Organismic." *American Zoologist* 4: 443–452.

——. 1973. *Genetic Diversity and Human Equality*. New York: Basic Books.

——. 1980. "The Birth of the Genetic Theory of Evolution in the Soviet Union in the 1920s." In *The Evolutionary Synthesis: Perspectives on the Unification of Biology*, edited by Ernst Mayr and William B. Provine, 229–242. Cambridge, MA: Harvard University Press.

Dobzhansky, Theodosius, and Gordon Allen. 1956. "Does Natural Selection Continue to Operate on Modern Mankind?" *American Anthropologist, New Series* 58 (4): 591–604.

Dobzhansky, Theodosius, and Ernest Boesiger. 1983. *Human Culture: A Moment in Evolution*. Edited and completed by Bruce Wallace. New York: Columbia University Press.

Dobzhansky, Theodosius, and M. F. Ashley Montagu. 1947. "Natural Selection and the Mental Capacities of Mankind." *Science, New Series* 105 (2736): 587–590.

Doyle, Arthur Conan. 1912. *The Lost World*. London: Hodder and Stoughton.

Dronamraju, Krishna Rao, ed. 1993. *If I Am to Be Remembered: The Life and Work of Julian Huxley with Selected Correspondence*. Singapore: World Scientific.

Dubow, Saul. 1995. *Scientific Racism in Modern South Africa*. Cambridge: Cambridge University Press.

van Dülmen, Richard, and Sina Rauschenbach, ed. 2004. *Macht des Wissens: Die Entstehung der modernen Wissensgesellschaft*. Cologne: Böhlau.

Dunbar, Robin Ian MacDonald. 1989. "Julian Huxley and the Rise of Modern Ethology." In *Evolutionary Studies: A Centenary Celebration of the Life of Julian Huxley. Proceedings of the Twenty-Fourth Annual Symposium of the Eugenics Society, London, 1987*, edited by Milo Keynes and G. Ainsworth Harrison, 58–79. Basingstoke: Macmillan.

Durant, John R. 1986. "The Making of Ethology: The Association for the Study of Animal Behaviour, 1936–1986." *Animal Behaviour* 34 (6): 1601–1616.

——. 1989. "Evolution, Ideology and World View. Darwinian Religion in the Twentieth Century." In *History, Humanity and Evolution: Essays for John C. Greene*, edited by James R. Moore, 335–373. Cambridge: Cambridge University Press.

——. 1992. "The Tension at the Heart of Huxley's Evolutionary Ethology."

In *Julian Huxley: Biologist and Statesman of Science. Proceedings of a Conference Held at Rice University 25–27 September 1987*, edited by C. Kenneth Waters and Albert Van Helden, 150–160. Houston: Rice University Press.

Durant, Will. 1926. *The Story of Philosophy: The Lives and Opinions of the Greater Philosophers*. New York: Simon and Schuster.

——. (1932) 2005. *On the Meaning of Life*. Frisco, TX: Promethean Press.

Edgar, Heather J. H. 2009. "Biohistorical Approaches to 'Race' in the United States: Biological Distances among African Americans, European Americans, and Their Ancestors." *American Journal of Physical Anthropology* 139 (1): 58–67.

Edwards, Anthony W. F. 1969. "Statistical Methods in Scientific Inference." *Nature* 222: 1233–1237.

——. 1974. "The History of Likelihood." *International Statistical Review* 42 (1): 9–15.

——. 1996. "The Origin and Early Development of the Method of Minimum Evolution for the Reconstruction of Phylogenetic Trees." *Systematic Biology* 45: 79–91.

——. 2003. "Human Genetic Diversity: Lewontin's Fallacy." *BioEssays* 25 (8): 798–801.

——. 2009. "Statistical Methods for Evolutionary Trees." *Genetics* 183: 5–12.

Edwards, Anthony W. F., and Luigi Luca Cavalli-Sforza. 1964. "Reconstruction of Evolutionary Trees." In *Phenetic and Phylogenetic Classification*, edited by V. E. Heywood and J. McNeill, 67–76. London: Systematics Association.

Egorova, Yulia. 2009. "De/geneticizing Caste: Population Genetic Research in South Asia." *Science as Culture* 18 (4): 417–438.

——. 2011. "DNA, Authentizität und historisches Gedächtnis." In *Biohistorische Anthropologie: Knochen, Körper und DNA in Erinnerungskulturen*, edited by Marianne Sommer and Gesine Krüger, 33–54. Berlin: Kadmos.

Eibl, Karl. 2010. "Literaturwissenschaft." In *Evolution: Ein Interdisziplinäres Handbuch*, edited by Philipp Sarasin and Marianne Sommer, 257–266. Stuttgart: Metzler.

Elton, Charles Sutherland. (1927) 2011. *Animal Ecology*. With an introduction by Julian Sorell Huxley and a new introduction by Mathew A. Leibold and J. Timothy Wootton. Chicago: University of Chicago Press.

——. 1958. *The Ecology of Invasions by Animals and Plants*. London: Methuen.

Entine, Jon. 2007. *Abraham's Children: Race, Identity, and the DNA of the Chosen People*. New York: Grand Central.

Erasmus, Zimitri. 2013. "Throwing the Genes: A Renewed Biological Imaginary of 'Race,' Place, and Identification." *Theoria* 60 (136): 38–53.

Erlingsson, Steindór J. 2006. "The Early History of the SEB and the BJEB." *Society for Experimental Biology Bulletin* (March): 10–11.

——. 2009. "The Cost of Being a Restless Intellect: Julian Huxley's Popular and Scientific Career in the 1920s." *Studies in History and Philosophy of Biology and Biomedical Sciences* 40: 101–108.

Erll, Astrid, Ansgar Nünning, and Sara B. Young, ed. 2008. *Cultural Memory Studies: An International and Interdisciplinary Handbook*. Berlin: de Gruyter.

Eshel, Ilan, and Luigi Luca Cavalli-Sforza. 1982. "Assortment of Encounters and Evolution of Cooperativeness." *Proceedings of the National Academy of Sciences* 79: 1331–1335.

Evans, David. 1992. *A History of Nature Conservation in Britain*. London: Routledge.

Fabian, Ann. 2008. "The West." In *A Companion to American Cultural History*, edited by Karen Halttunen, 125–138. Malden, MA: Blackwell.

Feldman, Marcus W. 1993. "Heritability, Race, and Policy." Paper no. 0051 presented at the "Race and Science" conference, Washington University, St. Louis, 11–12 Nov.

——. 2010. "The Biology of Ancestry: DNA, Genomic Variation, and Race." In *Doing Race: 21 Essays for the 21st Century*, edited by Hazel Rose Markus and Paula M. L. Moya, 136–159. New York: W. W. Norton.

Feldman, Marcus W., Luigi Luca Cavalli-Sforza, and Joel R. Peck. 1985. "Gene-Culture Coevolution: Models for the Evolution of Altruism with Cultural Transmission." *Proceedings of the National Academy of Sciences* 82: 5814–5818.

Feldman, Marcus W., and Richard C. Lewontin. 2008. "Race, Ancestry, and Medicine." In *Revisiting Race in a Genomic Age*, edited by Barbara A. Koenig, Sandra Soo-Jin Lee, and Sarah S. Richardson, 89–101. New Brunswick, NJ: Rutgers University Press.

Feldman, Marcus W., Sarah P. Otto, and Freddy B. Christiansen. 2000. "Genes, Culture, and Inequality." In *Meritocracy and Economic Inequality*, edited by Kenneth Arrow, Samuel Bowles, and Steven Durlauf, 61–85. Princeton, NJ: Princeton University Press.

Felsenstein, Joseph. 2004. *Inferring Phylogenies*. Sunderland, MA: Sinauer.

Fisher, Ronald Aylmer. 1925. *Statistical Methods for Research Workers*. Edinburgh: Oliver and Boyd.

——. 1930. *The Genetical Theory of Natural Selection*. Oxford: Clarendon.

——. 1956. "Blood-Groups and Population Genetics." *Acta genetica et statistica medica* 6 (4): 507–509.

Fleck, Ludwik. (1935) 1980. *Entstehung und Entwicklung einer wissenschaftlichen Tatsache: Einführung in die Lehre vom Denkstil und Denkkollektiv*. Frankfurt am Main: Suhrkamp.

——. 1983. *Erfahrung und Tatsache: Gesammelte Aufsätze*. Frankfurt am Main: Suhrkamp.

Fleckner, Uwe, ed. 1995. *Die Schatzkammern der Mnemosyne: Ein Lesebuch mit Texten zur Gedächtnistheorie von Platon bis Derrida*. Dresden: Verlag der Kunst.

Forbin, Victor. 1923. *Les fiancées du soleil*. Paris: Lemerre.

———. 1925. *Le secret de la vie*. Paris: Baudinière.

Ford, Edmund Brisco, and Julian Sorell Huxley. 1927. "Mendelian Genes and Rates of Development in *Gammarus chevreuxi*." *British Journal of Experimental Biology* 5: 112–134.

———. 1929. "Genetic Rate-Factors in Gammarus." *Development Genes and Evolution* 117 (1): 67–79.

Ford, Henry Jones. 1915. *The Natural History of the State: An Introduction to Political Science*. Princeton, NJ: Princeton University Press.

Fortun, Michael. 2008. *Promising Genomics: Iceland and deCODE Genetics in a World of Speculation*. Chicago: University of Chicago Press.

Fox, Robert. 2006. "Fashioning the Discipline: History of Science in the European Intellectual Tradition." *Minerva* 44: 410–432.

Fracchia, Joseph, and Richard C. Lewontin. 1999. "Does Culture Evolve?" *History and Theory* 38 (4): 52–78.

François, Olivier, and Eric Durand. 2010. "Spatially Explicit Bayesian Clustering Models in Population Genetics." *Molecular Ecology Resources* 10 (5): 773–784.

Franklin, Sarah. 2001. "Biologization Revisited: Kinship Theory in the Context of the New Biologies." In *Relative Values. Reconfiguring Kinship Studies*, edited by Sarah Franklin and Susan McKinnon, 302–322. Durham, NC: Duke University Press.

———. 2003. "Re-thinking Nature–Culture: Anthropology and the New Genetics." *Anthropological Theory* 3 (1): 65–85.

Franklin, Sarah, Celia Lury, and Jackie Stacey. 2000. *Global Nature, Global Culture*. London: Sage.

Franklin, Sarah, and Susan McKinnon, ed. 2001. *Relative Values: Reconfiguring Kinship Studies*. Durham, NC: Duke University Press.

Freedberg, David, and Vittorio Gallese. 2007. "Motion, Emotion and Empathy in Esthetic Experience." *Trends in Cognitive Sciences* 11 (5): 197–203.

Freud, Sigmund. (1913) 1918. *Totem and Taboo: Resemblances between the Mental Lives of Savages and Neurotics*. Translated by A. A. Brill. New York: Moffat, Yard.

Fried, Morton. 1975. *The Notion of Tribe*. Menlo Park, CA: Cummings.

Füßmann, Klaus, Heinrich Theodor Grütter, and Jörn Rüsen, ed. 1994. *Historische Faszination: Geschichtskultur heute*. Cologne: Böhlau.

Gallenkamp, Charles. 2001. *Dragon Hunter: Roy Chapman Andrews and the Central Asiatic Expeditions*. New York: Penguin.

Galton, Francis. 1883. *Inquiries into the Human Faculty and Its Development*. London: J. M. Dent.

———. (1883) 1907. *Inquiries into the Human Faculty and Its Development.* London: J. M. Dent.

Gannett, Lisa, and James R. Griesemer. 2004. "The ABO Blood Groups: Mapping the History and Geography of Genes in Homo Sapiens." In *Classical Genetic Research and Its Legacy: The Mapping Cultures of Twentieth Century Genetics*, edited by Hans-Jörg Rheinberger and Jean-Paul Gaudillière, 119–172. London: Routledge.

Gayon, Jean. 2003. "Do Biologists Need the Expression 'Human Races'? UNESCO 1950–51." In *Bioethical and Ethical Issues Surrounding the Trials of Nuremberg: Nuremberg Revisited*, edited by Jacques J. Rozenberg, 23–48. Lewiston: Edwin Mellen.

Gibbons, Ann. 1995. "When It Comes to Evolution, Humans Are in the Slow Class." *Science, New Series* 267 (5206): 1907–1908.

Gilbert, Walter. 1992. "A Vision of the Grail." In *The Code of Codes: Scientific and Social Issues in the Human Genome Project*, edited by Daniel J. Kevles and Leroy Hood, 83–97. Cambridge, MA: Harvard University Press.

Gillmore, Emma Wheat Hastings. 1932. *The How and Why of Life.* New York: Liveright.

Golinski, Jan. 1992. *Science as Public Culture: Chemistry and Enlightenment in Britain, 1760–1820.* Cambridge: Cambridge University Press.

Goodall, Jane. 1971. *In the Shadow of Man.* With photographs by Hugo van Lawick. Boston: Houghton Mifflin.

Goodman, Alan H., Deborah Heath, and M. Susan Lindee, ed. 2003. *Genetic Nature/Culture: Anthropology and Science beyond the Two-Culture Divide.* Berkeley: University of California Press.

Goodman, Morris. 1963. "Man's Place in the Phylogeny of the Primates as Reflected in Serum Proteins." In *Classification and Human Evolution*, edited by Sherwood L. Washburn, 204–234. Chicago: Aldine.

Gould, Stephen Jay. 1977. *Ontogeny and Phylogeny.* Cambridge, MA: Harvard University Press.

———. 1980. "G. G. Simpson, Paleontology, and the Modern Synthesis." In *The Evolutionary Synthesis: Perspectives on the Unification of Biology*, edited by Ernst Mayr and William B. Provine, 153–172. Cambridge, MA: Harvard University Press.

———. (1981) 1996. *The Mismeasure of Man.* Revised ed. New York: W. W. Norton.

———. 1989. "An Essay on a Pig Roast." *Natural History* 98 (1): 14–25.

———. 1995. "Ladders and Cones: Constraining Evolution by Canonical Icons." In *Hidden Histories of Science*, edited by Robert B. Silvers, 38–67. New York: New York Review of Books.

Grant, Madison. 1916. *The Passing of the Great Race or the Racial Basis of*

European History. With a preface by Henry Fairfield Osborn. New York: Charles Scribner's Sons.

Greely, Henry T. 2005. "Lessons from the HGDP?" *Science* 308: 1554–1555.

Green, Jens-Peter. 1981. *Krise und Hoffnung: Der Evolutionshumanismus Julian Huxleys*. Heidelberg: Carl Winter.

Greene, John C. 1981. *Science, Ideology, World View: Essays in the History of Evolutionary Ideas*. Berkeley: University of California Press.

——. 1990. "The Interaction of Science and Worldview in Sir Julian Huxley's Evolutionary Biology." *Journal of the History of Biology* 23 (1): 39–55.

Gregory, William King. 1927a. "Did Man Originate in Central Asia? (Mongolia the New World, Part V)." *Scientific Monthly* 24 (5): 385–401.

——. 1927b. "Hesperopithecus Apparently Not an Ape Nor a Man." *Science, New Series* 66 (1720), 579–581.

——. 1930. "A Critique of Professor Osborn's Theory of Human Origin." *American Journal of Physical Anthropology* 14 (2): 133–163.

——. 1934. *Man's Place among the Anthropoids*. Oxford: Clarendon.

——. 1937. "Biographical Memoir of Henry Fairfield Osborn 1857–1935, Presented to the Academy at the Autumn Meeting, 1937." *National Academy of Sciences Biographical Memoirs* 19: 53–119.

Gruenberg, H. 1939. "Men and Mice at Edinburgh: Reports from the Genetics Congress." *Journal of Heredity* 30 (9): 371–374.

Gruffudd, Pyrs. 2000. "Biological Cultivation: Lubetkin's Modernism at London Zoo in the 1930s." In *Animal Spaces, Beastly Places: New Geographies of Human-Animal Relations*, edited by Chris Philo and Chris Wilbert, 222–242. London: Routledge.

Hagen, Joel B. 2010a. "Datenbanken." *Evolution: Ein Interdisziplinäres Handbuch*, edited by Philipp Sarasin and Marianne Sommer, 175–179. Stuttgart: Metzler.

——. 2010b. "Waiting for Sequences: Morris Goodman, Immunodiffusion Experiments, and the Origins of Molecular Anthropology." *Journal of the History of Biology* 43 (4): 697–725.

Hailey, William Malcolm. 1938. *An African Survey: A Study of Problems Arising in Africa South of the Sahara*. London: Oxford University Press.

Halbwachs, Maurice. 1925. *Les cadres sociaux de la mémoire*. Paris: Alcan.

——. 1950. *La mémoire collective*. Paris: Presses universitaires de France.

Haldane, John Burdon Sanderson. 1924. *Daedalus, or Science and the Future*. New York: E. P. Dutton.

——. 1927. *Possible Worlds*. London: Chatto and Windus.

——. 1932a. *The Inequality of Man and Other Essays*. London: Chatto and Windus.

——. 1932b. *The Causes of Evolution*. New York: Harper and Brothers.

——. 1933. *Science and Human Life*. New York: Harper and Brothers.
——. (1932) 1937. *The Inequality of Man and Other Essays*. Harmondsworth: Penguin.
——. 1938. *Heredity and Politics*. London: Allen and Unwin.
Haldane, John Burdon Sanderson, and Julian Sorell Huxley. 1927. *Animal Biology*. Oxford: Clarendon.
Hamilton, Jennifer A. 2012. "The Case of the Genetic Ancestor." In *Genetics and the Unsettled Past: The Collision of DNA, Race, and History*, edited by Keith Wailoo, Alondra Nelson, and Catherine Lee, 266–278. New Brunswick, NJ: Rutgers University Press.
Hammack, David C. 1982. *Power and Society: Greater New York at the Turn of the Century*. New York: Columbia University Press.
Hammer, Michael F. 1995. "A Recent Common Ancestry for Human Y Chromosomes." *Nature* 378 (6555): 376–378.
Hammer, Michael F., and Stephen L. Zegura. 1996. "The Role of the Y Chromosome in Human Evolutionary Studies." *Evolutionary Anthropology* 5 (4): 116–134.
Haraway, Donna. (1989) 1992. *Primate Visions: Gender, Race and Nature in the World of Modern Science*. London: Verso.
——. 1991. *Simians, Cyborgs, and Women: The Reinvention of Nature*. New York: Routledge.
——. 1996. "Universal Donors in a Vampire Culture: It's All in the Family; Biological Kinship Categories in the Twentieth-Century United States." In *Uncommon Ground: Toward Reinventing Nature*, edited by William Cronon, 321–366. New York: W. W. Norton.
——. 1997. *Modest_Witness@Second_Millennium.FemaleMan©_Meets_Onco Mouse™: Feminism and Technoscience*. New York: Routledge.
Hardtwig, Wolfgang. 2005. "Geschichte für Leser: Populäre Geschichtsschreibung in Deutschland im 20. Jahrhundert." In *Geschichte für Leser: Populäre Geschichtsschreibung in Deutschland im 20. Jahrhundert*, edited by Wolfgang Hardtwig and Erhard Schütz, 11–32. Stuttgart: Franz Steiner.
Hardtwig, Wolfgang, and Alexander Schug, ed. 2009. *History Sells! Angewandte Geschichte als Wissenschaft und Markt*. Stuttgart: Franz Steiner.
Harries, Patrick. 2007. *Butterflies and Barbarians: Swiss Missionaries and Systems of Knowledge in South-East Africa*. Oxford: James Currey.
Harry, Debra, and Le'a Malia Kanehe. 2006. "Genetic Research: Collecting Blood to Preserve Culture?" *Cultural Survival Quarterly* 29 (4): 34.
Harth, Dietrich, ed. 1991. *Die Erfindung des Gedächtnisses: Texte zusammengestellt und eingeleitet von Dietrich Harth*. Frankfurt am Main: Keip.
Hartzog, Brooke. 1999. *The First Dinosaur Eggs and Roy Chapman Andrews*. New York: PowerKids Press.
Harvey, David C. 2007. "Heritage Pasts and Heritage Presents: Temporality,

Meaning and the Scope of Heritage Studies." In *Cultural Heritage: Critical Concepts in Media and Cultural Studies*. Vol. 1, *History and Concepts*, edited by Laurajane Smith, 25–44. London: Routledge.

te Heesen, Anke. 2006. *Der Zeitungsausschnitt: Ein Papierobjekt der Moderne*. Frankfurt am Main: Fischer.

Hellenthal, Garrett, George B. J. Busby, Gavin Band, James F. Wilson, Cristian Capelli, Daniel Falush, and Simon Myers. 2014. "A Genetic Atlas of Human Admixture History." *Science* 343 (6172): 747–751.

Hellman, Geoffrey. 1968. *Bankers, Bones and Beetles: The First Century of the American Museum of Natural History*. Garden City, NY: AMNH/Natural History Press.

Henn, Brenna M., Luigi Luca Cavalli-Sforza, and Marcus Feldman. 2012. "The Great Human Expansion." *Proceedings of the National Academy of Sciences of the United States of America* 109 (44): 17758–17764.

Hering, Ewald. 1870. *Über das Gedächtnis als eine allgemeine Function der organisirten Materie*. Vienna: K.K. Hof- und Staatsdruckerei.

Herrnstein, Richard J., and Charles Murray. 1994. *The Bell Curve: Intelligence and Class Structure in American Life*. New York: Free Press.

Higham, John. (1972) 1973. *Writing American History: Essays on Modern Scholarship*. Bloomington: Indiana University Press.

Hilgartner, Stephen. 1990. "The Dominant View of Popularization: Conceptual Problems, Political Uses." *Social Studies of Science* 20 (3): 519–539.

Hirschfeld, Ludwik, and Hanka Hirschfeld. 1919. "Serological Differences between the Blood of Different Races: The Results of Researches on the Macedonian Front." *Lancet* 194 (5016): 675–679.

Hochadel, Oliver. 2011. "Fossilien im Dienst der Nation: Atapuerca und der neue Beginn der spanischen Geschichte." In *Biohistorische Anthropologie: Knochen, Körper und DNA in Erinnerungskulturen*, edited by Marianne Sommer and Gesine Krüger, 134–163. Berlin: Kadmos.

Hogben, Adrian, and Anne Hogben, ed. 1998. *Lancelot Hogben: Scientific Humanist—An Unauthorised Autobiography*. Woodbridge: Merlin.

Hogben, Lancelot. 1932. *Genetic Principles in Medicine and Social Science*. New York: Alfred A. Knopf.

——. 1933. *Nature and Nurture: Being the William Withering Memorial Lectures*. London: Allen and Unwin.

——. 1938. *Science for the Citizen: A Self-Educator Based on the Social Background of Scientific Discovery*. London: George Unwin.

——. 1939. *Dangerous Thoughts*. London: Allen and Unwin.

——. 1943. *Interglossa: A Draft of an Auxiliary for a Democratic World Order, Being an Attempt to Apply Semantic Principles to Language Design*. Harmondsworth: Penguin.

——. 1966. "The Origins of the Society." In *The Origins and History of the So-*

ciety for Experimental Biology, edited by Michael Alfred Sleigh and James F. Sutcliffe, 5–11. London: SEB.

Hopwood, Nick. 2014. *Haeckel's Embryos: Images, Evolution, and Fraud*. Chicago: University of Chicago Press.

Hostettler, Otto. 2009a. "Heikle Gendaten werden illegal weitergegeben." *Beobachter*, 22 July. Accessed 2 July 2015. http://www.beobachter.ch/justiz-behoerde/artikel/ gentest_heikle-gendaten-werden-illegal-weitergegeben.

——. 2009b. "Staatsanwaltschaft pfeift Fedpol zurück." *Beobachter*, 2 Sept. Accessed 2 July 2015. http://www.beobachter.ch/justiz-behoerde/buerger-verwaltung/artikel/gentestch_staatsanwaltschaft-pfeift-fedpol-zurueck.

——. 2010. "Der Texas-Trick." *Beobachter*,17 Feb. Accessed 2 July 2015. http://www.beobachter.ch/justiz-behoerde/artikel/gentests_der-texas-trick.

Howe, Barbara J., and Emory L. Kemp. 1986. *Public History: An Introduction*. Malabar, FL: Krieger.

Hubback, David. 1989. "Julian Huxley and Eugenics." In *Evolutionary Studies: A Centenary Celebration of the Life of Julian Huxley. Proceedings of the Twenty-Fourth Annual Symposium of the Eugenics Society, London, 1987*, edited by Milo Keynes and G. Ainsworth Harrison, 194–206. Basingstoke: Macmillan.

Hughes, Lotte. 2006. *Moving the Maasai: A Colonial Misadventure*. Houndmills, UK: Palgrave Macmillan.

Hunter, Jane H. 2008. "Gender and Sexuality." In *A Companion to American Cultural History*, edited by Karen Halttunen, 327–340. Malden, MA: Blackwell.

Huntington, Ellsworth. 1924. *The Character of Races: As Influenced by Physical Environment, Natural Selection and Historical Development*. New York: Scribner.

Huxley, Aldous. 1932. *Brave New World*. London: Chatto and Windus.

Huxley, Julian Sorell. 1912. *The Individual in the Animal Kingdom*. Cambridge: Cambridge University Press.

——. 1914. "The Courtship-Habits of the Great Crested Grebe (Podiceps Cristatus); With an Addition to the Theory of Sexual Selection." *Proceedings of the Zoological Society of London* 84 (3): 491–562.

——. 1920. "Metamorphosis of Axolotl Caused by Thyroid-Feeding." *Nature* 104: 435.

——. 1923. *Essays of a Biologist*. London: Chatto and Windus.

——. 1924. "Constant Differential Growth-Ratios and Their Significance." *Nature* 2877 (144): 895–896.

——. 1926a. "On the History of Science." In *Essays in Popular Science*, by Julian Sorell Huxley, 163–169. London: Chatto and Windus.

——. 1926b. "The Annual Increment of the Antlers of the Red Deer (*Cervus elaphus*)." *Proceedings of the Zoological Society of London* 96 (4): 1021–1035.

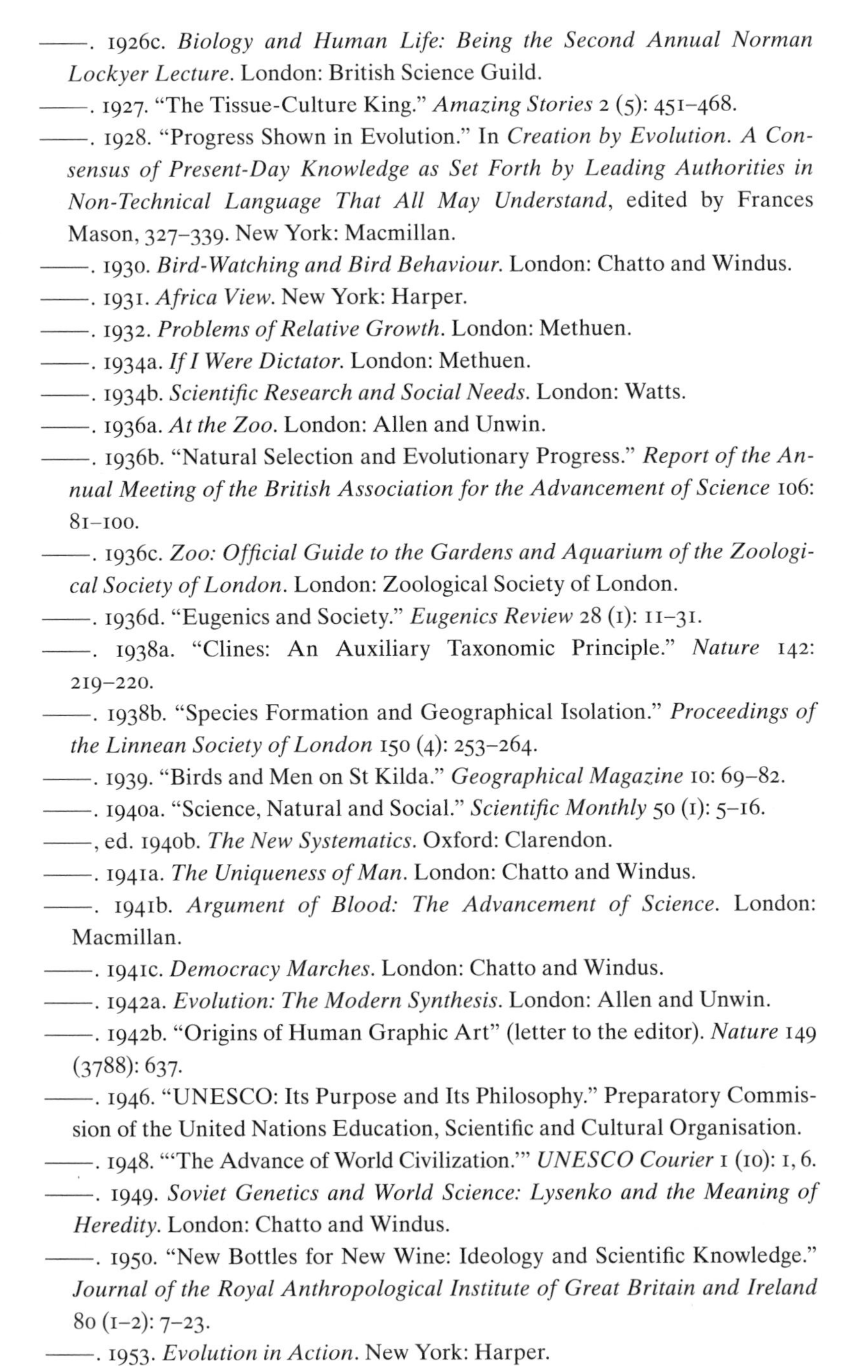

——. 1926c. *Biology and Human Life: Being the Second Annual Norman Lockyer Lecture*. London: British Science Guild.
——. 1927. "The Tissue-Culture King." *Amazing Stories* 2 (5): 451–468.
——. 1928. "Progress Shown in Evolution." In *Creation by Evolution. A Consensus of Present-Day Knowledge as Set Forth by Leading Authorities in Non-Technical Language That All May Understand*, edited by Frances Mason, 327–339. New York: Macmillan.
——. 1930. *Bird-Watching and Bird Behaviour*. London: Chatto and Windus.
——. 1931. *Africa View*. New York: Harper.
——. 1932. *Problems of Relative Growth*. London: Methuen.
——. 1934a. *If I Were Dictator*. London: Methuen.
——. 1934b. *Scientific Research and Social Needs*. London: Watts.
——. 1936a. *At the Zoo*. London: Allen and Unwin.
——. 1936b. "Natural Selection and Evolutionary Progress." *Report of the Annual Meeting of the British Association for the Advancement of Science* 106: 81–100.
——. 1936c. *Zoo: Official Guide to the Gardens and Aquarium of the Zoological Society of London*. London: Zoological Society of London.
——. 1936d. "Eugenics and Society." *Eugenics Review* 28 (1): 11–31.
——. 1938a. "Clines: An Auxiliary Taxonomic Principle." *Nature* 142: 219–220.
——. 1938b. "Species Formation and Geographical Isolation." *Proceedings of the Linnean Society of London* 150 (4): 253–264.
——. 1939. "Birds and Men on St Kilda." *Geographical Magazine* 10: 69–82.
——. 1940a. "Science, Natural and Social." *Scientific Monthly* 50 (1): 5–16.
——, ed. 1940b. *The New Systematics*. Oxford: Clarendon.
——. 1941a. *The Uniqueness of Man*. London: Chatto and Windus.
——. 1941b. *Argument of Blood: The Advancement of Science*. London: Macmillan.
——. 1941c. *Democracy Marches*. London: Chatto and Windus.
——. 1942a. *Evolution: The Modern Synthesis*. London: Allen and Unwin.
——. 1942b. "Origins of Human Graphic Art" (letter to the editor). *Nature* 149 (3788): 637.
——. 1946. "UNESCO: Its Purpose and Its Philosophy." Preparatory Commission of the United Nations Education, Scientific and Cultural Organisation.
——. 1948. "'The Advance of World Civilization.'" *UNESCO Courier* 1 (10): 1, 6.
——. 1949. *Soviet Genetics and World Science: Lysenko and the Meaning of Heredity*. London: Chatto and Windus.
——. 1950. "New Bottles for New Wine: Ideology and Scientific Knowledge." *Journal of the Royal Anthropological Institute of Great Britain and Ireland* 80 (1–2): 7–23.
——. 1953. *Evolution in Action*. New York: Harper.

——. 1955. "Evolution, Cultural and Biological." *Yearbook of Anthropology 1*: 2–25.

——. 1960a. "Huxley in Africa—1—The Treasure House of Wild Life." *Observer*, 13 Nov., 23–24.

——. 1960b. "Cropping the Wild Protein." *Observer*, 20 Nov., 23.

——. 1960c. "Wild Life as a World Asset." *Observer*, 27 Nov., 23, 25.

——, ed. 1961a. *The Humanist Frame: The Modern Humanist Vision of Life*. London: Allen and Unwin.

——. 1961b. *The Conservation of Wild Life and Natural Habitats in Central and East Africa: Report on a Mission Accomplished for Unesco, July–September 1960*. Paris: Unesco.

——. 1963. "Towards a Fulfilment Society." *New Scientist* 345: 712–714.

——. 1964. *Essays of a Humanist*. London: Chatto and Windus.

——. (1964) 1992. *Evolutionary Humanism*. Buffalo: Prometheus Books.

——, ed. 1965a. *Aldous Huxley 1894–1963: A Memorial Volume*. London: Chatto and Windus.

——. 1965b. *Ich sehe den künftigen Menschen: Natur und neuer Humanismus*. Munich: List.

——. 1966. *The Future of Man: Evolutionary Aspects*. N.p.: Ethical Culture Publications.

——. 1970. *Memories*. London: Allen and Unwin.

——. 1973. *Memories II*. London: Allen and Unwin.

Huxley, Julian Sorell, and Edward Neville da Costa Andrade, ed. 1935. *More Simple Science: Earth and Man*. Oxford: Blackwell.

Huxley, Julian Sorell, and Gavin Rylands de Beer. 1934. *The Elements of Experimental Embryology*. Cambridge: Cambridge University Press.

Huxley, Julian Sorell, and Alfred C. Haddon. 1935. *We Europeans: A Survey of "Racial" Problems*. London: Jonathan Cape.

Huxley, Julian Sorell, A. C. Hardy, and E. B. Ford, ed. 1954. *Evolution in Progress*. London: Allen and Unwin.

Huxley, Julian Sorell, and Henry Bernard Davis Kettlewell. 1965. *Darwin and His World*. New York: Viking Press.

Huxley, Julian Sorell, and Ludwig Koch. (1938) 1964. *Animal Language*. New York: Grosset and Dunlap.

Huxley, Julian Sorell, Gilbert Murray, and J. H. Oldham. 1944. *Humanism*. London: Watts.

Huxley, Julian Sorell, and Wolfgang Suschitzky. 1956. *Kingdom of the Beasts*. With texts by Huxley and photographs by Suschitzky. London: Thames and Hudson.

Huxley, Julian Sorell, H. G. Wells, John Burdon Sanderson Haldane et al. 1944. *Reshaping Man's Heritage: Biology in the Service of Man*. London: Allen and Unwin.

Huxley, Thomas Henry, and Julian Sorell Huxley. 1947. *Touchstone for Ethics, 1893–1943*. New York: Harper.

Huyssen, Andreas. 1995. *Twilight Memories: Marking Time in a Culture of Amnesia*. New York: Routledge.

Iggers, Georg G. 2007. *Geschichtswissenschaft im 20. Jahrhundert: Ein kritischer Überblick im internationalen Zusammenhang*. Göttingen: Vandenhoeck und Ruprecht.

Ingersoll, Ernest. 1883. *Knocking Round the Rockies*. New York: Harper.

———. 1900. *Nature's Calendar*. New York: Harper.

———. 1928. *Dragons and Dragon Lore*. New York: Payson and Clarke.

International Balzan Foundation, ed. 2009. *Luigi Luca Cavalli-Sforza: 1999 Balzan Prize for Science of Human Origins; Excerpts from the Publications; Balzan Prizes 1999 (Enlarged and Revised Edition, 2009); Meeting the Challenges of the Future; A Discussion between "The Two Cultures" (Leo S. Olschki Editore, 2003)*. Milan: Fondazione Internazionale Balzan. Accessed 2 July 2015. http://www.balzan.org/upload/EstrattoCavalli_ING.pdf.

James, William. 1899. *The Principles of Psychology*. New York: Holt.

Jensen, Arthur. 1969. "How Much Can We Boost IQ and Scholastic Achievement?" *Harvard Education Review* 39: 1–123.

Jobling, Mark A. 1994. "A Survey of Long-Range DNA Polymorphisms on the Human Y Chromosome." *Human Molecular Genetics* 3: 107–114.

Jobling, Mark A., and Chris Tyler-Smith. 2003. "The Human Y Chromosome: An Evolutionary Marker Comes of Age." *Nature Reviews Genetics* 4: 598–612.

Johnson, M. J., D. C. Wallace, S. D. Ferris, M. C. Rattazzi, and Luigi Luca Cavalli-Sforza. 1983. "Radiation of Human Mitochondria DNA Types Analyzed by Restriction Endonuclease Cleavage Patterns." *Journal of Molecular Evolution* 19: 255–271.

Jones, Frederick Wood. 1919. "The Origin of Man." In *Animal Life and Human Progress*, edited by Arthur Dendy, 99–131. London: Constable.

———. 1929. *Man's Place among the Mammals*. London: Edward Arnold.

Judson, Horace Freeland. (1979) 1996. *The Eighth Day of Creation: Makers of the Revolution in Biology*. Cold Spring Harbor: Cold Spring Harbor Laboratory Press.

———. 1992. "A History of the Science and Technology behind Gene Mapping and Sequencing." In *The Code of Codes: Scientific and Social Issues in the Human Genome Project*, edited by Daniel J. Kevles and Leroy Hood, 37–80. Cambridge, MA: Harvard University Press.

Juengst, Eric T. 1998. "Group Identity and Human Diversity: Keeping Biology Straight from Culture." *American Journal of Human Genetics* 63: 673–677.

Junker, Thomas, and Eve-Marie Engels, ed. 1999. *Die Entstehung der Synthe-*

tischen Theorie: Beiträge zur Geschichte der Evolutionsbiologie in Deutschland 1930–1950. Berlin: VWB.

Kaplan, Jonathan Michael, and Rasmus Grønfeldt Winther. 2014. "Realism, Antirealism, and Conventionalism about Race." *Philosophy of Science* 81 (5): 1039–1052.

Karafet, Tatiana M., Ludmila P. Osipova, Marina A. Gubina, Olga L. Posukh, Stephen L. Zegura, and Michael P. Hammer. 2002. "High Levels of Y-Chromosome Differentiation among Native Siberian Populations and the Genetic Signature of a Boreal Hunter-Gatherer Way of Life." *Human Biology* 74 (6): 761–789.

Kay, Lily E. 1993. *The Molecular Vision of Life: Caltech, The Rockefeller Foundation, and the Rise of the New Biology*. New York: Oxford University Press.

——. 2000. *Who Wrote the Book of Life? A History of the Genetic Code*. Stanford, CA: Stanford University Press.

Keith, Arthur. 1911. *Ancient Types of Man*. London: Harper.

——. 1915. *The Antiquity of Man*. London: Williams and Norgate.

——. 1927. "Darwin's Theory of Man's Descent as It Stands To-day." *Science, New Series* 66 (1705): 201–208.

Keller, Eva. 2015. *Beyond the Lens of Conservation: Malagasy and Swiss Imaginations of One Another*. New York: Berghahn.

Kelley, Robert. 1978. "Public History: Its Origins, Nature, and Prospects." *Public Historian* 1 (1): 16–28.

Kern, Stephen. (1983) 2003. *The Culture of Time and Space, 1880–1918*. Cambridge, MA: Harvard University Press.

Kevles, Daniel J. (1985) 1995. *In the Name of Eugenics: Genetics and the Uses of Human Heredity*. Paperback with new preface by the author. Cambridge, MA: Harvard University Press.

——. 1992a. "Huxley and the Popularization of Science." In *Julian Huxley: Biologist and Statesman of Science. Proceedings of a Conference Held at Rice University 25–27 September 1987*, edited by C. Kenneth Waters and Albert Van Helden, 238–251. Houston: Rice University Press.

——. 1992b. "Out of Eugenics: The Historical Politics of the Human Genome." In *The Code of Codes: Scientific and Social Issues in the Human Genome Project*, edited by Daniel J. Kevles and Leroy Hood, 3–36. Cambridge, MA: Harvard University Press.

Kevles, Daniel J., and Gerald L. Geison. 1995. "The Experimental Life Sciences in the Twentieth Century." *Osiris, 2nd Series* 10: 97–121.

Kevles, Daniel J., and Leroy Hood, ed. 1992. *The Code of Codes: Scientific and Social Issues in the Human Genome Project*. Cambridge, MA: Harvard University Press.

Keyes, C. R. 1936. "Henry Fairfield Osborn." *The Pan-American Geologist* 65 (3): 160–178.

Kimura, Motoo. 1968. "Evolutionary Rate at the Molecular Level." *Nature* 217: 624–626.

Kitcher, Philip. 1985. *Vaulting Ambition: Sociobiology and the Quest for Human Nature*. Cambridge, MA: MIT Press.

Klapisch-Zuber, Christiane. 2004. *Stammbäume: Eine illustrierte Geschichte der Ahnenkunde*. Munich: Knesebeck.

Klein, Kerwin Lee. 2000. "On the Emergence of *Memory* in Historical Discourse." In "Grounds for Remembering," special issue, *Representations* 69: 127–150.

Knight, Charles Robert. 1922. "Mural Paintings of Prehistoric Men and Animals." *Scribner's Magazine* 71: 279–286.

——. 1935. *Before the Dawn of History*. New York: Whittlesey House, McGraw-Hill.

——. 1946. *Life through the Ages*. New York: Knopf.

——. 1947. *Animal Drawing: Anatomy and Action for Artists*. New York: Dover.

——. 2005. *Autobiography of an Artist*. Edited by Jim Ottaviani. Ann Arbor: G. T. Labs.

Knoppers, Bartha Maria, Marie Hirtle, and Sébastien Lormeau. 1996. "Ethical Issues in International Collaborative Research on the Human Genome: The HGP and the HGDP." *Genomics* 34: 271–282.

Koch, Ludwig. 1955. *Memoirs of a Birdman*. London: Phoenix.

Kohler, Robert E. 1994. *Lords of the Fly: Drosophila Genetics and the Experimental Life*. Chicago: University of Chicago.

Kohlstedt, Sally Gregory. 1987. "International Exchange and National Style: A View of Natural History Museums in the United States, 1850–1900." In *Scientific Colonialism: A Cross-Cultural Comparison*, edited by Nathan Reingold and Marc Rothenberg, 167–190. Washington DC: Smithsonian Institution Press.

——. 2005a. "Nature, Not Books: Scientists and the Origins of the Nature-Study Movement in the 1890s." *Isis* 96 (3): 324–352.

——. 2005b. " 'Thoughts in Things': Modernity, History, and North American Museums." *Isis* 96 (4): 586–601.

——. 2010. *Teaching Children Science: Hands-on Nature Study in North America, 1890–1930*. Chicago: University of Chicago Press.

Kohlstedt, Sally Gregory, and Paul Brinkman. 2004. "Framing Nature: The Formative Years of Natural History Museum Development in the United States." In supplement 1 (2), *Proceedings of the California Academy of Sciences* 55: 7–33.

Kopec, Matthew. 2014. "Clines, Clusters, and Clades in the Race Debate." *Philosophy of Science* 81 (5): 1053–1065.

Kretschmann, Carsten, ed. 2003. *Wissenspopularisierung: Konzepte der Wissensverbreitung im Wandel*. Berlin: Akademie Verlag.

Krill, Karl-Heinz. 1968. “Die Gründung der UNESCO.” *Vierteljahreshefte für Zeitgeschichte* 16 (3): 247–279.

Krimbas, Costas B. 1994. “The Evolutionary Worldview of Theodosius Dobzhansky.” In *The Evolution of Theodosius Dobzhansky: Essays on His Life and Thought in Russia and America*, edited by Mark B. Adams, 179–193. Princeton, NJ: Princeton University Press.

Krüger, Gesine, Ruth Mayer, and Marianne Sommer, ed. 2008. *“Ich Tarzan”: Affenmenschen und Menschenaffen zwischen Science und Fiction*. Bielefeld: transcript.

Kühberger, Christoph, Christian Lübke, and Thomas Terberger. 2007. *Wahre Geschichte—Geschichte als Ware: Die Verantwortung der historischen Forschung für Wissenschaft und Gesellschaft*. Rahden: VML.

Kuklick, Henrika, ed. 2007. *A New History of Anthropology*. Malden, MA: Blackwell.

Kupper, Patrick. 2012. *Wildnis schaffen: Eine transnationale Geschichte des Schweizerischen Nationalparks*. Berne: Haupt.

Kuritz, Hyman. 1981. “The Popularization of Science in Nineteenth-Century America.” *History of Education Quarterly* 21 (3): 259–274.

Lamprey, Louise. 1921. *Long-Ago People: How They Lived in Britain before History Began*. Boston: Little, Brown.

——. 1927. *Children of Ancient Gaul*. Boston: Little, Brown.

Langford, George. 1920. *Pic, the Weapon-Maker*. New York: Boni and Liveright.

——. 1921. *Kutnar—Son of Pic*. New York: Boni and Liveright.

Latour, Bruno. 1986. “Visualisation and Cognition: Drawing Things Together.” *Knowledge and Society: Studies in the Sociology of Culture Past and Present* 6: 1–40.

——. 1987. *Science in Action: How to Follow Scientists and Engineers through Society*. Cambridge, MA: Harvard University Press.

——. 2008. “A Textbook Case Revisited—Knowledge as a Mode of Existence.” In *Handbook of Science and Technology Studies*, 3rd ed., edited by Edward J. Hackett et al., 83–112. Cambridge, MA: MIT Press.

Latour, Bruno, and Steve Woolgar. 1979. *Laboratory Life: The Social Construction of Scientific Facts*. Beverly Hills: Sage.

Lears, T. J. Jackson. 1981. *No Place of Grace: Antimodernism and the Transformation of American Culture 1880–1920*. Chicago: University of Chicago Press.

Lee, Richard B., and Irven DeVore, ed. 1968. *Man the Hunter*. Chicago: Aldine.

Lee, Sandra Soo-Jin, Joanna Mountain, Barbara Koenig, Russ Altman, Melissa Brown, Albert Camarillo, Luigi Luca Cavalli-Sforza, Mildred Cho, Jennifer Eberhardt, Marcus W. Feldman, Richard Ford, Henry T. Greely, Roy King, Hazel Markus, Debra Satz, Matthew Snipp, Claude Steele, and Peter A. Underhill. 2008. “The Ethics of Characterizing Difference: Guiding Principles on Using Racial Categories in Human Genetics (Open Letter)." *Genome Biology* 9 (7): 404.

Lefebvre, Henri. (1974) 1991. *The Production of Space*. Translated by Donald Nicholson-Smith. Oxford: Blackwell.

LeMahieu, Dan L. 1992. "The Ambiguity of Popularization." In *Julian Huxley: Biologist and Statesman of Science. Proceedings of a Conference Held at Rice University 25–27 September 1987*, edited by C. Kenneth Waters and Albert Van Helden, 252–256. Houston: Rice University Press.

Lenger, Friedrich. 2005. "Geschichte und Erinnerung im Zeichen der Nation: Einige Beobachtungen zur jüngsten Entwicklung." In *Erinnerung, Gedächtnis, Wissen: Studien zur kulturwissenschaftlichen Gedächtnisforschung*, edited by Günter Oesterle, 521–536. Göttingen: Vandenhoeck and Ruprecht.

Leslie, Mitchell. 1999. "The History of Everyone and Everything." *Stanford Magazine*, May/June. Accessed 2 July 2015. http://alumni.stanford.edu/get/page/magazine/article/?article_id=40759.

Lewenstein, Bruce V. 1992. "The Meaning of 'Public Understanding of Science' in the United States after World War II." *Public Understanding of Science* 1 (1): 45–68.

——. 1995. "Science and the Media." In *Handbook of Science and Technology Studies*, edited by Sheila Jasanoff, Gerald E. Markle, James C. Petersen, and Trevor Pinch, 343–360. Thousand Oaks, CA: Sage.

Lewin, Roger. 1985. "Molecular Clocks Scrutinized." *Science, New Series* 228 (4699): 571.

——. 1988a. "Conflict over DNA Clock Results." *Science, New Series* 241 (4873): 1598–1600.

——. 1988b. "DNA Clock Conflict Continues." *Science, New Series* 241 (4874): 1756–1759.

Lewontin, Richard C. 1972. "The Apportionment of Human Diversity." *Evolutionary Biology* 6: 391–398.

Lewontin, Richard C., Steven Rose, and Leon J. Kamin. 1984. *Not in Our Genes: Biology, Ideology, and Human Nature*. New York: Pantheon Books.

Li, Jun Z., Devin M. Absher, Hua Tang, Audrey M. Southwick, Amanda M. Casto, Sohini Ramachandran, Howard M. Cann, Gregory S. Barsh, Marcus W. Feldman, Luigi Luca Cavalli-Sforza, and Richard M. Myers. 2008. "Worldwide Human Relationships Inferred from Genome-Wide Patterns of Variation." *Science* 319 (5866): 1100–1104.

Lightman, Bernard. 2007. *Victorian Popularizers of Science: Designing Nature for New Audiences*. Chicago: University of Chicago Press.

Lindee, Susan M. 2003. "Provenance and the Pedigree: Victor McKusick's Fieldwork with the Old Order Amish." In *Genetic Nature/Culture: Anthropology and Science beyond the Two-Culture Divide*, edited by Alan H. Goodman, Deborah Heath, and M. Susan Lindee, 41–57. Berkeley: University of California Press.

——. 2013. "Map Your Own Genes! The DNA Experience." In *Genetic Explanations: Sense and Nonsense*, edited by Sheldon Krimsky and Jeremy Gruber, 186–200. Cambridge, MA: Harvard University Press.

Lotka, Alfred. 1925. *Elements of Physical Biology*. Baltimore: Williams and Wilkins.

Lowenthal, David. 1985. *The Past Is a Foreign Country*. Cambridge: Cambridge University Press.

——. 1998. *The Heritage Crusade and the Spoils of History*. Cambridge: Cambridge University Press.

Lucotte, Gérard. 1989. "Evidence for the Paternal Ancestry of Modern Humans: Evidence From a Y-Chromosome Specific Sequence Polymorphic DNA Probe." In *The Human Revolution: Behavioural and Biological Perspectives on the Origins of Modern Humans*, edited by Paul Mellars and Chris Stringer, 39–46. Edinburgh: Edinburgh University Press.

Lumsden, Charles J., and Edward Osborne Wilson. (1981) 2005. *Genes, Mind, and Culture: The Coevolutionary Process. 25th anniversary edition*. Singapore: World Scientific.

von Lünen, Alexander. 2009. "The Perfect Astronaut Would Be a Human without Legs—JBS Haldane and 'Positive Eugenics.' " In *Wie nationalsozialistisch ist die Eugenik?/What Is National Socialist about Eugenics?: Internationale Debatte zur Geschichte der Eugenik im 20. Jahrhundert/International Debates on the History of Eugenics in the 20th Century*, edited by Regina Wecker, Sabine Braunschweig, Gabriela Imboden, Bernhard Küchenhoff, and Hans Jakob Ritter, 127–138. Vienna: Böhlau.

MacEachern, Scott. 2000. "Genes, Tribes, and African History." *Current Anthropology* 41 (3): 357–384.

MacKenzie, John M. 1988. *The Empire of Nature: Hunting, Conservation and British imperialism*. Manchester: Manchester University Press.

Maltby, Clive. 2003. *The Journey of Man: The Story of the Human Species*. Tigress Productions. DVD.

Manias, Chris. 2015. "Building *Baluchitherium* and *Indricotherium*: Imperial and International Networks in Early-Twentieth Century Paleontology." *Journal of the History of Biology*. 48(2): 237–278.

Manni, Franz. 2010. "Interview with Luigi Luca Cavalli-Sforza: Past Research and Directions for Future Investigations in Human Population Genetics." *Human Biology* 82 (3): 245–266.

Marchal, Guy P. 2006. *Schweizer Gebrauchsgeschichte: Geschichtsbilder, Mythenbildung und nationale Identität*. Basle: Schwabe.

Marks, Jonathan. 1995. *Human Biodiversity: Genes, Race, and History*. New Brunswick, NJ: Transaction Publishers.

——. 2001. " 'We're Going to Tell These People Who They Really Are': Science and Relatedness." In *Relative Values: Reconfiguring Kinship Studies*, edited by Sarah Franklin and Susan McKinnon, 355–383. Durham, NC: Duke University Press.

——. 2012. "The Origins of Anthropological Genetics." In supplement 5, *Current Anthropology* 53: S161–S172.

Marrin, Albert. 2002. *Secrets from the Rocks: Dinosaur Hunting with Roy Chapman Andrews.* New York: Dutton Children's Books.

Matheka, Reuben M. 2008. "The International Dimension of the Politics of Wildlife Conservation in Keyna, 1958–1969." *Journal of Eastern African Studies* 2 (1): 112–133.

Matthew, William Diller. (1911) 1915. "Climate and Evolution." *Annals of the New York Academy of Sciences* 24: 171–318.

Mayer, Anna K. 2005. "When Things Don't Talk: Knowledge and Belief in the Inter-War Humanism of Charles Singer (1876–1960)." *British Journal for the History of Science* 38 (3): 325–347.

Maynard Smith, John, and N. Warren. 1982. "Models of Cultural and Genetic Change," review of *Cultural Transmission and Evolution: A Quantitative Approach*, by Luigi Luca Cavalli-Sforza and Marcus W. Feldman, and *Genes, Mind and Culture: The Coevolutionary Process*, by Charles J. Lumsden and Edward O. Wilson. *Evolution* 36 (3): 620–627.

Mayr, Ernst. 1942. *Systematics and the Origin of Species from the Viewpoint of a Zoologist.* New York: Columbia University Press.

——. 1961. "Cause and Effect in Biology." *Science* 134: 1501–1506.

——. 1962. "Accident or Design: The Paradox of Evolution." In *The Evolution of Living Organisms*, edited by G. W. Leeper, 1–14. Melbourne: Melbourne University Press.

——. 1963. *Animal Species and Evolution.* Cambridge, MA: Harvard University Press.

——. 1980a. "Prologue: Some Thoughts on the History of the Evolutionary Synthesis." In *The Evolutionary Synthesis. Perspectives on the Unification of Biology*, edited by Ernst Mayr and William B. Provine, 1–48. Cambridge, MA: Harvard University Press.

——. 1980b. "The Role of Systematics in the Evolutionary Synthesis." In *The Evolutionary Synthesis: Perspectives on the Unification of Biology*, edited by Ernst Mayr and William B. Provine, 123–136. Cambridge, MA: Harvard University Press.

——. 1988. *Toward a New Philosophy of Biology: Observations of an Evolutionist.* Cambridge, MA: Harvard University Press.

——. 1999. "Thoughts on the Evolutionary Synthesis in Germany." In *Die Entstehung der Synthetischen Theorie: Beiträge zur Geschichte der Evolutionsbiologie in Deutschland 1930–1950*, edited by Thomas Junker and Eve-Marie Engels, 19–29. Berlin: VWB.

Mayr, Ernst, and William B. Provine, ed. 1980. *The Evolutionary Synthesis: Per-*

spectives on the Unification of Biology. Cambridge, MA: Harvard University Press.

Mazumdar, Pauline M. H. 1992. *Eugenicists, Human Genetics and Human Failings: The Eugenics Society, Its Sources and Its Critics in Britain*. London: Routledge.

M'Charek, Amade. 2005. *The Human Genome Diversity Project: An Ethnography of Scientific Practice*. Cambridge: Cambridge University Press.

McGovern, Charles. 2008. "Consumer Culture and Mass Culture." In *A Companion to American Cultural History*, edited by Karen Halttunen, 183–197. Malden, MA: Blackwell.

McGregor, J. Howard. 1926. "Restoring Neanderthal Man." *Natural History* 26: 289–293.

McGucken, William. 1984. *Scientists, Society and State: The Social Relations of Science Movement in Great Britain 1931–47*. Columbus: Ohio University Press.

McNassor, Cathy. 2011. *Images of America: Los Angeles's La Brea Tar Pits and Hancock Park*. Charleston, SC: Arcadia.

Mellars, Paul, and Chris Stringer, ed. 1989. *The Human Revolution: Behavioural and Biological Perspectives on the Origins of Modern Humans*. Edinburgh: Edinburgh University Press.

Meloni, Maurizio. 2014. "How Biology Became Social, and What It Means for Social Theory." *Sociological Review* 62 (3): 593–614.

Milner, Richard. 2012. *Charles R. Knight: The Artist Who Saw through Time*. New York: Abrams.

Ministry of Town and Country Planning. 1947. *Conservation of Nature in England and Wales: Report of the Wild-Life Conservation Special Committee (England and Wales)*. London: H. M. Stationery Office.

Mitchell, Peter Chalmers. 1934. *Official Guide to the Gardens and Aquarium of the Zoological Society of London. 31st ed*. London: Zoological Society of London.

Mitchell, William John Thomas. 1998. *The Last Dinosaur Book: The Life and Times of a Cultural Icon*. Chicago: University of Chicago Press.

Mitman, Gregg. 1996. "When Nature Is the Zoo: Vision and Power in the Art and Science of Natural History." *Osiris, 2nd Series* 11: 117–143.

Morant, Geoffrey McKay. 1939. *The Races of Central Europe: A Footnote to History*. London: Allen and Unwin.

Moreno-Estrada, Andrés, Simon Gravel, Fouad Zakharia, Jacob L. McCauley, Jake K. Byrnes, Christopher R. Gignoux, Patricia A. Ortiz-Tello, Ricardo J. Martínez, Dale J. Hedges, Richard W. Morris, Celeste Eng, Karla Sandoval, Suehelay Acevedo-Acevedo, Paul J. Norman, Zulay Layrisse, Peter Parham, Juan Carlos Martínez-Cruzado, Esteban González Burchard, Michael L. Cuccaro, Eden R. Martin, and Carlos D. Bustamante. 2013. "Reconstructing the Population Genetic History of the Caribbean." *PLoS Genetics* 9 (11). doi: 10.1371/journal.pgen.1003925.

Morgan, Thomas Hunt. 1916. *A Critique of the Theory of Evolution: Lectures Delivered at Princeton University, February 24, March 1, 8, 15, 1916*. Princeton, NJ: Princeton University Press.

———. 1919. *The Physical Basis of Heredity*. Philadelphia: J. B. Lippincott.

———. 1923. "The Bearing of Mendelism on the Origin of Species." *Scientific Monthly* 16: 237–247.

Morier-Genoud, Eric. 2011. "Missions and Institutions: Henri-Philippe Junod, Anthropology, Human Rights and Academia between Africa and Switzerland, 1921–1966." *Schweizerische Zeitschrift für Religions- und Kulturgeschichte* 105: 193–219.

Morris, Alan. 1988. "Discussing Race in a Racist Society." *Anthropology Today* 4 (1): 3–5.

Morris, Desmond. 1967. *The Naked Ape: A Zoologist's Study of the Human Animal*. London: Cape.

Mountain, Joanna L., and Luigi Luca Cavalli-Sforza. 1994. "Inference of Human Evolution Through Cladistic of Nuclear DNA Restriction Polymorphisms." *Proceedings of the National Academy of Sciences* 91: 6515–6519.

———. 1997. "Multilocus Genotypes, a Tree of Individuals, and Human Evolutionary History." *American Journal of Human Genetics* 61: 705–718.

Mourant, Arthur Ernest. 1954. *The Distribution of the Human Blood Groups*. Oxford: Blackwell.

Muller, Hermann Joseph. 1961. "The Human Future." In *The Humanist Frame: The Modern Humanist Vision of Life*, edited by Julian Sorell Huxley, 399–414. London: Allen and Unwin.

———. 1965. "To the Editors, *The Scientific American*." *Eugenics Review* 57 (3): 101–104.

Nash, Catherine. 2007. "Mapping Origins: Race and Relatedness in Population Genetics and Genetic Genealogy." In *New Genetics, New Social Formations*, edited by Paul Atkinson, Peter Glasner, and Helen Greenslade, 77–100. London: Routledge.

———. 2012. "Genetics, Race and Relatedness: Human Mobility and Human Diversity in the Genographic Project." *Annals of the Association of American Geographers* 102 (3): 667–684.

———. 2015. *Genetic Geographies: The Trouble with Ancestry*. Minneapolis: University of Minnesota Press.

Nash, Roderick Frazier. 1966. "The American Cult of the Primitive." *American Quarterly* 18 (3): 517–537.

———. (1967) 1973. *Wilderness and the American Mind*. New Haven, CT: Yale University Press.

Nelkin, Dorothy. (1987) 1995. *Selling Science: How the Press Covers Science and Technology*. New York: W. H. Freeman.

Nelkin, Dorothy, and M. Susan Lindee. (1995) 2004. *The DNA Mystique: The Gene as a Cultural Icon*. Ann Arbor: University of Michigan Press.

Nelson, Alondra. 2008a. "The Factness of Diaspora: The Social Sources of Genetic Genealogy." In *Revisiting Race in a Genomic Age*, edited by Barbara A. Koenig, Sandra Soo-Jin Lee, and Sarah S. Richardson, 253–268. New Brunswick, NJ: Rutgers University Press.

——. 2008b. "Bio Science: Genetic Genealogy Testing and the Pursuit of African Ancestry." *Social Studies of Science* 38 (5): 759–783.

——. 2011. "Ruhestörung—Gräber, Knochen, DNA und eine neue afrikanische Ahnenforschung in den Vereinigten Staaten." In *Biohistorische Anthropologie: Knochen, Körper und DNA in Erinnerungskulturen*, edited by Marianne Sommer and Gesine Krüger, 54–77. Berlin: Kadmos.

——. 2012. "Reconciliation Projects: From Kinship to Justice." In *Genetics and the Unsettled Past: The Collision of DNA, Race, and History*, edited by Keith Wailoo, Alondra Nelson, and Catherine Lee, 20–31. New Brunswick, NJ: Rutgers University Press.

Newton, Maud. 2014. "America's Ancestry Craze: Making Sense of Our Family-Tree Obsession." *Harper's Magazine* (June): 29–34.

Nicholson, Edward Max. 1961. "The Place of Conservation." In *The Humanist Frame: The Modern Humanist Vision of Life*, edited by Julian Sorell Huxley, 385–397. London: Allen and Unwin.

——. 1987. *The New Environmental Age*. Cambridge: Cambridge University Press.

Nicholson, Edward Max, and Ludwig Koch. 1936. *Songs of Wild Birds*. London: Witherby.

——. 1937. *More Songs of Wild Birds*. London: Witherby.

Nietzsche, Friedrich. (1871) 1954. "Vom Nutzen und Nachteil der Historie für das Leben." In *Werke in drei Bänden*, 209–287. Munich: Carl Hanser.

Nuttall, George Henry Falkiner. 1904. *Blood Immunity and Blood Relationship*. Cambridge: Cambridge University Press.

Nye, Mary Jo. 2006. "Scientific Biography: History of Science by Another Means?" *Isis* 97 (2): 322–329.

Obermaier, Hugo. 1912. *Der Mensch der Vorzeit*. Berlin: Allgemeine Verlags-Gesellschaft.

Olby, Robert. (1974) 1994. *The Path to the Double Helix: The Discovery of DNA*. New York: Dover.

——. 1992. "Huxley's Place in Twentieth-Century Biology." In *Julian Huxley: Biologist and Statesman of Science. Proceedings of a Conference Held at Rice University 25–27 September 1987*, edited by C. Kenneth Waters and Albert Van Helden, 53–75. Houston: Rice University Press.

Oleson, Alexander, and John Voss, ed. 1979. *The Organization of Knowledge in Modern America, 1860–1920*. Baltimore: Johns Hopkins University Press.

Olson, Steve. 2002. *Mapping Human History. Genes, Race, and Our Common Origins*. Boston: Houghton Mifflin.

——. 2006. Review of *A Genetic and Cultural Odyssey: The Life and Work of L. Luca Cavalli-Sforza*, by Linda Stone and Paul F. Lurquin. *American Journal of Human Genetics* 78: 171–172.

Oppenheimer, Stephen. 2003. *Out of Eden: The Peopling of the World*. London: Constable.

——. 2004. *The Real Eve: Modern Man's Journey out of Africa*. New York: Basic Books.

——. 2006. *The Origins of the British: A Genetic Detective Story*. London: Constable.

Orth, Michael. 1986. "Utopia in the Pulps: The Apocalyptic Pastoralism of Edgar Rice Burroughs." *Extrapolation* 27 (3): 221–233.

Osborn, Henry Fairfield. 1884a. "Visual Memory." *Journal of Christian Philosophy* 3 (4): 439–450.

——. 1884b. "Illusions of Memory." *North American Review* 138 (330): 476–486.

——. 1890. "The Paleontological Evidence for the Transmission of Acquired Characters." *Science* 15 (367): 110–111.

——. 1891. "The Present Problem of Heredity." *Atlantic Monthly* 67: 353–364.

——. 1900. "The Geological and Faunal Relations of Europe and America during the Tertiary Period and the Theory of the Successive Invasions of an African Fauna." *Science, New Series* 11 (276): 561–574.

——. 1907. "Evolution as It Appears to the Paleontologist." *Science, New Series* 26 (674): 744–749.

——. 1908. "Coincident Evolution through Rectigradations and Fluctuations." *Science, New Series* 27 (697): 749–752.

——. 1910. *The Age of Mammals in Europe, Asia and North America*. New York: Macmillan.

——. 1912. "The State Museum and State Progress." *Science, New Series* 36 (929): 493–504.

——. (1913) 1924. *Impressions of Great Naturalists: Reminiscences of Darwin, Huxley, Balfour, Cope and Others*. New York: Charles Scribner's Sons.

——. 1915. *Men of the Old Stone Age: Their Environment, Life and Art*. New York: Scribner.

——. (1915) 1916. *Men of the Old Stone Age: Their Environment, Life and Art*. New York: Scribner.

——. 1917. *The Origin and Evolution of Life*. New York: Charles Scribner's Sons.

——. 1920. "Introduction." In *Pic, the Weapon-Maker*, by George Langford, xi–xii. New York: Boni and Liveright.

——. 1921a. "The Pliocene Man of Foxhall in East Anglia." *Natural History* 21 (6): 565.

——. 1921b. "The Second International Congress of Eugenics Address of Welcome." *Science, New Series* 54 (1397): 311–313.

——. 1921c. "Eugenics—The American and Norwegian Programs." *Science, New Series* 54 (1403): 482–484.

——. (1921) 1923. *The Hall of the Age of Man.* AMNH guide leaflet series 52. New York: AMNH.

——. 1922a. "*Hesperopithecus*, the First Anthropoid Primate Found in America." *American Museum Novitates* 37: 1–5.

——. 1922b. "Orthogenesis as Observed from Paleontological Evidence Beginning in the Year 1889." *American Naturalist* 56 (643): 134–143.

——. 1923. *Evolution and Religion.* New York: Charles Scribner's Sons.

——. 1924. "American Men of the Dragon Bones: Personal Impressions of a Field Trip to Mongolia with the Third Asiatic Expedition." *Natural History* 24: 350–365.

——. 1925. "The Cavemen Knew." *Collier's Weekly* 75 (21): 23.

——. (1925) 1947. *The Age of Man.* Revised to 1943 by William King Gregory and George Pinkley. New York: AMNH.

——. 1926a. "Giant Beasts of Three Million Years Ago." In *On the Trail of Ancient Man. A Narrative of the Field Work of the Central Asiatic Expeditions*, by Roy Chapman Andrews, 190–207. New York: Putnam.

——. 1926b. "The Origin of Species, 1859–1925." *Scientific Monthly* 22 (3): 185–192.

——. 1926c. "The Problem of the Origin of Species as It Appeared to Darwin and as It Appears to Us Today." *Science* 64: 337–341.

——. 1927a. *Creative Education in School, College, University, and Museum.* New York: Charles Scribner's Sons.

——. 1927b. *Man Rises to Parnassus: Critical Epochs in the Prehistory of Man.* Princeton, NJ: Princeton University Press.

——. 1927c. "Recent Discoveries Relating to the Origin and Antiquity of Man." *Science, New Series* 65 (1690): 481–488.

——. 1928a. "The Plateau Habitat of the Pro-Dawn Man." *Science, New Series* 67 (1745): 570–571.

——. 1928b. "The Present Status of the Problem of Human Ancestry." *Proceedings of the American Philosophical Society* 67 (2): 151–155.

——. 1929a. "The Revival of Central Asiatic Life." *Natural History* 29: 3–16.

——. 1929b. "Note on Geological Age of Pithecanthropus and Eoanthropus." *Science, New Series* 69 (1782): 216–217.

——. 1929c. *The Titanotheres of Ancient Wyoming, Dakota, and Nebraska.* 2 vols. Washington DC: Government Printing Office.

——. 1930a. "The Discovery of Tertiary Man." *Science, New Series* 71 (1827): 1–7.

——. 1930b. *Fifty-Two Years of Research, Observation and Publication, 1877–1929: A Life Adventure in Breadth and Depth*. New York: Charles Scribner's Sons.

——. 1930c. "Paleontology versus Genetics." *Science, New Series* 72 (1853): 1–3.

——. 1931a. "Arrest of Geologic, Archeologic and Paleontologic Work in Central Asia." *Science, New Series* 74 (1910): 139–142.

——. 1931b. "Paleontology versus Devriesianism and Genetics in the Factors of the Evolution Problem." *Science* 73 (1899): 547–549.

——. 1932. "Birth Selection versus Birth Control." *Science, New Series* 76 (1965): 173–179.

——. 1933. "Recent Revivals of Darwinism." *Science* 77 (1991): 199–202.

——. 1934a. "Aristogenesis, the Creative Principle in the Origin of Species." *Science, New Series* 7 (2038): 41–45.

——. 1934b. "The Dual Principles of Evolution." *Science, New Series* 80 (2087): 601–605.

——. 1936. *Proboscidea: A Monograph of Discovery, Evolution, Migration and Extinction of the Mastodonts and Elephants of the World*. Vol. 1. New York: American Museum Press.

——. 1942. *Proboscidea: A Monograph of Discovery, Evolution, Migration and Extinction of the Mastodonts and Elephants of the World*. Vol. 2. New York: American Museum Press.

Osborn, Henry Fairfield, and James McCosh. 1884. "A Study of the Mind's Chamber of Imagery." *Princeton Review* 13: 50–72.

Osborn, Henry Fairfield, and Chester A. Reeds. 1922. "Recent Discoveries on the Antiquity of Man." *Proceedings of the National Academy of Sciences of the United States of America* 8 (8): 246–247.

Otis, Laura. 1994. *Organic Memory: History and the Body in the Late Nineteenth and Early Twentieth Centuries*. Lincoln: University of Nebraska Press.

Pakendorf, Brigitte, and Mark Stoneking. 2005. "Mitochondrial DNA and Human Evolution." *Annual Review of Genomics and Human Genetics* 6: 165–183.

Pálsson, Gísli. 2007. *Anthropology and the New Genetics*. Cambridge: Cambridge University Press.

——. 2008. "Genomic Anthropology. Coming in from the Cold?" *Current Anthropology* 49 (4): 545–568.

——. 2012. "Decode Me! Anthropology and Personal Genomics." In supplement 5, *Current Anthropology* 53: S185–S195.

Parfitt, Tudor, and Yulia Egorova. 2005. "Genetics, History, and Identity: The Case of the Bene Israel and the Lemba." *Culture, Medicine, and Psychiatry* 29: 193–224.

——. 2006. *Genetics, Mass Media, and Identity: A Case Study of the Genetic Research on the Lemba and Bene Israel*. London: Routledge.

Patten, Robert L. 1992. "The British Context of Popularization." In *Julian Huxley: Biologist and Statesman of Science. Proceedings of a Conference Held at Rice University 25–27 September 1987*, edited by C. Kenneth Waters and Albert Van Helden, 257–262. Houston: Rice University Press.

Paul, Diane B. 1984. "Eugenics and the Left." *Journal of the History of Ideas* 45 (4): 567–590.

——. 1992. "The Value of Diversity in Huxley's Eugenics." In *Julian Huxley: Biologist and Statesman of Science. Proceedings of a Conference Held at Rice University 25–27 September 1987*, edited by C. Kenneth Waters and Albert Van Helden, 223–229. Houston: Rice University Press.

——. 1994. "Dobzhansky in the 'Nature-Nurture' Debate." In *The Evolution of Theodosius Dobzhansky: Essays on His Life and Thought in Russia and America*, edited by Mark B. Adams, 219–231. Princeton, NJ: Princeton University Press.

——. 1995. *Controlling Human Heredity, 1865 to the Present*. Atlantic Highlands, NJ: Humanities Press.

Peikoff, Kira. 2013. "I Had My DNA Picture Taken, with Varying Results." *New York Times*, 30 Dec. Accessed 2 July 2015. http://www.nytimes.com/2013/12/31/science/i-had-my-dna-picture-taken-with-varying-results.html?pagewanted=1&_r=0&hp.

Pennisi, Elizabeth. 2005. "Private Partnership to Trace Human History." *Science* 308 (5720): 340.

Pethes, Nicolas, and Jens Ruchatz, ed. 2001. *Gedächtnis und Erinnerung: Ein interdisziplinäres Lexikon*. Reinbek: Rowohlt Taschenbuch.

Piazza, Alberto, Paolo Menozzi, and Luigi Luca Cavalli-Sforza. 1981. "Synthetic Gene Frequency Maps of Man and Selective Effects of Climate." *Proceedings of the National Academy of Sciences* 78 (4): 2638–2642.

Porges, Irwin. 1975. *Edgar Rice Burroughs: The Man Who Created Tarzan*. 2 vols. New York: Ballantine Books.

Porter, Charlotte M. 1983. "The Rise to Parnassus: Henry Fairfield Osborn and the Hall of the Age of Man." *Museum Studies Journal* 1: 26–34.

Powers, Richard. (2006) 2007. *The Echo Maker*. London: Vintage.

Pratt, Mary Louise. 1992. *Imperial Eyes: Studies in Travel Writing and Transculturation*. London: Routledge.

Preston, Douglas J. 1986. *Dinosaurs in the Attic: An Excursion into the American Museum of Natural History*. New York: St. Martin's Press.

Pritchard, Jonathan K., Matthew Stephens, and Peter J. Donnelly. 2000. "Inference of Population Structure Using Multilocus Genotype Data." *Genetics* 155: 945–959.

Proctor, Robert N. 2003. "Three Roots of Human Recency: Molecular Anthropology, the Refigured Acheulean, and the UNESCO Response to Auschwitz." *Current Anthropology* 44 (2): 213–239.

Provine, William B. 1988. "Progress in Evolution and Meaning of Life." In *Evolutionary Progress*, edited by Matthew H. Nitecki, 49–74. Chicago: University of Chicago Press.

——. 1994. "The Origin of Dobzhansky's Genetics and the Origin of Species." In *The Evolution of Theodosius Dobzhansky: Essays on His Life and Thought in Russia and America*, edited by Mark B. Adams, 99–114. Princeton, NJ: Princeton University Press.

Pumfrey, Stephen, Paolo L. Rossi, and Maurice Slawinski, ed. 1991. *Science, Culture and Popular Belief in Renaissance Europe*. Manchester: Manchester University Press.

Rabinow, Paul. (1992) 1996. "Artificiality and Enlightenment: From Sociobiology to Biosociality." In *Essays on the Anthropology of Reason*, edited by Paul Rabinow, 91–111. Princeton, NJ: Princeton University Press.

Race, Robert Russell, and Ruth Sanger. 1950. *Blood Groups in Man*. Oxford: Blackwell.

Radick, Gregory. 2007. *The Simian Tongue: The Long Debate about Animal Language*. Chicago: University of Chicago Press.

Radin, Joanna. 2009. *From Anthropometry to Genomics: Reflections of a Pacific Fieldworker. By Jonathan Friedlaender, as Told to Joanna Radin*. Bloomington, IN: iUniverse.

——. 2013. "Latent Life: Concepts and Practices of Tissue Preservation in the International Biological Program." *Social Studies of Science* 43 (4): 483–508.

Rainger, Ronald. 1991. *An Agenda for Antiquity: Henry Fairfield Osborn and Vertebrate Paleontology at the American Museum of Natural History, 1890–1935*. Tuscaloosa: University of Alabama Press.

Rauhe, Simone. 2001. *Public History in den USA und der Bundesrepublik Deutschland*. Essen: Klartext.

Rayner, Sture. 1966. "Julian Huxley and His View on Eugenics in Evolutionary Perspective." *Hereditas* 56 (2–3): 207–212.

Reardon, Jenny. 2001. "The Human Genome Diversity Project: A Case Study in Coproduction." *Social Studies of Science* 31 (3): 357–388.

——. 2005. *Race to the Finish: Identity and Governance in an Age of Genomics*. Princeton, NJ: Princeton University Press.

——. 2008a. "Race without Salvation: Beyond the Science/Society Divide in Genomic Studies of Human Diversity." In *Revisiting Race in a Genomic Age*, edited by Barbara A. Koenig, Sandra Soo-Jin Lee, and Sarah S. Richardson, 304–319. New Brunswick, NJ: Rutgers University Press.

——. 2008b. "Race and Biology: Beyond the Perpetual Return of Crisis." *NTM Journal of History of Sciences, Technology, and Medicine* 16 (3): 373–377.

——. 2009. " 'Anti-Colonial Genomic Practice?' Learning from the Genographic Project and the Chacmool Conference." *International Journal of Cultural Property* 16 (2): 205–212.

Reardon, Jenny, and Kimberly TallBear. 2012. " 'Your DNA Is *Our* History': Genomics, Anthropology, and the Construction of Whiteness as Property." In supplement 5, *Current Anthropology* 53: S233–S245.

Reed, W. Maxwell. 1930. *The Earth for Sam: The Story of Mountains, Rivers, Dinosaurs and Men.* New York: Harcourt, Brace.

Regal, Brian. 2002. *Henry Fairfield Osborn: Race and the Search for the Origins of Man.* Aldershot: Ashgate.

Reitano, Joanne. 2006. *The Restless City: A Short History of New York from Colonial Times to the Present.* New York: Routledge.

Renfrew, Colin. 1987. *Archeology and Language.* Cambridge: Cambridge University Press.

Reumann, Miriam G., and Anne Fausto-Sterling. 2001. "Notions of Heredity in the Correspondence of Edwin Grant Conklin." *Perspectives in Biology and Medicine* 44 (3): 414–425.

Rheinberger, Hans-Jörg. 1992. *Experiment, Differenz, Schrift: Zur Geschichte epistemischer Dinge.* Marburg an der Lahn: Basilisken-Presse.

——. 1995. "Kurze Geschichte der Molekularbiologie." Preprint 24. Berlin: Max-Planck-Institut für Wissenschaftsgeschichte.

——. 1997. *Toward a History of Epistemic Things: Synthesizing Proteins in the Test Tube.* Stanford: Stanford University Press.

——. 2006. *Epistemologie des Konkreten: Studien zur Geschichte der modernen Biologie.* Frankfurt am Main: Suhrkamp.

Rheinberger, Hans-Jörg, and Staffan Müller-Wille. 2009. *Vererbung: Geschichte und Kultur eines biologischen Konzepts.* Frankfurt am Main: Fischer.

Rieppel, Lukas. 2012. "Bringing Dinosaurs Back to Life: Exhibiting Prehistory at the American Museum of Natural History." *Isis* 103 (3): 460–490.

Ritschel, Daniel. 1997. *The Politics of Planning: The Debate on Economic Planning in Britain in the 1930s.* Oxford: Oxford University Press.

Robinson, James Harvey. 1912. *The New History: Essays Illustrating the Modern Historical Outlook.* New York: Macmillan.

Roosevelt, Theodore. (1905) 1991. *The Strenuous Life.* Bedford: Applewood Books.

——. 1916. "How Old Is Man?" *National Geographic Magazine* 29 (2): 111–127.

Rose, Nikolas. 2007. *The Politics of Life Itself: Biomedicine, Power, and Subjectivity in the Twenty-First Century.* Princeton, NJ: Princeton University Press.

Rosenberg, Noah A., Jonathan K. Pritchard, James L. Weber, Howard M. Cann, Kenneth K. Kidd, Lew A. Zhivotovsky, and Marcus W. Feldman. 2002. "Genetic Structure of Human Populations." *Science* 298 (5602): 2381–2385.

Rouhani, Shahin. 1989. "Molecular Genetics and the Pattern of Human Evolution. Plausible and Implausible Models." In *The Human Revolution: Behav-*

ioural and Biological Perspectives on the Origins of Modern Humans, edited by Paul Mellars and Chris Stringer, 47–61. Edinburgh: Edinburgh University Press.

Ruhlen, Merritt. 1991. *A Guide to the Languages of the World*. Stanford, CA: Stanford University Press.

Ruse, Michael. 1988. "Molecules to Men: Evolutionary Biology and Thoughts of Progress." In *Evolutionary Progress*, edited by Matthew H. Nitecki, 97–126. Chicago: University of Chicago Press.

———. 1994. "Dobzhansky and the Problem of Progress." In *The Evolution of Theodosius Dobzhansky: Essays on His Life and Thought in Russia and America*, edited by Mark B. Adams, 233–245. Princeton, NJ: Princeton University Press.

———. 1999. *Mystery of Mysteries: Is Evolution a Social Construction?* Cambridge, MA: Harvard University Press.

Rüsen, Jörn. 1994a. *Historische Orientierung: Über die Arbeit des Geschichtsbewußtseins, sich in der Zeit zurechtzufinden*. Cologne: Böhlau.

———. 1994b. "Was ist Geschichtskultur?" In *Historische Faszination: Geschichtskultur heute*, edited by Klaus Füßmann, Heinrich Theodor Grütter, and Jörn Rüsen, 3–26. Cologne: Böhlau.

Sandage, Scott A. 2008. "The Gilded Age." In *A Companion to American Cultural History*, edited by Karen Halttunen, 139–153. Malden, MA: Blackwell.

Santos, Ricardo Ventura. 2003. "Indigenous Peoples, Changing Social and Political Landscapes, and Human Genetics in Amazonia." In *Genetic Nature/Culture: Anthropology and Science beyond the Two-Culture Divide*, edited by Alan H. Goodman, Deborah Heath, and M. Susan Lindee, 23–40. Berkeley: University of California Press.

Santos, Ricardo Ventura, Peter H. Fry, Simone Monteiro, Marcos Chor Maio, José Carlos Rodrigues, Luciana Bastos-Rodrigues, and Sérgio D. J. Pena. 2009. "Color, Race, and Genomic Ancestry in Brazil: Dialogues between Anthropology and Genetics." *Current Anthropology* 50 (6): 787–819.

Sanz, José Luis. 2002. *Starring T. Rex! Dinosaur Mythology and Popular Culture*. Bloomington: Indiana University Press.

Sarasin, Philipp, and Andreas B. Kilcher, ed. 2011. "Zirkulationen." *Nach Feierabend: Zürcher Jahrbuch für Wissenschaftsgeschichte. Zurich: diaphanes.*

Sarich, Vincent M., and Frank Miele. 2004. *Race: The Reality of Human Differences*. New York: Basic Books.

Sarkar, Sahotra. 1996. "Lancelot Hogben, 1895–1975." *Genetics* 142: 655–660.

Sarton, George. 1920. "The Faith of a Humanist." *Isis* 3 (1): 3–6.

———. 1924. "The New Humanism." *Isis* 6 (1): 9–42.

———. 1931. *The History of Science and the New Humanism*. New York: Henry Holt.

Schmidt-Salomon, Michael. (2005) 2006. *Manifest des evolutionären Humanismus: Plädoyer für eine zeitgemässe Leitkultur.* Aschaffenburg: Alibri.

Schneider, William H. 1995. "Blood Group Research in Great Britain, France, and the United States between the World Wars." *American Journal of Physical Anthropology* 38: 87–114.

——. 1996. "The History of Research on Blood Group Genetics: Initial Discovery and Diffusion." *History and Philosophy of the Life Sciences* 18: 277–303.

Schörken, Rolf. 1981. *Geschichte in der Alltagswelt: Wie uns Geschichte begegnet und was wir mit ihr machen.* Stuttgart: Klett-Cotta.

——. 1995. *Begegnungen mit Geschichte: Vom außerwissenschaftlichen Umgang mit der Historie in Literatur und Medien.* Stuttgart: Klett-Cotta.

Schwartz, Jeffrey H. 2006. "Molecular Systematics and Evolution." In *Encyclopedia of Molecular Cell Biology and Molecular Medicine.* Vol. 8, edited by Robert A. Meyers, 515–540. Weinheim: Wiley-VCH.

Schwarz, Angela. 2003. "Bilden, überzeugen, unterhalten. Wissenschaftspopularisierung und Wissenskultur im 19. Jahrhundert." In *Wissenspopularisierung: Konzepte der Wissensverbreitung im Wandel*, edited by Carsten Kretschmann, 221–234. Berlin: Akademie Verlag.

Schwarzenbach, Alexis. 2011. *WWF: Die Biographie*, Munich: Collection Rolf Heyne.

Scott, John Paul. 1950. "Foreword: Methodology and Techniques for the Study of Animal Societies." *Annals of the New York Academy of Sciences* 51: 1003–1005.

Scully, Marc, Turi King, and Steven D. Brown. 2013. "Remediating Viking Origins: Genetic Code as Archival Memory of the Remote Past." *Sociology* 47 (5): 921–938.

Secord, James A. (2001) 2003. *Victorian Sensation: The Extraordinary Publication, Reception, and Secret Authorship of Vestiges of the Natural History of Creation.* Chicago: University of Chicago Press.

——. 2004. "Knowledge in Transit." *Isis* 95 (4): 654–672.

Segal, Daniel Alan, and Sylvia J. Yanagisako, ed. 2005. *Unwrapping the Sacred Bundle: Reflections on the Disciplining of Anthropology.* Durham, NC: Duke University Press.

Segerstråle, Ullica. 2000. *Defenders of the Truth: The Battle for Science in the Sociobiology debate and beyond.* Oxford: Oxford University Press.

Seife, Charles. 2013. "23andMe Is Terrifying, but Not for the Reasons the FDA Thinks." *Scientific American*, 23 Nov. Accessed 2 July 2015. http://www.scientificamerican.com/article/23andme-is-terrifying-but-not-for-reasons-fda.

Selcer, Perrin. 2012. "Beyond the Cephalic Index: Negotiating Politics to Produce UNESCO's Scientific Statements on Race." In supplement 5, *Current Anthropology* 53: S173–S184.

Selvin, Paul. 1991. "The Raging Bull of Berkeley." *Science, New Series* 251 (4992): 368–371.

Semon, Richard. 1904. *Die Mneme als erhaltendes Prinzip im Wechsel des organischen Geschehens*. Leipzig: Engelmann.

Sepkoski, David. 2012. *Reading the Fossil Record: The Growth of Paleobiology as an Evolutionary Discipline*. Chicago: University of Chicago Press.

Sewell, James P. 1975. *UNESCO and World Politics: Engaging in International Relations*. Princeton, NJ: Princeton University Press.

Shapere, Dudley. 1980. "The Meaning of the Evolutionary Synthesis." In *The Evolutionary Synthesis: Perspectives on the Unification of Biology*, edited by Ernst Mayr and William B. Provine, 388–398. Cambridge, MA: Harvard University Press.

Shapin, Steven. 1980. "Social Uses of Science." In *The Ferment of Knowledge: Studies in the Historiography of Eighteenth-Century Science*, edited by George Sebastian Rousseau and Roy Porter, 93–139. Cambridge: Cambridge University Press.

Shapin, Steven, and Simon Schaffer. 1985. *Leviathan and the Air-Pump: Hobbes, Boyle, and the Experimental Life*. Princeton, NJ: Princeton University Press.

Shapland, Andrew, and David Van Reybrouck. 2008. "Competing Natural and Historical Heritage: The Penguin Pool at London Zoo." *International Journal for Heritage Studies* 14 (1): 10–29.

Shriver, Mark D., and Rick A. Kittles. 2008. "Genetic Ancestry and the Search for Personalized Genetic Histories." In *Revisiting Race in a Genomic Age*, edited by Barbara A. Koenig, Sandra Soo-Jin Lee, and Sarah S. Richardson, 201–214. New Brunswick, NJ: Rutgers University Press.

Silverman, Rachel. 2000. "The Blood Group 'Fad' in Post-War Racial Anthropology." In *Kroeber Anthropological Society Papers*, edited by Jonathan Marks, 11–27. Berkeley: University of California Press.

Silverman, Sydel. 2002. *The Beast on the Table: Conferencing with Anthropologists*. Walnut Creek, CA: AltaMira Press.

Simpson, Bob. 2000. "Imagined Genetic Communities: Ethnicity and Essentialism in the Twenty-First Century." *Anthropology Today* 16 (3): 3–6.

Simpson, George Gaylord. 1941. "The Role of the Individual in Evolution." *Journal of the Washington Academy of Sciences* 31 (1): 1–20.

———. 1944. *Tempo and Mode in Evolution*. New York: Columbia University Press.

———. 1947. "The Problem of Plan and Purpose in Nature." *Scientific Monthly* 64 (6): 481–495.

———. (1949) 1967. *The Meaning of Evolution: A Study of the History of Life and of Its Significance to Man*. New Haven, CT: Yale University Press.

———. 1951. *Horses: The Story of the Horse Family in the Modern World and Through Sixty Million Years of History*. New York: Oxford University Press.

——. 1960. "The World into Which Darwin Led Us." *Science, New Series* 131 (3405): 966–974.

——. 1961. *Horses: The Story of the Horse Family in the Modern World and through Sixty Million Years of History*. New York: Doubleday.

——. 1964. *This View of Life: The World of an Evolutionist*. New York: Harcourt, Brace and World.

——. 1964, 1966, 1967, 1969. *Biology and Man*. New York: Harcourt Brace Jovanovich.

——. 1966. "The Biological Nature of Man." *Science, New Series* 152 (3721): 472–478.

——. 1974. "The Concept of Progress in Organic Evolution." *Social Research* 41: 18–51.

——. 1977. "A New Heaven and a New Earth—and a New Man." In *Man's Place in the Universe: Changing Concepts*, edited by David W. Corson, 53–75. Tucson: College of Liberal Arts, University of Arizona.

Sims-Williams, Patrick. 1998. "Genetics, Linguistics, and Prehistory: Thinking Big and Thinking Straight." *Antiquity* 72: 505–527.

Singer, Charles. 1929. "Scientific Humanism." *Rationalist: A Journal of Scientific Humanism* 1 (1): 12–18.

Singer, Dorothea, and Charles Singer. 1957. "George Sarton and the History of Science," in "The George Sarton Memorial Issue," *Isis* 48 (3): 306–310.

Singh, J. P. 2011. *United Nations Educational, Scientific and Cultural Organization (UNESCO): Creating Norms for a Complex World*. London: Routledge.

Sluga, Glenda. 2010. "UNESCO and the (One) World of Julian Huxley." *Journal of World History* 21 (3): 393–418.

Smail, Daniel Lord. 2008. *On Deep History and the Brain*. Berkeley: University of California Press.

Small, Meredith F. 2006. "First Soldier of the Gene Wars: A Pioneer of Genetic Archaeology Maps the History of Human Migration." *Archaeology* 59 (3): 46–51.

Smith, Laurajane. 2007. "General Introduction." In *Cultural Heritage: Critical Concepts in Media and Cultural Studies*. Vol. 1, *History and Concepts*, edited by Laurajane Smith, 1–21. London: Routledge.

Smith, Roger. 2003. "Biology and Values in Interwar Britain: C. S. Sherrington, Julian Huxley and the Vision of Progress." *Past and Present* 178: 210–242.

Smocovitis, Vassiliki Betty. 1994. "Organizing Evolution: Founding the Society for the Study of Evolution (1939–1950)." *Journal for the History of Biology* 27 (2): 241–309.

——. 1996. *Unifying Biology: The Evolutionary Synthesis and Evolutionary Biology*. Princeton, NJ: Princeton University Press.

——. 1999. "The 1959 Darwin Centennial Celebration in America." *Osiris, 2nd Series* 14: 274–323.

——. 2012. "Humanizing Evolution: Anthropology, the Evolutionary Synthesis, and the Prehistory of Biological Anthropology, 1927–1962." In supplement 5, *Current Anthropology* 53: S108–S125.

——. (forthcoming). "The Unifying Vision: Julian Huxley, the Evolutionary Synthesis and Evolutionary Humanism." In *Pursuing the Unity of Science: Ideology and Scientific Practice between the Great War and the Cold War*, edited by Geert Somsen and Harmke Kamminga. London: Ashgate.

Smolenyak Smolenyak, Megan, and Ann Turner. 2004. *Trace Your Roots with DNA: Using Genetic Tests to Explore Your Family Tree*. New York: Rodale.

Smuts, Jan Christiaan. 1930. *Africa and Some World Problems*. Oxford: Oxford University Press.

Snow, Charles Percy. 1959. *The Two Cultures and the Scientific Revolution*. New York: Cambridge University Press.

Sober, Elliott. 1988. *Reconstructing the Past: Parsimony, Evolution, and Inference*. Cambridge, MA: MIT Press.

Soja, Edward W. 1996. *Thirdspace: Journeys to Los Angeles and Other Real-and-Imagined Places*. Cambridge, MA: Blackwell.

Sokal, Robert R., and Peter H. A. Sneath. 1963. *Principles of Numerical Taxonomy*. San Francisco: W. H. Freeman.

Sollas, William. 1911. *Ancient Hunters and Their Modern Representatives*. London: Macmillan.

——. (1911) 1915. *Ancient Hunters and Their Modern Representatives*. London: Macmillan.

Sommer, Marianne. 2000. *Foremost in Creation: Anthropomorphism and Anthropocentrism in National Geographic Articles on Non-Human Primates*. Berne: Peter Lang.

——. 2005a. "Ancient Hunters and Their Modern Representatives: William Sollas's (1849–1936) Anthropology from Disappointed Bridge to Trunkless Tree and the Instrumentalisation of Racial Conflict." *Journal of the History of Biology* 38 (2): 327–365.

——. 2005b. "How Cultural Is Heritage? Humanity's Black Sheep from Charles Darwin to Jack London." In *A Cultural History of Heredity III: 19th and Early 20th Centuries, edited by* Staffan Müller-Wille and Hans-Jörg Rheinberger, 233–253. Preprint 294. Berlin: Max-Planck-Institut für Wissenschaftsgeschichte.

——. 2006. "Mirror, Mirror on the Wall: Neanderthal as Image and 'Distortion' in Early 20th-Century French Science and Press." *Social Studies of Science* 36 (2): 207–240.

——. 2007a. *Bones and Ochre: The Curious Afterlife of the Red Lady of Paviland*. Cambridge, MA: Harvard University Press.

——. 2007b. "The Lost World as Laboratory: The Politics of Evolution between

Science and Fiction in Early Twentieth-Century America." *Configurations* 15 (3): 299–329.

———. 2008a. "History in the Gene: Negotiations between Molecular and Organismal Anthropology." *Journal of the History of Biology* 41 (3): 473–528.

———. 2008b. "Angewandte Geschichte auf genetischer Grundlage." *Nach Feierabend: Zürcher Jahrbuch für Wissenschaftsgeschichte* 4: 129–148.

———. 2009. "Angewandte Vorgeschichte: Das menschliche Gen zwischen Naturwissenschaft, Öffentlichkeit und Markt." In *History Sells! Angewandte Geschichte als Wissenschaft und Markt*, edited by Wolfgang Hardtwig and Alexander Schug, 20–30. Stuttgart: Franz Steiner.

———. 2010a. "From Descent to Ascent: The Human Exception in the Evolutionary Synthesis." *Nuncius* 25 (1): 41–67.

———. 2010b. "Human Tools of the European Tertiary? Artefacts, Brains and Minds in Evolutionist Reasoning, 1870–1920." *Notes and Records of the Royal Society* 65 (1): 65–82.

———. 2010c. "Seriality in the Making: The Osborn-Knight Restorations of Evolutionary History." *History of Science* 48 (161): 461–482.

———. 2010d. "(Net)Working a Stone into a Tool: The International Eoliths Controversy in the Light of New Approaches to the History of Archaeology." Keynote lecture, conference "New Historiographical Approaches to Archaeological Research," Berlin, 9–11 Sept.

———. 2010e. Review of *From Anthropometry to Genomics: Reflections of a Pacific Fieldworker. By Jonathan Friedlaender, as Told to Joanna Radin. American Journal of Human Biology* 22 (4): 567–568.

———. 2010f. "DNA and Cultures of Remembrance: Anthropological Genetics, Biohistories, and Biosocialities." In "BioHistories," edited by Soraya de Chadarevian, special issue, *BioSocieties* 5 (3): 366–390.

———. 2012a. " 'It's a Living History, Told by the Real Survivors of the Times—DNA': Anthropological Genetics in the Tradition of Biology as Applied History." In *Genetics and the Unsettled Past: The Collision of DNA, Race, and History*, edited by Keith Wailoo, Alondra Nelson, and Catherine Lee, 225–246. New Brunswick, NJ: Rutgers University Press.

———. 2012b. "Human Evolution across the Disciplines: Spotlights on American Anthropology and Genetics." *History and Philosophy of the Life Sciences* 34: 211–236.

———. 2012c. "'Do You Have Celtic, Jewish or Germanic Roots?'—Applied Swiss History before and after DNA." In *Identity Politics and the New Genetics: Re/Creating Categories of Difference and Belonging*, edited by Katharina Schramm, David Skinner, and Richard Rottenburg, 116–140. Oxford: Berghahn Books.

———. 2014. "Biology as a Technology of Social Justice in Interwar Britain: Arguments from Evolutionary History, Heredity, and Human Diversity." *Science, Technology, and Human Values* 39 (4): 561–586.

———. 2015. "Population-Genetic Trees, Maps and Narratives of the Great Human Diasporas." *History of the Human Sciences* 28(5) (first published online March): 108–145. doi:10.1177/0952695115573032.

Sommer, Marianne, and Gesine Krüger, ed. 2011. *Biohistorische Anthropologie: Knochen, Körper und DNA in Erinnerungskulturen*. Berlin: Kadmos.

Soodyall, Himla. 2003. "Reflections and Prospects for Anthropological Genetics in South Africa." In *Genetic Nature/Culture: Anthropology and Science beyond the Two-Culture Divide*, edited by Alan H. Goodman, Deborah Heath, and M. Susan Lindee, 200–216. Berkeley: University of California Press.

Speich Chassé, Daniel, and David Gugerli. 2012. "Wissensgeschichte: Eine Standortbestimmung." *Traverse* 1: 85–100.

Spencer, Quayshawn. 2014. "A Radical Solution to the Race Problem." *Philosophy of Science* 81 (5): 1025–1038.

Spuhler, James N. 1973. "Anthropological Genetics: An Overview." In *Methods and Theories of Anthropological Genetics*, edited by Michael H. Crawford and Peter L. Workman, 423–451. Albuquerque: University of New Mexico Press.

Squier, Susan M. 1994. *Babies in Bottles: Twentieth-Century Visions of Reproductive Technology*. New Brunswick, NJ: Rutgers University Press.

Stafford, Barbara. 2007. *Echo Objects: The Cognitive Work of Images*. Chicago: University of Chicago Press.

Stepan, Nancy Leys. 1982. *The Idea of Race in Science: Great Britain, 1800–1960*. London: Macmillan.

Stevens, Hallam. 2013. *Life Out of Sequence: A Data-Driven History of Bioinformatics*. Chicago: University of Chicago Press.

Stewart, Larry. 1992. *The Rise of Public Science: Rhetoric, Technology, and Natural Philosophy in Newtonian Britain, 1660–1750*. Cambridge: Cambridge University Press.

Stocking, George W., ed. 1988. *Bones, Bodies, Behavior: Essays on Biological Anthropology*. Madison: University of Wisconsin Press.

Stone, Linda, and Paul F. Lurquin. 2005. *A Genetic and Cultural Odyssey: The Life and Work of L. Luca Cavalli-Sforza*. New York: Columbia University Press.

Strasser, Bruno J. 2010. "Collecting, Comparing, and Computing Sequences: The Making of Margaret O. Dayhoff's 'Atlas of Protein Sequence and Structure,' 1954–1965." *Journal of the History of Biology* 43: 623–660.

———. 2011. "The Experimenter's Museum: GenBank, Natural History, and the Moral Economies of Biomedicine." *Isis* 102 (1): 60–96.

Strathern, Marilyn. 1992. *Reproducing the Future: Anthropology, Kinship, and the New Reproductive Technologies*. New York: Routledge.

Stringer, Chris, and Peter Andrews. 1988. "Genetic and Fossil Evidence for the Origin of Modern Humans." *Science* 239 (4845): 1263–1268.

Strobel, Gabrielle. 1993. "Human Genome Diversity Project Raises Serious Ethical Issues." *Stanford News Service*, 8 July. Accessed 2 July 2015. http://news.stanford.edu/pr/93/930608Arc3222.html.

Sturdy, Steve. 1998. "Reflections: Molecularization, Standardization and the History of Science." In *Molecularizing Biology and Medicine: New Practices and Alliances, 1910s–1970s*, edited by Soraya de Chadarevian and Harmke Kamminga, 273–292. Amsterdam: OPA.

Sturken, Marita. 2008. "Memory, Consumerism and Media. Reflections on the Emergence of the Field." *Memory Studies* 1 (1): 73–78.

Suárez, Edna. 2001. "Satellite-DNA: A Case-Study for the Evolution of Experimental Techniques." *Studies in History and Philosophy of Biological and Biomedical Sciences* 32 (1): 31–57.

Suárez-Díaz, Edna, and Victor H. Anaya-Muñoz. 2008. "History, Objectivity, and the Construction of Molecular Phylogenies." *Studies in History and Philosophy of Biological and Biomedical Sciences* 39 (4): 451–468.

Sulloway, Frank J. 1979. *Freud, Biologist of the Mind: Beyond the Psychoanalytic Legend*. New York: Basic Books.

Swetlitz, Marc. 1993. "Julian Huxley, George Gaylord Simpson and the Idea of Progress in 20th-Century Evolutionary Biology." PhD diss., University of Chicago, 1992.

——. 1995. "Julian Huxley and the End of Evolution." *Journal of the History of Biology* 28: 181–217.

Sykes, Bryan. 2001. *The Seven Daughters of Eve*. New York: W. W. Norton.

——. 2003. *Adam's Curse: A Future without Men*. London: Bantam.

——. 2006. *Blood of the Isles: Exploring the Genetic Roots of Our Tribal History*. London: Bantam.

Tabery, James. 2008. "R. A. Fisher, Lancelot Hogben, and the Origin(s) of Genotype-Environment Interaction." *Journal of the History of Biology* 41: 717–761.

Takezawa, Yasuko, Kazuto Kato, Hiroki Oota, Timothy Caulfield, Akihiro Fujimoto, Shunwa Honda, Naoyuki Kamatani, Shoji Kawamura, Kohei Kawashima, Ryosuke Kimura, Hiromi Matsumae, Ayako Saito, Patrick E. Savage, Noriko Seguchi, Keiko Shimizu, Satoshi Terao, Yumi Yamaguchi-Kabata, Akira Yasukouchi, Minoru Yoneda, and Katsushi Tokunaga. 2014. "Human Genetic Research, Race, Ethnicity and the Labeling of Populations: Recommendations Based on an Interdisciplinary Workshop in Japan." *BioMed Central Medical Ethics* 15 (33). doi:10.1186/1472-6939-15-33.

Taliaferro, John. 1999. *Tarzan Forever: The Life of Edgar Rice Burroughs, Creator of Tarzan*. New York: Scribner.

TallBear, Kimberly. 2007. "Narratives of Race and Indigeneity in the Genographic Project." *Journal of Law, Medicine and Ethics* 35 (3): 412–424.

Tang, Hua, Jie Peng, Pei Wang, and Neil J. Risch. 2005. "Estimation of Individual Admixture: Analytical and Study Design Considerations." *Genetic Epidemiology* 28 (4): 289–301.

Teilhard de Chardin, Pierre. (1955) 1959. *The Phenomenon of Man*. Translated by Bernard Wall, with an introduction by Julian Huxley. London: William Collins.

Terrell, John Edward, and Pamela J. Stewart. 1996. "The Paradox of Human Population Genetics at the End of the Twentieth Century." *Reviews in Anthropology* 26: 13–33.

Thacker, Eugene. 2005. *The Global Genome: Biotechnology, Politics, and Culture*. Cambridge, MA: MIT Press.

Thackray, Arnold, and Robert K. Merton. 1972. "On Discipline Building: The Paradoxes of George Sarton." *Isis* 63 (4): 472–495.

Thomas, Howard, ed. 1942. *The Brains Trust Book*. London: Hutchinson.

Thomson, Keith. 2008. *The Legacy of the Mastodon: The Golden Age of Fossils in America*. New Haven, CT: Yale University Press.

Tilley, Helen. 2003. "African Environments and Environmental Science: The African Research Survey, Ecological Paradigms, and British Colonial Development 1920–1940." In *Social History and African Environments*, edited by William Beinart and Joann McGregor, 109–130. Oxford: James Currey.

———. 2011. *Africa as a Living Laboratory: Empire, Development, and the Problem of Scientific Knowledge, 1870–1950*. Chicago: University of Chicago Press.

Tilney, Frederick. 1928. *The Brain from Ape to Man: A Contribution to the Study of the Evolution and Development of the Human Brain*. With a foreword by Henry Fairfield Osborn. New York: Paul B. Hoeber.

Toye, John, and Richard Toye. 2010. "One World, Two Cultures? Alfred Zimmern, Julian Huxley and the Ideological Origins of UNESCO." *History: The Journal of the Historical Association* 95 (319): 308–331.

Traweek, Sharon. 1988. *Beamtimes and Lifetimes*. Cambridge, MA: Harvard University Press.

Troeltsch, Ernst. 1922. "Die Krisis des Historismus." *Die Neue Rundschau* 33 (1): 572–590.

Underhill, Peter A. 2003. "Inferring Human History: Clues from Y-Chromosome Haplotypes." *Cold Spring Harbor Symposia on Quantitative Biology* 68: 487–493.

Underhill, Peter A., and Toomas Kivislid. 2007. "Use of Y Chromosome and Mitochondrial DNA Population Structure in Tracing Human Migrations." *Annual Review of Genetics* 41: 539–564.

Underhill, Peter A., Li Jin, Alice A. Lin, Syed Qasim Mehdi, Trefor Jenkins, Douglas Vollrath, Ronald W. Davis, Luigi Luca Cavalli-Sforza, and Peter J. Oefner. 1997. "Detection of Numerous Y Chromosome Biallelic Polymorphisms by Denaturing High-Performance Liquid Chromatography." *Genome Research* 7 (10): 996–1005.

Underhill, Peter A., Giuseppe Passarino, Alice A. Lin, Peidong Shen, Marta Mirazón Lahr, Robert A. Foley, Peter J. Oefner, and Luigi Luca Cavalli-Sforza. 2001. "The Phylogeography of Y Chromosome Binary Haplotypes and the Origins of Modern Human Populations." *Annals of Human Genetics* 65: 43–62.

Underhill, Peter A., Peidong Shen, Alice A. Lin, Li Jin, Giuseppe Passarino, Wei H. Yang, Erin Kauffman, Batsheva Bonné-Tamir, Jaume Bertranpetit, Paolo Francalacci, Muntaser Ibrahim, Trefor Jenkins, Judith R. Kidd, Syed Qasim Mehdi, Mark T. Seielstad, R. Spencer Wells, Alberto Piazza, Ronald W. Davis, Marcus W. Feldman, Luigi Luca Cavalli-Sforza, and Peter J. Oefner. 2000. "Y Chromosome Sequence Variation and the History of Human Populations." *Nature Genetics* 26: 358–361.

UNESCO. 2006. *Sixty Years of Science at UNESCO 1945–2005*. Paris: UNESCO Publishing.

Verne, Jules. 1864. *Voyage au centre de la Terre*. Paris: Hetzel.

——. 1873. *Around the World in Eighty Days*. Translated by George Makepeace Towle. Boston: Osgood.

Vetter, Jeremy. 2004. "Science along the Railroad: Expanding Field Work in the US Central West." *Annals of Science* 61: 187–211.

——. 2008. "Cowboys, Scientists, and Fossils: The Field Site and Local Collaboration in the American West." *Isis* 99 (2): 273–303.

Vollrath, Douglas, Simon Foote, Adrienne Hilton, Laura G. Brown, Peggy Beer-Romero, Jonathan S. Bogan, and David C. Page. 1992. "The Human Y Chromosome: A 43-Interval Map Based on Naturally Occurring Deletions." *Science* 258 (5079): 52–59.

Voss, Julia. 2007. *Darwins Bilder: Ansichten der Evolutionstheorie, 1837–1874*. Frankfurt am Main: Fischer.

Waddington, Conrad H. 1941. *The Scientific Attitude*. Harmondsworth, Middlesex: Penguin.

Wade, Nicholas. 2006. *Before the Dawn: Recovering the Lost History of Our Ancestors*. New York: Penguin.

——. 2014. *A Troublesome Inheritance: Genes, Race, and Human History*. New York: Penguin.

Wailoo, Keith, Alondra Nelson, and Catherine Lee. 2012. "Introduction: Genetic Claims and the Unsettled Past." In *Genetics and the Unsettled Past: The*

Collision of DNA, Race, and History, edited by Keith Wailoo, Alondra Nelson, and Catherine Lee, 1–10. New Brunswick: Rutgers University Press.

Wainscoat, James S., Adrian V. S. Hill, Swee Lay Thein, J. Flint, J. C. Chapman, David John Weatherall, J. B. Clegg, and Douglas R. Higgs. 1989. "Geographic Distribution of Alpha- and Beta-Globin Gene Cluster Polymorphisms." In *The Human Revolution: Behavioural and Biological Perspectives on the Origins of Modern Humans*, edited by Paul Mellars and Chris Stringer, 31–38. Edinburgh: Edinburgh University Press.

Wald, Priscilla. 2006. "Blood and Stories: How Genomics Is Rewriting Race, Medicine and Human History." *Patterns of Prejudice* 40 (4 and 5): 304–333.

Ward, David, and Olivier Zunz, ed. (1992) 1997. *The Landscape of Modernity: New York City, 1900–1940*. Baltimore: Johns Hopkins University Press.

Waters, C. Kenneth. 1992. "Revising Our Picture of Julian Huxley." In *Julian Huxley: Biologist and Statesman of Science. Proceedings of a Conference Held at Rice University 25–27 September 1987*, edited by C. Kenneth Waters and Albert Van Helden, 1–27. Houston: Rice University Press.

Weber, Marcel. 1998. *Die Architektur der Synthese: Entstehung und Philosophie der modernen Evolutionstheorie*. Berlin: de Gruyter.

Weber, Thomas P. 2005. *Darwin und die neuen Biowissenschaften: Eine Einführung*. Köln: DuMont.

Weindling, Paul. 2012. "Julian Huxley and the Continuity of Eugenics in Twentieth-Century Britain." *Journal of Modern European History* 10 (4): 480–499.

Wells, H. G. 1895. *The Time Machine*. London: Heinemann.

——. (1920) 1929. *The Outline of History: Being a Plain History of Life and Mankind*. New York: Literary Guild.

Wells, H. G., Julian Sorell Huxley, and G. P. Wells. 1929, 1930, 1931, 1934. *The Science of Life*. New York: Literary Guild.

Wells, Spencer. 2002. *The Journey of Man: A Genetic Odyssey*. London: Allen Lane.

——. 2006. *Deep Ancestry: Inside the Genographic Project: The Landmark DNA Quest to Decipher Our Distant Past*. Washington DC: National Geographic.

——. 2010. *Pandora's Seed: The Unforeseen Cost of Civilization*. New York: Random House.

Werskey, Paul Gary. 1971a. "British Scientists and 'Outsider' Politics, 1931–1945." *Science Studies* 1 (1): 67–83.

——. 1971b. "Haldane and Huxley: The First Appraisals." *Journal of the History of Biology* 4 (1): 171–183.

——. 1978. *The Visible College*. London: Allen Lane.

Whitley, Richard. 1985. "Knowledge Producers and Knowledge Acquirers: Popularisation as the Relation between Scientific Fields and Their Publics." In

Expository Science: Forms and Functions of Popularisation, edited by Terry Shinn and Richard Whitley, 3–28. Dordrecht: D. Reidel.

Wiener, Martin J. 1992. "The English Style of Huxley's Thought." In *Julian Huxley: Biologist and Statesman of Science. Proceedings of a Conference Held at Rice University 25–27 September 1987*, edited by C. Kenneth Waters and Albert Van Helden, 49–52. Houston: Rice University Press.

Wilson, Duncan. 2005. "The Early History of Tissue Culture in Britain: The Interwar Years." *Social History of Medicine* 18 (2): 225–243.

Wilson, Edward Osborne. 1975. *Sociobiology: The New Synthesis*. Cambridge, MA: Harvard University Press.

——. 1979. *On Human Nature*. Cambridge, MA: Harvard University Press.

Winther, Rasmus Grønfeldt. 2014. "The Genetic Reification of 'Race'?: A Story of Two Mathematical Methods." *Critical Philosophy of Race* 2 (2): 204–223.

Wöbse, Anna-Katharina. 2012. *Weltnaturschutz: Umweltdiplomatie in Völkerbund und Vereinten Nationen, 1920–1950*. Frankfurt am Main: Campus.

Wöbse, Anna-Katharina, and Mieke Roscher. 2010. "Zootiere während des Zweiten Weltkrieges: London und Berlin 1939–1945." *Werkstatt Geschichte* 56: 46–62.

Wolpoff, Milford H. 1989. "Multiregional Evolution: The Fossil Alternative to Eden." In *The Human Revolution: Behavioural and Biological Perspectives on the Origins of Modern Humans*, edited by Paul Mellars and Chris Stringer, 62–108. Edinburgh: Edinburgh University Press.

Worthington, E. B. 1938. *Science in Africa: A Review of Scientific Research Relating to Tropical and Southern Africa*. London: Oxford University Press.

Wright, Sewall. 1931. "Evolution in Mendelian Populations." *Genetics* 16: 97–159.

Wynne, Brian. 1995. "Public Understanding of Science." In *Handbook of Science and Technology Studies*, edited by Sheila Jasanoff, Gerald E. Markle, James C. Petersen, and Trevor Pinch, 361–388. Thousand Oaks: Sage.

YCC (Y Chromosome Consortium). 2002. "A Nomenclature System for the Tree of Human Y-Chromosomal Binary Haplogroups." *Genome Research* 12 (2): 339–348.

Yoxen, Edward. 1982. "Giving Life a New Meaning: The Rise of the Molecular Biology Establishment." In *Scientific Establishment and Hierarchies*, edited by Norbert Elias, Herminio Martins, and Richard P. Whitley, 123–143. Dordrecht: Reidel.

Zalasiewicz, Jan, Mark Williams, Alan Smith, Tiffany L. Barry, Angela L. Coe, Paul R. Bown, Patrick Brenchley, David Cantrill, Andrew Gale, Philip Gibbard, F. John Gregory, Mark W. Hounslow, Andrew C. Kerr, Paul Pearson, Robert Knox, John Powell, Colin Waters, John Marshall, Michael Oates, Peter Rawson, and Philip Stone. 2008. "Are We Now Living in the Anthropocene?" *GSA Today* 18 (2): 4–8.

Zuckerkandl, Emile. 1963. "Perspectives in Molecular Anthropology." In *Classification and Human Evolution*, edited by Sherwood L. Washburn, 243–272. Chicago: Aldine.

Zuckerkandl, Emile, and Linus Pauling. 1962. "Molecular Disease, Evolution, and Genetic Heterogeneity." In *Horizons in Biochemistry*, edited by Michael Kasha and Bernard Pullman, 189–225. New York: Academic Press.

——. 1965a. "Evolutionary Divergence and Convergence in Proteins." In *Evolving Genes and Proteins. A Symposium Held at the Institute of Microbiology of Rutgers University: New Brunswick—N.J., Sept. 17–18, 1964*, edited by Vernon Bryson and Henry J. Vogel, 97–166. New York: Academic Press.

——. 1965b. "Molecules as Documents of Evolutionary History." *Journal of Theoretical Biology* 8: 357–366.

Zvelebil, Marek. 1998. "Genetic and Cultural Diversity in Europe: A Comment on Cavalli-Sforza." *Journal of Anthropological Research* 54 (3): 411–416.

Index